Management for Professionals

The Springer series "Management for Professionals" comprises high-level business and management books for executives, MBA students, and practice-oriented business researchers. The topics cover all themes relevant to businesses and the business ecosystem. The authors are experienced business professionals and renowned professors who combine scientific backgrounds, best practices, and entrepreneurial vision to provide powerful insights into achieving business excellence.

The Series is SCOPUS-indexed.

Ben Vinod

Mastering the Travel Intermediaries

Origins and Future of Global Distribution Systems, Travel Management Companies, and Online Travel Agencies

 Springer

Ben Vinod
Grapevine, TX, USA

ISSN 2192-8096   ISSN 2192-810X   (electronic)
Management for Professionals
ISBN 978-3-031-51523-1   ISBN 978-3-031-51524-8   (eBook)
https://doi.org/10.1007/978-3-031-51524-8

This Springer imprint is published by the registered company Springer Nature Switzerland AG
The registered company address is: Gewerbestrasse 11, 6330 Cham, Switzerland

If disposing of this product, please recycle the paper.

The people who are crazy enough to think they can change the world are the ones who do.

—Steve Jobs

*This book is dedicated to the memory of
Dr. James W. Barany, Professor and
Associate Head at the School of Industrial
Engineering, Purdue University, West
Lafayette, Indiana. Professor Barany was
passionate about students, and he always
found the time in his busy work schedule to
provide valuable guidance during his seven
decades at Purdue. He was the consummate
advisor and mentor who taught me how to
succeed in all aspects of life.*

To my parents, who encouraged me.

To my wife, Ann, who has endured my obsession with the travel industry for over three decades.

Foreword

We can't solve today's problems with the mentality that created them.
Albert Einstein (1879-1955)

By way of introduction, I enjoyed the roles of CIO of American Airlines and CEO of what was then called The SABRE Group. Although I have held various other business positions at startups, as a professional Board Director, etc., it is mostly with the AA/SABRE experience in mind during the 70s to 90s that I write this foreword.

What occurred from a technology standpoint during that timeframe was nothing short of incredible as we pushed the then-existing capabilities to the limit in all aspects of technology—whether network, hardware, software, systems architecture, etc.

Our needs grew beyond available offerings from major technology suppliers at the time such as AT&T or IBM. In addition to creating proprietary solutions to automate the various functions of operating an airline and selling American airlines products, we had a dozen or more other large initiatives such as automating travel agencies around the globe or even hosting other airlines in our system as our purview. These activities required ever quicker response times, greater accuracy, and speedier implementations as competitors entered the marketplace. In short, the breadth of our business needs required thousands of concurrent solutions being developed at any one time and pushed the envelope into uncharted territories as we strove to update millions of fares each day in real time or to implement rather risky hardware architecture designs.

As I read the chapter summaries of Ben's book it took me on a walk down memory lane. Having been part of the development of the Sabre system, I was perpetrator of and/or witness to what was captured in his materials.

In retrospect, it was challenging, but also thrilling to be associated with so many "industry firsts" as we referred to them. Whether it was introducing the fully automated AAdvantage loyalty program for our airline customers or providing state-of-the-art robotics for use by our aircraft maintenance professionals, we juggled many balls and often learned as we forged ahead. We learned about the value and importance of timely, accurate data that could be monetized, and we learned that adopting new development tools and creating a laboratory where end users tested whether our requirements matched their needs could reduce our time to market.

After reading the particulars of what was occurring in the travel industry during the period, I hope you come away with two thoughts. One is associated with innovation and the other is leadership. The complexity of the global travel market-place, as well as the regulatory challenges and ever-changing technology environ-ment, can't be overstated. Being an intermediary in such a huge, global industry is not for the faint of heart; nor is creating a successful business to provide automation for them. As you wind your way to the final chapters of the book, you will find that the complexity and challenges have not lessened. As a supplier, the question of identifying who the traveler is and what their need is for any particular trip vs what you want to sell them and how you want to portray or differentiate your product only gets more complicated. It is fascinating to hypothesize what role Blockchain, GenAI, NDC, and other solutions will play in the travel industry going forward, as well as how they will interface with the still-existing legacy systems. Hopefully the use of the new tools will make it easier for the traveler to buy and have a great travel experience as a result!

President and CEO, The Sabre Group,   Kathy Misunas
Fort Worth, TX, USA

Chief Information Officer (1993–1995),
American Airlines, Fort Worth, TX,
USA

Founder and Principal, Essential Ideas,
New York, NY, USA
September 21, 2023

Foreword

Airline Distribution: Changing Market Structure, Performance, and Conduct

I met Ben Vinod when he was the Senior Vice President and Chief Scientist at Sabre, and I was an Advisor to the C-Suite at Sabre. Based on our discussions at Sabre and through reading his books, I am impressed on the depth and breadth of his insights on the travel industry. Now, in *Mastering the Travel Intermediaries*, Ben discusses in depth the origins and future of Global Distribution Systems, Travel Management Companies, and Online Travel Agencies in the context of some fundamental changes in the distribution space.

While "distribution" is hardly a new concept for airlines or intermediaries, the landscape is changing at a rapid pace for numerous reasons:

1. The rate at which consumers' demands, expectations, and preferences are changing.
2. The need for airlines to not only improve their financial position, particularly relating to distribution costs, but also develop flexibility and agility to market seats and ancillaries that go beyond the fees for bags, changes to bookings, and the choice of seats.
3. The emergence of powerful technologies (AI/Machine Learning, blockchain, cloud computing, and smart analytics, being just a few) that are empowering consumers to become more demanding and airlines to become more competitive.

Given these changes, what will be a successful distribution strategy in the air travel industry? What products, services, and experiences will different segments of travelers value and what technologies and data would be required to create and deliver this value? Ben offers not only a compelling view of both challenges and opportunities, but also some insights on some solutions that call for airlines and distributors to work together to create mutual value.

My career in the airline industry began at Trans World Airlines in the late sixties in network, fleet, and schedule planning that I still consider as the core product of airlines. And, although I did work for American Airlines from 1985 to 1987 in network/schedule planning and at that time yield management—now revenue management, Ben and my path did not cross. However, our paths did cross when we both worked for Sabre. We spent a lot of time brainstorming on the changing role of technology and the criticality of not only data and analytics, but what travelers really value and, in my opinion, even more important, what they are willing to pay. We now have exponentially more data on customers shopping patterns and behavior (accessible from airlines' websites and from intermediaries) and more computing power that can enable us to benefit from the use of technologies such as ChatGPT and other large language models. New technologies can even take into consideration travelers' experiences. However, such a holistic approach will require, as Ben points out, a realignment of the business models of all three major category of intermediaries—GDSs, OTAs, and TMCs—an alignment that would also require much greater cooperation to achieve mutual benefit.

There are many reasons why the business models of intermediaries need to change, and Ben understands and describes the dynamics of the change. Based on Ben's comprehensive analysis of the dynamics of change in the distribution space, he examines the future of travel distribution and the future of travel intermediaries. We both agree that the intermediaries do provide value-added services (using expertise, technologies, and infrastructure), for both airlines and their customers, given the ongoing complexity of air travel. However, using intermediaries does add costs not only for airlines, but, in some cases, for customers, too. I share Ben's perspectives and I have had numerous discussions with all three types of intermediaries as well as NDC aggregators on the need to adopt a bold new vision where suppliers of travel services produce, market, and deliver relevant products and intermediaries differentiate themselves from suppliers in retailing the products with seamless and frictionless experience. In some ways, both Ben and I have encouraged airlines and intermediaries to look over the horizon, while they deal with the challenges and opportunities of today.

I strongly recommend that practitioners in the airline sector and the distribution sector read this compelling book. It not only provides an incredible understanding of the historical aspects of all aspects of travel distribution and the impact of evolving technologies on the intermediaries, but also the potential strategies for the long-term survival of the intermediaries.

Technology and Business Strategy Advisor   Nawal Taneja
St. Petersburg, FL, USA
October 1, 2023

Preface

This is the third book in the series on marketing planning in the travel industry. While the first two books focused on the origins and advances in pricing and revenue management in the airline and hotel industries, this book is devoted to product distribution and the role of the intermediaries—past, present, and future. There are many types of intermediaries in the travel industry and the focus of this book is on Global Distribution Systems (GDS), Travel Management Companies (TMC), and Online Travel Agencies (OTA).

This book starts with an introduction to travel intermediaries and the role of Thomas Cook, who created the first travel agency. It continues with the early aviation pioneers; Congressman Clyde Kelly, C.R. Smith, Blair Smith, and the many individuals from American Airlines and IBM who influenced and participated in the development of the first airline reservation system. The creation of computerized airline reservation systems led to the creation of global distribution systems (GDS) by the airline community. GDS pioneers Robert Crandall and Max Hopper played critical roles that led to the launch of the first GDS, Sabre, in 1976 by American Airlines, that allowed travel agents to make automated airline bookings for the first time.

The remainder of the book is focused on the evolving business and technology landscape, challenges faced by the GDS, the role of travel management companies and the late entrant OTAs, the new channels that spawned from the Internet in the mid-1990s and grew to dominate hotel bookings, the role of the International Air Transport Association (IATA) with messaging standards for the airline industry and shifting pricing power from the GDS to the airline with the New Distribution Capability (NDC) initiative. Alternate commercial revenue models, the role of digital identity, and the potential impact of emerging technologies like blockchain are reviewed. The book concludes with a forecast of the future state of intermediaries.

Product distribution was already established as a dominant line of business when I joined American Airlines in June 1985. During my years at American Airlines and Sabre, I learned the airline planning and airline operations business followed by product distribution from supply aggregation and demand generation perspectives. In my role as SVP and Chief Scientist at Sabre (2008–2020), product distribution was a major area of focus along several dimensions such as air shopping algorithms, air shopping cache, air availability cache, air availability proxy, air ancillaries, offer management, advanced data analytics, hotel ranking, agency workflows, performance measurement, and many more. I also had the opportunity to make presentations and interact with several travel management companies, online travel agencies, airlines, and hotel chains at Sabre, customer site visits, and at travel conferences. The knowledge I gained through these experiences is reflected in this book.

I was compelled to write this book because I worked on both the airline side and GDS side of the travel value chain. Air product distribution is not well understood by many in the industry. It is a fiercely debated topic, and like a game of chess, there are several moves, and counter moves today in the product distribution space, especially air. As a management consultant and observer in the travel industry, my goal was to maintain a neutral perspective, compile the emerging trends, and provide my perspectives on product distribution.

This third book was the most challenging to write as I had to take many trips down memory lane when I started working at American Airlines (1985–1999) and later at Sabre (2004–2020) to piece together all the moving parts of the industry transition from a collaborative to contentious relationship between airlines and the intermediaries, to connect the dots into a timeline and describe the business of product distribution in a cohesive and succinct manner.

A special thanks to two established thought leaders in the travel industry, Kathy Misunas and Nawal Taneja, for providing valuable feedback and writing a foreword for this book.

Kathy Misunas of Essential Ideas was the first CEO of The Sabre Group (TSG) and was responsible for bringing together the various information technology groups into a single unit. She also held roles as CIO and SVP at American Airlines and CEO of Reed Travel Group.

Nawal Taneja is a well-known travel industry consultant, strategist, and trusted advisor in global aviation. He is also the author of several books that explore innovative best practices in the airline industry.

A special thanks to Phil Beck, Ross Darrow, and Ann Vinod, who patiently reviewed multiple iterations, offered their suggestions, and constructive feedback as I wrote this book.

Grapevine, TX, USA
October 30, 2023

Ben Vinod

Contents

Chapter 1
An Introduction to Travel Intermediaries

1.1 What Is an Intermediary?

In the travel value chain, there are many types of intermediaries that support travel suppliers by filling critical gaps in supplier business processes. For example, global distribution systems (GDSs), travel management companies (TMCs), and online travel agencies (OTAs) are **intermediaries,** entities that facilitate a transaction between a consumer (traveler) and a travel supplier (like an airline).

Travel intermediaries are middlemen that sit between travel suppliers and consumers and facilitate various types of transactions. Using intermediaries has an added cost to the travel supplier and, in some cases, the customer as well for the value-added services that they render. Intermediaries thrive due to the complexity of travel and take on roles that travel suppliers do not have the bandwidth to manage or outsource as a cost-effective strategy. They leverage a combination of technology and expertise to provide value to customers. Using travel intermediaries has its advantages for both consumers and suppliers. The travel industry has been traditionally characterized by its use of intermediaries.

1.2 Origins

The first intermediary in the travel business was Thomas Cook, an English businessman who founded Thomas Cook & Son, later Thomas Cook Group. This was the first travel agency as we know it today.

Thomas Cook was born in Melbourne, England in 1808. He was a baptist preacher and furniture maker. He was a strong advocate of the temperance movement in Great Britain that emerged in the 1820s and 1830s. He believed that alcohol abuse was one of the major problems in the Georgian era followed by the Victorian era. The temperance movement demanded political and economic reform. At a time

Fig. 1.1 The first organized tour by Thomas Cook. Source: https://ichef.bbci.co.uk/news/976/
cpsprodpb/99A3/production/_108913393_c2191392-1641-459c-8cf8-ebc1f46d39e2.jpg.webp

when drinking spirits was economically important and culturally unavoidable, the movement was dedicated to promoting moderation and, more often, complete abstinence with the use of intoxicating spirits.

Thomas Cook wanted to attend a rally in Loughborough. On July 5, 1841, he organized a group tour with the Midland Railway for a commission to transport 500 temperance supporters from Leicester to Loughborough to attend a demonstration. For many of the passengers who undertook the 11.5-mile round trip, it was their first railway trip in a third-class railway carriage. For one shilling, the passengers received a round trip train ticket, band entertainment, afternoon tea, and food. This was a milestone event. It was the first publicly advertised organized excursion that led to the creation of the travel and leisure industry. Thomas Cook was the first travel agency. In 1855, Thomas Cook organized the first international tour. The holiday package included travel, accommodation, and food. See Fig. 1.1.

By 1845, after the establishment of Thomas Cook & Son, he organized railway excursions for profit, and the following year he offered trips from England to Scotland.

In 1851, he gained prominence when he started organizing railway travel and accommodation for people from the provinces to attend Prince Albert's Great Exhibition in London. Over a six-month period, he transported over 150,000 people to London for this event. He rapidly expanded operations throughout Europe, North America, the Middle East, and the first world tour.

Thomas Cook & Son would go on to become one of the largest travel agencies in the world. It was later renamed Thomas Cook Group. Several bad deals in the 2000s eventually led to a debt burden of £1.7 billion, which resulted in the compulsory liquidation of the company in 2019. The successor was Thomas Cook Holidays, who bought the brand in 2019.

1.3   Types of Intermediaries

The travel industry relies on different types of middlemen, also known as intermediaries, to assist travelers and suppliers in various ways such as booking a trip, payment processing, customer service, etc. These intermediaries work with suppliers such as airlines, hotels, rental cars, and cruise lines to secure seats on specific itineraries based on availability. This section categorizes intermediaries based on their differences.

1.3.1   Global Distribution Systems

The global distribution system (GDS) is a digital two-sided marketplace that facilitates transactions between travel agencies and suppliers like airlines, hotels, rental cars, and cruise lines. The GDSs evolved from the airline computerized reservations systems for travel agents to serve their customers. Sabre, the first GDS, was launched in 1976. The GDS does not hold any inventory but provides travel agents with real time access to itineraries and fares that can be booked. It serves as the middleman but *does not* manage and control the point of sale. The GDS's have a vast and complex infrastructure for processing schedules and fares to promote offers and book travel at the various points of sale that are subscribers for their services.

Today there are three established GDS brands in the free world: Amadeus, Sabre, and Travelport. Travelport is a travel brand that owns three GDSs: Apollo, Worldspan, and Galileo International. The fourth major GDS brand is TravelSky, part of TravelSky Technology Limited, that operates in a regulated environment in China mandated by the Civil Aviation Authority of China (CAAC).

1.3.2   Travel Management Companies

Brick-and-mortar travel management companies (TMC) provide a value-added service to corporate and leisure customers. For corporations, they are in the business of managed travel, optimizing the travel budget, ensuring travel policy compliance for corporations, management reporting, and providing duty of care to ensure the safety of corporate employees when they travel. A TMC manages and controls the point of sale and access to one or more GDSs for airline schedules, fares, and availability to book travel for their customers.

The largest global TMCs today are American Express Global Business Travel, CWT (formerly Carlson Wagonlit Travel), BCD Travel (the owners wanted to register ABC Travel, but it was taken and settled for BCD Travel), and Flight Centre Travel Group. Beside the large global travel agencies, there are many mid-size and smaller travel agencies that make up the long tail.

In 2021, there were over 105,000 travel agencies worldwide, and sales from these agencies exceeded $833 billion. Approximately 65% of travel agencies in the U.S. have nine or fewer employees (Gitnux, 2021).

1.3.3   Online Travel Agencies

Online travel agencies (OTA) came into existence in 1996 after the creation of the Internet. They are predominantly focused on attracting leisure passengers to book travel. The online marketplaces created by OTAs allow customers to book travel products such as air, hotel, and car. Customers can also explore travel reviews before making a booking. An OTA controls the point of sale and access to the GDS for airline schedules and fares using the GDS supplied web services.

Booking Holdings, Inc., Expedia Group Inc., Airbnb Inc., Trip.com (formerly CTRIP) Group Ltd., and Tong Cheng Travel Holdings Ltd., are some of the largest OTAs in the world (GlobalData, 2021a).

1.3.4   Tour Operators

Tour operators specialize in promoting and selling packages. A package is a collection of travel products. Most commonly, the airfare and hotel products make up the package. Packages appeal to customers since the cost of purchasing the travel components together as a package is cheaper than buying the individual components of the package. But that is just the tip of the iceberg, since both travel and non-travel products like theatre tickets and destination activities can be included in a package. There also exists a hierarchy in the sales funnel. For example, a B2B tour operator may sell tours to a travel agent who can, in turn, sell them to customers.

Tour packages are very popular in Europe and the developing countries like Eastern Europe, China, and India. China Tourism Group Duty Free Corp Ltd. Expedia Group Inc., The Walt Disney Company, TUI AG, and REWE Group were the top five tour operators in the world in 2021 by revenue (GlobalData, 2021b). Of the top ten tour operators, five are based in China, three in Europe, and two in the U.S.

1.3.5   Airline Wholesalers

Flight wholesalers, also known as flight consolidators, buy airline seats in large quantities at discounted rates and then resell them to travel agencies. Consolidators emerged in the 1970s and 1980s before the Internet age. Since then, the need for consolidators declined, but they continue to thrive since many airlines use them to offload their inventory in less popular markets to fill empty seats. Consolidators negotiate net fares, also referred to as bulk or consolidator fares, directly with airlines. These discounts can be significant, ranging from 30% to 70% in certain markets. The term "net fares" signifies the amount owed to the airline by the consolidator who, then, increases the price of the net fares and sells them to travel agencies. The agencies markup the prices even more but ensure they are competitive by maintaining fare levels below publicly available fares.

While consolidators exist in domestic markets, they are the primary source of cheap international tickets. They negotiate the best deals with airlines at the lowest possible prices for business class and first class travel. With the decline in travel agency commissions since the mid-1990s, selling marked up net fares to travelers enables travel agencies to partially offset the loss from front-end commissions. The partnership between consolidators and travel agencies benefits both entities.

Large wholesalers subscribe to the GDSs, allowing smaller travel agencies access to the GDS using the credentials of the consolidator. This gives smaller travel agencies access to public fares, as well as private fares negotiated with the airlines by the consolidator. In addition, the partnership with consolidators allows travel agencies that are not affiliated with IATA/IATAN (International Airline Travel Agent Network) or ARC (Airline Reporting Corporation) to issue tickets on behalf of airlines. Figure 1.2 illustrates the role of the airline ticket consolidators and their relationship with travel agencies.

Some of the largest consolidators that specialize in international travel who have been around since the 1970s and 1980s are CENTRAV, DowntownTravel, GTT Global, Picasso Travel, and SKY BIRD Travel & Tours. Mondee is a relative newcomer, established in 2011, that focuses on leisure and business travel.

1.3.6   Hotel Wholesalers

Wholesalers negotiate deeply discounted rates with travel suppliers and, in turn, offer their travel products to travel agencies that sell directly to customers. Hotel wholesalers are also called consolidators.

Bed banks are wholesale operators that buy bulk hotel accommodation products at a deeply discounted fixed (static) price for future travel dates and resell them to third party sellers like tour operators, travel agents, OTAs, and loyalty membership programs rather than directly to the end consumer. Figure 1.3 illustrates how bed

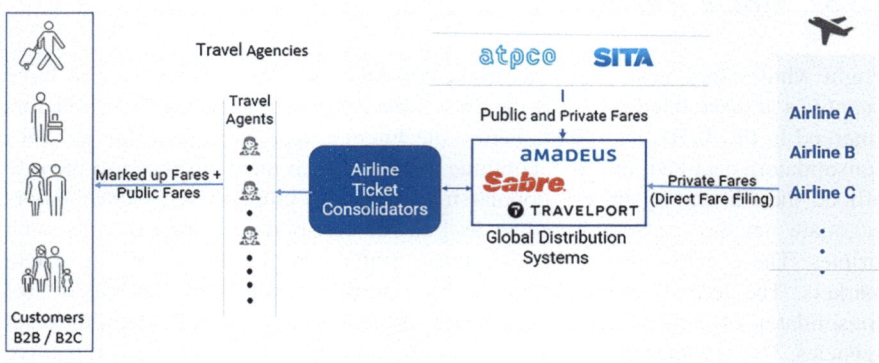

Fig. 1.2 Airline ticket consolidators and travel agencies

banks aggregate room inventory from hotels and resell to other businesses that manage the point of sale.

A bed bank marks up the hotel's deeply discounted negotiated static rate and promotes the property through its extensive network of resellers. The reseller, in turn, adds a markup or commission when they sell rooms and packages to travelers. With bed banks, multiple intermediaries exist, and each intermediary takes a cut since they facilitate the transaction. Bed banks thrive in the lodging industry due to the fragmented nature of the hotel industry, with many small hotel chains and independents that make up the long tail. Despite margin erosion, the primary benefit of bed banks to accommodation providers is increasing occupancy by distributing the product, thereby providing wide exposure to travelers worldwide through a network of travel agencies, tour operators, OTAs, and other niche resellers. Besides lodging, many bed banks also offer other products such as local activities, tickets, transfers, and car rentals.

Beds online, part of HOTELBEDS, is the leading booking engine for travel agents. HOTELBEDS, the leading B2B bed bank, provides an extranet for hotel suppliers to list their properties and manage the distribution. HOTELBEDS acquired Tourico Holidays and GTA Travel in 2017 and now has inventory for over 180,000 hotels in 185 destinations worldwide. Other bed banks are WebBeds, HPro Travel, Bonotel, Travco, GRNconnec, and MG Bedbank.

1.3.7   Metasearch

Metasearch engines are popular with leisure travelers. A significant proportion of online visitors to metasearch sites do not have a fixed destination or travel date in mind. Metasearch sites provide the capability for leisure customers to explore

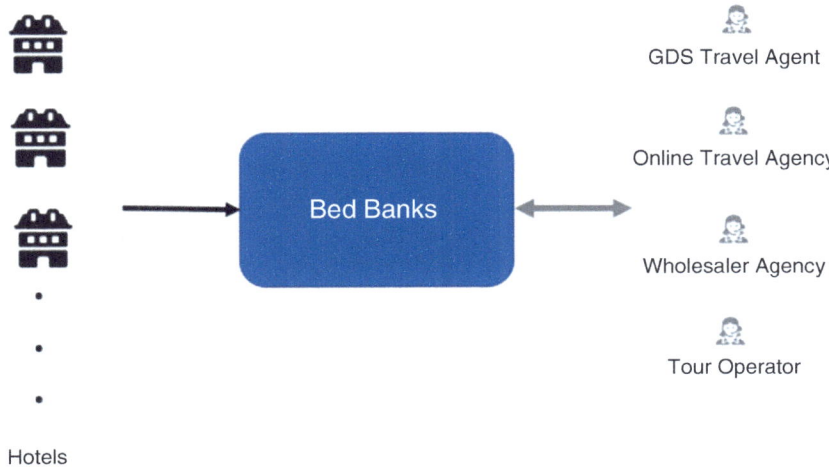

Fig. 1.3   Aggregation of hotel room inventory by bed banks

destinations and deals that are available. Metasearch engines aggregate content from supplier sites (e.g., airlines, hotels) and OTAs and present the results to an online shopper. Metasearch engines do not capture a booking or have access to room inventory. When an online user makes a booking, the metasearch engine delivers the booking to the source from where the content was aggregated.

The largest metasearch engines are Kayak, Skyscanner, Google Flights, and Tripadvisor.

1.3.8   Airbnb

Airbnb, founded in 2008 in San Francisco, is a home sharing service. It is a disruptor in hospitality since it takes market share with credible hotel inventory in key markets, targeting leisure and extended stay customers. It is a community-based online platform for listing and renting local properties. It is a marketplace and serves as an intermediary that links renters (hosts) and travelers and facilitates the process of booking a short-term or long-term home rental. The host includes all information about the property including the price in the Airbnb hosting service including high-definition photos of the property. When a guest makes a booking, a payment is made to Airbnb for the daily rate, cleaning fee, additional guest fees, and the Airbnb booking fees. After the stay is completed, Airbnb pays the rent to the host minus the

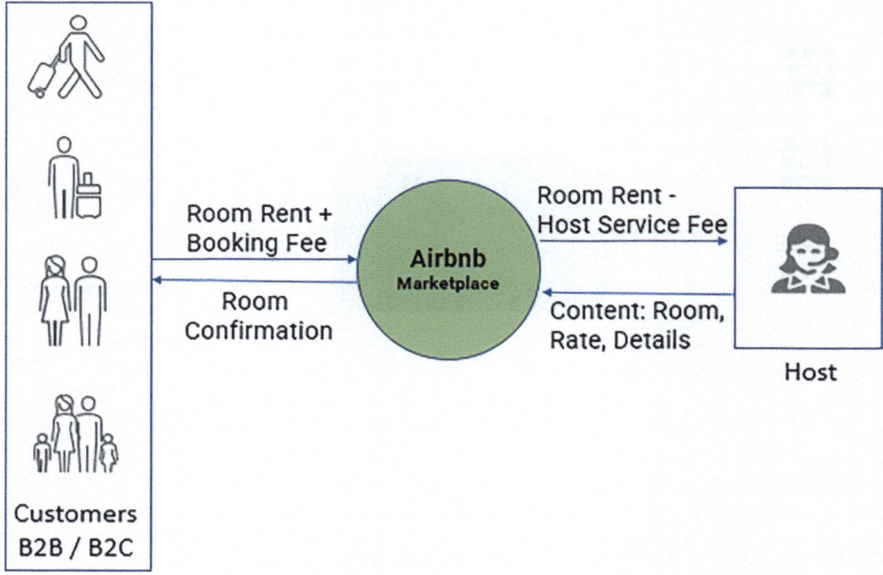

Fig. 1.4 The Airbnb business model

commission. Both the host and guest can rate the experience, which can benefit future guests.

The aggregator business model of Airbnb is like Uber and eBay that facilitates a transaction between two entities. Figure 1.4 illustrates the role of Airbnb as a facilitator of a marketplace for individuals (hosts) to rent their rooms to customers. These intermediaries can get away with high transaction fees from both the buyer and the seller since they are not corporate entities with bargaining power.

The Airbnb business model is made for a mass market geared for exponential growth.

1.3.9   Destination Management Companies

Destination management companies (DMCs) are specialized intermediaries that provide the local expertise, coordination of activities, and events in a target destination. DMC's sell tours, local events, local activities, ground transportation, and travel packages to tour operators. In addition, they work with local suppliers to provide value-added services like entertainment guides, interpreters, and theme-based packages based on the type of group such as corporations, conferences, and incentive groups.

1.3.10 Tourism Boards

Tourism boards are funded by public or private sources and develop tourism in a specific destination, region, or country. They are multi-functional and address marketing, branding, setting policy, education, regulation, and collaborate with tour operators and travel agencies to create and distribute tourism-related products and services.

1.3.11 NDC Aggregators

The NDC aggregators are relatively new and only came into existence after the launch of IATA's New Distribution Capability (NDC) initiative in 2012. They are viewed as disruptors in air product distribution since their fundamental goal is to aggregate NDC content from airlines and deliver it to a point of sale. When a booking is made through an NDC aggregator, the GDS never sees the booking. Bypass of the GDS for air bookings is called disintermediation.

1.3.12 ARC and BSP

The Airline Reporting Corporation (ARC) is a travel agency accreditation organization that provides billing and settlement in North America. It serves as an intermediary between carriers and U.S. based travel agencies. For travel agencies outside the U.S., IATA's Billing and Settlement Plan (BSP) serves as the international accreditation organization and simplifies the selling, reporting, and remitting procedures of IATA Accredited Passenger Sales Agents. ARC and BSP serve as the single point of remittance and settlement of finances between travel agents and airlines. In the absence of ARC and BSP, travel agents would have to connect to each airline for financial settlement.

ARC and BSP support two primary workflows.

When the travel agent is the merchant of record, the agent sells tickets, receives payment that goes into their merchant account, and then pays the airline with the pertinent clearing bank.

When the airline is the merchant of record, a traveler uses a payment gateway on the travel agent's desktop to make a payment for the ticket directly to the airline. In this situation, commissions from airlines are paid to a travel agent by IATA or ARC after a period.

Besides the two primary workflows, there are several variations such as the IATA EasyPay proprietary payment system.

An agency debit memo (ADM) is a notification sent by an airline to a travel agency, requesting payment for a specific amount due to an error or wrongdoing on the agency's part. In the U.S., ARC serves as the intermediary between carriers and domestic travel agencies, issuing debit memos to the travel agencies. For travel agencies located outside the U.S., IATA fulfills this role, sending debit memos on behalf of the airlines.

1.3.13   Hotel Commissions Payment

There are third party companies like Onyx CenterSource, Travel Agency Commission Payment (TACP) from NTT Data, and CTS Systems that enable payments of commissions from a hotel to the travel agencies. These commission management and reporting entities can also be classified as intermediaries that provide a service to hotel chains and independents to pay commissions to travel agencies. These intermediaries support multiple forms of payment, such as paper check, automated clearing house (ACH), and wire transfer. Hotel commissions are typically paid out within eight days after the customer checks out. These intermediaries also provide a portal for travel agencies to view the status of their commission payments.

1.3.14   Crew Accommodations and Passenger Handling During Flight Disruptions

An example of a highly specialized intermediary is crew accommodations for pilots and flight attendants. There are vendors (e.g., Accommodation Plus International, TA Connections, and Value Group) that provide a full-service end-to-end turnkey solution for airline operations and during disruptions to the airline schedule due to weather and other unpredictable factors.

In this scenario, the intermediary serves two distinct functions.

First, they handle the accommodation of airline crews in hotels during normal airline operations. They also handle the sourcing and contracting for crew bulk rates from hotel chains and independents, negotiate rates with ground transportation providers, electronically distribute hotel and ground transportation schedules to all suppliers, a crew database to validate qualification for the negotiated hotel, a crew member mobile app for alerts and hotel cancellations, and invoicing crew expenses to the airlines. These systems also interface with the various airline crew tracking systems to receive up-to-date information on crew status and schedule changes.

Second, when there is a disruption to the airline schedule and flights are cancelled, passengers that are affected are managed by this entity by accommodating displaced passengers in hotels and arranging ground transportation.

1.3.15   Medical Tourism

Another example is medical tourism, where a patient travels to another country for various reasons such as lower cost for treatment, specialized medical care, and unavailable or unapproved procedures in the home country. According to the Centers for Disease Control and Prevention (CDC), the most common procedures that people undergo on medical tourism trips are dental care, surgery, cosmetic surgery, fertility treatments, organ and tissue transplantation, and cancer treatment. Medical intermediaries, commonly called medical travel agents or facilitators, provide a complete package that includes documents, permits, travel arrangements for air, hotel, ground transportation, and aftercare accommodations.

1.4   Why Did Intermediaries Come into Existence?

Intermediaries came into existence to fill a void in the business process and to support travel suppliers. Travel intermediaries such as GDSs, TMCs, and OTAs came into existence since suppliers needed them to fill seats on flights. These intermediaries are similar to regional airlines that provide valuable higher net yield connecting passengers to fill seats on the mainline fleet.

Travel is also an information-intensive industry. Intermediaries have embraced advances in information technology, and the Internet has promoted transparency and unbiased marketplaces. First, the GDSs provided unparalleled access to schedules and fares from airlines for travel agents to book travel. Since 1996, the Internet enabled online travel agencies (OTAs) to provide the same transparency to online users. The metasearch engines followed by aggregating the web search results for online customers.

Today, transaction cost economics is the primary driver of disintermediation of intermediaries by suppliers. The larger question remains: *does disintermediation improve the efficiency of product distribution?*

1.5   Intermediary Bookings

All bookings made by an intermediary are termed *indirect* bookings. These intermediaries are travel management companies, online travel agencies, wholesalers, and tour operators. Instead of booking an air itinerary or a hotel room directly with the travel supplier, these intermediaries facilitate the booking process. The intermediary channels provide transparency for a wide range of airlines, hotels, resorts, as well as tour packages. They are compensated by the travel suppliers for every reservation made through their booking platform. The compensation varies over a wide range from around 2.2% for airline bookings made through a GDS to 25% or higher for hotel bookings made through an OTA.

1.6   The Travel Agency Revenue Model

In recent decades, travel agencies have relied on revenue streams such as GDS incentives, supplier commissions, and fees for managing corporate accounts to cover both fixed and variable operational costs. However, the industry is evolving, and, in the future, travel agencies will be less reliant on GDS incentives and supplier commissions. Instead, their primary source of revenue will originate from value-added services paid for by the traveler.

This change will transform travel agencies into an entity more focused on the traveler rather than the supplier.

1.7   What Is This Book About?

The book focuses on the origins and future of three major travel intermediaries and their crucial role in the travel value chain, the challenges they confront, and what lies ahead as technology and the business landscape continue to evolve.

The book focuses on GDSs, TMCs, and OTAs, with an emphasis on air travel. Over the past decade, the role of these intermediaries and the cost of acquiring customers have generated widespread discussions, debate, and controversy in industry forums.

Before we can talk about the current state of the union of these key travel intermediaries and their prospects for the future, we need to start at the beginning, to put it all in perspective.

Chapter 2
Origins of the Global Distribution Systems (1925–1983)

2.1 Origins of Civil Aviation

The first significant achievement in the modern era of aviation occurred when Orville Wright made the first, sustained, powered flight on December 17, 1903. The plane was built by Orville and his elder brother Wilbur. This led to the development of the first practical airplane in 1905.

The first known scheduled airmail service took place in the United Kingdom between Hendon, North London, and Windsor, on September 9, 1911. The effectiveness of aircraft in combat was first tested during The Great War (World War I: 1914–1918). The initial airmail service demonstrated an aircraft's potential for commercial use. The first scheduled airmail service in the U.S. took place from Washington D.C. to New York City with an intermediate stop in Philadelphia on May 15, 1918.

The Contract Air Mail Act or the Kelly Act, named after Representative Clyde Kelly of Pennsylvania, was passed on February 2, 1925. This act required private contractors to bid for routes to transport mail during the early days of commercial aviation. Representative Kelly was dubbed as the "father of the air mail" for his pioneering initiative to revolutionize mail service in the U.S. The Kelly Act was a significant legislation that aimed to liberate the airmail from the complete control of the U.S. Post Office. It enabled the Postmaster General to negotiate autonomous contracts with private companies for mail delivery.

Airmail planes had one seat that was available for sale to a commercial passenger. To make a reservation, passengers would contact the airmail company in the departure city. If a seat were available on the requested departure date, they could secure it.

The Kelly Act was responsible for the emergence of a profitable commercial airline industry. Scheduled commercial passenger service was started by Pan American Airways, Western Air Express, and Ford Air Transport Service. By the 1930s, four major U.S. domestic airlines came into existence, and they dominated air travel for most of the twentieth century. These airlines were United Airlines, American

B. Vinod, *Mastering the Travel Intermediaries*, Management for Professionals, https://doi.org/10.1007/978-3-031-51524-8_2

Airlines, Eastern Airlines, and Transcontinental and Western Air (TWA). The other major airline, Delta Air Lines, gained a foothold in commercial aviation during the Second World War.

When airlines began carrying more than one passenger on a flight, automation was required to handle tasks such as determining seat availability, accepting reservations, and updating inventory counts. Before the advent of computers, airlines would process reservations manually. This involved utilizing a centralized board that showcased the flight schedule for the upcoming thirty days. Employees would field phone calls for reservations, jot down passenger details, and subsequently allocate seats on the preferred flights.

In the 1930s, most airlines utilized a *request and reply* system (Copeland, 1995). Seat inventory was controlled at the departure city of a flight, which meant that the process for making changes to a reservation on behalf of a customer involved communication between the airline agent and the inventory management agent at the departure city. For example, a flight from Chicago to New York would be managed by a single office. Each scheduled service had a flight card, which served as an index card.

Airlines previously employed groups of operators solely dedicated to handling reservations. These operators would sit at round tables with numerous index cards, each representing a different flight, stored on a rotating shelf, an industrial version of a "Lazy Susan" shown in Fig. 2.1.

Fig. 2.1 Flight cards on a "Lazy Susan". Source: C.R. Smith Museum, Fort Worth, Texas

The index cards were colored, and they resembled the colors of Tiffany lamps' glass shades. This approach to handling reservations was informally referred to as the "Tiffany Card" and "Tiffany System".

To secure a seat, the operator would need to locate the corresponding index card, indicate the booked seat, and manually write out the flight ticket. This entire process would typically take 90 min per reservation. While this workflow served its purpose, it became increasingly burdensome as air travel gained popularity. With the growing number of flights and simultaneous bookings, the reservation system became a significant bottleneck. Due to the limited seating capacity around the reservations table, only eight operators could work at a time. Consequently, airlines faced significant challenges in managing their increasing fleets and processing multiple reservations simultaneously.

Initially, these offices were typically located at airports. However, as time went on, they became more centralized at major airports or at the switching office of the telephone company to minimize the need for additional phone lines. To book a flight for a customer, a sales agent would contact the appropriate booking office and inquire about availability for a specific flight. The booking agent would then retrieve the corresponding flight card from a filing cabinet and inform the sales agent about seat availability. Once the sale was confirmed, the agent would record the passenger's information, such as the itinerary, name, and contact details, on a passenger name record (PNR) card. This card would then be transmitted to the inventory system via teletype or telephone.

In 1939, American Airlines replaced their *request and reply* system with a *sell and report* system. Agents at the Boston office discovered that they could freely sell seats until the flight was nearly full, reducing the number of phone calls and speeding up customer requests, resulting in increased agent productivity. Once the flight bookings reached a certain threshold, agents would receive a "stop sale" message and return to the *request and reply* system. Despite being an improvement, this system still had inefficiencies and lacked automation. One major issue arose when a flight was completely booked, requiring the booking agent to inform the sales agent that no seats were available. The customer would then request another flight, leading the booking agent to retrieve a new flight card from the cabinet and repeat the process. During peak demand, multiple booking agents would often need to retrieve the same flight card, causing delays in the booking process.

2.2 Origins of the Airline Reservations System

In the modern era, the host computer reservations system (host CRS), commonly referred to as an airline reservations system, consists of several key domains. These domains include flight schedules, schedule change, inventory, passenger name record (PNR), pre-reserved seats, ticketing, and departure control system (DCS). DCS is also called check-in system to onboard passengers and transmit APIS (Advanced Passenger Information System) data for international flights when the flight is closed. A reservations system is necessary for customers to book flights according to an airline's schedule. The inventory control domain of the host CRS serves as an execution component for accepting or rejecting bookings based on seat availability.

Despite the expansion of commercial aviation after World War II, the reservation process was outdated and lacked automation. This deficiency resulted in reduced airline agent productivity. The *sell and report* system utilized by airlines was inefficient. To avoid oversales, airlines had to maintain seats in reserve so that the last few seats could be processed by the slower *request and reply* system. Hence, any office could sell tickets for a flight that was below a threshold, thereby expediting the reservations process. Once the threshold, usually 75%, was exceeded, it reverted back to the *request and reply* system.

The introduction of interactive real time computing technology for the fledgling airline industry was still a few years away.

2.2.1 The Reservisor

As technology progressed, reservation systems underwent significant advancements. Shortly after the conclusion of World War II, American Airlines introduced the Reservisor, which marked the first automated reservation system.

Cyrus Rowlett Smith, also known to his friends as "C.R." in the industry, took over as the President of American Airlines in 1934 when he was just 35 years old. He then went on to lead the airline for the next 34 years, during which he played a significant role in shaping the entire airline industry. In the 1940s, C.R. Smith aimed to automate the reservations processing of the airline with the help of electrome-chanical devices. However, none of the manufacturers of business equipment during that period could fulfill American's requirements for reservations processing completely. These requirements included finding flight availability, recording book-ings or cancellations, retaining processed transactions on the agent's device until they were manually cleared, and automatically notifying other stations about the flight's status. Additionally, the data processing had to be cost effective, with fewer errors than the existing manual system.

Charles E. Ammann, an employee of American Airlines who worked in the Advanced Process Research Department, was one of the early pioneers that contrib-uted toward the automation of the airline reservation process. After the Second World War, Charles Ammann, who had earlier developed the *sell and report* system for American Airlines, devised an electromechanical system that he called the Reservisor. He broke reservations processing into three logical steps: checking availability, updating the seat inventory when a seat was booked or canceled, and recording the passenger data such as name, address, and telephone number. The three logical steps were solved in stages with the Reservisor, the Magnetronic Reservisor, and the Resewriter.

The Reservisor was an electromechanical version of the flight boards used for the sell and report system (Eklund, 1994). The core of the machine consisted of a matrix where the rows represented the flights, and the columns represented the flight departures for the next ten departure days. When a flight reached its predefined threshold of 75% or higher, a relay was inserted into the board to short-out the lines when they were energized.

After C.R. Smith approved the initiative, Charles Ammann worked with the Teleregister Company based in Stamford, Connecticut. Teleregister was founded in 1948 as an independent company by the Western Union. With its extensive knowledge in communication, Teleregister played a significant role in merging basic data processing and data transmission. Teleregister developed the device to Charles Amman's specifications. The company also manufactured display units specifically designed for stockbrokers.

The first electromagnetic system designed to determine seat availability was the Boston Reservisor, named after the American Airlines agent office, where the testing took place. Testing of the Boston Reservisor started at American's Boston office on February 2, 1946, to replace the physical index card and card files to determine seat availability. Because of the time it took to reconcile the passenger's name record against seat inventory, it required a buffer of a few seats to avoid denied boardings. The Reservisor enabled the Boston office to process an incremental 200 passengers a day with 20 fewer operators.

2.2.2   The Magnetronic Reservisor

Encouraged by the performance of the Reservisor, Charles Ammann now focused his attention on not just availability but the actual inventory of seats. American and Teleregister decided to use a drum memory for storage, used by the Harvard Mark III computer, to directly manipulate the number of seats available. The name, Magnetronic Reservisor, was because the system used magnetic drums random access memory to store seat inventory.

Figure 2.2 illustrates a block diagram of the Reservisor (Eklund, 1994). It consisted of two machines, each built around its own magnetic drum memory which worked on the same data simultaneously. They compared signals, identified discrepancies, and printed exception messages for an agent to review and resolve. The duty cycle of the machine was 22 h, with 2 h of scheduled maintenance each day.

In 1952, the Magnetronic Reservisor was installed at American's booking office in LaGuardia airport in New York which could answer availability queries for 1000 flights for 10 days into the future in less than 1.2 seconds. With this system, many operators could look up information simultaneously and notify ticket agents over the telephone whether a seat was available. It was still considered inefficient because an operator and agent had to communicate at each end of the phone line. In addition, the

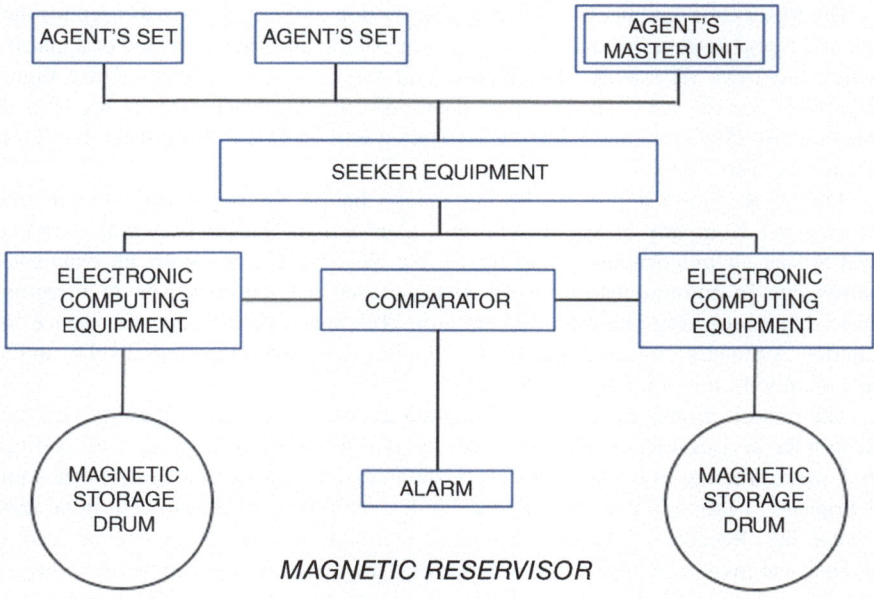

Source: C.R. Smith Museum, Fort Worth, Texas

Fig. 2.2 The Magnetic Reservisor

issue of reconciling the passenger name record with seat inventory after each booking continued to persist.

The deployment was a success and before long there were Teleregister systems at Braniff, National, Northeast, Pan American, United, and Western.

An early model of the Magnetic Reservisor on display at the C.R. Smith Museum is shown in Fig. 2.3 and an enlarged version of the Teleregister label on the Magnetronic Reservisor in Fig. 2.4.

2.2.3   The Reserwriter

While the Magnetronic Reservisor solved the availability and booking problems and improved the accuracy of seat inventories, recording the details of the passenger after the sale had to be addressed to reconcile passenger name records to seat counts.

Charles Ammann worked with IBM to add special programming and storage devices to an IBM Model 858 Card-A-Type to create a Data Organizing Translator. Called the Reserwriter, it was tested in American's Buffalo booking office in 1956. Passenger data was typed on a punched card for storage. The punched card was then processed into a paper tape and sent to the ticketing offices over American's existing

Fig. 2.3 The Magnetic Reservisor (early version). Source: C.R. Smith Museum, Fort Worth, Texas

Fig. 2.4 The Teleregister label on the Magnetronic Reservisor. Source: C.R. Smith Museum, Fort Worth, Texas

teletype network to print the paper tickets with complete routing information that determined if itinerary information had to be sent to a downline station and whether those stations were with American or another carrier. The Reserwriter fulfilled the final gap in Ammann's logical three-step vision and allowed remote offices to directly book and cancel flights while simultaneously recording passenger information. Resewriters were installed at American's larger booking offices by 1958.

The Reservisor systems developed by Teleregister were a significant milestone in the history of transaction processing. It was a precursor to the larger IBM and UNIVAC mainframe computer environments that succeeded it. Unfortunately, Teleregister was a small company with fewer than 200 employees and was no match for IBM and UNIVAC who had deep pockets for research and development. IBM went on to develop the fully automated Sabre reservations system that supported high-volume transaction processing. Teleregister continued in business, supplying computing and communications equipment to the banking industry and brokerage firms. They were acquired by the Thompson Ramo Woolridge (TRW) group in 1965.

2.2.4   The Chance Encounter

While the Reservisor, Magnetronic Reservisor, and Reserwriter went a long way to improve productivity in the booking offices of American, it was still highly dependent on manual input and prone to human error. About 8% of all bookings contained errors. In addition, the process of booking a flight was still manually intensive, requiring input from as many as twelve different individuals and took up to 3 h. At the dawn of the Jet Age, Pan American was the launch customer for the Boeing 707. In 1952, American ordered its first 707s, to complement the existing Douglas DC-7 prop liner fleet. The Boeing 707 jets had a cruising speed of 600 miles/h, compared to 365 miles/h for the DC-7. The Boeing 707s went into service for American in 1958. The faster new jets allowed more flights per aircraft per day and delivered more passengers faster than the booking systems could sell tickets. C.R. Smith realized the magnitude of the operational problem that could increase the cost of processing a reservation.

The chance encounter between American Airlines CEO C.R. Smith and a young IBM International Business Machines Corporation (IBM) salesman R. Blair Smith on a Los Angeles to New York flight in 1953 revolutionized airline reservations processing (Copeland, 1995) on an unprecedented scale. Blair Smith was on his way to an IBM training session in New York.

Blair Smith[1] later recalled the conversation with C.R. Smith: "I told [C.R. Smith] I was going back to study a computer that had the possibility of doing more than just

---

[1] IBM Archive: Sabre: The First Online Reservations System, https://www.ibm.com/ibm/history/ibm100/us/en/icons/sabre/s

keeping availability. It could even keep a record of the passenger's name, the passenger's itinerary, and, if you like, his phone number. Mr. C.R. Smith was intrigued by this. He took out a card and wrote a special phone number on the back. He said, 'Now, Blair, ... when you get through with your school, our reservation center is at LaGuardia Airport. You go out there and look it over. Then you write me a letter and tell me what we ought to do.'"

During the training session, Blair Smith updated IBM's CEO, Thomas J. Watson Jr., regarding his discussion with C.R. Smith. Watson emphasized the importance of fulfilling C.R. Smith's requests, which included conducting a tour of the reservation center, composing a recommendation letter, and sharing a copy of the letter with Watson. Blair Smith proposed a collaborative endeavor between IBM and American Airlines to develop a computerized reservations system. This initiative, named SABER (Semi-Automatic Business Environment Research), aimed to explore the technical feasibility of automating the reservations process by linking passenger names with seat reservations. Upon the completion of the project in 1958, American Airlines signed a new contract with IBM to design the functional specifications for the world's first passenger name record (PNR) system, enabling the seamless matching of passengers to seats.

2.2.5   Semi-Automatic Ground Equipment (SAGE)

The year 1951 saw the establishment of project SAGE (Semi-Automatic Ground Environment) by IBM, acting on behalf of the United States Air Force. The project aimed to create a real-time computer system for air traffic control. SAGE, partially developed by IBM, was an air defense system promising interactive real time computing. In 1959, SAGE was declassified by the federal government. The project made several technical advances such as magnetic core memory, active- standby dual processors, modems for digital communications over voice-band channels, time sharing the central processor, input–output control with memory cycle stealing and branch, and index instructions (Copeland & McKenney, 1988; Astrahan & Jacobs, 1983). Additionally, the IBM SAGE computer, also known as AN/FSQ-7, had a communication front end that received real time data from tracking devices over communications lines. Knowledge gained from SAGE proved invaluable for developing the world's first computerized airline reservations system.

IBM discovered that instead of transmitting messages from radars to interceptor aircraft, they could employ the same system for transmitting messages from travel agents to airline ticketing offices. This system would have the capability to inform agents about seat availability, handle their bookings, and even generate tickets, all without the need for human intervention on the receiving end of the telephone line.

2.2.6   The IBM Reservations System

IBM considered the Sabre partnership to be a risky venture and enlisted the consulting firm Arthur D. Little to assess its feasibility. At that time, IBM focused solely on selling or leasing hardware and relied on customers to create the software. To avoid failure, IBM made an investment to support program development. Roger Burkhardt and Wilfred (Fred) Plugge, both from American, were the main driving forces behind this initiative. To expand the programming team, American hired mathematician Mal Perry, who played a pivotal role in the development of the reservations system. The deployment of the system was planned to occur in phases, starting in 1961. The joint IBM/American team relocated to American Airlines' headquarters at 99 Park Avenue in Manhattan, New York.

Bill Elmore was responsible for coding the first PNR demonstration on an IBM 650. This demonstration, consisting of 25,000 instructions, served as an early indicator of how powerful the Sabre system could be. Unfortunately, a software bug in the program resulted in the 650's drum memory unexpectedly resetting to zero. As a result, doubts and skepticism surrounding the system were further amplified (Head, 2002).

By 1960, both Pan American and Delta had signed contracts with IBM, and the joint project for all three carriers was named SABER. This prompted American to seek a different name for its system, which became known as SABRE (Semi-Automated Business Research Environment).

IBM made errors in their proposals to Pan Am, American, and Delta by proposing incompatible hardware and not recognizing that the main challenge was application programming in assembly language. American utilized the binary IBM 7090 computers, while Delta's DELTAMATIC used the IBM 7070, and Pan Am's PANAMAC utilized the IBM 7080 computers, which were decimal machines (Copeland & McKenney, 1988) that executed operations on decimal numbers and addresses without the need for conversion to a binary representation. These computers differed in terms of their throughput capabilities and processors (Siwiec, 1977). In hindsight, to reduce development costs by distributing them among the three airlines, they should have standardized on the more expensive binary IBM 7090 computers.

The Sabre central reservations system site was in Briarcliff Manor in 1962. Robert V. Head, who began working with the Sabre team in October 1959, had an interesting experience. While visiting Bill Elmore's office, which had a view of the exercise yard at the Sing Correctional Facility, Elmore explained to Head the difference between prisoners at Sing and Sabre programmers: the prisoners knew when they would be released. After his time on the Sabre project, Robert V. Head became a prolific writer. He contributed articles to banking, automation, and computer journals and authored several books on information systems and management.

Sabre was launched in 1964 after experiencing several cost overruns and delays. This system was the first fully operational computerized reservations system. DELTAMATIC and PANAMAC followed suit and became operational in 1965.

The Sabre system was equipped with two IBM 7090 mainframe computers, one for real-time processing and the second, for backup and batch jobs. That were set up in a state-of-the-art data center situated in Briarcliff Manor, New York. This cutting-edge computing environment allowed American to check for availability efficiently, update inventory, and generate passenger name records (PNRs) automatically at a reasonable cost. As a result, American's reservation error rate decreased to less than 1%, and reservations could be swiftly processed in a matter of seconds. The new system boasted an impressive capacity of handling 7500 reservations per hour, a significant improvement compared to the previous manual card system where it took an average of 90 min to process a single reservation.

The online transaction processing (OLTP) system was the first of its kind in the world. At the time of its launch, Sabre was the world's largest private real time commercial data processing system, second only to the U.S. government. Raytheon cathode ray tube terminals (CRT) were used by agents to access the mainframe reservations system, which were completely reliant on the host. These terminals were dubbed "dumb terminals" by the industry because they had very limited capabilities on their own.

In 1964, IBM introduced the System/360, the first of its computers to use interchangeable software and peripheral equipment. Based on the knowledge gained from developing the Sabre system, IBM developed a generalized version of the reservations system called Programmed Airline Reservations System (PARS) for midsized carriers. PARS used a specialized operating system, the Airline Control Program (ACP), which was designed for handling a large volume of inputs, fast terminal response, system availability, reliability, and recoverability. It could run on System/360 models 40 through 75, though Model 65 was the most common configuration (Siwiec, 1977). IBM took orders for the processors for deployment in 1968 from Braniff, Continental, Delta, Northeast, and Western.

In 1965 Eastern decided to deploy PARS rather than pursue internal development. Eastern Airlines deployed PARS on the powerful System/360 computers which became operational in 1968. Eastern's System One was an important milestone, as it was based on PARS and became the new standard for reservations processing. TWA and United were allied with hardware vendors Burroughs and Univac but encountered technical difficulties due to their lack of experience with teleprocessing systems, unlike IBM's work on SAGE. Both carriers ultimately decided independently to purchase Eastern's software with installation support from IBM as an alternate solution (Copeland & McKenney, 1988).

By 1971, PARS became the industry standard with the application software for passenger name reservation and seat inventory. United introduced the Apollo system in 1971 based on PARS.

2.2.6.1   Sabre Upgrade to PARS

From 1971 to 1973, Sabre underwent a series of upgrades to the PARS-based system. These upgrades included the replacement of the IBM Selectric terminals with cathode ray tubes in 1973. After the upgrade to a PARS-based system in 1971, nine out of the top ten major U.S. airlines were also using PARS (with IPARS being the international version). These systems, which were developed in TPF, were transaction oriented and capable of processing a high volume of messages per second. They supported various reservations functions such as schedules, inventory management, passenger name records, shopping, itinerary pricing, ticketing, and departure control. The introduction of ACP and later TPF high-performance operating systems brought about a significant revolution in transaction processing, greatly improving the efficiency of airline passenger operations and ultimately contributing to airline profitability. The generic name for IBM-based reservations systems is Passenger Service System (PSS).

2.2.6.2   ACP and TPF

The operating system for the reservations systems was called the Airline Control Program (ACP). ACP was developed by IBM in the mid-1960s to support reservations processing for airlines in North America and Europe. It was a low-level programming language in Assembly and code was written in 128-byte, 381-byte, 1 K, or 4 K blocks. If the code could not be contained in a block, the blocks were chained for continuity. Remarkably, ACP has survived conceptually unchanged into the twenty-first century.

In 1979, IBM introduced the transaction processing facility (TPF) as a substitute for ACP. TPF was specifically designed to handle high volumes of transactions and accommodate a very large number of simultaneous users, while maintaining rapid response times. The TPF Database Facility (TPFDF) is a high performance specialized database. There is also the Airline Control system (ALCS) monitor developed by IBM that integrates TPF services to operate in a MVS (Multiple Virtual Storage) environment. Several airlines operate ALCS-based reservations systems.

Common examples of high-volume transaction processing applications include airline reservations, banking, retail, healthcare, and credit card processing. These companies depend on the computing power of mainframes to handle numerous applications that process vast amounts of data, enabling them to cater to hundreds of thousands of users at the same time.

Today, prominent customers that utilize TPF include American Airlines, United Airlines, Delta Air Lines, Sabre, Travelport, American Express, Discover Financial Services, JPMorgan Chase, and Visa Inc. TPF runs on a traditional IBM System/370 assembly language environment.

The TPF 4.1 release was replaced by z/TPF v1.1 in September 2005 with 64-bit addressing and mandates use of the 64-bit GNU development tools.

2.2.7   *Current Airline Reservations Environment*

Today, Amadeus' Altea Customer Management Suite, Sabre's *SabreSonic* Customer Sales and Service and Navitaire New Skies reservations systems host many airlines. Navitaire, a wholly owned subsidiary of Accenture, was acquired by Amadeus in 2015. Radixx, a travel technology company founded in 1993 that provides reservations processing for LCCs, was acquired by Sabre in 2019. Other vendors that provide hosting services for reservations to a few airlines are Shares, SITA Horizon Customer Sales and Service, Hitit Computer Services, ameliaRES (owned by InteliSys Aviation), AeroCRS, ACCELaero (owned by Information Systems Associates FZE), Mercator (Accelya) and iFly Res (IBS). UniSys Aircore® in a partnership with TravelSky provides hosted reservations services for carriers based in China. The Civil Aviation Administration of China (CAAC) mandates that all Chinese carriers should be hosted for reservations in China. While the industry has been consolidating and using the services of CRS vendors that provide hosted services, about 10% of the world's airlines continue to use proprietary systems for reservations processing.

PSS market share is measured based on passengers boarded worldwide. Excluding China and the CAAC carriers, Amadeus (Altea and Navitaire) has approximately a 46% share of the market, followed by Sabre at 15%. SITA has a declining market share that is about 5%, and the balance is made up of the smaller vendors Mercator, Radixx (owned by Sabre), Unisys (outside China), Shares, Hitit Computer Services (Crane Passenger Service System), and others.

Figure 2.5 shows the evolution of airline reservations systems based on IBM hardware and software (Airline Control Program).

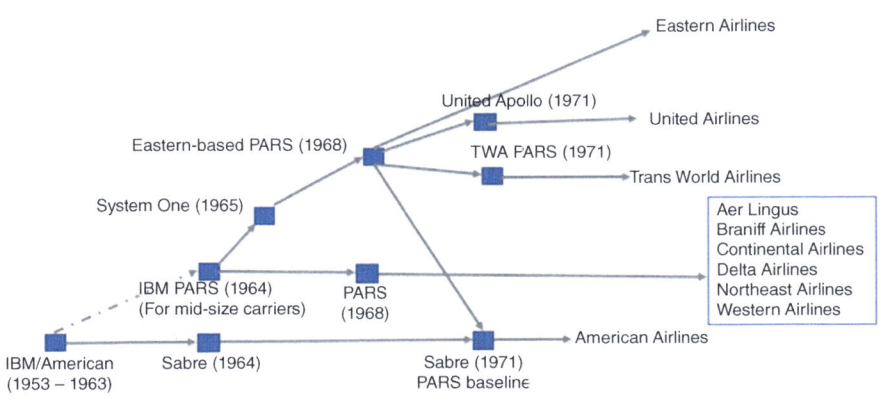

■ Airline Reservations System(CRS) Milestone

Fig. 2.5 Origins of the IBM-based Airline Reservations Systems

2.3   The British Pioneering Work with UNIVAC and IBM

Technically, the UNIVAC (Universal Automatic Computer) Air Lines Reservations System was the first computerized airline reservation system that was launched in 1958.

In the 1960s, British European Airways (BEA) was the fifth-largest airline in the world and carried over seven million passengers per year. In contrast, British Overseas Airways Corporation (BOAC) (formed from the merger of Imperial Airways and British Airways Ltd. in 1939) carried a million passengers per year and was ranked 35th. Beacon was BEA's online computer network that was developed between 1963 and 1967 as a full-scale passenger reservations system. The Beacon reservations system (Knight, 2008) could handle over 10 transactions per second and was connected to all BEA reservations offices in Europe. The Beacon hardware consisted of two UNIVAC 490 mainframe computers.

In 1965, IBM made inroads at BOAC which had a complex international route network consisting of many segments on the same flight number. IBM promised a turnkey reservations system and hardware based on their experience with AA's Sabre, Pan Am, and Delta. Peter Hermon, who was responsible for the development of the new airline reservations system called BOADICEA, selected IBM because of their Programmed Airline Reservations System (PARS) software. BOADICEA was the name of a British queen who lived in AD 60. PARS did not have all the features required for an international carrier like BOAC. IBM collaborated with BOAC to enhance the application with international features, leading to the creation of IPARS (International PARS).

The creation of IPARS with BOAC input to IBM was probably the single greatest achievement in reservations processing on an international scale. BOAC became the first European carrier to adopt a computer reservations system with reservations, inventory control, passenger name records, and interline messaging. Interline messaging was important; 60% of BEA's reservations came from other airlines. The interline messages were based on IATA's AIRIMP telex message that were sent over the airline-owned SITA network. The original hardware deployed was two IBM 360/Model 50 computers with channel switching for peripherals to be switched between processors. For the second version, a third interchangeable processor was added, and the hardware was upgraded to the IBM 360/Model 65a, each with a 256k core and a processing speed of 650,000 instructions per second (0.65 MIPS). Over 1000 BOAC terminals worldwide were connected to this environment and maintained connections to hundreds of airlines (Harris, 1993).

On April 1, 1974, BEA and BOAC merged to create British Airways. Beacon did not survive the merger and BOAC's rival system, BOADICEA, in partnership with IBM hardware became the new standard. The merger of Beacon and BOADICEA resulted in the creation of BABS (British Airways Booking System). BOAC and its successor British Airways were dominant in the 1970s, 1980s and early 1990s marketing versions of BABS to airlines that included the hosting of reservations processing.

This is a print advertisement (Fig. 2.6) for Boadicea from 1969 that I purchased at Camden Market, London.

Fig. 2.6 BOAC print advertisement of Boadicea

2.3.1 The SAS Technology Platform

Scandinavian Airline System (SAS) has a long, checkered history of customizing reservations system environments to fit their business needs. Their initiatives also helped Univac (and later Unisys) gain a foothold in the reservations market to compete against IBM (Grimstad, 2015).

SAS Data was established in 1958 when SAS acquired its first computer. Electronic data processing at SAS consisted of three systems—Space Availability System (SPAS) in 1958, Space Availability Control (RAMAC) in 1960, and load control and check-in (ZEBRA) in 1961. SPAS was built by Standard Electric Lorenz, RAMAC was running on an IBM 305 and ZEBRA on a STANTEC computer in 1961.

SPAS was a special purpose computer and required 8000 transistors. It could instantly tell a local booking terminal if space was available on a flight. The Standard Electric Lorenz "Agent Set" was used in sales offices and travel agencies worldwide. Once availability on an aircraft was known, reserving a seat required a confirmation message to SPAS with a telegram from the booking office.

SPAS was unaware of the seat capacity on an airplane. The lack of visibility to the exact number of available seats posed a limitation, as an aircraft could take off with unoccupied seats. To address this, a space control system was implemented on an IBM 305 RAMAC, which was connected to SPAS.

To address weight and balance and check-in of passengers, a British built and tube-based STANTEC computer, nicknamed Zebra was introduced. SPAS, RAMAC, and ZEBRA were connected to each other.

In 1965, SAS created SASCO 1, a new version of all three inter-connected systems and the semi-automatic Telegram Center running on an IBM 1410. The Telegram Center in Copenhagen, which processed 25,000 telegrams a day was replaced by the Message Switching Computer (MESCO) in Stockholm in 1967 followed by Copenhagen and Oslo on UNIVAC 418 computers. SAS deployed a front-end Data Switching Computers (DASCO) at all three locations for the incoming traffic from the agent sets.

When IBM failed to deliver the new IBM 1410 hardware with the promised capacity of twice the 1965 system, SAS deployed the Univac 494 as application computers to replace the IBM 1410 for the new SASCO 2. The new Univac system was operational in January 1969. In 1977, the Univac 418 computers used for communications were replaced by Collins Communications in Canada and called TELCON 1. SAS acquired a Collins system with 2500 terminals. TELCON 1 also replaced both MESCO and DASCO.

To complement the Univac application computers, SAS acquired IBM mainframes in 1974 and established a parallel development and operating environment with Unisys. To run real-time applications on the IBM computers, an Information Management System (IMS)[2] hierarchical database for high-performance online

---

[2]IMS is available on IBM's z/OS.

transaction processing was deployed with an interface for SASALFA terminals to communicate with the developed system. To support more terminals. TELCON 1 was replaced by TELCON 2 in 1977, a Univac system based on Sperry 3760 and Sperry DCP40 processors. TELCON 3 with DCP40 dual communication nodes from Unisys was deployed in 1981. It consisted of an IBM mainframe front end, a Unisys mainframe front end, and a concentration of terminals in the SAS cities— Copenhagen, Stockholm, and Oslo.

In 1982, SAS Data Services became a division of SAS. In the 1980s and 1990s, the system went through various enhancements, such as Direct Access and Higher Connectivity with the EDIFACT protocol. The SAS technology platform in the 1990s was comprised of UniSys, IBM, Tandem, Unix, AS/400, PC's, and local area networks (LAN) with TCP/IP communications. The SAS reservations system was called ResAid. After 45 years (1958–2004), SAS Data was acquired by Computer Sciences Corporation (CSC) in 2004 and renamed CSC Airline Solutions. For economic and strategic reasons, SAS migrated to the Amadeus Altea Customer Management Suite in 2013.

2.3.2 UNIVAC and Airline Reservations Systems

Besides BEA and Scandinavian Airline System, Northwest Airlines, Capitol Airlines, Eastern Airlines, Iberia, Lufthansa, Air France, and many airlines invested in the UNIVAC airline reservations system. Burroughs Corporation acquired Sperry Corporation in a hostile takeover in 1986 to create Unisys and had a plan to compete with IBM in the mainframe market. However, with the decline in mainframe computing, they changed direction. Unisys continued to develop and market its reservation services under the acronym USAS (Univac Standard Airline Systems) and manufacture enterprise class computers with the Unisys 1100-series, 2200-series, and ClearPath server lines.

The USAS applications for passenger reservations (USAS*RES) and cargo (USAS*CGO) were written in the early 1970s in Fortran, though some applications were written in COBOL, the Unisys 1100/1200 assembly language, and Logic and Information Network Compiler, a fourth-generation programming language (LINC 4GL). USAS had many large airline customers including Lufthansa, Iberia, Cathay Pacific, Scandinavian Airline System, Air France, the CAAC carriers in China (China Southern, Air China, China Eastern are the largest), and others. At Air France, the reservation system was customized and launched as Alpha 3 (Alpha trois) in 1970. Air France also hosted several airlines on Alpha3 (Regional, Brit Air, City Jet, VLM Airlines, and CCM Airlines).

The Unisys AirCore reservations system is the new generation open source passenger reservations system. In 2005, AirCore partnered with Lufthansa Systems to launch Future Airline Core Environment (FACE). The objective of FACE was to provide airlines with a range of passenger service solutions, including expanded electronic distribution channels, and Unisys established a partnership to reduce

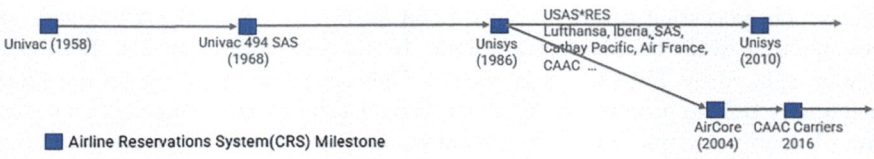

Fig. 2.7 Univac and SAS

airline distribution costs. It was subsequently abandoned in 2007 for commercial reasons. AirCore is used in China by the CAAC carriers which collectively transported 417,255,845 passengers, down 36.74% from 2019 (659,629,070 passengers)[3] caused by the COVID-19 pandemic (CAAC, 2020). Unisys AirCore is a replacement for the legacy passenger reservations system USAS*RES.

In 1987, four leading UNIVAC airline reservations system users founded Amadeus. When Amadeus was launched in 1992, a version of SYSTEM ONE was used by Amadeus to manage passenger name records for airlines and travel agents. At that stage, Amadeus did not have inventory control, ticketing or departure control and relied on USAS, for the founding airlines of Amadeus for these services. Amadeus did not begin development of inventory control, seat maps, ticketing, and departure control until British Airways and Qantas signed a contract in 2000 to replace BABS with the new Amadeus Altea Customer Management Suite. Today, Altea is the dominant reservations system provider to the full-service airlines.

Figure 2.7 illustrates significant milestones of Univac at SAS and beyond.

2.4   The SITA Reservations System

The SITA passenger reservations system is used by several smaller airlines. Gabriel was the SITA derivation of the Sperry USAS*RES system that Sperry had sold to Aeroflot in the 1960s. The SITA Gabriel reservations system focused on second- and third-world countries in Africa, India, Eastern Europe, and the former Soviet Union.

SITA invested in and launched a new version called Horizon Passenger Management and Distribution System, which is a reservations system for both low-cost carriers (LCC) and full-service carriers (FSC). FSCs and LCCs are discussed in detail in Sect. 2.16.

In 2003, SITA decided to migrate Gabriel reservation system customers to the new Unisys Aircore platform (FlightGlobal, 2003).

---

[3] Air transport, passengers carried—China, World Bank data, Air transport, passengers carried - China | Data (worldbank.org).

2.5 Key Organizations that Shaped the Airline Industry

There were a few key organizations that helped formulate policy for the fledgling airline industry during the early years of commercial aviation.

2.5.1 IATA and ICAO

The first incarnation of the International Air Transport Association (IATA) occurred on August 25, 1919. It was established as an organization of airline carriers, not of international governments. Membership was voluntary and new membership applications were approved by a majority of existing members. IATA focused on collaboration and sharing of information on air safety, standardization of aircraft design and construction, weather navigation, flight communications, weight and balance, landing strip requirements, aircraft deicing, and high altitude flights between member airlines. The last meeting of the first IATA was held in 1938 before the start of World War II.

By 1944, many nations were interested in formulating a universal international air transport policy. Based on invitations sent to nations by the United States, 54 nations sent representatives to the Chicago Civil Aviation Conference in 1944. At the conference, there were two distinct points of view. The U.S. aviation industry was far more mature than that of European nations. While the U.S. wanted open and free competition, the other nations, the most vocal being the United Kingdom, wanted an agency to control capacity, frequency, and fare levels (Taneja, 1976) on international routes as a protectionist measure. The convention ended in a stalemate over the economic aspects of regulating international air travel. However, at the Chicago Convention, several milestones were achieved. The Five Freedoms of the Skies (Vinod, 2021a, Appendix A) were articulated,[4] aircraft registration, air safety, and the issuance of licenses. The Chicago Convention was signed on December 7, 1944, by 52 states and the Provisional International Civil Aviation Organization (PICAO) was created as an intergovernmental body to address technical matters such as air safety, pending final ratification from the 26 member States (Chicago Convention, 1944). It functioned in a provisional capacity until March 5, 1947, when the last signature for ratification was received. ICAO, the International Civil Aviation Organization, became operational as an official entity on April 4, 1947.

In the spring of 1945, representatives from 31 scheduled airlines met in Havana, Cuba, to organize their own international governance body and resurrect IATA. The Canadian delegation was in favor of the establishment of an organization. IATA was created and incorporated by an Act of the Canadian Parliament on December

---

[4]ICAO only recognizes the first five freedoms as recognized by international treaties. In total, there are nine traffic freedoms, which also addresses various types of cabotage, several of which are governed by bilateral treaties.

18, 1945. The following year, delegates from the U.S. and United Kingdom met in Bermuda to provide a stamp of legitimacy to the fledgling organization and to address trans-Atlantic fare regulation. At Bermuda, the stalemate from the Chicago Convention was overcome when the United Kingdom withdrew its demands on frequency and capacity restrictions in return for the U.S. acceptance of direct control of air fares that were subject to government approval. IATA was thus given the power to schedule Traffic Conferences (TC) to set international fares, while the routes themselves were subject to mutual bilateral inter-governmental agreements (Pillai, 1969). The objective of the cartel or syndicate was to ensure that fares were protected to avoid relentless competition, while ensuring that they were as low as possible, in the interest of the traveling public. Strange as this may sound, IATA had two distinct functions: the standard organizational function to support the airline community and the semiautonomous rate-fixing syndicate through the Traffic Conferences that regulated prices, schedules, and routes in the interest of the profit of the carriers. The fares were always subject to final government approval.

2.5.2 The Civil Aeronautics Board

To ensure a federal focus on aviation safety, President Franklin D Roosevelt signed the Civil Aviation Act in 1938. This important piece of legislation established the independent Civil Aeronautics Authority (CAA). The CAA had a three-member Air Safety Board that conducted accident investigations and made recommendations to prevent them in the future. The legislation also expanded the government's role in civil aviation and the CAA had the authority to regulate airline fares, and routes served by individual carriers. In 1940, President Roosevelt split the CAA into two agencies, the Civil Aeronautics Administration which is now part of the Department of Commerce, and the Civil Aeronautics Board (CAB). The Civil Aeronautics Administration retained responsibility for Air Traffic Control (ATC), airman and aircraft certification, safety enforcement, and airway development for the future. The responsibilities of the CAB included accident investigation, safety rulemaking, and economic regulation of the airlines.

As a fledgling industry, the U.S. Congress granted CAB the power to exempt airlines from the antitrust laws. CAB was given regulatory power over civil aviation including airline tariffs, airmail rates, mergers, and competition. CAB virtually controlled the performance of the U.S. airlines until the late 1970s by regulating prices for each route and governing which carrier could operate on which route.

2.5.3 Air Transportation Association of America (ATA)

The Air Transport Association (ATA) was founded in 1936 in Washington D.C. ATA played a key role in major government decisions on the U.S. aviation industry, including the creation of CAB, establishment of air traffic control system, and airline deregulation. ATA collaborates with the U.S. administration and Congress to create a tax and regulatory environment that enables U.S. airlines to provide the services needed to compete globally.

ATA changed its name to A4A (Airlines for America) in 2011 with the tagline "We Connect the World." IATA and A4A publish a wide range of manuals for all aspects of the airline industry. IATA published AIRIMP (Air Interline Messaging Procedures), the reservations interline message procedures passenger manual. A4A publishes the Standard Interline Passenger Procedures, known as SIPP. The AIRIMP telex messaging was the backbone of inter-airline communications between reservations systems and GDS connectivity in 1976. It continues to be used today, even after EDIFACT was introduced in 1987/1988.

Members of A4A move over 2.3 million passengers across 80 countries and 66,000 tons of cargo across 220 countries every day.

2.5.4 Federal Aviation Agency (FAA)

In May 1958, A.S. "Mike" Monroney, the Democratic Senator from Oklahoma and chairman of the Aviation Subcommittee of the Senate Commerce Committee, introduced the Federal Aviation Act of 1958 in Congress to create an independent Federal Aviation Agency to oversee the safe and efficient use of the national airspace. On August 23, 1958, President Dwight Eisenhower signed the Federal Aviation Act, which transferred the Civil Aviation Authority's functions to a new independent Federal Aviation Agency that was responsible for civil aviation safety. Known for his contributions to civil aviation, Mike Monroney was called "Mr. Aviation" in the Senate.

2.5.5 Department of Transportation (DOT)

President Lyndon Johnson was an advocate of creating a single department that was responsible for developing and carrying out comprehensive transportation policies and programs across all modes of transportation. In 1966, Congress authorized the creation of a cabinet level position that covered all aspects of transportation. The new Department of Transportation (DOT) became fully operational on April 1, 1967, and the Federal Aviation Agency became one of several modal organizations within DOT. It was renamed the Federal Aviation Administration (FAA). As part of the reorganization, CAB's accident investigation function was transferred to the newly created National Transportation Safety Board.

2.5.6   *American Society of Travel Advisors (ASTA)*

The American Society of Travel Agents was founded in 1931, and its members represent 80% of all travel sold in the U.S. through the travel agency distribution channel, the GDS. They also have hundreds of internationally based travel agencies. In 2018, it was renamed the American Society of Travel Advisors, to reflect the changing role of the travel agent over the past few decades. It is the largest trade organization for the travel industry and represents more than 2600 travel agencies and supplier companies.

2.5.7   *Global Business Travel Association (GBTA)*

The Global Business Travel Association is the world's leading forum for business travel. The business travel and meetings organization is based in Washington, D.C. with over 9000 members that manage more than $345 billion of global business travel.

2.5.8   *Business Travel Association (U.K.)*

The Business Travel Association (BTA) serves as the voice for the business travel industry in the United Kingdom. The BTA's primary focus is to foster engagement and advancement within the industry through conferences, industry discussions, educational sessions, networking opportunities, and promoting interaction among individuals.

2.5.9   *Travel Management Coalition*

The Travel Management Coalition (TMC) consists of influential leaders in the travel sector who have been severely affected by the COVID-19 outbreak. Its main objective is to ensure a complete recovery of the industry and its customers and employees which suffered significant damage due to the pandemic. As a collective, the TMC represents thousands of travel agents based in the United States and serves millions of corporate and leisure travelers who have been greatly impacted by the pandemic.

2.5.10 Folatur

Based in Latin America, Folatur promotes the development of the distribution channel for the tourism industry, represented by travel agencies and tour operators, all of which promote transparent competition.

2.6 Airline Paper Tickets

In 1909, DELAG (*Deutsche Luftschiffahrts-Aktiengesellschaft*, translates as "German Airship Travel Corporation") introduced paper tickets for the first scheduled air flights in Frankfurt, Germany. The world's first commercial airline operated a fleet of Zeppelins, which were manufactured by the Luftschiffbau Zeppelin Corporation, and offered sightseeing flights starting in June 1910. Later that same year, inter-city passenger operations were added to their services. The Zeppelin was a rigid lighter-than-air airship that achieved flight by creating buoyancy or lift by using a gas that is lighter than air.

Inaugurated on January 1, 1914, the St. Petersburg-Tampa Airboat Line operated the first heavier-than-air passenger air service using a fixed-wing aircraft. Abram C. Pheil purchased the first ticket for the 23-mile journey, making him the world's first airline passenger. It was an open-air wooden craft that rarely exceeded an altitude of five feet. The Line holds the distinction of being the first scheduled passenger and commercial freight airline. It operated for three months until March 31, 1914, and carried 1205 passengers without any incidents.

In the early days of commercial aviation, airline tickets were like train tickets and were created by the operator who provided the service. In 1930, the Traffic Committee of IATA (International Air Transport Association) developed the first standards for handwritten tickets, which allowed for up to four segments in each itinerary. These standards served the airline industry until the 1970s.

The airline tickets were like miniature works of art, colorful, and symbolic of the airline's identity, both cultural and economic. The book Tickets Please! (Schmitz, 2020) features over 150 ticket designs.

IATA's Billing and Settlement Plan (BSP) for travel agents, launched in Tokyo in 1972, was the second significant milestone in ticketing after the introduction of the IATA standards in 1930. The IATA neutral paper ticket was created as a result. These tickets featured the IATA logo on their cover and were valid for travel on a wide range of airlines across the globe, accessible by any travel agent. The flight coupons worked like cheques, with travel agents writing them, IATA reconciling them, and airlines cashing them.

The year 1983 marked the third significant milestone in ticketing with the introduction of a magnetic stripe on the back of ticket coupons to aid automation. This enabled electronic storage of all travel details on the ticket, which could double as a boarding pass. Though paper tickets were still prevalent, the digital transformation of ticketing, called E-Ticketing, had begun.

The first paperless E-Ticket emerged in 1994. Despite the emergence of E-Tickets, customers who booked on an airline website still had to visit the local airport or city ticket office to make a payment and receive a paper ticket.

2.7 Origins of the GDS

From the mid-1960s to the mid-1970s, travel agencies and airlines worked together to promote a reservations system that would standardize workflows and improve the productivity of agents (Vinod, 2009). This concept was referred to as a neutral industry-wide reservations system. In 1967, the Reuben H. Donnelly Corp. began developing the Donnelly Official Airline Reservations System (DOARS) on a Univac platform. DOARS was not successful due to insufficient funding from the 21 airline participants involved in the project.

The Automated Travel Agency Reservations System (ATARS), a system that was exclusively for airlines and travel agents based on PARS, sought approval from the Civil Aeronautics Board (CAB). The U.S. Justice Department interpreted the exclusivity feature as a per se violation of anti-trust laws, which in turn triggered a CAB investigation. The ATARS agreement was modified to meet the requirements but was abandoned before CAB could reach a verdict.

The contract to develop a common integrated travel agency system between the American Society of Travel Agents (ASTA) and Control Data Corporation (CDC) in 1973 also failed when the airline constituents could not accept a computer vendor controlling access to travel agents.

Max Hopper joined American in 1972, the same year that the Sabre system migrated to the IBM System/360 mainframe computers at a new underground facility in Tulsa, Oklahoma. Robert Crandall, the dynamic detail-oriented former TWA executive, was hired as Chief Financial Officer of American Airlines in 1973, Crandall also had responsibility for data processing based on his experience at TWA.

In 1974, Robert Crandall and Max Hopper proposed the development of the Joint Industry Computerized Reservations System (JICRS) to gain control of the distribution channel. Their proposal involved the creation of a joint industry task force comprising airlines, ASTA, and hardware vendors. Crandall and Hopper played crucial roles in specifying the functional scope, benefits, and costs of JICRS. A technical evaluation team concluded in July 1975 that JICRS was feasible and would produce significant cost savings for participating airlines. However, United, as the largest domestic U.S. carrier at that time, found the financing terms unacceptable since they were tied to passenger volumes. This meant that United would become the largest investor, which was not acceptable to them. United announced its withdrawal from the JICRS proposal in January 1976 and announced its intention to promote its PARS-based Apollo system to travel agents and commercial accounts within nine months. United then notified the industry that it planned to accept Apollo orders in May 1976 for delivery starting in September, resulting in the failure of JICRS.

Carriers like American and TWA were compelled to adopt similar strategies, sparking a competition to create, improve, and actively endorse these systems within the travel agency community. ASTA made a final effort to establish a joint system called the Multi-Access Agent Reservations System (MAARS), but it failed when CAB rejected antitrust immunity (ATI) due to concerns about exposing other airline CRSs being promoted to travel agencies.

Following United's announcement, American made swift progress and installed the first Sabre terminals for travel agents in April 1976 through its Travel Agency Automation Program. By the end of the year, Sabre had been installed in 130 travel agencies. It was the first CRS to offer agency features to travel agencies and later became known as the GDS.

In the 1970s, attempts to establish a unified system for travel agents failed. As a result, airlines began promoting their own reservations systems to travel agents as a means of selling their products. Additional features were incorporated into these reservations systems, and the enhanced versions used by travel agents identified during the JICRS study became known as Global Distribution Systems (GDS). These systems expanded their support for agency point-of-sale functions worldwide, while the original airline system, which stored seat inventory, is now referred to as host CRSs or simply CRSs. The implementation of a single system for multiple airlines quickly followed. American's Sabre system took the lead ahead of United's Apollo and TWA's PARS because it focused on incorporating the agency features specified in the JICRS study. When United and TWA realized that travel agents preferred American's Sabre system, they also prioritized adding these agency features, but were a year behind Sabre.

MAARS Plus, a Multi-Access Airline Reservations System developed by ITT in 1977, stands out for its distinctive attributes. Unlike other systems, MAARS Plus was not owned by airlines or travel agencies. Its key feature is the ability to establish direct connections with the reservations systems of all participating carriers. This direct access functionality allows for a tailored reservations session workflow specific to each airline's CRS. Familiarity with the command line entries of each system was necessary to navigate through the booking process. Additionally, MAARS Plus stored reservations directly on the airline reservations systems. Each reservation was indexed according to its source, facilitating easy navigation for travel agents to the appropriate airline system.

MAARS Plus falsely claimed to be the sole unbiased system. However, the screens were clearly biased toward the reservation systems of the individual airlines. While ITT received backing from Eastern Airlines and a few smaller carriers, their initiative ultimately faltered in both operational and financial aspects due to Eastern Airlines' withdrawal of support to focus on marketing its own reservations system called System One Direct Access (SODA) in 1981. The failure of this system can be attributed to two main factors. First, the absence of a common language made it difficult for travel agents to understand the various command line entries and codes from the individual airline systems and, second, the revenue model was flawed because investors could not recover subscription fees from the travel agents.

2.7.1   Other Multi-Access Initiatives

Besides MAARS Plus, there were other initiatives as well to establish a multi-access reservations system. In 1976, Travicom was an initiative by partners British Airways, British Caledonian, hardware vendor Videcom and back-office systems supplier CCI. It was created in 1977 and was the first multi-access system used by travel agents with high transaction volumes. Like MAARS Plus, the system had access to multiple travel suppliers with a real time link and the travel agent made transactions in the host reservations system of the airline. It had the same issues for users who were required to know the cryptic native commands of each linked airline reservations system. British Airways took control of Travicom and by 1987 it handled over 90% of bookings in the United Kingdom. It was replaced by Galileo in 1988. Travicom changed its trading name to Galileo UK, the distribution company that sold the Galileo GDS to travel agents and provided technical service and customer support. In 2000, Galileo International acquired Travel Automation Services that include Galileo UK from British Airways (Travel Weekly, 2000).

Another initiative was the Sabre multi-access environment called SMART (Sabre Multi-Access Reservations Terminal). SMART was an improvement over existing multi-access reservations systems since it had a common language for input, but the responses were specific to the individual host CRSs. Multi-access systems benefited flag carriers who had a monopoly in their home market. They were eventually replaced by the airline owned GDSs that entered these markets and, to appease the flag carriers, offered a direct access capability to travel agents who preferred to transact directly with a host CRS using its native commands.

2.7.2   The First GDS

By December 1976, there were 130 travel agencies that subscribed to Sabre. Max Hopper, the chief visionary of the Sabre system owned by American Airlines, was the first to market with a product for travel agents. American aggressively marketed the reservations system as an extension of the airline sales force, to sign up new subscribers. It was perceived by competitors that the product display of flight schedule and availability favored American over other carriers. It was the first mass-market software for travel agencies. Max Hopper is widely credited as being the father of the Sabre reservations system.

United's Apollo was introduced to the travel agency community a few months later. Apollo did not include the travel agency features, identified in the JICRS study, which Sabre supported and United quickly realized that travel agents preferred the

Sabre system. Trans World Airlines (TWA) introduced their system (TWA PARS) shortly after Apollo. They also introduced PARS II with limited features, which found a niche with small travel agencies, followed by the Trans World Airlines (TWA) PARS system. The common perception was that the airline-owned proprietary reservations systems gave these airlines an unfair competitive edge over airlines that were not promoting a host CRS of their own.

From a product development and capability perspective, Sabre was a year ahead of the other CRSs. Most of the CRSs had similar marketing strategies. In return for the hardware and training that was provided at no cost, the agencies agreed to pay a subscriber fee based on the volume of bookings made by the agency. Agencies offset these expenses with the standard front-end commissions paid by the airlines, about 10% on the net price of the ticket. Hence, if the federal excise tax was 8%, the effective commission on the face value of a ticket was 9.2%. In addition, they also received override commissions to encourage bookings on specific airlines, based on thresholds achieved for key performance indicators (KPIs). The volume of business at travel agencies expanded rapidly, with the agency channel providing most of the bookings for the airlines.

The flight listings on the Sabre system did not adhere to the OAG (Official Airline Guide) format. In the OAG format, nonstop flights were displayed first, followed by direct flights, online connections, and interline connections. On the other hand, the Sabre display was more advanced and considered total travel time (also known as elapsed time), deviation from the requested time, and carrier preference. The differences in the display order between OAG and Sabre created a perception that the display favored American Airlines. Interestingly, these three schedule parameters play a significant role in consumer choice model calibrations today using the multinomial logic model (MNL). They help revenue management analysts understand consumer behavior and how they select an itinerary from a set of choices (Vinod 2021a, Chap. 4).

When the GDSs were created in 1976, they facilitated business-to-business (B2B) commerce. Automating the sales reservations process for travel agents enhanced their productivity, and they served as an extension of the airline's sales force. In 1976, EDIFACT had not yet been developed. AIRIMP teletype standards were used for all inter-system communications.

Figure 2.8 illustrates the evolution of airline reservation systems (1953–1986) based on IBM hardware and software that led to the creation of the GDS with agency features.

2.8 Airline Deregulation

In 1976, President Gerald Ford initiated the process toward deregulation of the U.S. airline industry when he released a report of the Economic Policy Board Task Force on International Air Transportation Policy (Ford, 1976). An economic policy

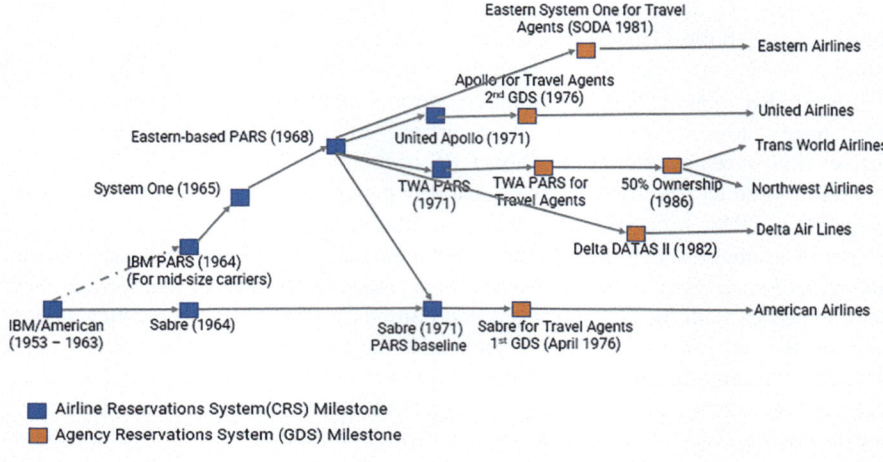

Fig. 2.8 Origins of the Global Distribution Systems (1953–1986)

based on competition was initiated with bilateral and multilateral agreements. The new air transport policy embraced five principal objectives:

1. Rely on competitive market forces.
2. Provide low price air transportation.
3. Generate sufficient profits for U.S. air carriers.
4. Foster national defense and foreign relations.
5. Encourage safety and efficacy for the airways and environment.

The report also acknowledged the possibility of excess capacity and predatory pricing and emphasized the need for restrained competition. The report sought route competition among carriers and recommended a continued role for IATA to negotiate international air tariffs for international flights. Airline members of IATA were warned to allow competitive market factors to influence the creation of innovative and flexible fares, routes, and schedules for *individual* carriers.

Prior to deregulation, airlines operated in a tightly regulated environment; regulated by governments and self-regulated through organizations such as CAB and IATA. To set the stage for airline deregulation, President Jimmy Carter, who succeeded President Gerald Ford, appointed an economist, Alfred E. Kahn as chairman of the CAB. The Airline Deregulation Act was signed into law by President Carter on October 24, 1978. A powerful voice in favor of deregulation of the airline industry, economist Alfred E. Kahn, was the chief architect and promoter of deregulating the airline industry in the United States despite opposition from labor unions and industry executives. He believed that air transportation would thrive under competitive market forces based on the fundamental economics of supply and demand. He is acknowledged worldwide as the "Father of Airline Deregulation". Airline deregulation resulted in increased service, a wider choice of

air services to choose from and increased competition. Airline deregulation disman-
tled a comprehensive system of government controls (Kahn, 1988a, b).

By January 1, 1983, a year after the termination of the CAB's route authority, the
act required the complete deregulation of all fares. This deregulation of the airline
industry brought about what is now known as pricing and revenue management. For
the first time in the history of commercial aviation, airlines were able to adjust their
routes and fare structure in response to customer demand and competitive pressures.
Today, airlines worldwide are experiencing various stages of deregulation, making
the investment in new technology and solutions essential to survive in a highly
competitive environment. Since deregulation, each major airline has implemented
flight scheduling with fleet assignment, fare management, and some form of pricing
and revenue management, although with varying degrees of sophistication. The
pricing and revenue management process is the most critical factor in determining
airline profitability. This practice has matured on multiple fronts, such as the
adoption of best business practices, decision-support capabilities for advanced
forecasting, publishing profitable schedules, response to competitive fare changes
with timely fare management, revenue optimization, inventory control recommendations for execution in the host CRS, and cost-effective product distribution.

The airline industry saw a significant and immediate change under deregulation
(Crandall, 1995; Dempsey & Gesell, 1997). During the regulated era, competition
was artificially limited, and airlines were almost guaranteed profitability. The Civil
Aeronautics Board approved routes and set fares that ensured a 12% return on flights
that were only 55% full. Deregulation transformed airline profitability, by increasing
competition and productivity, and ushered in an era of survival of the fittest. The
U.S. Domestic Airline Deregulation Act mandated the end of route restrictions by
December 31, 1981, and rate regulation by January 1, 1983. The CAB accelerated
this process and ended route regulation in 1979 and rate regulation in 1980. CAB
was dissolved on January 1, 1985, because of deregulation.

Airline deregulation also had its negative impacts on the flying public. Several
smaller cities lost commercial service which in turn resulted in the movement of
people toward the larger cities at the expense of the smaller cities. At that time this
had an adverse impact on the general health of the U.S. economy.

The end of government-mandated fare levels through the CAB's Standard Industry Fare Level (SIFL) came with deregulation (Kretsch, 1995) of the domestic
U.S. market, leading to a proliferation of fares from competing airlines, ranging
from full fare to discounted and deeply discounted fares. This made air travel
accessible to a wider segment of the population that had never flown before.
Between 1976 and 1990, after adjusting for inflation, average yields per passenger
mile decreased 30% (Kahn, 1988a). Deregulation also impacted the global airline
industry, not just the domestic U.S. market.

In 1978, the deregulation of the U.S. airline industry caused the government to
lose control, leading to a surge in fare filings and new markets served by airlines. The
intense competition that followed caused several airlines to cease operations, including Braniff Airlines (1982), Continental Airlines (1983), Air Vermont (1984), Air
Florida (1984), Capitol Air (1984), PEOPLExpress (1986), Eastern Airlines (1991),

Midway Airlines (1991), and many more. The rapid growth that resulted from deregulation also led to a decline in airline service. Labor unions claimed that airlines used bankruptcy to break up union contracts, The number of airline consolidations increased post-deregulation.

2.8.1 CAB and the Imperfect Cartel

The attitudes of the different nations on protectionism versus free trade differed based on factors like their reliance on air transport services, negotiation power, geographical location, and ability to generate traffic (Taneja, 1976).

When President Carter appointed economist Dr. Alfred E. Kahn as the Chairman of CAB in 1977. Kahn believed that competitive market forces would shape the airline industry in the decades ahead and he supported the following key policy changes:

1. The Air Cargo Deregulation Act of 1977
2. The Airline Deregulation Act of 1978
3. The June 9, 1978, Order to Show Cause on IATA Conference Agreements
4. The International Air Transportation Competition Act (IATCA).

The primary purpose of the "Order to Show Cause" was to inform IATA and the airlines that CAB planned to withdraw antitrust immunity from the rate-fixing traffic conferences. A hearing was scheduled to determine why such antitrust immunity should continue. Even before the Order to Show Cause, market competitive forces had already begun to undermine the effects of IATA's cartel activities. The stable price structure established by IATA began to collapse because of changes in both public and private policy. The British Government ordered BOAC to avoid IATA. In 1977, Sir Freddie Laker launched a long haul "no frills" low-cost carrier to carry transatlantic passengers from London, Gatwick to New York's JFK at lower prices than the IATA cartel. Competing airlines introduced cheap standby and advance purchase fares. Despite these cracks in the armor, CAB wanted to make sure that IATA did not gain an upper hand and control the cartel price structure.

The airline deregulation act also called for the dissolution of CAB over the next seven years. Immediately after deregulation, President Carter appointed Marvin S. Cohen to succeed Alfred E. Kahn as the Chairman of the CAB. Cohen was hand-picked by Kahn as his successor when he became the President's top counselor on inflation policy. Cohen continued Kahn's deregulation policy under President Ronald Reagan, accelerated the deregulation of the U.S. domestic market and influenced the deregulation of international markets.

Since the creation of IATA for the second time, CAB gave immunity from U.S. antitrust laws for U.S. and foreign flag carriers that participated in the rate fixing tariff conferences hosted by IATA. This protection continued for three decades. IATA was not a cartel in the traditional sense since it did not control total industry output, dictate prices, or allocate markets or profits to its member airlines.

However, the very existence of IATA provided a framework for its member airlines to negotiate and set prices in markets which leads to collusive behavior.

Kahn believed that IATA was a cartel since it organized its Traffic and Tariff Conferences. Kahn labeled IATA a *smoothly oiled price-fixing cartel* and the participants in IATA's multilateral negotiations *as protectionists and cartelizers*. In his opinion, IATA's role stifled innovation, competition, and fare efficacy amongst the airlines. After more than three decades of protectionism, in 1978, CAB openly questioned the need for antitrust immunity enjoyed by IATA in a deregulated airline industry (Peterson, 1990). Subsequently, a hearing was scheduled, and airlines and IATA were invited to justify why the antitrust immunity should not be revoked. The formal declaration of this order was postponed multiple times at the request of the Reagan Administration. It was finally ordered in May 1985, despite resistance from European flag carriers and smaller airlines that had been losing money. The CAB policy of aggressively pursuing deregulation continued even after its functions were absorbed by the Department of Transportation (DOT).

After more than four decades of deregulation in the U.S. airline sector, the overwhelming consensus is that it has proven to be a triumph. This is especially evident in the reduction of average ticket prices, a greater number of available flights, improvements in operational effectiveness for carriers and a razor-sharp focus on profitability. In addition, safety standards have remained commendable throughout this period. However, it is important to acknowledge the downside, which includes several airline bankruptcies, significant fluctuations in profitability with losses often outweighing gains by an order of magnitude, and a decline in customer service.

2.9   Origins of the Frequent Flyer Programs

In 1979, Texas International Airlines established the first mileage-based frequent flyer program that rewarded passengers based on miles flown. However, the program was terminated within a year when it merged with Continental Airlines in 1982. Western Airlines ("*the only way to fly*") launched the Travel Bank program in 1980, which offered a US$50 travel certificate for every five trips, but it was not based on miles traveled.

On May 1, 1981, American Airlines introduced the AAdvantage frequent flyer program, which was created by Robert Crandall to boost the brand among repeat customers, monitor frequent flyer spending habits, and provide targeted incentives for the most loyal customers. The idea for the program evolved from William Bernbach, the founder, and CEO of Doyle Dane Bernbach, American's advertising agency, who drew inspiration from commercial banks that were rewarding their top clients with free products such as toasters and electric blankets. Bernbach recommended introducing a "loyalty fare" for frequent flyers that never materialized but eventually led to the development of the mileage based AAdvantage program.

Table 2.1 Frequent Flyer Programs in 1981

Airline	Frequent Flyer Program
American Airlines	AAdvantage
United Airlines	Mileage Plus
Trans World Airlines	Frequent Flight Bonus/Aviators
Delta Air Lines	Frequent Flyer Program (later called SkyMiles)
Western Airlines	Air Pass II
Braniff Airlines	Travel Bonus Bonanza
Continental and Eastern Airlines	Frequent Traveler/OnePass
Republic Airways	Frequent Flyer
Northwest (Orient) Airlines	Free Flight Plan/WorldPerks
Air Canada	Altitude

United Airlines launched Mileage Plus a week after American Airlines launched AAdvantage. The first wave of mileage-based frequent flyer programs, all launched in 1981 after American's AAdvantage program, is summarized in Table 2.1.

2.9.1   The Ill-Fated AAirPass Program

The deregulation of the airline industry led to intense competition and reduced ticket prices. American Airlines experienced a financial setback in 1980, with an annual loss of $76 million. To secure funds, the company introduced a discount program called AAirpass in 1981, exclusively available to frequent flyers.

The AAirpass program had various options which included passes for two 5000 coach miles a year for 5, 10, or 15 years at a cost of $19,900, $39,500, and $59,900, respectively (New York Times, 1981). The AAirpass program also offered lifetime passes for people aged 52 and older providing 25,000 miles of travel a year for a one-time fee of $66,000. An unlimited mileage pass with no age limit was offered for $250,000 and offered purchasers the privilege of unlimited lifetime first-class travel on American Airlines flights. Additionally, for an extra $150,000, pass holders could acquire a companion pass, enabling them to bring a travel companion on their flights.

American incurred financial losses from the program despite the limited sale of only 65 of the unlimited mileage passes. This was primarily due to the excessive travel undertaken by certain pass holders. The AAirPass policy, which was quite lenient, permitted pass holders to cancel their trips at the last minute without facing any penalties. In addition, the airline was still obligated to cover per-passenger taxes and fees imposed by airports and various countries.

In 1990, American recognized its mistake and decided to increase the price of an unlimited AAirpass (with a companion) to $600,000. The airline did not want to eliminate the program, so it further raised the cost to $1.01 million in 1993. Eventually, in 1994, the airline discontinued the sale of unlimited passes altogether.

However, American brought it back as a one-time offer in the 2004 Neiman Marcus Christmas catalog, priced at $3 million. Unfortunately, there were no buyers for this revived offer.

The airline also terminated some passes due to what they considered fraudulent activity. The fact that the unlimited AAirpass program was available for purchase for 13 years, until 1994, is ultimately surprising. However, even though the airline managed to sell only 65 of these passes, it is evident that the termination of the program indicated a financial setback for the company.

No other airline has ever tried to imitate American Airlines and introduce lifetime passes. However, there have been various instances of pass books with specified durations and flight coupon books that could be exchanged for travel. These programs have never generated any profit for the airlines because they forget the fundamentals of revenue management. Pass holders often chose to redeem their passes on peak flights, causing high-valued revenue passengers to be displaced.

2.9.2   The Runaway Success of Loyalty Programs

Airline loyalty programs have proven to be highly successful for airlines. These programs generate new demand, establish brand value even though the end product is essentially a commodity. These programs cultivate long-term relationships with loyal customers who are dedicated to the frequent flyer program. For the airlines, they also serve as an additional source of revenue through collaborative marketing initiatives with credit card companies, telephone companies, and various promotions with packaged goods, and retail companies (Vinod, 2011b).

A quote that has been circulating for at least ten years states that *airlines are essentially credit card companies that happen to also fly planes.* From modest beginnings, airline loyalty programs have evolved into an incredibly lucrative venture. In fact, airlines generate more revenue from their credit card operations than they do from their actual flights. Airlines are a lot like central banks (Locke, 2020) since the airlines issue frequent flyer miles like central banks and treasuries print money. By selling their miles to credit card companies, airlines can rake in billions of dollars through their frequent flyer programs. Airlines sell these miles at a discount to banks. Companies like JP Morgan, Barclays, and American Express are major frequent flyer point buyers. Airlines also sell miles to raise cash during lean periods. These miles are typically earned through credit card transactions, primarily by business class passengers. Many argue that the core business of airlines lies in selling miles, rather than selling seats on their planes. The profitability of selling air miles is remarkable, with airlines making a three to one profit ratio. This makes it a highly profitable venture compared to running an airline, which is asset intensive and involves significant fixed costs and variable costs such as fuel and employee expenses. As a result, airlines collaborate with banks and credit card companies to sell air miles.

The largest airlines in the world boast memberships of more than 20 million and partnerships with hundreds of companies. In 2023, the three largest frequent flyer programs were United Mileage Plus with 110 million members, Delta SkyMiles with 109 million members, and American AAdvantage with 100 million members.

Redemptions of miles or points for travel were initially available only through airline call centers. After the arrival of the Internet and the online consumer direct channel, most redemptions are handled through airline websites. Travel agencies that subscribe to a GDS do not redeem miles or points for flights except in rare cases like the relationship between Amadeus and Air France.

Loyalty programs have a significant impact on revenue management. Every frequent flyer redemption displaces revenue passengers, which represents a cost to the airline. Hence, frequent flyer redemption policy is influenced by its impact on revenue management. Frequent flyer redemption bookings are capacity controlled. While the launch of loyalty programs was an important milestone in the history of commercial aviation, it has had little impact on the GDS. Travel agents, unless expressly authorized by an airline, do not process frequent flyer redemptions for their customers.

To reward profitable behavior, Delta followed by United, converted their frequent flyer programs from a mileage-based award system to a revenue-based system in 2014. American followed in 2016 after its merger with US Airways. Hence, instead of earning frequent flyer miles based on the distance traveled, mileage is accrued based on the price of the ticket, and frequent travelers in high standing receive a multiplier to accelerate their accumulation of miles. Instead of earning miles based on the flown distance, members of these frequent flyer programs earn between 5 and 11 miles per dollar spent. The new system is not without its flaws since revenue without considering the route does not capture profitability. For example, someone who spends $300 on a flight from Dallas/Fort Worth to Omaha is probably more profitable for the airline than someone who spends the same amount on a transcontinental flight, say Los Angeles to Boston.

2.10   GDS Poised for Growth with Co-Host Agreements

The rapid growth of the GDS model provided automation to travel agencies. The five airlines that owned the five largest CRSs: American Airlines, United Airlines, Trans World Airlines, Delta Air Lines, and Eastern Airlines saw the significance of this initiative as a competitive differentiator. The benefit of CRS (now called GDS) ownership was widely referred to as the *halo effect*.

By the end of 1976, only a few hundred travel agencies had been automated by Sabre and Apollo. But the race was on to add travel agency subscribers, and by June of 1978, several thousand travel agencies were automated.

After airline deregulation, and facing stiff competition, airlines with vested interests in agency reservations systems started investing in back-office systems. Back-office accounting is required for travel agents to monitor sales performance and commissions that they are owed by the airlines. In 1978, United Airlines bought a license to use the software from Agency Data Systems (ADS), a Florida-based mini-computer accounting systems company for travel agencies owned by Terry Jones. Triggered by United's acquisition of the software, American bought ADS outright, retaining the company's employees. Later, Terry Jones went on to become President of Travelocity and Sabre CIO. United marketed the system as Apollo Business Systems but no longer had access to the developers of ADS. Eventually, all the CRSs developed their own comprehensive back-office accounting systems to track both individual and aggregate travel agency productivity, agent sales reports, and commission payments.

By 1978, the CRS vendors had automated several thousand travel agencies and the competition to add new travel agency subscribers was intense. In addition, travel agencies wanted access to broad airline content, not just the content from a single airline. This led to the co-host agreements. Co-hosting is selling space to other airlines on an airline's own system. The co-host agreements between airlines were an important milestone to expand the geographic reach of an airline where it had no presence. The CRS vendor airlines only negotiated co-host agreements with airlines that did not compete.

American, concerned that it was falling behind United due to its larger network, introduce the co-host concept whereby other airlines were given preferential screen display of their schedules for a fee. American and Western Airlines signed the first co-host agreement which enabled Western to market the system in the western U.S. thereby preventing United's Apollo from dominating the region. Travel agency contracts, the commitment to use a specific CRS, were multi-year and the importance of the first mover advantage cannot be understated. American had the first mover advantage and secured co-host agreements with five airlines.

As part of the co-host agreements, the co-host airline paid the CRS vendor airline for favorable placement in content displays in city pair availability. The CRS vendor airline also charged the co-host a booking fee for each reservation. In turn, a co-host could also charge an airline a fee for bookings made through an agency terminal that they sponsored. Remarkably, it was strategic marketing to gain market presence in unserved markets and overall market share of airline bookings. Though not the primary intent, it was also a way for American to leverage significant economies of scale and recoup their initial investment on the CRS with the monthly cash payment received for this service.

On October 1, 1982, American, the second largest carrier, and Delta Air Lines, the fourth largest carrier, signed a co-hosting agreement (Carmichael, 1982). With this agreement, American and Delta were essentially co-hosts in each carrier's system, which meant that their flights would receive priority listing on their respective computer systems. At that time, Sabre was the most sophisticated in the travel industry with 7000 installations used by 40% of the travel agents. For this reason, Delta paid a higher fee for preferential display on Sabre than American paid to be listed on DATAS II, which had not yet been rolled out. This left only three airlines— Ozark, Piedmont and United that were not listed on the Sabre system. By the end of

1982, United was the only airline not listed on the Sabre system. It was listed on the second largest system, Apollo.

Delta's DATAS II was billed as the *world's first totally unbiased* automated reservations system (Shifrin, 1982). DATAS II used the OAG format which used elapsed time as the deciding factor for competing flights departing at the same time. The OAG type display showed every direct flight available between two cities strictly by departure time, giving Delta no advantage. If two or more flights had the same departure time and elapsed time, the listing was in alphabetical order. This contrasted with competing reservation systems that exhibited inherent favoritism toward their own flights, granting them priority listings. Additionally, other airlines secured secondary preference by paying additional fees to be designated as "co-hosts" within these systems.

Marketing DATAS II in this way was based on an investigation launched by the Justice Department's Antitrust Division in June 1982 to determine if airlines with CRSs biased their displays to harm competitors. The investigation was triggered after a federal grand jury in Texas started looking into allegations that American had used illegal tactics, such as manipulating the displays, that resulted in Braniff Airlines filing for bankruptcy.

By 1982, the five major CRS providers had all secured extensive co-host agreements and approximately 70% of all US domestic travel agencies were automated, which accounted for 80% of domestic air travel.

Table 2.2 illustrates the co-host agreements with the 5 major CRSs in 1982.

Table 2.2 Airlines with co-host agreements with CRS vendors in 1982

American Airlines Sabre	United Airlines Apollo	Eastern Air Lines SODA	Trans World Airlines PARS	ITT MARS PLUS
Air Cal	Air Florida	American	American	Eastern
Air Florida	Alaska Air	Continental	Continental	Northwest
Continental	Continental	NY Air	Delta	Pan Am
Delta	Delta	Ozark	Eastern	TWA
Eastern	Frontier	Pan Am	Mississippi Valley	
Frontier	Mississippi Val-ley	Piedmont	Pan Am	
Golden West	Ozark	Republic	United	
Northwest	Pan Am	TWA	USAir	
Ozark	Republic	USAir	Western	
Pacific South-west	TWA	Western		
Pan Am	USAir			
Piedmont				
Republic				
TWA				
USAir				
Western				

Source: CAB report to Congress

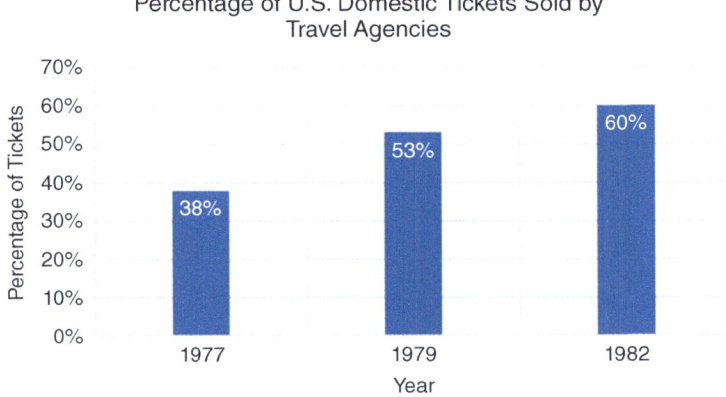

Fig. 2.9 Growth of the travel agency channel. Source: U.S. Department of Transportation: Airline Marketing Practices, Travel Agencies, Frequent Flyer Programs, and Computer Reservations Systems, 1999

2.10.1   Co-Host Agreements and the Barrier for Participation for Smaller Carriers

Airlines that signed co-host partnerships with the airline CRS vendors early in the sales process had gained an advantage. CRS vendors maintained different tiers of booking fees. For example, Western Airlines had a minimal booking fee by virtue of being the first to sign a co-hosting agreement with American. Airline CRS vendors estimated the level of competition from airlines that wanted to sign the co-host agreements. Airlines that offered fierce competition after deregulation paid an exorbitant booking fee of $2 to $3 per segment. Smaller carriers were also subject to higher segment booking fees. Many airlines found these fees to be too high to participate in a CRS, which effectively cut them out of the travel agent distribution channel and the more lucrative corporate bookings.

The growing importance of the travel agency channel is shown in Fig. 2.9 from data collected by the Department of Transportation.

2.11   CRS Market Power and Origins of the CRS Rules

In the absence of any governance to control anti-competitive behavior, owners of the major CRSs gained an advantage over their competitors by biasing the display of schedules in their favor. Various studies during this period, conducted by both airlines and government agencies confirmed that 90% of all bookings were made off the first screen and 50% to 60% of bookings were made off the first line in the display to travel agents. The incremental revenues attributed to the first line on the first screen was very significant and the airlines that owned the CRSs devised ways to ensure that their flight occupied the first line. Other airlines could pay for a better screen presence, and this was typically offered as part of the co-host agreements.

For example, the algorithm used by Sabre operated by assigning a time penalty that represented carrier preference. American Airlines, as the host, received no penalty. Co-hosts typically receive a penalty of 30–45 min and other airlines receive a penalty of 60–75 min. As a direct result of the screen manipulation, American occupied the strategically first position on city pair availability displays, even though it was not deserved from a schedule perspective.

In June 1983, a House subcommittee opened hearings (Lammi, 1983) on allegations that the major airlines were leveraging their computer reservations systems (the GDSs) to gain an unfair competitive advantage over the smaller airlines. Simply put, it was an investigation of the *haves* versus *the have-nots.*

Dan McKinnon, Chairman of the Civil Aeronautics Board, testified before the Public Works and Transportation panel's aviation subcommittee. He acknowledged that there was "*bias*" in the reservations system screen displays. He said, "*All of these airlines have developed and implemented computer programs that give themselves a priority in the way flight are displayed to travel agents*" and added that the real question was "*How far can a computer reservation system be biased before it becomes anticompetitive?*"

CAB and the Justice Department submitted a report of their analysis and McKinnon acknowledged "*We don't claim to have all the answers at this point.*" In 1983, more than 80% of the travel agents in the United States had automated reservations systems, dominated by American Airlines and United Airlines that provided 75% of these systems followed by TWA with 15%.

To support the ongoing investigation, in September 1983, CAB requested the five major airlines that marketed their GDSs—American Airlines, United Airlines, Trans World Airlines, Delta Air Lines, and Eastern Air Lines to submit detailed information on the workings of their systems within 30 days. A DOT study in 1988 concluded that 11% of American's bookings made through Sabre were the direct result of display bias.

When complaints from the smaller airlines and travel agencies grew louder, CAB identified four anti-competitive behaviors that were prevalent in 1983. The Code of Federal Regulations (CFR) was introduced on November 11, 1984, which prohibited anti-competitive behavior from airlines that owned the dominant CRSs. They were

1. Carrier preferencing in displays was eliminated and all displays were required to display results which best met the query parameters submitted by a travel agent. Hence, on United's Apollo, for a specific market, an American flight may be ranked first and appear on the first line of the display.
2. Capabilities available in the CRS for a specific airline should be universally available to all participating airlines.
3. Discrimination between CRS providers was also banned which ensured that all airlines that owned and marketed a CRS to travel agencies participated in all the

CRSs. Hence an airline that participated in one CRS had to participate in all of them. It was also mandated that travel agency booking data collected by a CRS were required to be made available to the competing CRSs for a fee. This is now referred to as MIDT (Marketing Information Data Tapes).
4. Booking fees charged by the CRSs to participating carriers were required to be non-discriminatory. Before this ruling, price discrimination prevailed, and some carriers paid no segment booking fees whereas others did. Booking fee billing data previously provided on microfiche or paper is provided as BIDT (Billing Information Data Tapes) by the CRS vendors.

The CRS rules went into effect on November 11, 1984, though it did not dispel the debate over anti-competitive behavior.

2.11.1   Marketing Information Data Tapes (MIDT)

Marketing Information Data Tapes is a dataset based on bookings made by travel agents that subscribe to a GDS. The data is collected before a ticket is issued. In its raw form, almost all the data stored in the passenger name record (PNR) is available. This booking information is extremely valuable to an airline to assess its competitive position in key markets, project booking trends, and act if required. This data is also used to measure travel agency productivity from a historical perspective. The collection of this data began in 1984 and the CRS vendors started actively marketing and selling this data since 1987. MIDT only has GDS (indirect) bookings.

An alternative to MIDT is a competing product called the IATA Direct Data Solution (DDS). This dataset is superior to MIDT since it includes visibility into direct and indirect (agency) ticket sales across all geographic regions. This dataset was created in partnership with ARC, Cirium, and airlines who contribute their sales data.

2.12   The GDS Platform

The very nature of the GDS marketplace that facilitated transactions by bringing buyers and sellers together was one of the first indications in the business world about the power of platforms.

Before the Internet era, there were only channels that were classified as offline channels and limited to travel agencies, airline call center agents, airport ticket offices (ATO), city ticket offices (CTO), and gate agents. Figure 2.10 illustrates the two offline channels that existed from the pre-Internet era.

Between 80% and 85% of all airline bookings flowed through the GDS channel, classified as the offline indirect channel. It was indirect since the bookings were made through an intermediary, the GDS, and not directly on the airline's

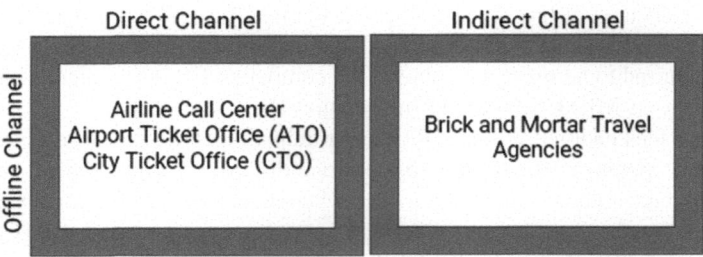

Fig. 2.10 Offline direct and indirect channels pre-Internet

reservations system. The remaining 15% to 20% of airline bookings were made through airline call centers, airport ticket offices (ATOs), and city ticket offices (CTOs). These channels were classified as offline direct since an agent made the booking on behalf of a customer by directly accessing the airline reservations system.

2.12.1 SabreTalk

SabreTalk, a programming language developed jointly by American Airlines, Eastern Airlines, and IBM, is known as PL/TPF (Programming Language for TPF) in the industry. It was introduced as a version of PL/1 in the late 1960s. The SabreTalk compiler was designed specifically to generate programs that operate within TPF. The compiler creates re-entrant IBM System 370 Assembler language programs, which can handle multiple requests while residing in the main storage as a single program image. As a higher level language, SabreTalk offers an efficient approach to application programming, allowing programmers to focus on the logic of their programs rather than intricate machine-specific details, ultimately boosting productivity. Another advantage of using SabreTalk is its simplicity for program modification. This simplicity arises from the reduced number of coding statements within a given segment and the high level of readability associated with these statements. Furthermore, if programmers incorporate descriptive names and comments, SabreTalk can become almost self-explanatory in many cases. Lastly, the language's clarity, along with the comprehensive error flagging routines integrated into the compiler, significantly reduces the time required for program debugging. Typically, SabreTalk applications have longer path lengths compared to assembler applications but are shorter than those written in other high-level languages (IBM, 2003).

Several airlines, including British Airways and Delta Air Lines, relied on SabreTalk programs until the early 2010s. However, after the original developers stopped providing support, companies had to resort to hiring translators to convert SabreTalk programs into different programming languages like C, C++, Java, and C#.

2.13   The Role of the Travel Management Company

The early pioneers who introduced the concept of a travel agency date back to the nineteenth century. Thomas Cook, a Baptist preacher, negotiated a deal with the Midland Railway to transport 540 members of his temperance society in 1841. The train ticket cost 1 shilling which included lunch for which Cook received a commission from the railway. It is acknowledged that Thomas Cook was the first documented travel agent who created excursion packages. In 1845, he arranged packages excursions for 165,000 without the aid of modern technology—telegraph and phones.

George C. Taylor became the fourth president of American Express in 1914. He expanded foreign remittance operations and inaugurated the travel services with a travel department in 1915. As one company executive wrote to George Taylor in 1915, *"Already, we supply travelers with the tickets for their European tours; we receive and forward their mail; we provide reading and writing rooms for their convenience; we store and forward their baggage and packages; we engage their return steamship accommodations. In fact, we are doing already for travelers practically everything except that which is most remunerative to ourselves, namely furnishing eastbound steamship tickets to Europe: providing hotel accommodations and conducting small parties desiring such a service."*

Since the early days, travel agencies have evolved from printing paper tickets, delivering the ticket to a home or office, and collecting payment from customers. Today, their role has been transformed to that of a travel advisor and performs many functions such as optimizing travel spend for their corporate customers, providing duty of care, and ensuring compliance of corporate employees to the corporate travel policy.

Travel agencies that subscribe to a GDS also specialize in leisure travel, but corporate travel is the primary source of bookings. For example, large travel management companies like Flight Center Travel Group (FCTG) have a strong leisure focus and have business units like FCM Travel and Executive Travel Group that specialize in corporate travel.

However, the GDS channel primarily serves the corporate managed travel segment. Travel agents that specialize in corporate travel subscribe to a GDS to book air, hotel, car, and rail. Corporations want visibility into employee travel spend and stipulate that air, hotel, and car bookings should be made through a travel management company (TMC) that is responsible for managing the corporate travel budget.

Larger travel agencies that have global reach and serve corporate customers from many points of sale are called travel management companies (TMCs). GDSs and TMCs typically have 5-year contracts. The TMCs work with the travel managers at corporations who provide travel guidelines and policy for employee corporate travel. The TMCs provide a range of travel-related services for employees of a corporation.

There was a time when corporate travel departments were viewed as profit centers since travel agencies shared the front-end commissions received from the airlines. As front-end commissions dwindled over time, corporations pay the agency a management fee, and a fee per transaction based on how it was booked such as via online corporate booking tool or call center.

Historically, leisure passengers booking through a travel agency had been a free service since travel agencies were compensated with commissions from airlines and incentives from GDSs. Corporate travel is managed travel, and corporations pay a fee to a TMC for services rendered. However, with declining commissions from airlines, travel agencies are forced to provide value-added services for which customers are willing to pay.

The challenge faced by travel agencies, large and small, is that they must continuously innovate and identify ways to add value to both the airlines and their customers, both corporate and leisure. This is required to augment the travel agency's position as a valuable intermediary in the travel value chain.

The range of services provided by a travel agency to a corporate customer include:

Corporate Booking Tools
The migration of corporate travel bookings from travel agents and call centers to self-service online corporate bookings started in the late 1990s and will continue despite being an inferior product. Corporate online booking tools (OBT) have archaic workflows and have not kept pace with online booking engines offered by the OTAs for leisure travelers. The leading corporate booking tools are Concur Travel, Travelport's Deem, Sabre GetThere, Navan, Happay, TravelBank, and Egencia.

TMCs support a range of online booking tools including proprietary and third party solutions. Corporate buyers select the OBT that fits their business specific travel needs and preferences.

Expense Reporting
Expense reporting systems enable automated submission of traveler expenses and receipts for approvals and reimbursement. Expense reporting is an integral part of corporate travel and corporate booking tools integrate to multiple expense reporting tools such as SAP Concur, Emburse Chrome River, Expensify, Coupa, and many more.

Optimizing the Travel Budget
Working closely with their clients, many TMCs support the year-over-year analysis of the amount spent on corporate travel and recommend changes to the travel policy to tighten the travel budget. For example, the thresholds can be revised for the price of a booked itinerary versus the lowest fare in the market, advance purchase mandates for international trips, hidden city pricing for cost savings, mandate ticketing by the ticketing deadline to avoid price jumps, and other travel policy adjustments related to changes in fare rules.

Minimize Revenue Leakage
Selling fares based on airline seat inventory availability can go up or down. Predicting airfares for future dates is called fare forecasting. Many factors influence the selling fare in a market such as the inventory control recommendations from the

airline's revenue management system, response to competitor fare changes and the introduction of promotional fares. The GDS air shopping data can be used to develop machine learning algorithms such as Q-Learning, to predict when fares would increase and when they would decrease. Fare prediction models can be used to automate the rebooking of a customer's itinerary to realize cost savings. Sophisticated automated rebooking vendors use fare forecasting to trigger a cancel and rebook to realize cost savings. Reinforcement learning techniques are effective in predicting if selling fares will go down or up (Vinod, 2013a). Most of the automated rebooking vendors use a brute force periodic evaluation of selling fares to determine if savings can be realized. Corporations expect their TMCs to use an automated rebooking tool to realize cost savings for their corporate customers.

There are several vendors in the airfare and hotel price assurance space such as Coupa (Yapta was acquired by Coupa Software in 2020), FairFly, Trappit, and others. Many TMCs offer this feature to their corporate customers through a technology partner or their own internally developed tools.

Itinerary Management for Corporate Employees

Time management of employees is an important function of corporate travel agents. Organizing an international business trip and even domestic trips can take considerable time. Corporate travel agents manage the entire itinerary for a corporate traveler that includes air travel, accommodations, and local transportation.

Mobile-enabled itinerary management tools do not have the capabilities to book travel but enable travelers to organize their total trip even if part of the trip were booked outside the managed travel program. This can include air bookings, hotel, car rental, restaurants, etc. There are many competing vendors in this space such as Sabre's TripCase and Concur Travel's Tripit. Many TMCs provide mobile apps for managing itineraries. They may be proprietary itinerary management apps or third party apps like TripCase and Tripit. Travelers can use these apps to check the status of a flight, receive disruption alerts, and access check-in, as well as destination information. Furthermore, mobile apps are increasingly allowing users to easily communicate with live agents for in-trip support through chat, text, or phone.

Compliance with Corporate Travel Policy

Corporate policy is defined by the travel manager at a corporation. Conformance to corporate travel policy is required in the context of managed travel. Examples of corporate policy established by corporate travel managers include:

1. Travel booked on domestic trips must be within 25% of the lowest fare. This is sometimes referred to as the *lowest logical fare* to minimize the cost of travel.
2. Business travel is only applicable when the elapsed time exceeds 8 h.
3. Carriers that are not allowed.
4. The purpose of the trip also determines if travel is permitted. Examples are conferences, corporate office visits, customer visits, sales activity, etc.

When an employee does not book travel that is complaint with corporate policy, an exception alert is sent to the employee's supervisor for approval or denial.

Duty of Care
It is mandatory for TMCs to oversee corporate travel programs on behalf of corporations to ensure the safety of employees when they travel. Managing the travel risk based on the travel agent's knowledge of the market is of immense value. For corporate customers, knowledge of the employee itinerary and location of the employee is critical for duty of care. It is mandatory or TMCs to be able to locate and communicate with employees when the need arises. If an employee is traveling in a conflict zone or unpredictable events occur at the destination, the travel agency reaches out to the employee to make sure they are safe and accounted for. A Duty of Care solution that monitors global security events and subsequent identification of affected employees is built into the travel agency desktop to assist TMCs responsible for overseeing travel programs on behalf of corporations.

Gift Cards
Some TMCs in collaboration with corporate buyers provides incentives like gift cards from retailers or items of monetary value to employees for booking travel under budget. In some cases, they are rewarded for simply abiding by the corporate travel policy. Providing incentives to promote behavioral change in employees is frowned upon by many corporations, who expect employees to exhibit cost conscious behavior when they are on a business trip.

Centralized Data Reporting
Reporting gives corporations insight into the travel spend by week, month, or quarter, insights into the effectiveness of the corporate travel compliance program, carbon footprint, and traveler wellness.

Support Corporate Sustainability Initiatives
Corporate sustainability initiatives are defined by the corporate entities that engage with a TMC. The objective of a corporate travel agency is to help corporations to spend less on their business and be more sustainable.

IATA has pledged to be carbon neutral by 2050, but there is no well-defined path to achieve that goal. Electric airplanes for commercial aviation are several decades away. Sustainable fuels have only seen limited production and hydrogen as an alternative fuel for conventional airplanes holds some promise. What can be done today is recommend sustainable eco-friendly hotels on corporate itineraries.

After years of consolidation, the largest TMCs are American Express Global Business Travel, BCD Travel, CWT (formerly Carlson Wagonlit Travel), Travel Leaders Group, American Express Travel, Corporate Travel Management, Flight Centre Group, and Fareportal.

2.13.1 TMC Pricing Model

TMCs have relied on transaction pricing, which dominates the industry, and is unlikely to change for a variety of reasons. Alternate billing models such as subscription per traveler, periodic management fees, performance-based pricing, and incentives based on travel program savings have never gained traction. Transaction-based pricing provides corporate clients with a pay-as-you-use model that can be defined in TMC-corporate buyer contracts. Equally important, it can be monitored, reconciled, and compared against competing TMC bids.

Not all transactions are created equal; they vary based on the degree of human interaction. The greater the interaction, the higher the fees. For example, online bookings without agent assistance are the cheapest, followed by online bookings with agent assistance, followed by telephone support. A few TMCs charge a flat fee for each trip made by a corporate traveler, regardless of the degree of interaction (online or offline) to complete the booking.

To ensure trust between a corporation and a travel agency, a key to the future is providing transparency by sharing data for all transaction types and the monitoring of a client's revenue performance. Experimentation with blockchain-based contracts to govern terms and conditions and data access has already begun.

2.14 Operating a GDS

The infrastructure to operate a GDS has evolved in leaps and bounds. In 1976, when the GDS was introduced by American Airlines to travel agents, it supported a single supplier with call center, and gate agents who could access the system to book and modify existing reservations. This is illustrated in Fig. 2.11.

Since then, it has evolved to support multiple suppliers—airlines, hotels, rental car, cruise lines, rail, etc. and multiple points of sale such as call center, online travel agencies, city ticket offices, airport ticket offices, gate agents, and airline websites. This is illustrated in Fig. 2.12.

The GDS is housed in a data center with a variety of mainframe and mid-range systems, a telecommunications network, and an array of servers for sellers and buyers to access the system. Travel agencies access the system mostly with intelligent workstations operating on a local area network, though remote parts of the world may still be using dumb terminals. GDSs used to manage their own network

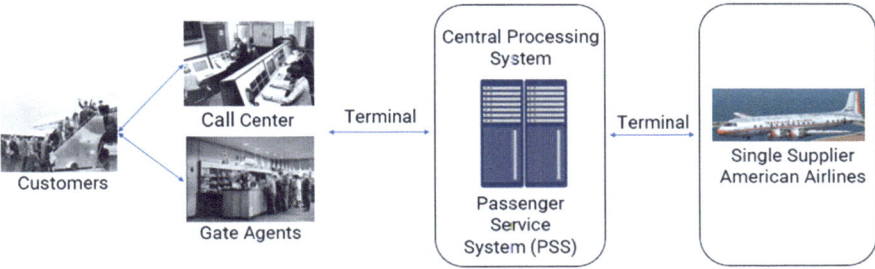

Fig. 2.11 Initial deployment of the GDS in 1976

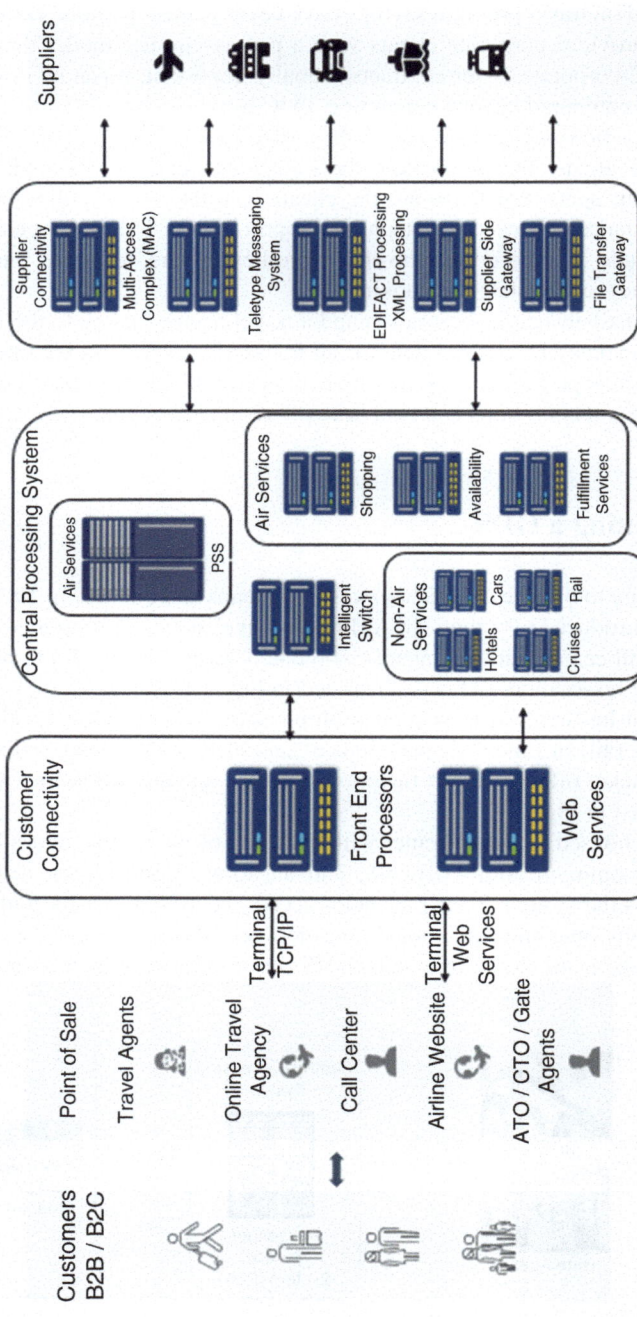

Fig. 2.12 Current Deployment of the GDS

services. Today, most of the network services are provided by SITA, AT&T, and Equant (merged with French Telecom's Global One in 2001), which are more cost effective. With Internet connectivity, international travel agencies must no longer wait to be connected over dedicated telecommunication circuits, which used to be the standard practice. The pricing infrastructure that maintains the complex fare rules and determines the prices of itineraries is a very critical function. It must be accurate to avoid airline debit memos when the passenger itinerary is priced incorrectly and ticketed. Typical response times for travel agency queries such as city pair availability, shopping, and booking are less than 3 seconds, and the system operates with an uptime reliability of 99.96%.

Besides the vast technical infrastructure, GDS also has dedicated sales and service teams to add new subscribers and provide technical support such as training, onsite support and a 24x7 help desk to address software and hardware issues, communications, workflow, and procedural issues.

2.15 Reservations System and GDS Basic Terminology

This chapter ends with a discussion on some of the core terminology used in the context of airline reservation systems and GDSs in the context of B2B contracts. This is required to understand how the GDS, and airline reservation system vendors bill their customers.

Figure 2.13 illustrates a sample airline network. All routes in this network are flown by the same airline. The nodes in the network represent the airports LAX (Los

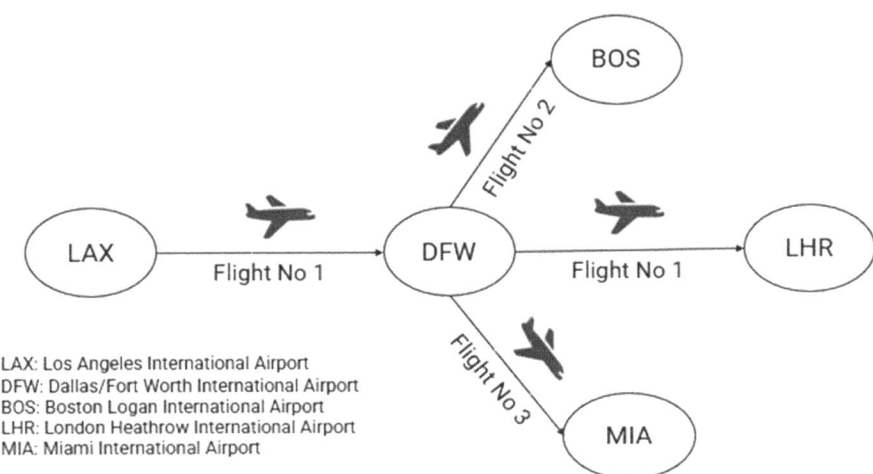

Fig. 2.13 Sample network

Angeles), DFW (Dallas/Fort Worth), BOS (Boston), MIA (Miami), and LHR (London Heathrow).

A flight leg represents a nonstop city pair. The flight legs in this network are LAX-DFW, DFW-LHR, DFW-BOS, and DFW-MIA.

A flight number represents a line of flight. It is a direct flight without a change in equipment. A change of gauge is a fraudulent airline practice that inconvenience passengers. It is an airline marketing term used to designate that a specific flight number changes aircraft, part way through the direct flight. For example, the first leg is operated by a narrowbody aircraft, and the second leg on the same flight number is operated by a widebody aircraft. The Department of Transportation (DOT) permits this inherently fraudulent practice and expects the travel agent to explain the situation to a customer.

The origin and destination pairs on a flight number are the segments. The segments on flight number 1 are LAX-DFW, DFW-LHR, and LAX-LHR.

A market is a collection of all O&D services that have the same origin and destination of the market. Airline fares are typically filed by market, though there may be flight specific fares.

An origin and destination (O&D) service can be a nonstop or a connecting market with specific departure, arrival and connect points. A connecting itinerary has more than one flight number. In this example, LAX-BOS and LAX-MIA are O&D services with a connection over DFW. O&Ds are important for schedule development and revenue management.

Passengers boarded means passenger enplanements. Passengers boarded, referred to as PBs, is a segment flown by a passenger. In some cases, the PBs can be higher than segments flown that are discussed in the examples below.

When a booking is made and a Passenger Name Record (PNR) is created, it may include flights (one or more), hotel, rental, car, etc. that reflects the various services booked for the journey.

2.15.1 GDS Air Segment Bookings

Segment counts identify the pricing of services by a GDS. When a travel agent makes a booking through the GDS, travel suppliers pay a segment booking fee to the GDS.

For air bookings, the number of segments is calculated by multiplying the number of passengers with each individual flight segment, identified by a separate flight number, in a PNR. Typically, a flight booking consists of 2.5 segments.

Consider the following examples.

Example 1 A one way flight from DFW to BOS for one person, one ticket.

Number of Segments: 1
Number of Legs: 1
Passengers Boarded: 1

Number of Passengers: 1
Passenger Name Record (PNR): 1

Example 2 A round trip flight from DFW to BOS for one person, one ticket.
Number of Segments: 2
Number of Legs: 2
Passengers Boarded: 2
Number of Passengers: 1
Passenger Name Record (PNR): 1

Example 3 A round trip flight from DFW to LHR for a family of 2, one ticket.
Number of Segments: 2
Passengers Boarded: 4
Number of Passengers: 2
Passenger Name Record (PNR): 1

Example 4 A round trip flight from LAX to LHR (direct flight, not non-stop) for a family of two, one ticket.
Number of Segments: 4
Passengers Boarded: 4
Number of Passengers: 2
Passenger Name Record (PNR): 1

Example 5 A round trip flight from LAX to BOS (connecting over DFW) for a family of two, one ticket.
Number of Segments: 8
Passengers Boarded: 8
Number of Passengers: 2
Passenger Name Record (PNR): 1

Example 6 Consider the Eva Air flight from New York (JFK) to Taipei (TPE) which makes a technical fuel stop in Anchorage, Alaska (ANC). The round trip is for one passenger, one ticket. The PB count depends on whether the passengers must deplane at the refueling station. If the passengers must deplane and re-enplane, it will count as 2 PBs.
Number of Segments: 2
Passengers Boarded: 4
Number of Passengers: 1
Passenger Name Record (PNR): 1

2.15.2   GDS Segments for Other Lines of Business

GDSs also have segment definitions for hotels, car rentals, cruise lines, and tour bookings.

For hotel bookings, each hotel segment represents a separate reservation processed by a GDS, regardless of the number of rooms, party size, or length of stay.

For rental car bookings, a rental car segment represents a separate reservation processed by a GDS, regardless of the number of cars, party size, or length of rental.

For cruise line bookings, a cruise line segment represents each separated cabin reservation processed by a GDS, regardless of the number of the number in party or the length of the cruise.

For tour bookings, each tour segment represents a separate reservation processed by a GDS, regardless of the party size or duration of the tour.

2.16   Full-Service Carriers and Low-Cost Carriers

In the context of product distribution, there are significant differences between full-service carriers and low-cost carriers.

Full-service carriers (FSC), which are also referred to as legacy carriers, offer a range of service classes (first, business, and coach), loyalty programs, operate hub-and-spoke (Peterson, 1986) connection networks, grant access to exclusive lounges for top-tier customers, provide on-board catering, baggage allowance, inflight entertainment, and various other amenities. Additionally, they primarily operate from major airports rather than secondary airports. FSCs recognize the importance of corporate clients and actively participate in the Global Distribution Systems (GDSs).

Most low-cost carriers (LCCs) primarily offer point-to-point direct flights and operate in a no-frills setting with only one class of service (coach). They typically depart from and arrive at secondary airports, which have lower landing fees. LCCs focus on selling additional services, have fewer comfort amenities, minimal loyalty programs, and prioritize online sales directly to consumers. It is uncommon for most LCCs to participate in the Global Distribution Systems (GDSs).

The ultra-low-cost-carriers (ULCC) constitute another category within the airline industry. They cater to travelers seeking a budget-friendly experience, particularly attracting leisure customers. ULCCs employ a business model centered around low costs, with their marketing strategy prioritizing the provision of significantly cheaper airfares compared to established legacy carriers. These airlines offer unbundled fares, which means that services such as seat assignments, check-in or carry-on baggage fees, and in-flight meals are not included in the base fare. Passengers can choose to pay for these services separately, allowing them to customize their travel experience according to their preferences and budget. One key distinction between ULCCs and traditional low-cost airlines lies in their minimalistic approach. While low-cost airlines also prioritize cost savings, ULCCs take it a step further by offering fewer inclusions within the base fare and instead implementing a higher number of additional fees for optional services. ULCCs focus on direct online sales, and do not participate in the indirect GDS channel.

Chapter 3
GDS: Platform Power (1984–1995)

3.1 Intermediaries and Why They Came into Existence

The Global Distribution Systems are intermediaries since they facilitate a transaction between a seller (airlines, hotels, rental car, cruise lines, rail, etc.) and a customer (corporate and leisure travelers). They came into existence in 1976 when airlines started promoting their reservations systems with agency features to travel agencies to sell their product. They served as an extension of airline sales. The GDSs made it easier for travel agents to comparison shop for itineraries in a market across carriers and simplified the entire booking and ticketing process. They were so successful that 80% to 85% of total air travel was booked through the GDSs until 1996, when the Internet arrived. The remaining bookings were made at the three primary physical locations: the airline call center, airport ticket office (ATO) and city ticket office (CTO).

The next twelve years (1984–1995) was a relatively calm period in which airlines that launched the various GDSs reaped the benefits of free cash flow. The CRS business was also more profitable than running an airline. Airline owners of the CRSs had a clear advantage over the smaller airlines that did not have an ownership in a GDS.

3.2 Regulation to Control Screen Bias

The air shopping travel agent entry that returns priced itineraries from one or more carriers did not exist when the GDS was launched in 1976. It was not until the mid-1980s that all GDSs supported rudimentary air shopping capabilities. Sabre introduced Bargain Finder[SM] in 1984 which returned nine itineraries for a shopping request. It was the industry's first automated low fare search capability. These shopping algorithms required an itinerary to be booked by a travel agent before

© The Author(s), under exclusive license to Springer Nature Switzerland AG 2024
B. Vinod, *Mastering the Travel Intermediaries*, Management for Professionals,
https://doi.org/10.1007/978-3-031-51524-8_3

new itineraries were priced and displayed, hence it is called the WPNI (Will Price New Itineraries) entry. It is an agency product that quotes the lowest available fare in the market for a booked itinerary and displays alternate itineraries with lower fares which could be a different route or carrier. Other GDSs introduced similar capabilities. The Apollo/Galileo flight search product was called Best Buy Quote ($BBQ™), and Worldspan's was called Power Pricing (Worldspan Power Shopper®).

Until the development of the shopping capability in the GDS, the single most important entry for a travel agent was the city pair availability. This entry returned a set of one way itineraries across carriers with availability by booking class for the requested city pair and date. Agents could sell a seat off a line of the city pair availability display and repeat the process for the return and complete the booking, followed by ticketing. Until the introduction of EDIFACT in 1988, the numeric availability attached to each booking class was based on teletype. With EDIFACT, subject to an airline's participation level for connectivity, the availability displayed could be based on true last seat availability returned from an airline's reservations system. This is based on the request (PAOREQ) and response (PAORES) transactions. EDIFACT also provides instant confirmation for sell transactions, called secure sell transactions since availability, is queried in the airline reservations system before the sell is completed, with the request (ITAREQ) and response (ITARES) transactions.

The regulatory period started in 1984 with the scrutiny associated with display bias. Smaller airlines and travel agencies were concerned with the bias associated with the display of itineraries by carrier for city pair availability. Based on these concerns, CAB identified four anti-competitive behaviors that were prevalent in 1983. The Code of Federal Regulations (CFR) was introduced and made effective on November 11, 1984, which prohibited anti-competitive behavior from airlines that owned the dominant CRSs. The four anti-competitive behaviors are discussed in Chap. 2, Sect. 2.11.

Even though the CRS rules went into effect on November 11, 1984, it did not dispel the debate over anti-competitive behavior.

Besides the regulatory concerns, it was a relatively calm and prosperous period where the GDSs collectively enjoyed an 80% share of total airline bookings until the arrival of the Internet on a large scale in 1996. The remaining 10% to 20% of bookings were made by wholesalers, airline call centers, airport ticket offices (ATO), and city ticket offices (CTO).

The landscape for bookings changes dramatically with the arrival of the Internet in 1996. Travelocity was the first Online Travel Agency (OTA) to enter the market ahead of airline direct channels.

3.3   Checks and Balances with Negotiated Contracts

GDSs negotiate contracts with airlines for access to their content. They also maintain contracts with travel management companies and smaller travel agencies that are subscribers to the GDS for content access to sell the airline products.

3.3.1   Airline: GDS Full Content Agreements (FCA)

The term "full content agreements" (FCA) is a contractual provision that requires an airline to provide access to its schedules, fares, and availability to the online and offline channels of a GDS that they provide to other sales and distribution channels, inclusive of their own websites. The GDS position is that they offer a balanced solution for the efficient marketplace with the well-publicized full content renewal agreements that they negotiate with full-service carriers (FSC), most notably the U.S. majors, global pure play, and international flag carriers. The FCA provision in airline-GDS contracts has been central to almost every dispute for over two decades, including the antitrust lawsuits made by American against Sabre and Travelport in the past.

The full content renewals promoted by GDSs are based on six tenets for participating carriers. They are:

1. Full content (schedules and fares). This ensures that the fares available on an airline consumer direct website, inclusive of web fares, are available in the GDS.
2. Long-term agreements. These contracts are typically three years or five years in duration.
3. Service fee protection.
4. Parity treatment, here an airline agrees to give a GDS access to the same content that is available through other GDSs.
5. Flexibility for travel agents to choose the tools they want.
6. Flexibility to pay incentives based on value.

Hence, airlines that sign the full content renewals cannot publish lower fares on their websites that are not available through the GDS channel.

An agreement to display an airline's schedules and fares begins with a participating carrier agreement (PCA). GDSs like Sabre have a Direct Connect Availability (DCA) agreement that is negotiated with a participating carrier. For a percentage reduction in booking fees for the duration of the contract, which is typically three to five years, the contract commits the airline to the highest level of participation. This highest level of participation, also referred to as connectivity, is called direct connect availability (DCA), which is also known as seamless availability. This provides travel agents with accurate information to book travel such as numeric true last seat availability by booking class in a city pair availability display. In return, a participating airline guarantees access to all published fares, promotions, and services that are available on their own consumer direct channel to the online and offline channels of the GDS. The term "DCA contract" is a U.S. centric term not used by GDSs in Europe and Asia. The term originated in contracts since the full content agreements came with the highest level of connectivity so that travel agents would not be subject to "unable to confirm (UC) at sell" or "price jumps".

The FCAs do not apply to the sale of ancillary products and services sold by airlines. This means that airlines do not have to distribute ancillary products to ATPCO and GDSs. The introduction of IATA's new distribution capability (NDC) will eventually render the full content agreements obsolete since pricing power, the power to price an itinerary in the GDS environment, shifts from the GDS to the airline.

Since 2006, travel agencies have been charged 80 cents per segment by the GDSs for access to full content from participating carriers. This is called the GDS content access fee (see Sect. 7.3.1), and it is an opt-in program. This fee is deducted from the GDS incentives paid to travel agents. Each GDS has a name for this fee. They are

1. Amadeus: Amadeus Content Plus
2. Sabre: Efficient Access Solution
3. Travelport: Content Continuity Program

However, the wording in GDS—travel agency contracts is ambiguous (Pestronk, 2018). The GDSs only stipulate that content provided by a participating carrier will be made available through the GDS. However, if the participating carrier withholds any fares, the GDS is not liable for absence of full content. The provisions of these contracts stipulate the 80 cents per segment fee, even if the travel agency does not have access to full content.

Sections 4.6 and 7.3.1 discuss the context under which the GDSs were forced to charge travel agents 80 cents per segment for access to full content from participating carriers.

3.3.2   Most Favored Nation (MFN) Clause in GDS: Airline Contracts

The term most favored nation (MFN) originated in international trade relations where a country grants the same rights, privileges, and immunities to a country that it has already granted MFN status.

The word "nation" is used because the term comes from international trade agreements between nations. This clause requires nation A to provide nation B with all the concessions that nation A has already granted to nation C.

An MFN clause exists in contracts between airlines and the GDSs, but it is misleading. It has nothing to do with nations and does not guarantee price equality in booking fees. The MFN clause mitigates risk where the airline is guaranteed to get the best deal now and in the future. What it simply means is that if the GDS grants a better deal to its favorite airline, then the same deal also applies to the airline negotiating the contract. Contrary to common belief, the MFN clause is not over-arching. In an MFN clause the "best deal" is unbiased display of an airline's content on a travel agency desktop. It does not mean that the segment booking fees are the same, since GDS fees are influenced by the size of the carrier and expected booking volume. At best, it is a *feel good* clause in airline—GDS contracts.

3.3.3   Most Favored Nation (MFN) Clause in TMC Contracts

When a corporate client of a TMC requests an MFN clause in the contract, it should be clear to the TMC to phrase the contract carefully with the required caveats to protect themselves. An MFN clause in this scenario means that the TMC must agree to provide the requestor (for example, Corporation A with the same prices as those offered to Corporation B if Corporation B gets the lowest price from the TMC. In a corporate travel management contract, the MFN clause is as follows (Pestronk, 2019).

> The Parties intend that Client shall have the status of a most-favored customer with respect to matters of pricing and contract terms for the Deliverables provided hereunder. If a travel management company (TMC) offers more favorable prices or contract terms at any time during the Term to any of its other customers, TMC shall immediately notify Client, and Client shall be entitled to the more favorable prices and contract terms for all Deliverables provided hereunder after the date of such offer. Upon Client's request, TMC shall certify to Client in writing that it is in compliance with this provision.

Like the GDS-airline contracts, the MFN clause between a TMC and a corporation is not meaningful since no two corporations are equal with similar service levels and itinerary complexity. Hence, from a TMC perspective, this is not a viable clause in a contract, though some corporations demand MFN status during contract negotiations. The TMC fee is a function of transaction volume and complexity of the itineraries booked. From a volume perspective, the lowest fees are rendered to large corporations, and government/military agencies. The second part of the equation is itinerary complexity. Even if the volume of bookings between two corporations is the same, more time is spent in servicing a corporation that books complex international itineraries a year than a corporation that books the same number of U.S. domestic trips. TMCs, to protect their business, must be careful to add qualifying conditions so that only accounts that are similar in volume and complexity are grouped together. During negotiations, TMCs can also demand agency exclusivity. Like the airline-GDS contracts, at best it is a *feel good* clause in TMC—corporation contracts.

3.4   Yield Management: The Early Period

Yield management originated in the airline industry with scheduled air carriers that operate under a Federal Aviation Regulations (FAR) 121 certificate issued by the Federal Aviation Administration (FAA).

Before airline deregulation, airlines operated within a strictly regulated environment that was overseen by governments and controlled internally through entities like the Civil Aeronautics Board (CAB) and the International Air Transport Association (IATA).

Key economic, regulatory, and competitive market factors led to the creation of the discipline of yield management (Cross, 1998). In the 1940s, the Civil Aeronautics Board (CAB) had the power to regulate tariffs, air mail rates, and competition in the airline industry. Airlines followed United's example in 1940 by introducing coach class, a cheaper alternative to first class. In 1948, charter passenger airlines began offering lower fares than scheduled airlines, prompting Capital Airlines to request permission from the CAB to offer discounted fares as well. This marked the beginning of the discount travel era, which resulted in various coach class restricted fares, including off-peak travel, no stopovers, and no refunds. Discounted fares expanded in the 1960s, with the introduction of qualified fares such as youth fares, family fares with discounts for children, clergy fares (which required identification cards), and Discover America excursion fares. The 1970s saw the proliferation of discounted fares, with an even wider range of options available to passengers.

In 1949, Southern Airways conducted the first "planned overbooking/controlled oversales" experiment (Cross, 1998). In the mid to late 1950s and 1960s, airlines began practicing overbooking as a means of compensating for cancellations and no-shows. Prior to 1961, U.S. airlines discreetly practiced overbooking, though it was not publicly acknowledged until Marvin Rothstein (Rothstein, 1985), Director of Operations Research at American Airlines, brought it to light. In that same year, the CAB acknowledged that one in every 10 passengers on the twelve leading airlines at that time did not show up for their flights, which caused economic distress to the airlines. The CAB did not officially approve the practice of overbooking but mandated a no-show penalty of 50% of the ticket value to passengers and put in place a reciprocal policy wherein airlines paid a penalty equal to 50% of the ticket value for passengers who were left at the gates on overbooked flights. These penalties were abandoned in 1963. A subsequent study conducted by the CAB in 1965–1966 determined that there were 7.69 denied boardings for every 10,000 passengers boarded (CAB Economic Regulation Docket 16563, 1967). In 1967, the CAB concluded that airlines could accommodate more passengers by practicing controlled overbooking. Overnight, overbooking became an accepted practice. CAB raised the penalty for passengers left at gates to 100% of the ticket value. The involuntary denied boardings, reported as the number of denied boardings for every 10,000 passengers boarded, was monitored by the CAB, and is still reported today by its successor organization, the U.S. Department of Transportation (DOT).

Kenneth Littlewood from the British Overseas Airways Corporation (BOAC) proposed the first discount allocation model (Littlewood, 1972) before airline deregulation. Littlewood suggested that airlines should prioritize maximizing revenue rather than passenger occupancy on flights with perishable seat inventory. Littlewood's rule for two booking classes can be extended to multiple booking classes. In 1980, William Swan from American Airlines' Operations Research (O.R.) department extended Littlewood's model with the logit approximation to the normal distribution, assuming demand is normally distributed. This method was first implemented by American Airlines in 1982 (Smith et al., 1992).

Development of the first yield management system began as a rule-based system (Smith, 2007). It was later improved to a decision support system that relied on marginal revenues. The system went through five iterations, starting with the Multiclass Optimization Modeling System (MOMS) in 1979, followed by the Discount Allocation Decision System (DADS) in 1980, the City Allocation Reporting System (CARS) in 1981, and the Super City Analysis and Reporting System (SCARS) in 1982. These iterations eventually led to DINAMO (Dynamic INventory Allocation Maintenance Optimizer) in 1985, which introduced overbooking and multi-class discount allocation controls. DINAMO utilized an economic overbooking model (Smith, 1982) to assess the tradeoffs between denied boarding costs and the cost of an unused seat. The discount allocation model was later known as $EMSR_B$ (Belobaba, 1992), extended Littlewood's Rule (Littlewood, 1972) to multiple booking classes. It determined the booking limits by generating joint protection levels for higher valued booking classes relative to the lower valued booking class. This was a computationally viable heuristic to optimal booking limits (Curry, 1990) that requires the evaluation of multidimensional numerical quadrature (convolution integrals) and more computing power. The seat inventory controls in this first-generation yield management system were based on leg/segment controls. These controls involved nested inventory controls for booking classes by leg, segment close indicators, and segment limit sales by booking class.

The initial versions of these applications were implemented on IBM mainframe computers using the MVS operating system. The SCARS reports were generated from a MARK IV program and provided to the analysts for examination. When DINAMO was introduced, the yield management analysts utilized an IBM mainframe 3270 terminal emulator to access data and reports.

During his tenure as President of Texas International Airlines, Donald Burr implemented "Peanut Fares" in select markets with low load factors, after receiving approval from the CAB. Following suit, American Airlines introduced SuperSAAver™ fares, which were applicable only on roundtrips, came with a 45% discount on coach fares, required a 30-day advance purchase, and were restricted within a seven to 45 day return window—all approved by the CAB. American Airlines cited the competition posed by transcontinental charter flights from New York to Los Angeles and New York San Francisco as the reason for this introduction. In 1978, the CAB introduced a "free zone," legalizing fares that fell between a 70% discount over coach fare formula and a 10% increase over the coach fare formula.

The term "yield management" is credited to Robert L. Crandall during his tenure as Senior Vice President at American Airlines (Cross, 1995). Crandall assumed the role of President at American Airlines in 1980 (Serling, 1985), at a time when airline deregulation was already in progress. Under Crandall's leadership, American Airlines became increasingly concerned about the aggressive price reductions implemented by competitors in an unregulated market, which posed a threat to profitability. As a result, the airline industry experienced a surge of innovation. Carriers operating within a deregulated environment quickly recognized the significance of optimizing passenger mix to enhance yields and the importance of maximizing the number of revenue passengers on each flight. This was because the cost of accommodating an additional passenger on a flight was minimal, resulting in nearly pure profit.

Yield management emerged on a large scale a few years following the deregulation of the U.S. airline industry in 1979 during the term of President Jimmy Carter. American Airlines, under CEO Robert Crandall, developed the first yield management capability in 1986 to combat the competitive threat from PEOPLExpress (Smith et al., 1992; Cook, 1999; Cummings, 2007, Vinod 2016a), a rapidly growing deep discounter. The fundamental premise of Yield Management is that seat inventory has a limited shelf life and not all customers are equal (Cross, 1995, Vinod 2021a). In the mid-1980s, PEOPLExpress posed a considerable threat to network carriers like American Airlines by offering deeply discounted tickets. Revenue Management became a strategic and tactical tool for American to match PEOPLExpress's fares, but it was capacity controlled. Analysts at American could control the availability of deeply discounted fares, while PEOPLExpress sold all their seats at the discounted price, which was unsustainable in the long run to support the airline's cost structure. By September 1986, PEOPLExpress was in financial trouble and ceased operations as an independent airline on February 1, 1987, when they merged with Continental Airlines. Donald Burr realized the importance of yield management too late and was quoted saying (Bryan, 1989), "*What you don't know about Revenue Management could kill you!*"

There is another aspect of what American did during this period. When the deeply discounted Ultimate SuperSAAver™ fares were introduced in 1985, prices were discounted 70% and ranged between $39 and $129, but these deeply discounted fares were capacity controlled. These new Ultimate SuperSAAver™ fares bought air travel within reach of an entirely new segment of the population that had never flown before.

Yield management maximizes revenue from *perishable inventory* based on the fundamental premise that *all customers are not created equal*. Seat inventory is perishable, since once the flight departs, the unsold seat inventory is lost forever. *Yield management is the process of selling the right seat to the right customer at the right price at the right time*. Yield management is called "revenue management" today since it is revenue and not yield (revenue per revenue passenger mile or revenue per revenue passenger kilometer) that is maximized. The transition happened officially in 1993 when the IATA Conference on Yield Management, which was organized in 1988, was rebranded as the IATA Conference on Revenue

Management. Besides yield, the core metrics monitored by revenue management analysts are the revenue per available seat mile (RASM) or revenue per available seat kilometer (RASK) and load factor.

Today revenue management is a central nervous system of the airline that influences decisions in flight scheduling, fleet assignment, fare management, reservations and inventory control, screen display on airline websites, and airline operations (Vinod, 2015a, Vinod, 2021a Chapter 10). The final products of revenue management are inventory control recommendations in the host CRS that determine booking class availability for travel agents to book flights.

3.4.1   Impact on Other Travel Verticals

When Bob Crandall introduced the term "Yield Management" into the airline vocabulary, no one realized at that time the far-reaching future impacts of this new concept. Fundamentally, it ushered in a new era of competitiveness in the airline industry.

Revenue management is one of several applications that was classified as strategic operations research (Bell et al., 2003) since it creates a sustainable competitive advantage. For his many contributions that revolutionized the airline industry, shortly before his retirement in 1998 after 25 years at American Airlines, Scott McCartney of the Wall Street Journal stated that Robert Crandall was "*the man who changed the way the world flies*" (McCartney, 1998).

Revenue management in the airline industry is widely acknowledged to contribute between 3% and 8% in incremental revenues. Business process maturity and level of sophistication of the inventory controls play a key role in the value of revenue management in an organization as a mission critical function (Donovan, 2005; Vinod, 2016a) with significant growth opportunities for management.

The lodging industry followed, and senior executives embraced and committed to the adoption of the concept, application software, and business process for competitive advantage. Early adopters were Marriott, Hilton, and Holiday Inn (Leven, 1994; Vinod, 2004).

Besides airlines and hotels, revenue management has found wide acceptance in travel verticals that have perishable inventory (Lieberman, 2010). While revenue management started in the airline industry and the hotel industry were fast followers, other modes of transportation that have perishable inventory like passenger rail companies, rental car, cruise lines, and ferry lines also saw the successful adoption of revenue management.

In the rental car industry, early adopters include Hertz (Carroll & Grimes, 1995), Avis and Budget. National Car Rental attributes revenue management to saving the company from bankruptcy in 1995 (Geraghty & Johnson, 1997). Among cruise lines, early adopters in the early 1990s were Admiral, Carnival, and Royal Caribbean (Fisher & Mongalo, 1993) to manage the sale of reservation requests and upsell to higher-valued cabins based on value.

A range of industries beyond travel like high-tech manufacturing, consumer electronics, and retail (Vinod, 2005a) have seen adoption of new approaches to solve their specific problems. Revenue management spawned active research and innovation from industry and academic institutions. Rapid improvements based on business model changes, modernization of data science and technology continues unabated to this day for competitive advantage.

Academic focus was introduced in the late 1980s (Belobaba, 1987, 1989), the 1990s (Lee, 1990; Weatherford, 1991; Weatherford & Bodily, 1992; Williamson, 1992; Chatwin, 1993; Kärcher, 1996), and 2000s (Li, 2008). With the growing interest in this discipline, several books have appeared that focus on both the practice and the theory (Daudel & Vialle, 1989; Cross, 1997; Ingold et al., 2000; Yeoman & McMahon-Beattie, 2004, 2011; Talluri & van Ryzin, 2004; Phillips, 2005; Boyd, 2007; Ng, 2008; Mauri, 2012; Garrow, 2016; Taneja, 2017; Gallego & Topaloglu, 2019; Szende, 2020; Vinod, 2021a, 2022a).

3.5   Airline Revenue Management: Leg/Segment Controls and O&D Controls

The airline seat is a perishable asset. When a flight departs, an empty airline seat is lost forever. Besides, there are differences in the preferences for business travelers and leisure travelers. Customers who book an airline ticket may be schedule sensitive (business) or price sensitive (leisure), leading to the creation of multiple fares for the same flight.

In the airline industry, *yield management is the process of selling the right seat to the right customer at the right price at the right time to maximize revenues.* Yield management was renamed revenue management in 1993, since it is revenue and not yield (revenue per revenue passenger mile) that is maximized. Yield management has a concept of *reading days*, which are predefined points in time before flight departure when demand forecasts and inventory controls are updated.

Control of seat inventory in the airline industry is either by flight leg, flight segment, or by origin and destination (O&D).

To explain this distinction, Fig. 3.1 illustrates a sample airline network. The nodes in the network represent the airports SEA (Seattle), SFO (San Francisco) LAX (Los Angeles), DFW (Dallas/Fort Worth, the hub), BOS (Boston), LHR (London Heathrow), and MIA (Miami).

A flight leg represents a nonstop city pair. Hence, the flight legs in this network are SEA-DFW, SFO-DFW, LAX-DFW, DFW-LHR, DFW-BOS, and DFW-MIA.

A flight number represents a line of flight. The origin and destination pairs on a flight number are the segments. The segments on flight number 1 are SFO-DFW, DFW-LHR, and SFO-LHR.

An origin and destination (O&D) service can be a nonstop or a connecting market with specific departure, arrival, and connect points. A connecting itinerary has more

Fig. 3.1 Sample network

than one flight number. In this example, SEA-DFW, SFO-DFW, and LAX-DFW are inbound flights into DFW. DFW-BOS, DFW-LGR, and DFW-MIA are outbound flights from DFW.

There are many O&D connecting services in this network. For example, LAX-BOS, LAX-LHR. LAX-MIA, SEA-BOS, SEA-LHR, SEA-MIA, SFO-BOS, and SFO-MIA are O&D services with a connection over DFW.

A market is a collection of all O&D services that have the same origin and destination of the market.

Overbooking by cabin is an important feature to compensate for the effects of cancellations and no-shows (Smith, 1982; Rothstein, 1985). No-shows are customers who hold a reservation at departure and do not show up for a flight. Go-shows are passengers who show up without a confirmed reservation for a flight or those that show up with a confirmed reservation for which no reference is found in the carrier's host reservations system. The latter category is referred to as no record passengers or NOREC's. Mis-connects are passengers who missed their flight connections due to flight delays.

Spoilage represents the number of empty seats on closed out flights. Oversales represent passengers who were denied boarding. The risks associated with oversales are passenger compensation, loss of goodwill, and higher operating costs at the airports. The smaller the spoilage factor, the more effective the overbooking policy is.

Overbooking is a balancing act with two objectives, minimizing spoilage *and* minimizing oversales, which should be satisfied *simultaneously* to maximize onboard revenues.

There are two types of overbooking models: the economic overbooking model and the more conservative oversale rate constrained model (expressed in oversales per 10,000 boarded). Overbooking ensures that more passengers can book on their first choice (Crandall, 1991) of travel.

There are two categories of revenue management systems, which vary depending on how an airline controls its seat inventory for the control of discounted fares. They are called leg/segment controls and O&D controls. These controls influence the city pair display of booking class availability for travel agents.

The control of reservation requests by origin and destination is a vast improvement in inventory control technology (Feldman, 1995; Gallacher, 1996) over leg/segment controls. It is applicable for airlines that operate a hub-and-spoke network to control the flow of local and connecting traffic.

This book only discusses airline revenue management in the context of booking class availability that is required to support city pair availability and air shopping for GDS bookings. Air shopping is also required for airline direct bookings. In the future with New Distribution Capability or NDC (see Chap. 6), booking classes are no longer required, simplifying the availability determination logic for O&D carriers. Simplifying the availability determination process in the absence of RBDs for O&D carriers is discussed in Chap. 9.

3.5.1 Leg/Segment Controls

Initially, all airlines worldwide utilized leg/segment revenue management, which establishes nested controls for inventory based on flight legs or segments using variations of the expected marginal seat revenue calculation (Belobaba, 1987, 1989). Leg/segment revenue management systems forecast the unconstrained demand by leg class or segment class and determine the nested inventory controls which are stored on the inventory system of the host CRS. The optimization is at a flight number level to determine the inventory controls. These controls often involve segment close indicators (open/close by booking class) or segment limits (numeric limits by booking class) in conjunction with nested inventory controls by leg or segment. They are commonly used by carriers that primarily operate point-to-point routes without substantial connecting traffic. This approach continues to be widely employed in the industry.

3.5.1.1 Calculation of Leg/Segment Availability

For every city pair availability request from a travel agent, the airline host CRS calculates the availability by booking class. For leg/segment carriers, determining seats available by booking class is straightforward since numeric seat availability is stored on the inventory detail record on the airline's host CRS. With a multiple serial nesting structure, seat availability is typically calculated based on net nesting or threshold nesting (Vinod, 2006). In addition, segment close indicators or segment limit sales are used to selectively control through traffic on the flight number.

With threshold nesting, a booking in any booking class impacts all the booking classes in the booking class hierarchy, thus maintaining the protection level across all classes, except for those that are closed, regardless of what booking class was sold. The calculation of seat availability in a cabin in booking class i is as follows:

$$\text{Seats Available}_i = \text{Authorization}_i - \text{Total Seats Sold}; \quad i = 1, 2, \ldots, N$$

Authorization_i is the booking limit for class i and Total Seats Sold is the total seats sold across all booking classes in the cabin.

$$\text{Total Seats Sold} = \sum_{k=1}^{N} \text{Seats Sold}_k$$

With net nesting, inventory action is taken in the requested class and all classes *above* the requested class in the hierarchy, up to the base class (usually Y). The calculation of seat availability in a cabin is as follows:

$$\text{Seats Available}_i = \text{Authorization}_i - \sum_{k=i}^{k=N} \text{Seats Sold}_k \; ; \quad i = 1$$

Seats Available_i

$$= \text{Min}\left[\text{Availability}_{i-1}, \left(\text{Authorization}_i - \sum_{k=i}^{k=N} \text{Seats Sold}_k\right)\right] \; ; \quad i = 2, \ldots, N$$

Authorization_i is the authorization limit for class i, Seats Sold_k is the seats sold in class k and $\text{Availability}_{i-1}$ is the seat availability in the class immediately above i, in the class hierarchy.

As the calculation indicates, net nesting requires booking counts by booking class while with threshold nesting, only the total seats sold in a nested hierarchy is required.

Table 3.1 illustrates seat availability based on threshold nesting and net nesting calculation method.

The authorization level for the highest booking class in the hierarchy Y is the overbooking limit. There are two types of models to calculate the overbooking limit: the economic overbooking model or the more conservative quality of service constrained model. The authorization levels for all booking classes are updated daily by the revenue management system which reflects the current seats available.

As observed in Table 3.1, the inventory detail record on the airline reservations system stores information by flight leg or flight segment the booking limits, bookings, and seats available by booking class.

When a direct connect availability request for booking class availability is received from a GDS, the host CRS returns numeric availability to the GDS with a

Table 3.1 Booking class availability on the inventory detail record

Booking class	Authorization	Bookings	Seats available Threshold nesting	Seats available Net nesting
Y	420	5	165	165
B	385	10	130	135
M	320	20	65	80
H	260	35	5	40
V	210	50	−45	25
Z	180	55	−75	25
Q	60	80	−195	−20
Total bookings		255		

Flight Leg: DFW-BOS Coach Capacity: 300 seats Flight Number: 1
Current Bookings by Booking Class: Y = 5, B = 10, M = 20, H = 35, V = 50, Z = 55, Q = 80

simple read-only lookup of the inventory detail record. For city pair availability requests from a GDS, numeric availability is usually capped at seven or nine.

If the connectivity is limited to teletype between the GDS and the host CRS, then booking class availability in the GDS is based on availability status (AVS) messages, received from the host CRS, that is stored in an internal database.

Regardless of the level of connectivity, AVS is the backup default for booking class availability in situations when there is a time out when availability is requested. Hence, the transmission of AVS messages and AVS generation and recap during schedule changes is a core capability in the host CRS.

3.5.2 O&D Controls

Network airlines operate hub-and-spoke networks. The typical hub economics consists of 30% of local traffic and 70% of connecting traffic. O&D revenue management systems forecast the unconstrained demand by service class and the network optimization model calculates the optimal inventory controls which are stored on the inventory system of the host CRS. Network optimization is either at an *optimization group* level, which is an arrival-departure complex at a hub, or for the entire schedule for a future departure date. To manage seat availability based on origin and destination, airlines utilize a technique known as continuous nesting (also known as bid price controls). The bid price is generated by a network optimizer in the airline's revenue management system by flight leg. For leg j, the bid price is denoted by γ_j. The bid price is the opportunity cost of not having an incremental seat on a flight. Adoption of bid price controls by network carriers started in the late 1990s and continues to this day. Two of the largest airlines, American Airlines and US Airways, migrated to bid price controls in November 1998 (Cook, 1999). Approximately 30 to 50 of the major network carriers have adopted origin and destination revenue management.

3.5.2.1 Calculated of O&D Seat Availability

The calculation of seat availability with O&D controls consists of two steps: physical availability followed by financial availability.

The physical availability is simply the difference between the authorized capacity for the cabin and the total seats sold in the cabin to determine if a booking is allowed.

For financial availability, the bid prices are additive when multiple legs are involved to cover an itinerary. The net contribution for each service class is calculated by subtracting the total bid price from the market class fare value (MCFV) for each booking class. If it is positive, then a booking can be made in the booking class (Vinod, 2006). The MCFV are calculated periodically by aggregating fares considering fare qualification rules, which can be determined when an availability request is made. It is an average of historical and future fares weighted by flown traffic by point of commencement. This level of precision is required to ensure that the MCFV used for availability determination is as close as possible to the ticketed fare, though they will never be the same.

The net contribution (financial availability) for service s, class c, is calculated as follows, where S. In the collection of services in the network.

$$\text{Net Contribution}_{sc} = \text{MCFV}_{sc} - \sum_{j \in S} \gamma_j$$

Figure 3.2 illustrates the calculation of availability with O&D controls on the host CRS for a sample network.

For O&D controls, a bid price curve that is a function of seats sold (or seats available) is stored on the inventory detail record. The bid price curve is calculated by the revenue management system because the value of the bid price to seats sold on a flight leg and cabin is not a linear function.

Direct connect availability requests for city pair availability between the GDs and the host CRS is a computational burden, since the net contribution calculation must be applied iteratively to find the precise number of seats available by booking class. For city pair availability requests from a GDS, numeric availability is usually capped at seven or nine by the host CRS.

If the connectivity is limited to teletype between the GDS and the host CRS, then booking class availability in the GDS is based on availability status (AVS) messages, received from the host CRS, that is stored in an internal database. The AVS messages generated by the host CRS are based on O&D availability of the local leg. Availability status messages by O&D do not exist in the GDS since it will lead to excessive volumes of data that will have to be transmitted from the host CRS to the GDS.

Regardless of the level of connectivity, AVS is the backup default for booking class availability in situations when there is a time out when availability is requested. Hence, the transmission of AVS messages and AVS generation and recap during schedule changes is a core capability in the host CRS.

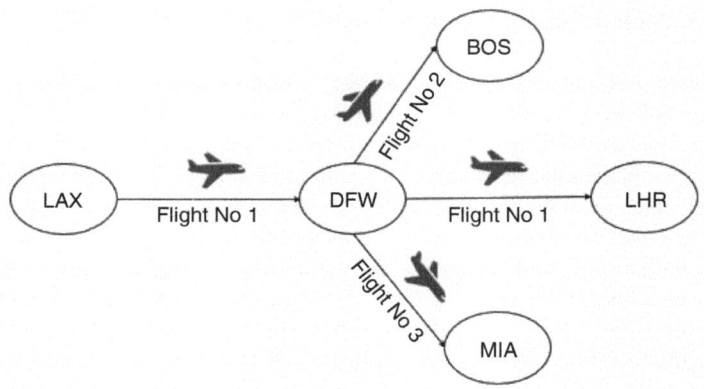

Origin	Destination	Class	Fare Value	Total Bid Price	Net Contribution	Open/Close	Flight Leg	Bid Price
LAX	DFW	Y	$600	$300	$300	Open	LAXDFW	$300
LAX	DFW	B	$500	$300	$200	Open	DFWBOS	$450
DFW	BOS	Y	$550	$450	$100	Open	DFWLHR	$700
DFW	BOS	B	$450	$450	$0	Closed	DFWMIA	$350
DFW	LHR	Y	$800	$700	$100	Open		
DFW	LHR	B	$700	$700	$0	Closed		
DFW	MIA	Y	$300	$350	–$50	Closed		
DFW	MIA	B	$250	$350	–$100	Closed		
LAX	BOS	Y	$900	$750	$150	Closed		
LAX	BOS	B	$750	$750	$0	Closed		
LAX	LHR	Y	$1,200	$1,000	$200	Open		
LAX	LHR	B	$1,050	$1,000	$50	Open		
LAX	MIA	Y	$600	$650	–$50	Closed		
LAX	MIA	B	$540	$650	–$110	Closed		

Fig. 3.2 Calculation of booking class availability for O&D control

3.5.2.2   Legacy System Enhancement Versus Open System for O&D Booking Class Availability

Enhancing a legacy mainframe reservations system to support O&D controls is an expensive proposition. It requires an expansion of the inventory detail records, development of the new O&D booking class availability logic, followed by the extensive testing of all the applications that access the inventory records in the reservations system. A better alternative is to develop a satellite co-operative processor that is channel attached to the host CRS to provide the O&D availability logic for inventory control.

Sabre collaborated with Air France in 1997 to address the limitation of enhancing legacy TPF-based reservations systems for O&D inventory control. Together, they developed the *Availability Processor*™, a Unix-based satellite co-operative processor. This processor was connected to Alpha3, Air France's host CRS, to handle availability and sell processing for O&D control (Smith et al., 1997; Vinod et al., 1997). Nowadays, the use of satellite processors for availability and sell processing in an open systems environment is widely adopted. Examples of such processors include *SabreSonic* Inventory (an enhanced version of the Availability Processor™), Altea Inventory/RAAV (Revenue Availability with Active Valuation), and PROS RTDP. Google/ITA also utilizes a read-only availability proxy, known as DACS (Dynamic Availability Calculation System), for processing availability requests in their shopping engine.

3.5.3   The Last Frontier: Individual Seat Pricing

Advances in revenue management have always focused on more granular inventory controls to maximize revenues. Revenue management of individual seats is the most granular level of inventory control by customer segment or individual customer. Pricing of seats can be by seat type (e.g., aisle, window, bulkhead, exit row, etc.) or even at the individual seat level. The inventory control of individual seats represents the last frontier in revenue management (Vinod, 2021d). Pricing of seats by type or individual in a GDS environment requires a request for seat pricing from the airline host CRS. It is far easier to control inventory at the seat level in an NDC environment when the airline is in control of the offer, eliminating the need for additional messages between the GDS and the airline host CRS.

3.5.4   Value Generated by Revenue Management

In addition to discount allocation controls, airlines also employ overbooking as a strategy to account for cancellations and no-shows. Unlike hotels or cruise lines, overbooking risk is minimized when there are multiple flights serving the same market, thereby ensuring that passengers on less popular flights during peak travel times are accommodated. Various studies have demonstrated that revenue management can generate additional revenues ranging from 3% to 8%, depending on the level of sophistication of inventory controls and the maturity of the airline's business process.

   Full-service carriers, also known as network carriers, started to migrate to origin and destination controls (O&D) to control the flow of connecting traffic versus local traffic in the late 1980s with virtual nesting controls (Smith, 1986; Vinod, 1989) and later with bid price controls in the 1990s (Vinod, 1995). Virtual nesting relies on virtual buckets and booking limits by bucket to enable O&D controls. For example, when a virtual bucket is closed, all service classes that are indexed or mapped to the bucket are automatically closed for sale. Only a few large network carriers such as American Airlines, United Airlines, Delta Air Lines, KLM Royal Dutch Airlines, and SAS used virtual nesting controls. Today, continuous nesting control based

O&D controls, also known as bid price controls, are the most popular form of O&D controls that is used by several network carriers. American, for example, migrated from virtual nesting controls to bid price controls in November 1998 (Cook, 1999).

Another aspect of revenue management is the control of group traffic to ensure that the minimum acceptable fare negotiated with a group does not displace higher valued individual passengers (Yuen, 2002, 2003; Vinod, 2013b). Group bookings are negotiated and booked directly by the airline without involving travel agents.

Over the past three decades, revenue management has matured as a discipline with well-defined repeatable business processes. Revenue management has played a larger role and been a key influencer in decision making across airline planning, airline operations and airline marketing (Vinod, 2015a).

With the growing importance of the sale of ancillary products and services as a revenue stream, traditional revenue management of the base fare is transitioning to offer management (Chap. 9).

3.5.4.1   First, Second and Third Order Network Effects

The objective of origin and destination revenue management is to determine the right mix of short-haul, medium-haul and long-haul demand that should be accommodated on an airline's global route network to maximize total revenues subject to capacity constraints. By optimizing the network, instead of a flight leg or flight segment, the model simultaneously evaluates trade-offs of the network effects simultaneously.

The objective of the O&D network optimization model is to capture the first, second and third order effects in the network and determine the optimal inventory controls to maximize network revenues. Many of the GDS connectivity products such as seamless (interactive) sell, seamless (interactive) availability, married segments, journey, and married to journey based on EDIFACT (see Sect. 3.5) were specifically developed for O&D network carriers.

For airlines on leg/segment inventory controls, seamless sell and seamless availability are required if the host CRS inventory control system has business rules that influence the availability by booking class for each booking request which cannot be reflected in the standard AVS/AVN messages.

To explain the first-, secon, and third-order effects in the airline network, consider the 3 × 3 hub-and-spoke network shown in Fig. 3.1.

If there is a surge in demand from SEA to DFW because of a promotional fare filing, the first-, second-, and third-order impacts on traffic flows are summarized in Table 3.2.

The narrative for the positive and negative higher order network effects is as follows:

1. *First-Order Effect:* When a promotional fare is introduced from SEA to DFW, the local SEA-DFW market will see a surge in demand (+)

Table 3.2 First-, second-, and third-order network effects

First-order effects (+)	Second-order effects (−)	Third-order effects (+)
SEA-DFW	SEA-BOS	SFO-BOS
	SEA-LHR	SFO-LHR
	SEA-MIA	SFO-MIA
		LAX-BOS
		LAX-LHR
		LAX-MIA

2. *Second-Order Effect:* Because of the increased demand from SEA to DFW (local market), availability will be restricted to the connecting markets that have the SEA-DFW leg in common (−)
3. *Third-Order Effect:* Because availability is restricted for the connecting markets originating out of SEA (the second-order effect), there will be greater availability for markets that do not originate in SEA, and hence they will see increased traffic (+).

3.6   The EDIFACT Protocol

Electronic Data Interchange for Administration, Commerce and Transport (EDIFACT) is a global standard set of syntax rules defined by the United Nations for the inter-company electronic data exchange between two or more business partners. The EDIFACT EDI standard was created by the United Nations in 1985. The syntax was designed to keep files as compact as possible. In 1987, the UN/EDIFACT Syntax Rules were approved by the International Organization for Standardization as ISO 9535. The standards are called UNSM (United Nations Standard Messages).

The standardized, computer-readable format for transforming data enables the automation of commercial transactions and is widely used in international businesses, multiple industries such as financial institutions, insurance providers, food and pharmaceutical suppliers, retailers, automotive manufacturers, and travel. It is a collection of 200 messages that have been adapted across industries. When it was introduced, it was acknowledged as a common, uniform language through which computers can communicate for fast and efficient online transaction processing.

In the airline industry, EDIFACT was introduced in 1988. The original UN/EDIFACT standard was not suited for real-time interactions. It was the airline industry, acting through ATA and IATA that developed the interactive EDIFACT architecture. In travel, the interactive request and response fall under the TIQREQ (Travel, tourism, and leisure Information Inquiry Request) and TIQRSP (Travel, tourism, and leisure Information Inquiry Response) categories. It was leveraged for interactive availability, interactive sell, married segment controls, journey controls, married-to-journey, interactive seat maps, bid price exchange between operating carrier and marketing carrier, and many more transaction flows. The introduction of EDIFACT for travel-related transactions such as availability, booking, and ticketing was a quantum leap in message processing between GDSs and airline reservations systems (CRS) with instant confirmation.

EDIFACT Is a rigid, compact format and has limitations for today's retailing needs of airlines. The airline version of interactive EDIFACT supports high-volume transaction processing. Scalability remains to be proven and demonstrated by the NDC/XML-based messaging standard.

With the growth of the Internet in the early 2000s. XML became the more widely supported file syntax. EDIFACT has a rigid syntax which many low-cost carrier reservations systems do not support.

The new XML-based message, introduced as part of the NDC initiative, for exchanging data between an airline and a GDS will over the next 10 to 15 years (2033–2038) replace the EDIFACT protocol which has been the mainstay of GDS interactive connectivity since its introduction in 1988.

3.7 Advances in Air Connectivity

With the introduction of EDIFACT, there was significant innovation on the connectivity front. Major features such as journey controls and married to journey controls were introduced in the early 1990s well before the introduction of origin and destination (O&D) revenue management.

Air connectivity products are fundamental to reservations processing on the GDS. They automate the workflow to enable the GDS subscribers to request and sell an airline's product. Airlines support various air connectivity products at varying degrees of sophistication using teletype, EDIFACT, and XML messaging standards to receive and respond back to requests.

All GDSs have a unique global identifier for sending and receiving transactions that are to be processed. For example, the identifier for Sabre is 1S, Amadeus is 1A, and Worldspan is 1P.

3.7.1 GDS Connectivity Product Features

The product features for GDS connectivity include both teletype and EDIFACT protocols. Advanced revenue management capabilities such as origin and destination controls require the newer EDIFACT messaging for availability and sell transactions.

In the world of airline connectivity, the term "direct connect" is frequently used. It should not be confused with the usage of the term "direct connect," which is used in the context of airline distribution, which is facilitating an airline booking without using a GDS.

3.7.1.1   AVS/AVN

Availability status messages, also known as AVS, are messages sent through teletype to inform airlines and GDSs whether seats on a particular flight by booking class are available for sale or not. These messages can be inbound or outbound and may include information such as flight closure, waitlist availability, flight cancellation, or flight reopening. An AVS message is an open/close status message, and AVN is numeric status based on the number of seats available.

The GDS stores AVS and/or AVN messages based on the agreement with the airline. Unless an airline sends a close message, all flight/segment/departure date/classes are considered open for sale by the host airline. The GDS handles all availability and sales processing and sends a teletype message after recording the sale activity. However, if an airline posts AVS status in the "request" status, travel agents must sell the seats pending confirmation from the airline.

3.7.1.2   Basic Booking Record (BBR)

The Basis Booking Record (BBR), also known as Basic Booking Request, is the lowest level of participation that an airline can have in a GDS. At the end of a transaction (end transaction—ET), the airline sends a teletype sell request, and upon receipt of the message, confirms the booking. BBR allows travel agencies to make reservations with airlines that have basic internal systems. The availability request for city pairs only shows the class of service available, without any numerical value, and travel agents will not know if the seat is confirmed until the airline responds after the transaction is completed. There are no travel agency incentives paid by the GDS to the travel agency for BBR bookings.

3.7.1.3   Direct Access Interactive (DAI)

GDS agents can request availability and sell against an airline's host inventory resident on the host CRS. The GDS communicates with the host CRS through a pool of airline LNIATAs (LiNe Interchange Address Terminal Address) allocated as connections to a GDS. The airline views these transactions as if they were performed by an airline agent. To carry out this process, travel agents must request access to the airline's inventory system.

3.7.1.4   Seamless Sell and Seamless Availability

Seamless connectivity between the host CRS and the GDSs that provide a significant percentage of total bookings is a *requirement* for origin and destination inventory control. GDSs offered seamless sell transaction processing before seamless availability was offered in the early 1990s. Seamless sell is also called direct connect sell (DCS), and seamless availability is also called direct connect availability (DCA).

3.7.1.5   Direct Connect Sell (DCS)

Direct connect sell (DCS), also known as seamless sell, is a necessary level of participation for an airline with a GDS for O&D control. In this situation, AVS data stored on the GDS database or direct access is used to display availability. Sales are made against the airline inventory system. DCS allows a GDS travel agent to sell an airline segment from the GDS as if the agent were selling directly from the host CRS. The GDS instantly and transparently checks the airline's internal database to provide the travel agent with true last seat availability. DCS enables full usage of host CRS O&D controls to compute availability before the booking is made.

Upon completion of the booking process on a DCS airline participant, the record locator is returned to the travel agent as an end transaction (ET) confirmation and a means of documenting the booking within the GDS. Messages transmitted through this form of connectivity conform to the EDIFACT standard data packet and encompass comprehensive POS (point of sale) information at the agent level. DCS is mandatory for airlines who are using O&D-based inventory control.

3.7.1.6   Direct Connect Availability (DCA)

Direct connect availability (DCA), also known as seamless availability, is a necessary level of participation with a GDS for O&D inventory control. It allows GDS travel agents to receive accurate last seat availability in the city pair availability (CPA) display based on the availability processing logic that is resident in the host CRS. An airline's last seat availability is integrated into the standard CPA display in the GDS. A DCA sales indicator is appended to segments within the CPA display to differentiate the airline's product.

DCA does not require special formats or additional keystrokes. With DCA, a GDS subscriber can view an airline's availability as seen by an airline reservation agent. With no special formats or additional keystrokes, an availability request is received from the GDS as if the request originated from the host CRS. The GDS "seamlessly" retrieves availability from the host CRS database to provide the GDS agent with up-to-date true last seat availability.

Direct connect availability (DCA) offers a key benefit: it enhances the visibility of available inventory for sale by directly querying the airline's availability package on the host CRS and displaying it on the GDS. The selling process for confirmed bookings remains consistent for both DCS and DCA, as the airline's availability processing logic is used to determine what can be sold.

The airline's host CRS will receive messages, under this method of connectivity, in an EDIFACT standard data packet containing full POS information at the travel agent level. Direct connect availability (DCA) cannot be supported without the presence of direct connect sell (DCS). DCA is a requirement for O&D control.

Seamless availability extends the host CRS's inventory management system through the GDSs to the subscriber. In addition, seamless availability interacts with low fare search products offered by the GDSs. The host CRS will receive messages, under this method of connectivity, in an EDIFACT standard data packet containing full POS information at the agency level. Hence, if the market class fare adjustment (MCFA) table (also called the market adjustment table (MAT)) in the host CRS used for O&D control supports POS-based fare adjustments, inventory can be controlled at the appropriate POS level.

Without seamless availability agreements, booking class availability on a GDS is decided by the AVS/AVN present on the local GDS database. These controls can be either leg based or segment based. However, with seamless connectivity, a GDS agent can directly access host CRS availability, enabling them to provide true last seat availability and full origin and destination inventory controls.

3.7.1.7 Market Restricted Flights

Airlines sometimes publish market restricted "MR Flights" on a GDS to reduce the impact of availability requests that must be handled by the host CRS. MR Flights are flights that are restricted to a certain market. When an O&D availability request includes an MR Flight, it is sent to the host CRS for true last availability based on the O&D inventory control logic on the host CRS. If a flight is not market restricted, the availability response may be based on AVS or AVN status stored locally on the GDS.

With a few exceptions, most airlines hosted on Amadeus pay an additional polling fee, which can vary by territory, when the availability request is sent to the airline's inventory system for true last seat availability, in addition to the traditional segment booking fees charged by GDSs. Polling fees can be avoided if availability is returned based on AVS/AVN, which defeats the purpose of O&D controls. Revenue management systems can recommend flights that should be placed on market restricted (MR) status based on booked load factor, days to departure, and expected proportion of connecting traffic.

There are some fundamental differences in the flow of availability transactions for airlines when they originate from an Amadeus travel agency versus Sabre and Travelport. Also, the impact of incremental polling fees is greatest for airlines hosted on Amadeus for reservations. The polling has implications for airlines on O&D inventory control.

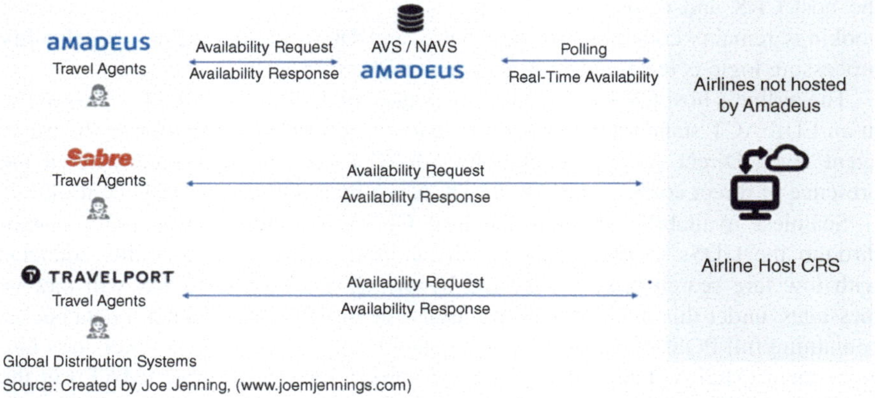

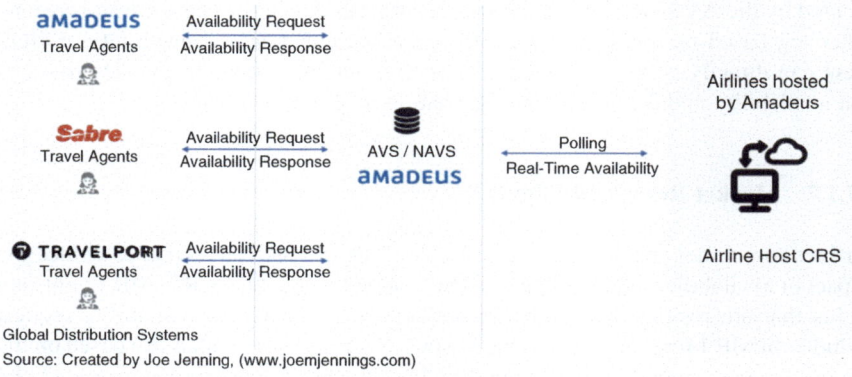

Fig. 3.3 Availability request/response for non-amadeus carriers

Fig. 3.4 Availability request/response for amadeus carriers

Figure 3.3 illustrates the availability request and response workflow for airlines that are not Amadeus carriers. The highest level of participation for connectivity between an airline and a GDS is seamless availability or direct connect availability, which is interactive. For O&D control, the availability processing logic resident in the host CRS must be invoked to determine true last seat availability. However, airlines not hosted on Amadeus are required to pay the polling fee for requests originating from an Amadeus travel agency to query the host CRS for true last seat availability. The alternative is to return availability based on AVS/NAVS, which dilutes the value of O&D controls.

Figure 3.4 illustrates the availability request and response workflow for airlines that are Amadeus carriers, also referred to as Amadeus System Users.

In this scenario, since Amadeus is the receiving system (identifier 1A) for airlines hosted by Amadeus for reservations, all channels inclusive of Sabre and Travelport are subject to a polling fee that the airline must pay unless they resort to an AVS/NAVS response for availability. Airlines use the MR flight status capability to reduce the polling fees paid to Amadeus.

Flights are classified MR by an airline's revenue management system, based on the connecting traffic carried on board these flights, which requires accurate POS-based O&D availability. The number of flights that are posted as market restricted is a function of the percentage of total reservations provided by a GDS and the underlying transaction cost structure for seamless availability. O&D revenue management must recommend a set of flights by date that must be on market restrict status based on two key factors: the bid price relative to the fares flowing over the segment, and the expected number of connecting passengers forecast over the segment.

The MR capability is also useful for airlines when they cutover from leg/segment controls to O&D controls with a phased migration when markets are incrementally enabled with O&D controls.

The System User concept is unique to Amadeus. It is powerful since it offers protection from supplier direct distribution since Amadeus serves as the receiving id and booking engine on the websites of the Amadeus carriers (System Users). Hence, regardless of where the booking originates: the airline website, an Amadeus travel agent or an Amadeus carrier ticket counter, Amadeus is firmly entrenched in the value chain and receives revenue for the booking. The System User concept also enhances airline alliance capabilities of Amadeus by providing instant access to passenger profiles, alliance availability, fare quote, and frequent flyer data. Airlines benefit from this capability because they do not have to enhance their core reservations system.

3.7.1.8   Married Segments

Married segment logic in the airline's reservations inventory system is a requirement for O&D control. Married segment logic allows segments to be married or "joined" as a connection from an O&D-based availability display and treated as a single unit through the booking, pricing, and ticketing process. This protects market-specific inventory from unauthorized, partial cancellations. This logic ensures that inventory determined to be available on an origin and destination basis is sold *as* an origin and destination *with* the applicable fare. A common practice when seats are not available on a segment is to make a reservation for a connecting service that includes the segment followed by the subsequent cancellation of the upline or downline segment to fulfill the original request. When flight segments have been married as an origin and destination service, cancellation of *any* segment in the service will result in *all* segments within the service to be canceled. Sell from availability transactions involving connecting flight segments should be *married* to indicate the origin and destination sold. Similarly, when segments are sold from the availability display of a GDS, all connecting segments in the service should be sold as an origin and destination to conform with the way their value and availability levels have been calculated.

Participating airlines decide which segments to marry, and the control of these segments is applied with the sell request seamlessly. Up to three segments can be married and any reservation modification or partial cancellation must be authorized by the airline. Travel agencies realize that the sell is valid and legitimate, thus knowing that neither airline action nor debit is forthcoming due to inaccurate or illegal booking of flights.

3.7.1.9   Journey Data

Journey is a requirement for O&D control. Journey data provides additional information to the airline about the rest of the passenger itinerary. The airline can evaluate individual segments relative to the entire journey. Journey control data can be sent with Direct Connect Availability, Direct Connect Sell, and Direct Access Interactive messages and follows the actionable travel segments (TVL). All the segments or a specified number of segments requested by an agent in the AAA/PNR will be included in the journey data. It does not support past date/flown segments or cancel segment processing and the number of journey data segments sent to the travel agent before and after an availability or sell query will depend on the carrier.

Journey data enables the evaluation of a flight segment relative to a passenger's entire journey, ensuring the product is priced based on its true value. Journey data when used in conjunction with married segment control generates incremental revenues for the airline.

3.7.1.10   Married to Journey

Married to Journey Data is an optional feature that allows segments not sold at the same time to be married together. Seamless Availability, Seamless Sell, Married Segments, and Journey Data must be activated by the airline to access this feature. Married segment control allows participating airlines to marry newly sold segments in a single sell entry. However, some airlines find this restrictive, since they are unable to marry newly sold segments to one that was previously sold. Married to journey provides the distinct capability to marry previously sold segments with a new sell request.

GDS connected travel agencies realize in a world of airline partnerships and market partnerships that their booking is seen in its entirety and valued as such. The airline can see the whole picture in their evaluation to confirm the request. Journey data are a requirement for O&D control.

3.7.1.11 Interactive Seat Maps

Interactive seat maps provide a live representation of individual seat availability for each cabin on a flight and can be seamlessly integrated into graphical seat maps. This functionality reduces the need for travel agents to contact an airline reservations center, enabling them to prioritize revenue-generating inquiries.

3.7.1.12 Interactive Pre-Reserved Seats

Interactive pre-reserved seats enable travel agents to promptly request a specific seat assignment and receive immediate confirmation from the airline. A travel agent can make a seat request with or without utilizing a seat map.

Both Interactive Seat Maps and Interactive Pre-Reserved Seats serve as the basis for airline merchandising initiatives like paid seats.

3.7.1.13 Point of Sale

Point of sale (POS) is defined as the location at which bookings are sold. It provides information about the location of the agency. POS automatically comes with direct connect EDIFACT products like DCA and DCS. It is optional with teletype products. Extended POS data provides additional fields, and the sequence of elements is based on industry standards. Unlike an online direct airline distribution request for availability, information about the user inquiring can be immediately given when requesting availability.

For seamless availability requests, airlines that operate with an origin and destination (O&D) revenue management system respond with availability for booking classes by O&D based on the logic resident in the host CRS inventory. It could be based on the POS or point of commencement (POC). POC is preferred by international carriers who offer the same fare regardless of the POS. If flights are already booked and active in the agency's AAA (agent assembly area, an area in memory in which items of data are stored temporarily during the shopping or booking process), journey data will be used to determine availability for the new segment based on the entire itinerary.

The POS is a many-to-one hierarchy where rules can be applied at multiple levels to control availability. The hierarchy begins with region, country, city group, city, agency group down to the ARC IATA number of the individual travel agency. An example of a POS hierarchy is shown in Fig. 3.5.

In cases where the POS is in different countries, it is possible for the unit of currency to vary. In the context of international markets, it is crucial to consider currency fluctuations and the differentiation of traffic freedoms (Vinod, 2021a, Appendix A), which include third, fourth, fifth, and sixth freedom traffic. These factors play a significant role in managing the flow of traffic within the airline network.

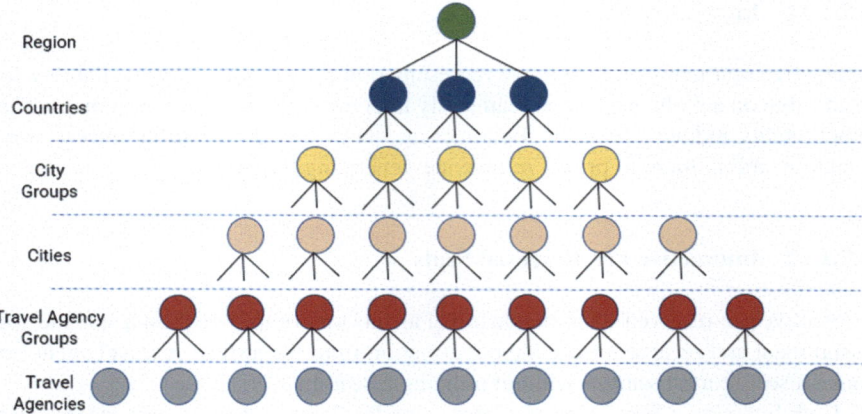

Fig. 3.5 Sample POS hierarchy

3.7.1.14   Point of Commencement

Point of commencement (POC) is offered with direct connect availability and journey data. POC uses a journey's starting point information in addition to POS information to determine availability. The GDS definition for POC is based on the PADIS/IATA standard, and it constitutes the ORG (origination) of the chronological first segment (obtained by sorting segments by departure date and time) in the PNR or shopping request, regardless of the interline or marketing airline for that segment. However, from an airline's inventory control perspective, this information is not always useful. O&D carriers expect the inventory system to determine the *chronological first host airline segment* from journey data when an interline itinerary is involved. For O&D inventory control, airlines prefer POC to prevent abuse by travel agents and OTAs who can switch between different points of sale to benefit from more favorable seat availability.

3.8   Participation Levels in Airline-GDS Contracts

The participation levels defined the sophistication of the connectivity between an airline reservations system and the GDS. The segment booking fees vary by level of connectivity. The four primary levels of connectivity are:

Basic Booking Request. Also called Basic Booking Record, is a teletype sell request sent at end transaction (ET) where the airline confirms upon receipt of the teletype message. Participating carriers cannot send AVS messages and hence the GDS cannot show a numeric value in the city pair availability display and the travel agent will not know if the seat is confirmed until the airline responds back after end transaction. This participation level uses AVS messages to show classes of service, inventory is not displayed, and schedules are accepted 189 days in advance.

Full Availability. This level of connectivity is not ideal for carriers with moderate to complex business models. It supported common GDS Teletype (TTY) communications that were ideal for carriers that require greater functionality than BBR. It is a teletype sell request sent at end transaction (ET) where the airline confirms upon receipt of the teletype message. Participating carriers send AVS messages to open/ close flights. The AVS messages were used to show inventory availability, and schedules were accepted 330 days in advance.

Direct Connect Sell. This connectivity tier is ideal for carriers with simple schedules but complex operations. It utilizes AVS messages to show inventory availability, but provides an interactive, real-time link during the actual booking process (interactive sell or seamless sell processing). When a sell request is submitted, the GDS transparently checks for true last seat availability in the host CRS. The airline returns the record locator after end transaction. It is required for carriers using O&D revenue management controls.

Direct Connect Availability. This is the highest level of participation. It is ideal for carriers with complicated schedules, many classes of service, or complex operations. It did not use AVS messages and provided an interactive, real-time link to the host CRS during the city pair availability display (also known as interactive availability or seamless availability) and booking processes (interactive sell or seamless sell). It is required for carriers using O&D revenue management controls to determine true last seat availability and secured sell based on the value of the O&D reservation request.

Besides these core connectivity options, there are several additional participation options. All GDSs support them, though the names can be different.

Numeric Availability (NAV) that allows a carrier to send AVS message with inventory count by class of service. It provides travel agents with more accurate information than normal AVS messages.

Carrier Specific Display (CSD) provides the capability for travel agents to sort the display by carrier, it is also an opportunity for carriers to increase sales through a GDS channel.

Answerback (AB) is a feature where a carrier's reservations system provides a PNR/Record Locator which is appended to the GDS booking. It provides travel agents with a higher comfort level of assured reservations.

Direct Access (DA) allows a travel agent to view a carrier's own reservations system in response to availability and other functions. It is an opportunity for carriers to increase sales through the GDS channel and to sell inventory in a manner consistent with their own reservations staff procedures.

Claim it (CL) allows travel agent to gain control over a partially completed PNR within the carrier's reservations system. It reduces calls from travel agents to the carrier for handling incomplete reservations.

Group Management Tool (GMT) allows travel agents to create and manage blocks of group inventory through the GDS.

3.9   Hotel and Rental Car Connectivity

A direct connection between a hotel CRS and all GDSs is not necessary. Due to the slow delivery schedule for connectivity provided by GDSs to hotel CRSs in the 1980s, a technology start-up called THISCO (The Hotel Industry Switch Company) was established in December 1988 by sixteen hotel companies and Reed Travel Group (Davis, 2002). The first president of THISCO was John F. Davis III. Davis had an interesting anecdote about his conversation with Max Hopper, the chief technologist and founding father of the Sabre reservations system. When Davis informed Hopper that Sabre would have to connect to THISCO,'s UltraSwitch, which would then connect to the different hotel CRSs, Hopper responded with the statement, *"Sabre would connect to THISCO over my dead body!"*

Sabre became the first GDS to connect to THISCO in December 1989, despite his remark. By 2002, the switch was processing over 300 million transactions per month. In 1995, Pegasus Systems was established as the parent company of THISCO, the Hotel Clearing Corporation (HCC), and TravelWeb. HCC, founded by John F Davis III in 1992, allowed travel agents to consolidate their hotel commissions into a single check for a fee. TravelWeb, launched in 1994, was the first comprehensive Internet catalog of hotel properties worldwide. It later introduced a real-time booking engine for hotels. In 1999, Pegasus Systems acquired REZsolutions, which was formed through the merger of Utell International and Anasazi, Inc. Pegasus Systems was then renamed Pegasus Solutions in 2001, offering services such as travel agency commission payment through Hotel Clearing Corporation, content management, and connectivity to OTAs and hotel websites (Burns, 2015). The distribution division of Pegasus Solutions, operating separately from reservations services, was spun off and renamed DHISCO (Distribution Hospitality Intelligent Systems Company). As of 2018, DHISCO processes over 9 billion transactions per month for more than 110,000 hotels.

The switch's relevance in the industry has diminished (Hoare, 2010) due to the significant number of hotel bookings coming directly from major OTAs. These OTAs have their own direct connection solutions to hotel CRS and property management system (PMS) platforms, which utilize XML interfacing standards developed by the Open Travel Alliance (OTA), Hotel Technology New Generation (HTNG), and pre-packaged push messages like FastRez, created by Open Travel. However, DHISCO can still be utilized to target online travel agencies that serve niche markets. In 2018, DHISCO was acquired by RateGain, an Indian travel technology company.

In 1987, AVIS Rent a Car launched a competing product called Wizcom's ResAccess. This product was designed to connect AVIS locations to the GDSs, allowing travel agents to access real-time rates and availability from the rental car company's reservations system. Wizcom's customer base for electronic distribution includes rental car companies, hotels, and tour operators.

3.10   Continued Growth of the GDSs

By the 1990s, SABRE was neither a proprietary competitive weapon for American Airlines nor a general distribution system for the airline industry (Hopper, 1990). The GDSs were electronic travel supermarkets that linked suppliers of travel and related services such as package tours and theatre bookings to travel agents and corporate travel departments. Airlines that owned the GDSs were not treated differently; they paid booking fees to the GDSs like all the other airlines in the system.

During this period, a notable feature launched by the Sabre GDS was Sabre Traveler Automation Records (STARS) that eliminated the manual customer contact list maintained by travel agents. The CRSs came to be known as the GDSs as the systems developed agency point-of-sale support worldwide, whereas the airline system where seat inventory was stored is known today as host CRSs, or simply CRSs. Hosting multiple airlines in a single system followed quickly.

When airlines realized the value of GDS ownership, several new GDSs appeared in the market in the 1980s. Eastern Airlines launched System One Direct Access (SODA) in 1981. It became operational in 1982 and was based on PARS. Delta Air Lines launched DATAS II in 1982 based on PARS technology, terminating the joint marketing agreement with United's Apollo system.

To gain control of the computer reservations market, Galileo was founded in 1987 by British Airways, Swissair, KLM Royal Dutch Airlines, Alitalia, Olympic Airways, Sabena, Air Portugal, Austrian Airlines, and Aer Lingus.

Galileo International was created in 1992 when Covia, a wholly owned subsidiary of United Airlines that controlled Apollo acquired Galileo and merged it with Apollo. It was owned by 11 North American and European airlines: Aer Lingus, Air Canada, Alitalia, Austrian Airlines, British Airways, KLM Royal Dutch Airlines, Olympic Airlines, Swissair, TAP Air Portugal, United Airlines, and US Airways. In 1997, Galileo International bought Apollo from United Airlines, US Airways, and Air Canada through an initial public offering (IPO).

Amadeus Global Travel Distribution was created in 1987 as a joint venture between Air France, Lufthansa, Iberia, and Scandinavian Airlines System (SAS). System One, developed by Eastern Airlines was the baseline for the Amadeus reservations system (passenger name record) code running on TPF and the pricing engine was from Air France, running on Unisys. The system became operational in 1991 by integrating four national reservations systems, Esterel in France, Savia in Spain, Smart in Sweden and START in Germany (Kärcher, 1996). These national systems were controlled and (partly) owned by the founding airlines of Amadeus.

Abacus, based in Singapore, was founded in 1987 by Cathay Pacific Airways, Singapore Airlines and Thai Airways International PLC. Thai Airways later dropped out of the partnership and four Asian carriers joined the partnership. They were All Nippon Airways, China Airlines, Malaysia Airlines, and Royal Brunei. Philippine Airlines joined a year later followed by Garuda Indonesia, EVA Airways, Hong Kong Dragon Airlines, and Silk Air. Abacus acquired a 10% share of PARS which served as the booking system. The agreement signed between PARS and Abacus created the foundation for Worldspan.

In Australia, there were two competing groups who wanted to establish Asia-Pacific computer reservations systems to fend off rival booking systems from Europe and the U.S. which were coming into the Asia-Pacific region. Qantas Airways, Ansett Airlines, and Australian Airlines wanted to establish an Asia-Pacific GDS to offer a complete booking service to their customers. Qantas' subsidiary Asia Pacific Distribution Limited was aligned with Sabre and called its offering Fantasia. Ansett Airlines and Australian Airlines established the Southern Cross Distribution System (SCDS), a joint marketing and distribution company with Galileo.

In 1989, All Nippon Airways and Japan Airlines participated in a feasibility study for Fantasia. However, All Nippon Airways pulled out of Fantasia in favor of rival Abacus in Singapore (Harrington, 1989a), and Japan Airlines did not commit to Fantasia. Eventually, Qantas scaled back its ambitious plans to develop a system for Asia-Pacific carriers due to its failure to attract airline equity partners on the initiative and focused on signing travel agencies like the United Travel Agents Group, the largest independent agency body in Australia (Harrington, 1989c). SCDS fared better and signed several domestic airlines in Australia and the South Pacific including Air Pacific the national airline of Fiji that was 20% owned by Qantas (Harrington, 1989b). Clearly, the center of gravity in the Asia-Pacific region had shifted toward Abacus, which was successful in attracting equity investment from many international carriers in the region.

Since 1998, Sabre had maintained a 35% stake in Abacus. Sabre acquired the remaining portion for $411 million in 2015 from the 11 Asian airlines who held a majority share. At the time of the acquisition, Abacus served over 100,000 travel agents across the Asia Pacific region's 59 markets. With a focus on Asia Pacific, Abacus developed local relationships with airlines and hotels. Equally important Abacus acquired low-cost carrier content in the region and content from the Chinese carriers. In addition, Sabre also acquired the Abacus national marketing company (NMC), Abacus Distribution System, that was based in Hong Kong.

Travelport was created in 2001 by Cendant following the acquisition of Galileo GDS and Cheap Tickets. Travelport acquired Orbitz in 2004. Travelport was sold to the Blackstone Group in 2006. Travelport acquired Worldspan in 2007.

A wave of consolidations resulted in just four major GDSs. These are Sabre, Apollo, Worldspan, and System One/Amadeus. Sabre was partially spun off from AMR Corp (17.8%), parent of American Airlines, in June 1996 which raised 627 million dollars. PARS and DATAS II merged to create Worldspan in February 1990 and was initially owned by Delta Air Lines, Northwest Airlines, and Trans World Airlines.

Besides airlines, GDSs also have access to content from hotels, rental cars, rail operators, and cruise lines. It is used by traditional brick-and-mortar travel agents, consolidators, wholesalers, and online travel agencies. The GDS is viewed as a marketplace that links buyers to the suppliers. It is also known as an aggregator of content or intermediary that facilitates transactions between buyers and sellers.

Today, the major global GDSs are Amadeus, Sabre, and Travelport.

3.11   DOT's 1992 Revisions to the CRS Rules

In 1992, the Department of Transportation (DOT) that inherited CAB's duties determined that the CRS rules were still required to promote competition and prevent CRS owners from biasing screen displays in favor of an airline that owned the CRS.

The three primary requirements to ensure that each owner airline and its CRS would treat other airlines equitably are

1. Flight display screens cannot favor one airline over another.
2. For the same level of service, the prices for bookings must be the same for all airlines, including the owner airlines. This eliminated differences in booking fees for cohost or subscriber airlines.
3. The mandatory participation rule required airlines with a 5% ownership interest or more in a CRS to participate in competing systems at the same level at which it participates in its own system.

Figure 3.6 illustrates the ticket distribution relationship under the revised CRS Rules.

DOT also included a sunset date of December 31, 1997, for the CRS Rules. It was subsequently extended to January 2004.

3.12   Revenue Streams for Travel Agents

How do travel agents make money? As intermediaries between airlines, GDSs, and customers, they are the only ones who receive payments from all three parties. Their customers are either corporations (B2B) or leisure travelers (B2C). To book travel for their customers, travel agents subscribe to a GDS and pay a subscription fee to access schedules and fares. Travel agents have four main sources of revenue,

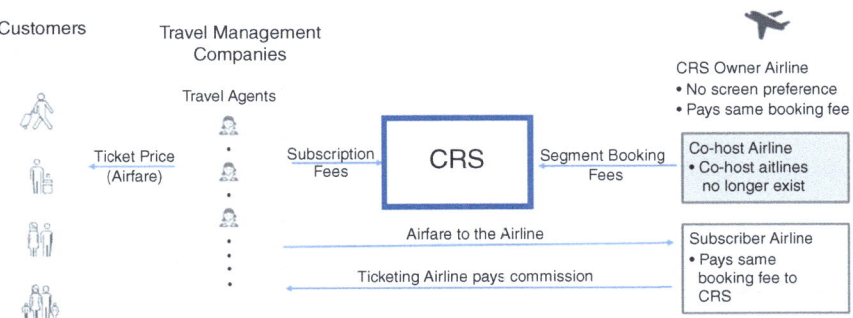

Fig. 3.6 Ticket distribution relationship under CRS Rules

including GDS incentives, front-end commissions from airlines, override or back-end commissions from airlines, and service fees from corporate or leisure customers.

3.12.1 Agency GDS Incentives

The GDS gives two types of incentives to travel agencies: a one-time payment for signing a contract and an ongoing productivity-based fee. The first can be for a contract renewal or converting from a competing GDS. The second incentive rewards agencies based on performance such as the number of bookings made. The incentives are a percentage of the segment booking fee that the GDS collects from airlines. Over the past 20 years, incentives have increased and currently average about 53% for brick-and-mortar travel agencies. It is the single largest marketing cost borne by the GDSs.

GDSs face intense competition for agency incentives because most major travel agencies use a dual GDS strategy, utilizing at least two and sometimes three GDSs for booking transactions. These incentives have a detrimental effect on GDS revenues, as they often exceed 50% of the segment booking fees. Additionally, financial incentives such as signing, and growth bonuses may be included.

Travel agency incentives paid by GDSs have been increasing at an alarming rate. According to a report by the General Accounting Office (GAO, 2003), the average amount of incentives paid by GDS companies to travel agencies between 1995 and 2002 increased from $22.3 million to $233.4 million, which is over a 900% increase.

Agency incentives paid by GDSs in the 2000s which were in the 10% range hover around 53% of the segment booking fees in 2023.

This information is shown in Fig. 3.7.

In 2006, GDSs started charging 80 cents per segment full content fee to travel agents for bookings made for airlines that participate and guarantee full content in the GDS (see Sects. 3.3.1 and Sect. 4.6).

The world of travel agency incentives will change with the adoption of IATA NDC, discussed in Chaps. 6 and 14.

3.12.2 Front-End Commissions

On February 9, 1995, Delta Air Lines rescinded the 10% travel agent commissions on U.S. domestic fares and put a cap on commissions of $50 for round-trip fares over $500. For one-way fares above $250, commissions were capped at $25. Other airlines also followed Delta's decision, which led to the American Society of Travel Agents (ASTA) filing a lawsuit against the airlines who imposed the cap, claiming that they illegally conspired to do so. The lawsuit was settled in 1996 (McDowell, 1996), but the cap on commissions remained, leading to a trend of declining travel agency commissions. By 2002, agents received a $10 commission for a one-way

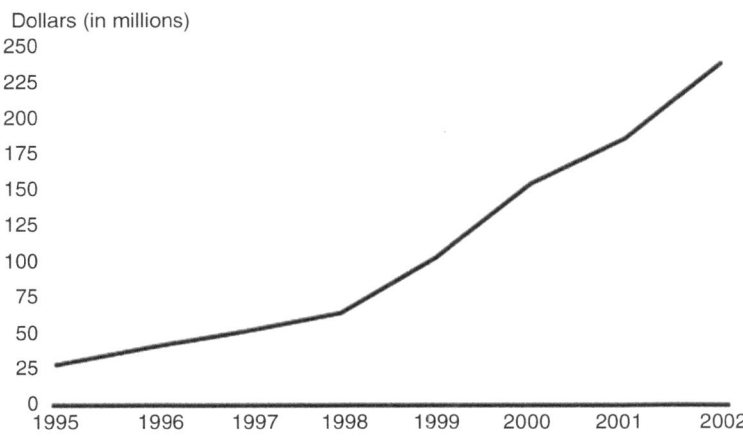

Source: GAO analysis of data provided by domestic GDSs.

Note: Amounts shown are in nominal dollars.

Fig. 3.7 Average payments to U.S. travel agents by each GDS, 1995–2002

ticket and $20 for a round trip ticket (Alexander, 2002). After the tragic terrorist attacks on September 11, 2001, Delta Air Lines announced a plan to end all commissions to travel agents (Isidore, 2002), with the expectation that travel agents would charge customers for their service. This move, combined with reduced commissions and the rise of the Internet, did not lead to the disintermediation of travel agents. Instead, it resulted in bankruptcies and mergers to achieve economies of scale.

In the current U.S. domestic markets, airlines restrict front-end commissions to certain markets where they intend to increase ticket sales. International markets tend to be more profitable for travel agencies as they receive front-end commissions.

Although front-end commissions for U.S. domestic markets have decreased, they are still an important source of income for travel agents who specialize in selling luxury tour packages and cruises. It is a misconception that commissions are no longer significant for travel agents.

3.12.3   Backend (Override) Commissions

Performance-based incentives, also known as override commissions or back-end commissions, are earned by travel management companies. If a travel agency meets the negotiated performance targets set by the airline, such as a certain number of bookings per quarter in a specific market or market entity (collection of markets), they will receive a predetermined lump sum payment from the airline.

3.12.4 Net Fare Markup

Travel management companies negotiate discounted net fares with airlines and discounted rates with hotels, which they can sell at a marked-up price for a profit. Travel agencies tend to sell these discounted fares to their customers as they can earn greater profits per transaction.

3.12.5 Service Fees

Revenue is generated through service fees charged to customers for the services provided. Two types of customers are served: B2B corporate customers and B2C leisure customers.

Corporate customers typically require fulfillment of various functions such as spend optimization, corporate policy compliance, and duty of care. The fee structure varies by TMC and may be either a lump sum payment or an a la carte payment structure, which includes fees for airline bookings, hotel bookings, change fees, agent touch fees, and emergency contact fees.

Leisure travelers may be charged a service fee for services rendered. Travel agents are likely to find cheaper fares than OTAs such as Booking.com and Expedia for complex international trips. Many customers, especially millennials, are unaware of the value provided by travel agents, who have access to better seats, upgrades, rooms, and can make bookings for ground transportation and local attractions.

Figure 3.8 illustrates the logical flow of payments in the travel distribution market.

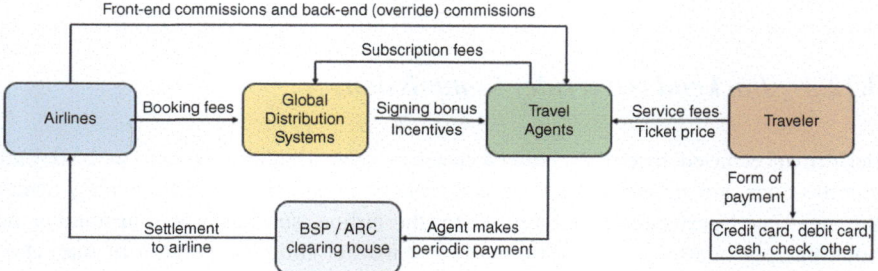

Fig. 3.8 Flow of payments in the travel distribution market

3.13 Revenue Management and Airline Reservations

The relevance of revenue management in the context of airline reservations systems is very significant. This is because the inventory control component of airline reservations serves as the execution component to accept and reject passenger requests based on the optimal inventory controls recommended by revenue management. The inventory control capability in an airline's reservations system can be basic or advanced depending on how the seats are managed by an airline. The inventory control can range from leg-based controls to segment-based controls to origin and destination (O&D) based controls. In addition, GDS connectivity dictated by the level of participation between an airline and a GDS can add sophistication to inventory controls. Examples of GDS connectivity features that can enhance the inventory control capabilities are seamless availability, seamless sell, married segment controls, journey controls, point of commencement, and married to journey (Vinod, 2021a, Chap. 4).

3.13.1 Hosted Reservations and the Perceived Halo Effect

In the 1970s and early 1980s, the travel agencies that accessed a CRS (later called GDS) was owned by a handful of airlines. The CRS owners were able to extend their market reach beyond the traditional ATO, CTO, and central reservations offices. They also controlled the display order of the flights of airlines that participated in a CRS. This benefit to owner airlines was frequently referred to as the "*halo effect.*"

The term "*halo effect*" that is erroneously mentioned today by GDSs that also host airline reservations, in an entirely different context. From a revenue generation perspective, the single most important component of an airline reservations system is inventory control, which determines in real time the availability of seats by booking class on a flight leg, flight segment, or service (origin and destination). Inventory control is the execution component of recommendations made by the revenue management system. Seat availability is the end product of the revenue management process to maximize airline network revenues by determining the right mix of short haul, medium haul, and long haul customers to accept based on value. Inventory controls come in many flavors such as leg controls. Segment controls and O&D controls depending on the sophistication of the revenue management system (Vinod, 2004).

The halo effect of having a GDS that is inter-operable with an airline CRS with the same or similar codebase has been a selling point for GDS vendors like Sabre and Amadeus who host many airlines for reservations processing.

Over the past two decades many full-service carriers, to contain development and maintenance costs of their home grown internal CRSs, have migrated to a multi-hosted PSS environment from Amadeus, Sabre, SITA, Navitaire (for LCC carriers), and many smaller vendors. Historically, Sabre has been hosting airlines since

American Airlines developed the first reservations system. Besides American, some of these customers in the 1990s included Alaska Airlines, Hawaiian Airlines, Gulf Air, and many more. Amadeus did not have a complete reservations system with reservations, inventory, ticketing, and departure control until the 2000s. After British Airways and Qantas decided to sunset their British Airways Booking System (BABS), they signed an agreement to migrate to Amadeus' Altea Customer Management Suite in 2001. This led to the development of Altea Inventory and Altea Departure Control System (DCS) to complement Altea Res which was a derivative of System One. Unlike Altea Res that was mainframe dependent for processing transactions, Altea Inventory and Altea DCS were developed on open systems. Migration to the new system occurred a decade later. In 2011, All Nippon Airways switched to Amadeus. Cathay Pacific and Singapore followed in 2012. The GDSs use their marketing power to convince airlines to become users of their reservations system (CRS) with the claim of a better booking experience due to system compatibility, accuracy of true last seat availability and matching PNR records between the GDS and the CRS. This is, however, not true but serves primarily as a marketing tool to win share in the reservations marketplace.

For carriers that are hosted by Amadeus and Sabre, it can be argued that articulating the value of end-*to-end inventory visibility* is of strategic importance. The halo effect does not apply toward a neutral display. It can be argued that the halo effect, if it does exist, is in the context of carrier specific preferential city pair availability and shopping displays to a travel agent. Hosted carriers *will benefit from the halo effect* of hosted inventory, the ability to *sort and control the line order of the display based on actual bookings to re-direct demand to low load factor flights and preserve market share*. This, however, requires changes to the direct connect availability and shopping responses from the airline CRS to the GDS, and is not practiced in the industry today.

For carriers that are *not* hosted by a GDS like Amadeus and Sabre, the halo effect will be *limited to carriers that have deployed an availability proxy, that mirrors the availability processing logic in the host airline CRS*.

In summary, the perceived benefits of a hosted CRS and a GDS from the same vendor are debatable at best.

3.14 Governance, Standards, and Partners

This section reviews the governance, industry standards, and partners of the GDSs.

3.14.1 Industry Standards and Governance

To participate in a GDS, airlines and GDSs have a participating carrier agreement (PCA). The Full Content Agreement (FCA) is a provision in the PCA for an airline to provide the same content through the GDS that they provide to their consumer direct website, call centers, or any other channel. The definition of full content is access to airline schedules, fares, and seat availability. Sabre called these contracts Full Content DCA agreements. DCA stands for direct connect availability between the GDS and the airline host CRS for true last seat availability via EDIFACT and represented the highest level of participation from a connectivity perspective.

There are two primary entities that define how the airline CRS and GDS work with each other, IATA and ATA (Air Transport Association of America). ATA changed its name to A4A (Airlines for America) in 2011 with the tagline "We Connect the World."

IATA, the International Air Transport Association, represents, leads, and serves the airline industry comprised of over 250 airlines worldwide. A4A, Airlines for America, previously known as the Air Transport Association of America, Inc. (ATA), is the premier trade organization of the principal U.S. airlines. A4A members transport over 90% of all U.S. airline passenger and cargo traffic.

IATA and A4A publish a wide variety of manuals for all aspects of the airline industry. IATA published AIRIMP (Air Interline Messaging Procedures), the reservations interline message procedures passenger manual. A4A publishes the Standard Interline Passenger Procedures, known as SIPP.

The standards define several messaging protocols.

The teletypewriter message, referred to as teletype, is a simple asynchronous message. In the 1920s, airlines recognized the need to communicate reservation activity between themselves and started sending messages to printers using teletype. Messaging standards were established after World War II as the airline industry flourished. These teletype standards were used as the foundation of system-to-system communications. IATA standardized teletype messages in the airline industry. There are many message types, including name messages for sell, update, and cancel reservations, availability status messages advising when flights are open/closed for sale, airline schedule updates, ticketing, and special formats to support codeshare agreements.

Interactive messaging using EDIFACT (Electronic Data Interchange for Administration, Commerce and Transport) was defined by IATA Passenger and Airport Data Interchange Standards (PADIS). This is also referred to as Type A messaging. The Type A EDIFACT message formats are governed by PADIS (Passenger and Airport Data Interchange). These messages are interactive and have an immediate application response with timeout handling at the application level. Message delivery is not guaranteed. With Type B messages such as teletype, delivery is guaranteed, and the response takes longer. AIRIMP governs the Type B message formats.

The Open Travel Alliance (OTA) is a nonprofit standards body that has developed messaging standards for interoperability to disparate systems across all verticals (air, hotel, car, etc.) in the travel industry. XML (extended markup language) message standards are being defined by the Open Travel Alliance (OTA) and Open AXIS using EDIFACT as a reference. IATA is also involved in developing the new XML and JSON (Java Script Object Notation) messaging standard for the new distribution capability (NDC). These new message standards make it easier for new types of standard and rich content from airlines and service providers to be transmitted to a GDS.

All GDSs participate in the IATA forums as non-voting members, alongside airlines and other service providers.

3.14.2   Communications Partners

SITA (Société Internationale de Télécommunications Aéronautiques) was founded in 1949 by 11 airlines to combine their individual communication networks to realize cost efficiencies. The 11 original airlines were Air France, KLM Royal Dutch Airlines, Sabena, Swissair, TWA, British European Airways Corporation (BEAC), British Overseas Airlines Corporation (BOAC), British South American Airways (BSAA), Swedish A.G. Aerotransport, Danish Air Lines, and Norwegian Air Lines. SITA manages complex communication solutions for its air transport, government and GDS customers. It has one of the world's largest, fastest, and most reliable messaging services backbone and supports the world's largest teletype messaging community and routinely exchanges over 25 million messages a day. SITA's international data network operates in over 200 countries and territories. The GDSs and online service providers use the SITA network to supplement their own.

ARINC (Aeronautical Radio, Inc) was chartered by the U.S. Federal Radio Commission, which later became the Federal Communications Commission (FCC), to serve as the airline industry's single licensee and coordinator of radio communications outside of the government. The airline industry uses teletype messages over ARINC or SITA networks to communicate between reservations systems. ARINC also introduced ACARS (Aircraft Communications Addressing and Reporting System) in 1978 to support aircraft to ground communications via aircraft band radio or satellite. ARINC was acquired by Rockwell Collins in 2013 and has been a part of Collins Aerospace since 2018.

3.14.3   Settlement Partners for Airlines and Agencies

Interline travel uses the services of settlement partners to distribute payment to the appropriate parties. The IATA Clearing House (ICH) enables the world's airlines and airline-associated companies to settle their interline billing. There are over 475 members and associates and settles over 50 billion in interline and service transactions annually. The scope of clearance is further expanded through its inter-clearance agreement with the U.S.-based Airline Clearing House (ACH) which settles over $12 billion in receivables on behalf of 91 airlines.

There are two settlement partners for travel agencies: BSP and ARC. Both are owned and governed by the airlines. Travel agents issue one sales report and remit one amount to a central point and airlines receive one settlement covering all agents. The Billing and Settlement Plan (BSP) facilitates the selling, reporting, and remitting procedures for IATA accredited passenger sales agents. All IATA accredited agents in the BSP country of operation are automatically eligible to participate in a BSP. General Sales Agents (GSA) and Airport Handling Agents (AHA) may also participate in a BSP based on nomination from the airline they represent. The Airline Reporting Corporation (ARC) serves the travel industry with financial services, ticket distribution, and settlement for ARC accredited travel agency locations and corporate travel departments in the United States, Puerto Rico, and the U.S. Virgin Islands.

The financial settlement process supports two major workflows and their variants. First is the scenario where the travel agent is the merchant of record and, second, where the airline is the merchant of record. When the travel agent is the merchant of record, the travel agent collects payment from the traveler, and then submits the payment to BSP/ARC clearing houses for settlement to the airline. When the airline is the merchant of record, the traveler accesses a payment gateway on the agent's system to pay for the ticket directly to the airline.

3.14.4 Industry Partners for Airline Fares

The Airline Tariff Publishing Company (ATPCO) and SITA are the main publishers of airfares and rules. Prior to airline deregulation, airfares were printed on the backs of airline timetables. A few airlines continue to publish timetables, but do not publish airfares due to their volume and complexity.

The Air Traffic Conference of America which was part of the Air Transport Association of America (ATA) was founded in 1945 to publish passenger tariffs. In 1965, they divested from ATA as an independent company and continued to publish passenger tariffs to travel agents. ATPCO was established in 1975 by 11 founding airlines (Air Canada, Air France, Air Nippon Airways, American Airlines, British Airways, Delta Air Lines, Hawaiian Airlines, KLM Royal Dutch Airlines, LATAM Airlines, Lufthansa Airlines, and United Airlines and was branded as ATPCo and later changed to ATPCO.

These fare aggregators collect fare information from over 400 airlines and distributes it to GDSs and CRSs both domestically (U.S.) and internationally on varying frequencies. All GDSs also have the capability to receive fares directly, used primarily for private fares, thereby bypassing ATPCO and SITA.

Before the digital era, ATPCO collected and distributed fares, rules, and routes in large bound books. In 2022, ATPCO donated a historical airfare collection consisting of over 1500 volumes of fare, rule, and route books leading up to the dawn of the Internet to the Library of Congress.

3.14.5   Industry Partners for Airline Schedules

The "Official Aviation Guide of the Airways" has been publishing airline schedules since February 1929 in the U.S., listing 35 airlines offering a total of 300 flights. After a merger with a rival publication, it was officially called OAG, the Official Airline Guide. The ABC World Airways Guide was first published in the United Kingdom in 1948. OAG published its Pocket Flight Guide in 1970 and is still published today in multiple regional versions. A competing product was the American Express FlightGuide.

OAG participated with IATA to develop the IATA Standard Schedules Information Manual (SSIM) for the transmission of airline schedules data in 1972. SSIM continues to be the primary source for schedule data in the airline industry. ABC and OAG brands merged when Reed Elsevier, which already owned ABC, acquired OAG in 1993 and retained the OAG brand name.

Innovata evolved from Dittler Brothers who published travel schedules since 1923 and emerged as a competitor to OAG in 1998. Innovata was acquired by FlightGlobal, part of Reed Elsevier Group in 2014. Cirium is the new identity for the FlightGlobal data and analytics business. OAG and Cirium (successor to Innovata) are two of several vendors that distribute airline schedules, Standard Schedule Information Manual (SSIM), to the GDS. Airlines submit their schedules to schedule aggregators for worldwide distribution. The fare aggregators also distribute connect point and minimum connect time (MCT) information. IATA standard MCT are set by airport authorities and airlines can file exceptions to the standards.

3.14.6   GDS and Collaborative Entities

The primary role of the GDS to support these entities is content sourcing. The GDSs have relationships with travel management companies (TMC), corporations, online travel agencies (OTA), and leisure travel agencies. The Computerized Airline Sales and Marketing Association (CASMA) is an industry conference that brings GDSs, airlines, and vendors in travel distribution together to discuss current trends and the state of the industry.

TMCs support corporate travel programs and compete on service, price, technology, duty of care, and reporting. Examples are American Express Global Business Travel (AmEx GBT), Carlson Wagonlit Travel (CWT), BCD Travel, and Corporate Travel Management (CTM).

Corporations can contract directly with a GDSs corporate booking tool or indirectly with a TMC for travel technology and services to support their corporate travel program. Corporations can also source content directly from suppliers, though this is an exception than the norm. Corporations can be Fortune 500 companies as well as smaller companies.

OTAs provide travelers with access to leisure content and online self-service purchase options. OTAs rely on GDSs for shopping and booking APIs. The older APIs are SOAP (Simple Object Access Protocol) based while the newer APIs are REST (Representational State Transfer) APIs. REST APIs provide simpler methods of accessing web services, are JSON compliant, and can benefit mobile applications. Examples of OTAs are Booking Holdings, Expedia Group, Trip.com. MakeMyTrip, EaseMyTrip, Despegar, Lastminute.com, eDreams, CheapOair, Svenska Resegruppen AB, etc.

The leisure travel agencies, both online and brick-and-mortar, provide end-to-end travel service to customers based on their goals, finding the best options, and booking travel. Examples are dnata, American Express, and Flight Centre Travel Group.

Beyond these entitie,s there are other technology providers that may use products and services offered by GDSs, for example, Google Flights, farecompare, Bing, Yapta, and TripBam.

3.14.7   Government Oversight

The travel and transportation industries are controlled by many government bodies. There are transportation boards such as the U.S. Department of Transportation (DOT), which mandates on-time reporting in the U.S. The European Commission (EU) publishes a GDS code of conduct that governs many aspects of the GDS business such as air availability displays, contract terms, and conditions. Various governmental bodies dictate data privacy standards. Immigration and security organizations dictate policy. A GDS must comply with all governmental requirements in countries where they do business.

3.15   Airline Agent Adoption of QIK

QIK, an acronym for Qantas Intelligent Keypad, is an airline agent application that was initially developed in the late 1980s by Qadrant, a subsidiary of Qantas Airways. It served as a front end tool for mainframe computer reservation systems. QIK was developed specifically for the airline's reservation call centers, to enhance airline agent productivity. The name QIK was chosen because the application utilized a separate keypad in addition to the regular keyboard. These additional keys on the keypad acted as function keys. However, in later versions, the physical keyboard was replaced with a standard QWERTY[1] keyboard, including function keys from F1 to F12. The 1988 version, called QIK-1 was DOS-based.

The applications were marketed under the brand names QIK, QIK-RES, and QIK-CHEK. They effectively encapsulated the airline's business rules within a smart application based on personal computers. Consequently, the application could send the necessary transactions to the airline's mainframe or host for further processing. By implementing QIK, the training time required for an airline agent could be significantly reduced from six weeks to just two weeks. This is important because of the high turnover of airline agents who were paid hourly wages at airline call centers. Besides, the automation of host transactions played a key role in eliminating format entry errors. As a result, the need to resend transactions was greatly reduced, leading to a noticeable decrease in the mainframe usage costs for airlines.

During the early 1990s, Qantas established a partnership with DMR Consulting Group based in Montreal, Canada to promote QIK and other IT solutions. They named this joint venture Qadrant International. In 1997, DMR Consulting acquired the remaining 49% of Qadrant's stocks from Qantas Airways, making them the sole owner of the company (Aviation Week, 1997). Qadrant then collaborated with Sabre Decision Technologies (SDT), a subsidiary of AMR, which was also the parent company of American Airlines, to enhance QIK under the leadership of Tom Cook. This collaboration resulted in the development and release of QIK-II in 1992, which expanded the platform from DOS to OS/2 & Windows. QIK-3 was designed for a heterogeneous computing world that used Sun Microsystems and its Java applications for portability. Eventually, QIK-II was migrated to SITA's Common Use Airport platform CUTE/OS, maintaining the ongoing partnership between the companies. These QIK applications are utilized by over 70 airlines globally.

Contrary to common belief, *Turbo Sabre*, the online travel agent assisted reservations tool for high-volume agencies, did not deploy QIK, but a competing application developed by Sabre Computer Services (SCS) called Insight. This version of QIK had a graphical user interface instead of the 4x4 keypad attached to the computer.

3.16 Electronic Tickets

In 1994, the first paperless E-Ticket was introduced. The digital version of the paper tickets was stored on the reservations systems making the ticketing process cheaper. However, even with the advent of E-Tickets, customers who reserved flights on an airline's website were still required to physically visit a nearby airport or city ticket

---

[1]In 1874, Remington and Sons introduced the first commercial typewriter, called Remington Number 1. It was invented by Christopher Sholes who implemented the QWERTY keyboard on it. The top row of keys begins with Q, W, E, R, T, and Y.

office to finalize payment and acquire a physical ticket. Global standards for E-Ticketing were established by 1997. However, the adoption of E-Ticketing was slow and paper tickets flourished into the 2000s and peaked in 2005 when an astounding 285 million IATA neutral paper tickets were issued.

On June 1, 2008, IATA ceased distribution of paper ticket stock to travel agencies worldwide. According to IATA, the elimination of paper tickets resulted in an annual savings upside of $3 billion. The cost of issuing a ticket has been drastically reduced from $10 to $17 to $1 or less with electronic tickets.

To receive an electronic ticket, a passenger must have a virtual coupon record (VCR) number that is associated with their booking. VCR information can be shared between the airline and their interline partners, ground handlers, and GDSs.

Electronic ticketing has several benefits.

Foremost is real-time revenue recognition and immediate access to revenue data for balance sheets. Previously, airlines would send tickets to offshore low cost processing centers where each ticket had to be manually entered into a computer before revenue could be recorded.

There are additional benefits to paperless ticketing. Notably, the tickets cannot be misplaced and there is no need to visit a travel agent or airline ticket office to make changes or obtain a new ticket. With the prevalence of mobile phones and mobile apps, E-Ticketing serves as the forefront of mobile booking and self-service check-ins.

Despite the rise of electronic tickets, the ticket number remains a crucial element in resolving various travel-related problems. If an airport agent is unable to locate a customer's E-Ticket, providing the ticket number can simplify the check-in process and result in the issuance of a boarding pass. A common issue is the crediting of frequent flyer mileage from partner airlines, particularly if the frequent flier ticket is lost by the partner airline within the airline alliance which can be resolved with a ticket number.

There are lessons to be learned from the ticketing transformation from paper airline tickets to E-Tickets. The transition process took fourteen years, from 1994 to 2008. A comprehensive conversion to E-Ticketing only took place after IATA issued a mandate and stopped issuing the paper tickets in 2008. The NDC program for Offers was initiated in 2012 and One Order in 2015. The specifications went through a few iterations and started stabilizing in 2018. Perhaps the NDC program will achieve mass adoption in fourteen years, by 2032. Unlike E-tickets, it is unlikely that IATA can issue a NDC adoption mandate to airlines worldwide due to the high level of investment required to enable NDC.

3.17   Airline Divestiture

In the late 1990s, the airlines started divesting their interests in the GDSs. They determined that ownership of a GDS was no longer strategic in a regulatory environment that mandates unbiased screen displays. Government regulation neutralized the strategic advantage of owning a GDS.

Galileo acquired Apollo Travel Systems from United Airlines, US Airways, and Air Canada for 700 million dollars and went public in 1997 as an independent company.

Michael Durham, who was the chief financial officer of AMR Corp, moved to Sabre in 1995 to lead the partial spinoff of Sabre. At that time, he also oversaw Travelocity.com, the leading travel site on the Internet at that time. He resigned three years later, before Sabre completely separated from American in 2000 and became an independent company.

Amadeus took back ownership from the founding airlines in the early 2000s and is listed on the Madrid stock exchange.

3.18   The Origins of Air Shopping

Air shopping is the single largest application and infrastructure supported by a GDS. It is the agent entry that returns a range of priced itineraries from which the customer selects a flight.

When the GDS was introduced in 1976, the concept of a shopping entry did not exist. Travel agents would submit a city pair availability request for the outbound flight that specifies the city pair, date, and approximate time of departure. The GDS displays booking classes that are available for sale and the agent can book an itinerary by selecting a line from the city pair availability response that is displayed. The same process is repeated for the return flight. Next the agent submits a pricing entry to determine the total fare for the itinerary. The price can be further refined by submitting a new class pricing entry that determines the lowest fare for the itinerary if a different combination of open booking classes on the outbound and inbound flights produced a lower fare subject to fare combinability rules. There was no mechanism to search for alternate itineraries that may be more desirable or produce a lower price.

When the GDS was introduced in 1976, the display of nonstops and connections in the city pair availability display was not dynamic and relied on the Sabre Sales Guide (SSG) which pre-built nonstops, single-connects, and a limited number of double-connects in TPF several times a week to address schedule changes.

Sabre introduced the industry-first Bargain Finder[SM] in 1984 which returned nine lowest priced itineraries for a shopping request. The earlier versions did not have the ability to dynamically build connecting itineraries. Instead, the connections were prebuilt nightly based on the latest airline schedules. The algorithm in the earlier versions was rudimentary and based on simple heuristics to determine the shopping response. The shopping request includes the departure date for a one-way trip and a departure date/return date for a round trip. After the introduction of Bargain Finder in the 1980s all GDSs supported rudimentary air shopping capabilities. For example, the Apollo/Galileo flight search product was called Best Buy Quote and Worldspan's was called power pricing. Until the late 1990s, Sabre maintained a fare pricing complex (FPC) running on TPF with mainframes to support shopping and pricing.

These shopping algorithms required an itinerary to be booked by a travel agent before new itineraries were priced and displayed. In 1993, an enhanced version, called Bargain Finder Plus^{SM}, returned nineteen itineraries. It was popular with travel agents, though the shopping results lacked adequate diversity. When the OTAs arrived in 1996, demand for air shopping and related requirements increased dramatically. OTAs required a shopping service that had to return a larger number of itineraries (200–1000+) for each shopping request without having to book an itinerary first. Hence, results had to be based on the standard shopping parameters of origin, destination, number in party, departure date and return date. Fare-led algorithms were augmented with schedule-led algorithms to ensure diversity of itineraries. To support the growing demands for air shopping from Travelocity and other OTAs who depended on Sabre's web services for shopping services, an open systems version of shopping and pricing was initiated by Sabre's then Chief Technology Officer Craig Murphy in 2001 and launched in 2004, called the Air Travel Search Engine (ATSE). It was later renamed Air Travel Shopping Engine. Dynamic schedules (DSS), a component of ATSE, replaced SSG. DSS did not rely on the pre-built static list of non-stop, single, and double-connect flights. Instead, itineraries were created dynamically with no limits imposed on the multi-connect flights. This capability increased the sales opportunities for subscribers by offering an expanded number of flights in the city paid availability and schedule displays. ATSE went through significant enhancements with the estimated seat value (ESV) algorithm for Travelocity in 2008 (Benzinger et al., 2008) followed by the high-performance shopping engine (Steeb & Sohn, 2006), where data are organized by VTCR-Fare Class (Vendor, Tariff, Carrier, Rules, Fare Class). ITA Software (de Marcken, 2003) provides an alternative to GDS shopping. It was deployed when Orbitz was launched and is used today by several airline websites and Google Flights.

Chapter 5 provides a more in depth view of air shopping and why scale matters for GDS transaction processing.

3.19   Legacy Technology

Host CRS and GDS technology is legacy mainframe operating system and was based on IBM's Airline Control Program (ACP) which evolved to TPF (Transaction Processing Facility), also a low-level assembly programming language. TPF was introduced by IBM in 1969 and it went through a series of enhancements. TPF 4.1

was released in 2014. Reservations systems on TPF 4.1 migrated to IBM's latest Z/ TPF version over the past decade. The z/Architecture, a 64-bit instruction set introduced in 1991, is the successor to the 32-bit System 360 architecture. With the modernization initiative by IBM, TPF is now a mature 64-bit operating system.

When ACP and later TPF were conceived by IBM, these operating systems were designed for high-volume transaction processing. Besides reservations processing for airlines and hotels, it is also used for high-volume transaction processing in banking for credit card transactions and financial transactions such as payments and money transfers.

Many major organizations still use TPF for their mission-critical applications on mainframe systems. This is because TPF is known for its stability and reliability. Besides, TPF provides a considerable level of security and scalability, both of which are crucial for large enterprises. Besides travel, some of the largest TPF users include American Express, JP Morgan Chase, Visa Inc., and Discover Financial Services.

Based on the recognition that total cost of ownership (TOC), flexibility, and time to market are critical, all the GDSs have a plan to migrate reservations processing from a mainframe TPF environment to a service-based architecture based open systems technology platform. A key argument is that an open systems architecture that scales vertically with the transaction volume can be processed on commodity hardware, which lowers the cost of transaction processing. It is a phased migration strategy for all components of transaction processing—reservations, inventory, departure control shopping, pricing, schedules, and ticketing. Over the years, it has become increasingly difficult to recruit programmers who are well versed in TPF, which provides an added incentive to migrate to an open systems environment that use modern programming languages. Amadeus was the first to complete the migration to open systems successfully in 2018. They however had an advantage that only their core PNR processing was in TPF, on the System One baseline, while inventory and departure control were developed directly on open systems.

All interactive availability and sell transactions between a GDS and an airline are handled by EDIFACT, a stable, but rigid and restrictive format. The new XML-based "standard" promoted by IATA is called NDC, which allows airlines to deliver rich content, dynamic pricing, and air ancillary products directly to GDSs, TMCs, and OTAs. The new communication protocol will replace the EDIFACT protocol, which has been around since the 1980s.

Since it was conceived in 2012, the first set of NDC messaging standards were released in September 2015. It was meant to change how airlines ticket and distribute their product and allows airlines to deliver personalized content directly to OTAs, TMCs, and metasearch engines bypassing the GDS or through a GDS. NDC has the potential to break the GDS oligopoly that has existed since 1976.

3.20   What Happened to the Pioneer Airline Central Reservation Systems?

The fate of the airline reservations systems that spawned the GDSs is intertwined in the airline ownership of these systems, airline bankruptcies, and the sustained growth of the GDSs. The major reservations systems that came into existence are reviewed in this section.

3.20.1   Northwest Airlines Stake in PARS

In 1986, TWA PARS was the third largest airline-owned computer reservations system for travel agencies behind American's Sabre and United's Apollo. Of the approximately 22,000 U.S. travel agents that used these systems to book flights, about 40% used Sabre, 30% used Apollo and 12% used TWA PARS. Carl Icahn, the investment banker who had taken control of TWA earlier that year, wanted to sell all or part of PARS to generate cash flow for TWA that was in deep financial straits because of fare wars, declining load factors and a costly flight attendants strike. Northwest Airlines acquired a 50% stake in TWA PARS. Under the terms of the agreement, PARS was split into a services company and a marketing company, and each airline owned 50% (LA Times Archives, 1986).

In April 2008, Delta Air Lines and Northwest Airlines announced a merger agreement, which at that time was the largest airline in the world with 786 aircraft. With the merger the Delta Air Lines brand survived and so did the Delta reservations system, DELTAMATIC. Northwest Airlines schedules and flights were merged into DELTAMATIC by 2010.

3.20.2   System One

Frank Lorenzo was the Chairman, President, and Chief Executive Officer of Texas Air Corporation. He acquired Continental Airlines in 1981 and Eastern Airlines in 1986. Texas Air purchased System One from its bankrupt subsidiary, Eastern Airlines for $100 million. At that time System One owned about 20% of the travel agency market.

In 1990, Texas Air Corporation sold one half stake in System One to Electronic Data Systems (EDS), which was a subsidiary of General Motors at that time, for $250 million. By selling System One's Airline Services Division, EDS gained a foothold in the passenger reservations processing business, which was subsequently renamed SHARES. Continental Airlines and Eastern Airlines were hosted on SHARES. As part of the deal, Texas Air Corp. awarded EDS with a 10-year contract to provide both Eastern Airlines and Continental Airlines with data and technology services.

In 1998, Amadeus acquired 100% ownership of System One by buying out the 1/3rd interest owned by Continental Airlines and the 1/3rd interest owned by EDS.

At that time Amadeus only had a PNR capability for GDS processing of travel agency transactions and relied on the Amadeus carriers, Iberia, Lufthansa, Air France, and SAS, to provide reservations inventory control, departure control and ticketing capabilities with their existing Unisys USAS systems. The landscape changed in 2000 when British Airways and Qantas, who were on the British Airways Booking System (BABS) which evolved from BOADICEA, contracted with Amadeus for a complete end to end reservations system with inventory control, departure control and ticketing capabilities. This led to the creation of Altea Customer Management Suite with Altea Reservations, Altea Inventory, and Altea DCS. The Altea system was rolled out in phases, starting with Altea Inventory in 2005 followed by Altea DCS in 2012.

3.20.3   Apollo

The Apollo reservations system was used by United Airlines, hosted out of a data center in Denver. On October 1, 2010, United Airlines and Continental Airlines merged in an $8.5 billion all-stock merger of equals. After the merger, a decision was made to retain SHARES, the Continental Airlines reservations system, as the reservations system for the new combined airline and the Apollo reservations system was subsequently sunset in March 2012.

Apollo, the GDS is still used by Travelport for travel agencies in the U.S., Canada, Mexico, and Japan. However, Travelport has embarked on the Travelport + initiative and the plan calls for the sunset of the Apollo GDS and Worldspan GDS in 2023 after all subscribers have migrated to Galileo International.

3.20.4   The Braniff "Cowboy" System

Braniff International Airways, a major airline prior to airline deregulation, contracted with an airline school called Atlantic Airlines School in Kansas City, Missouri for access to PARS. The system was called "Cowboy" by Braniff and was introduced on July 12, 1969, at a computer data center in Dallas, Texas in the Brookhollow Industrial District near Love Field in Dallas, Texas. The Braniff International Computer and Accounting Center housed the Braniff Cowboy Computer System.

After airline deregulation in 1978, Braniff collapsed in May 1982, Southwest Airlines bought the Braniff Cowboy system and used an upgraded version for over three decades. Southwest was a simple no frills point to point carrier. The reservations system was renamed SAAS (Southwest Airlines Automated System) and Southwest customized the application. As an airline with a simple LCC model, they did not enhance the system as the airline industry evolved to support international flying, code sharing, schedule flexibility, overnight flights, interlines, full E-Ticket number in the GDS reservations at ticketing, and the sale of ancillary products.

These changes support the smaller TMCs that book Southwest Airlines corporate travel through the GDS. However, larger TMCs typically book Southwest Airlines either through BookingBuilder software or SWABiz, the free corporate online booking portal offered by Southwest Airlines for business travel. BookingBuilder is a link and not a screen scraper and maintains the integrity and display of Southwest's fares and schedules. It can access Southwest.com and enable travel agents to automate the booking as part of the GDS workflow.

Southwest Airlines replaced SAAS in May 2017 with the Amadeus Altea reservations system. The transition from SAAS to Amadeus took three years, with international routes followed by domestic routes.

3.20.4.1 Ticketless Travel from Morris Air

Interestingly Southwest Airlines is often credited with inventing ticketless travel. It was not Southwest, but Morris Air that invented e-ticket (or ticketless, which is distinct from E-Ticket) travel where ticket coupons are produced in an electronic format and updated as the passenger's status changes through the airport handling process. Morris Air was an LCC based in Salt Lake City, Utah that started operations in 1992. Morris Air was acquired by Southwest Airlines in December 1993.

While IATA E-Ticket standards were developed to eliminate paper ticket stock, it is sometimes incorrectly called ticketless. The term ticketless solutions are associated with LCCs who opt to partner with travel providers that are not affiliated with the Airline Reporting Corporation (ARC) or the Billing & Settlement Plan (BSP). These airlines operate in a ticketless environment, eliminating the need for physical or electronic tickets. To secure the Passenger Name Record (PNR), airlines settling with non-ARC or non-BSP affiliated providers require guaranteed payment. These airlines have several alternatives available to them for processing credit card payments outside of the BSP/ARC system. Once the booking process is finalized, the PNRs must be manually reviewed to verify that the vendor has confirmed the booking and to rectify any errors communicated by the vendor after the host CRS responds to the initial booking.

3.20.5 Navitaire

Accenture, the management consulting company, founded PRA Solutions in 1993. Initial products offered by PRA Solutions were airline revenue accounting and revenue integrity software. PRA Solutions acquired the Open Skies reservations system from Hewlett Packard (HP) in December 2000 that ran on HP3000 hardware. The company was renamed Navitaire in 2001. Under Accenture, Navitaire became the leading no frills reservations system for low-cost carriers. When HP discontinued the HP3000 hardware in the early 2000s, Navitaire developed a replacement reservations system called New Skies in 2015. Today, Navitaire hosts more than 60 LCCs and high-speed rail operators worldwide. In January 2016, Amadeus acquired Navitaire from Accenture for $830 million.

3.21 Platform Power

Today, in the digital era, there are many powerful platforms. Many retailers are dependent on Google to refer customers to their site. The Amazon platform is powerful because a vast number of products can be searched for and bought on the platform.

Like the GDS model, the power of the platform has a seismic impact since it captures a disproportionate share of the value a company creates (Edelman, 2014).

The Global Distribution Systems that came into existence in 1976 well before the Internet and digital transformation were the first examples of the power of digital platforms. The GDS platforms support the sale of travel products across all lines of business such as air, hotel, car, cruise lines, rail, ferry, and ground transportation.

To launch a GDS requires anchor tenants (suppliers) and subscribers (travel agents). Every GDS, when they were launched had an anchor tenant, a fundamental requirement for success. For example, Sabre had American Airlines, Apollo had United Airlines, PARS had Trans World Airlines, and Amadeus had Air France, Lufthansa, Iberia, and SAS. The travel agents that accessed the system to book travel were the subscribers.

The GDS business is a B2B transaction between a seller and a travel agent on behalf of a customer, which could be a corporate customer or a leisure customer. In a B2B context, *platform power* implies business entities become increasingly dependent on the platform to transact business at scale. The GDS is an oligopoly, where Amadeus, Sabre and Travelport have 90% of the market share for indirect bookings. The platforms of these GDSs are powerful, and there is a very high barrier for entry for a new entrant to become a major player. This is precisely what the airlines want to break up with the NDC initiative, but it is going to be a slow process due to *platform power*.

The key traits of "platform power" in the case of GDSs are:

3.21.1 Scalability

In the online transaction processing business of travel, scale matters. The GDSs facilitate billions of transactions daily to book air, hotel, rental car, rail, cruise lines and ferry lines. The value proposition for the GDS is well known. Travel suppliers are the major beneficiaries. With significant investments in air shopping, the travel agency desktop and high volume online transaction processing it is economical and difficult to replicate at the scale and efficiency. The cost of distribution, specifically air, is about 2.2%, which is a fraction of the merchant fees and commissions charged by OTAs for hotel bookings.

In the case of the GDS, it was the unique capability to search for flights across carriers and to book and ticket travel on behalf of a customer. The GDS workflow is so efficient that a booking can be made in under 30 seconds. This is because, in the traditional GDS model (not the NDC model), has access to airline schedules from schedule aggregators, fares filed by airlines through fare aggregators, and the GDS only queries an airline's reservation system for true last seat availability to ensure that the itineraries presented on a travel agent's display are bookable. Of all the steps in the booking and ticketing process, air shopping is the most expensive computationally and it takes less than five seconds to execute a shopping request and display bookable itineraries on a screen.

In the travel industry scale matters, which is the ability to support high volume transaction processing. For example, in 2019, before the COVID-19 pandemic, the Sabre GDS processed over 249 billion shopping requests and the peak shopping requests exceeded 28,000 per second. The look to book ratios originating from OTAs that are processed by the GDS can sometimes exceed 20,000:1 in certain markets. Besides, historically, shopping volumes have increased by 60% year over year. In addition, the itineraries returned during shopping exceeded 2.5 trillion per month (Vinod, 2020a). Amadeus and Travelport support similar volumes as well.

3.21.2 Travel Agency Productivity

The GDSs provide a travel agency desktop product, which is the customer facing agency workflow for the GDS product. The agency desktops are frequently updated with new versions with a focus on simplifying the booking process for travel agents. These enhancements ensure travel agents can complete a high volume of bookings daily. Another reason for stickiness to the GDS platform that it brings buyers and sellers together to transact business is because of buy-in from all major airlines (sellers) and travel agencies (buyers) who participate in these marketplaces for speed, efficiency, and convenience.

3.21.3   Standards for Connectivity

Connectivity is a critical component for a successful digital marketplace. GDSs, by their very role as intermediaries, connect sellers and buyers together to transact business. GDSs provide support for Type A and Type B messages.

Type A EDIFACT refers to the real-time interactive message formats that are regulated by PADIS (Passenger and Airport Data Interchange Standards). IATA PADIS covers all the standard passenger interactions, such as availability, sell, flight check-in updates, passenger list, boarding pass reprints, baggage transfers, booking, itinerary pricing, ticketing, and many more. These EDI messages are based on the EDIFACT syntax used by the GDSs.

Type B messages originated from teletype technology and preceded Type A messages. Teleprinters were invented in the 1920s to facilitate message transmission without the need for operators trained in Morse code. This technology, which is still widely used in the aviation industry, communicates over ARINC and SITA networks. The AIRIMP documents govern the Type B message formats.

Since the early 2000s, the GDSs have provided the capability for connectivity between a travel agency desktop and a LCC's host reservations system that did not process EDIFACT transactions. With XML (eXtensible Markup Language) connectivity, travel agencies can view real-time availability, flight information, and seat maps from LCCs. In addition, IATA has developed the new messaging standard for the New Distribution Capability (NDC), which will utilize XML and JSON (Java Script Object Notation), also supported by the GDSs.

The governing body that developed these standards were IATA, member airlines and A4A (Airlines for America).

3.21.4   Supplier Connectivity and Inter-Operability

Supplier connectivity is a core competency of the GDS. It supports current tiers of connectivity for air and connectivity to suppliers in other lines of business such as hotels, car rentals, and cruise lines. Traditional connectivity is with EDIFACT. GDSs also support XML interfaces to low-cost carrier reservation systems that do not support EDIFACT for shopping, pricing, and selling seats.

3.21.5   Security

Security encompasses authentication, authorization, and viewer permissions. Employee profile records (EPRs) are a crucial aspect of GDS security that regulates the capabilities of agents who utilize the GDS. The EPR grants travel agents access to the GDS and determines which functions they can or cannot perform within the system.

EPRs consist of various components, such as the agent ID for signing in, the agent sign which identifies each individual agent, the office code for categorizing travel agents based on their job function, duty codes that allow users to execute specific tasks, keywords that facilitate specific functionalities, and passcodes (passwords) for securing user sign-ins and system access.

3.21.6 Payment

GDS platforms facilitate various payment methods including bank transfers and credit card processing. Moreover, GDS systems enable virtual payment for the settlement of payment transactions between travel agencies and travel suppliers.

Chapter 4
GDS: The Internet, New Channels and Transparency (1996–2011)

4.1 Introduction

Adoption of the Internet for online direct and indirect sales of travel products changed airline product distribution economics. Airline investment in the direct channel was more grounded in emotion than costs. The GDS is a low-cost channel through which airlines do not have to spend any time marketing their product. The direct channel requires a significant investment in marketing dollars to enhance the brand and attract traffic to the airline's website.

The year 2004 was a very important year in airline distribution from several perspectives – the arrival of the GDS new entrants (GNEs, pronounced GeNiEs), GDS deregulation and the network airlines start to mimic the low-cost carriers by selling air ancillary products to increase the total price of the ticket.

4.2 The Origins of Online Channels

Well before the Internet era, in 1985, American launched eAAsy Sabre, a green screen interface for a home user to access schedules, availability, and fares with any modem-equipped computer. It was an important first step in the online era, but it was also mostly symbolic. This was when Max Hopper was Senior Vice President of Information Services at American Airlines (1985–1993). At that time, Kathy Misunas was at American Airlines, and played a role in the creation of eAAsy Sabre, well before she held the joint appointment as CIO at American Airlines and the first CEO of The Sabre Group (TSG) (1993–1995). Consumers had the ability to reserve flights, but the ticketing process had to be handled by either airlines or travel agents. The launch of eAAsy Sabre was a delicate balancing act for Sabre. As the leading GDS for travel agents, it risked the wrath of travel agents who were unhappy

© The Author(s), under exclusive license to Springer Nature Switzerland AG 2024   119
B. Vinod, *Mastering the Travel Intermediaries*, Management for Professionals,
https://doi.org/10.1007/978-3-031-51524-8_4

with the notion of consumers and airlines engaging in direct transactions, thus excluding them from the selling and booking process.

Market penetration was low and fewer than 150,000 customers booked and purchased tickets using eAAsy Sabre (Gutis, 1989). In 1990, the three largest online service providers that subscribers could access for a wide range of services such as news, weather, banks, and travel were Compuserve with 550,000 members followed by Prodigy and Nynex (Lavin, 1990). Using an online service provider, customers could access one or more of the three largest airfare services: eAAsy Sabre, the Official Airline Guide (OAG) electronic edition travel services, and Travelshopper from TWA and Northwest Airlines.

American and Sabre were the first to realize the potential of the Internet and launched Travelocity, the world's first online travel reservations system on March 12, 1996, even before the arrival of airline websites. Sabre's former CIO, Terry Jones, was the first CEO of Travelocity and responsible for developing the booking engine. From its early origins, the OTAs transformed themselves into web supermarkets to serve as a one-stop shop for air, hotels, rental car, and cruise lines. The OTA is an indirect channel. They use a GDS as the back end to manage bookings and a fulfilment agency for ticketing, customer service, and accounting.

With the Internet's introduction in 1996, the way online customers accessed schedules and fares was revolutionized. This had a significant impact on the share of total bookings made through the GDS. The Internet gave rise to two new channels: the consumer direct online channel which bypassed the GDS, and the consumer indirect online channels. These indirect online channels, known as OTAs, utilized the GDS as a backend for various services, including air shopping and bookings. Although the GDS still captured bookings made through the OTA channel, the margins were lower since the OTAs demanded higher incentives than TMCs.

Expedia was founded by Rich Barton and launched by Microsoft in late 1996 and began offering online services on the Microsoft Network. Microsoft later sold Expedia to media conglomerate USA Networks, Inc. in 2002. Priceline was founded in 1997 by Jay S. Walker. Priceline introduced the novel reverse auction "name your price" model in 1998 for which they were granted a patent (Walker et al., 1998). This model reversed the buyer–seller relationship since the buyer submits a guaranteed price that they are willing to pay to travel to a destination. Priceline aggregates requests from buyers and submit them to sellers for a response. Airlines participate in the reverse auction model to get rid of surplus inventory. However, the model has its flaws; it does not offer the identity of the seller or the travel schedule, that may be less than appealing, until after the purchase has been made. The popularity of this model is limited to leisure travelers who are seeking deep discounts.

The OTAs provided market transparency to travelers on an unprecedented scale. Travelers could view schedules and fares on the Internet without having to talk to a brick-and-mortar travel agent. Giving information and choice directly to consumers was a significant improvement over traditional distribution. This market transparency has led to innovations in revenue management, called competitive revenue management (Vinod, 2021a, Chap. 7), where competitor selling fares and schedules were used as input into the revenue management systems to set inventory controls.

A few years after Expedia, Travelocity, and Priceline had consolidated their market positions in the online space, Orbitz was launched in 2001 with investment from American, Continental, Delta, Northwest, and United. Unlike the other OTAs, Orbitz supported a lower-cost model with direct supplier links and a percentage of their online bookings bypassed the Worldspan GDS and were booked directly in the airline CRSs. Travelocity's former Chief Executive Sam Gilliland (Hansell, 2002a) remarked on the direct link to suppliers *"If you have a direct connection, you are just pushing costs from one place to another. In the end, it will have very little economic benefit to the airlines."* Orbitz was acquired by Cendant Corporation in 2004. In 2006, Travelport, the travel distribution business of Cendant was acquired by The Blackstone Group and Orbitz was later acquired by Expedia in 2015. Expedia also acquired Travelocity from Sabre in 2013. Today, the largest OTAs are Expedia (Expedia Holdings) and booking.com (Booking Holdings). The Priceline Group, which owned booking.com, was renamed Booking Holdings in 2018.

Niche OTAs Hotwire.com, Hotels.com, and CheapTickets.com are owned by Expedia Holdings. TripAdvisor.com, the online travel company known for its user generated content (e.g., hotel reviews, restaurant reviews, etc.) was founded in 2000. It was owned by Expedia from 2005 to 2011 and it is now an independent company.

The most recent additions to the online realm are metasearch engines. Kayak and Skyscanner made their debut in 2004, followed by Google Flights in 2011. Metasearch sites are widely used today. A metasearch site examines multiple sources, including supplier websites and online travel agencies (OTAs) and presents a display based on the itineraries obtained from these sources. It also provides direct links to these sites for booking. While the metasearch site receives a referral fee, the booking itself is not guaranteed as the customer may choose to abandon the shopping cart after being directed to the supplier site. In the case of metasearch, if an airline does not participate in a Global Distribution System (GDS), the metasearch engine can determine the lowest fare directly from the supplier site, if available. However, metasearch engine sites may have their preferences and, in some instances, only search sites that offer them a commission for referring customers.

Most low-cost carriers (LCCs) do not participate in GDSs and their consumer direct channel serves as the only way to book an airline seat. Advances in technology and hosting services have made it easier for LCCs, whose primary segment is the leisure traveler, to bypass the GDS and travel agencies in favor of direct distribution to lower the cost of customer acquisition.

4.3 Platform Power Revisited

Today many retailers advertise on Google, and manufactured goods are available through Amazon. The eBay platform sells a range of products that are not available on other sites and fittingly they market eBay as *"seriously, we have everything."*

These platforms, such as eBay, Google, and Amazon are expensive for users of these systems or for the suppliers. For example, the eBay fees paid by a seller can range from 10% to 20% of the total transaction.

These intermediaries also threaten the existence of businesses since if they do not agree with their terms and fees, they will no longer be allowed to participate in the marketplace. Strategies to counter this threat are a high priority to control the cost of online advertising (Edelman, 2014).

Google Flights is reshaping the metasearch industry. The foundation of Google Flights is ITA Software, acquired by Google in 2010. There are notable distinctions between the two. Unlike ITA Software's QPX shopping engine that powers airline websites and relies on real-time shopping results, Google Flights combines live and cached shopping results. While Google Flights has traditionally provided direct links to supplier websites, it has only recently started offering the same for OTA sites outside of the U.S. By offering referrals to airlines and OTAs at no cost, Google's recent decision (Sullivan, 2020) is expected to have a considerable impact on other metasearch players that heavily rely on referral revenue. To compensate for the loss in revenue, Google plans to introduce new ad formats that provide suppliers and travel partners with alternative options for promoting their products. The rationale behind this move is to prioritize flights based on their relevance to users, considering factors such as price and convenience.

Compared to these platforms, the GDS platform is relatively inexpensive. The cost of air distribution is approximately 2.2%.

4.3.1   Redefining the Distribution Channels

Until the late 1980s and early 1990s, the sale of airline products was strictly offline direct and indirect. Travel agencies were the offline indirect channel and the offline direct channel consisted of airline call center agents, airport ticket offices (ATO), and city ticket offices (CTO). The Internet led to the creation of two new channels, the online direct and online indirect through which products could be distributed and sold. These new channels offered end consumers instant access to unprecedented transparency of airline schedules and available selling fares. The online direct channels are the airline.com branded channels and these bookings bypass the GDSs.

The direct channels transact directly with an airline's host CRS for booking and ticketing activity. Air shopping is a separate service to display priced itineraries. For the first time, the GDS share of the market was going down as the direct channels attracted more site traffic and conversions from leisure passengers. The airline direct bookings were predominantly leisure bookings.

The indirect channels typically transact with a GDS for schedules, fares, and availability. Although there has been a shift from offline to online over the past decade, the indirect channels contribute over 50% of the bookings worldwide in 2022. The online indirect bookings were made predominantly by online leisure customers through the OTAs. The OTAs use the GDS as a backend for retrieving

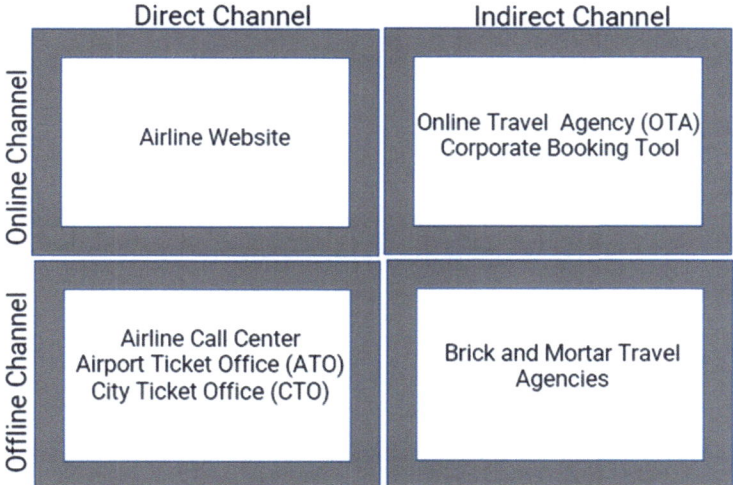

Fig. 4.1 The new reality in 1996 after Internet adoption

shopping results using web services provided by the GDS and to store the passenger name records after the booking has been created. The online indirect channels augmented or at least protected the GDS booking share, though at a reduced margin because of the higher incentive they demanded compared to the brick-and-mortar TMCs.

Figure 4.1 illustrated the new reality of distribution channels after adoption of the Internet.

Each distribution channel represents a storefront through which an airline's products are displayed for sale. What is displayed for sale is the output of the airline's revenue management process. The fundamental challenge is that the same content—schedules, fares, and availability, should be displayed across all channels of distribution. This is made possible with full content agreements between airlines and GDSs, and interactive (seamless) connectivity to ensure availability is identical across all channels.

4.4   The Travel Value Chain

The travel value chain is shown in Fig. 4.2.

A customer can access and book travel through direct channels or indirect channels. The direct channels are those that access schedules, fares, and availability from an individual supplier's reservations system (host CRS). All bookings that originate from brick-and-mortar travel agencies and OTAs are termed indirect bookings and the GDS serves as an intermediary between the OTA/TMC and the airline.

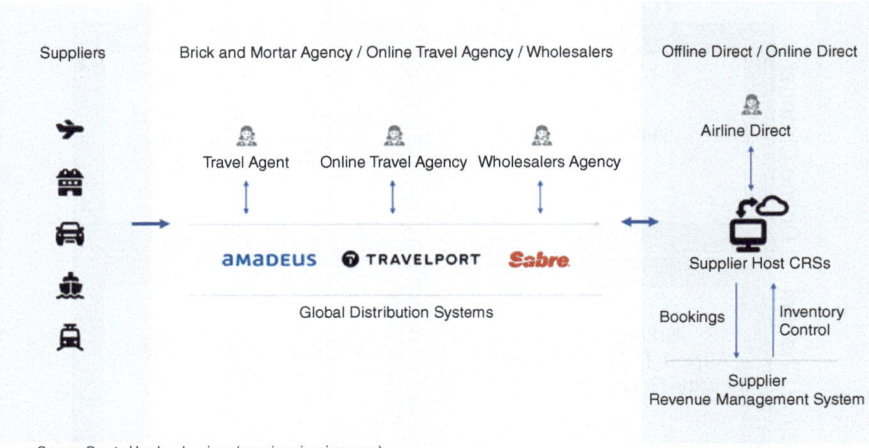

Source: Created by Joe Jennings (www.joemjennings.com)

Fig. 4.2 The travel value chain

A travel agent subscribes to one or more GDSs to access schedules, fares, and availability to facilitate the booking of an itinerary. The most common entry used by travel agents is the short sell or the long sell to create a booking. A short sell is a sell request from the city pair availability or schedule display. A long sell is a direct sell of a constructed itinerary with individual segments. The shopping entry came later and was launched in 1984 by Sabre, and it was called Bargain Finder. The Bargain Finder entry returned nine priced itineraries. This was followed by Bargain Finder Plus which returned 19 itineraries while Bargain Finder Max returned 60 or more itineraries depending on the parameter configuration.

An online travel agency provides access to schedules, fares, and availability to online users. An OTA typically accesses travel content from a GDS using web services and the shopping entry is used to return priced itineraries to an online user.

A wholesaler is an airline consolidator that specializes in bulk ticket purchases from airlines. They acquire capacity on routes from airlines in bulk at a discount and resell them to travel agencies and individual customers. Wholesalers acquire unpublished net fares from the airlines. The net fares are discounted, and savings can exceed 30%. Travel agents acquire the net fares from wholesalers and mark them up for sale to customers, which will be still lower than the public fares in the market. There are variations to the wholesaler model. Some acquire the inventory in bulk as risk inventory from the airline. However, many wholesalers do not buy in bulk but on consignment by accessing the available inventory by booking class at the discounted net fare levels. In many parts of the world, and especially markets that are not mature, wholesalers thrive and fill empty seats on airlines and promote brand awareness with the price conscious customer. In international markets, airline consolidators are also called bucket shops or tour operators.

Revenue management is an advanced decision support application that recommends optimal inventory controls which are implemented in the airline's host reservations system. Inventory control in the host CRS serves as the execution component to accept and reject reservation requests, based on inventory control recommendations from the revenue management system.

Today's GDS is more than just air supplier content, but links travel suppliers like airlines, hotels, rental cars, cruise lines and rail with travel agencies. It is a business-to-business (B2B) model. The GDS is considered a very efficient and cost-effective channel to distribute and sell supplier inventory. However, today, it is a legacy platform with limitations on how suppliers want their content displayed and sold. For travel agencies, it provides a platform for comparison shopping across suppliers so that they can select the supplier of choice to make a booking based on a customer's preferences. A GDS has connectivity to supplier reservations systems to access availability. In the future in a New Distribution Capability (NDC) world, suppliers will also price itineraries and ancillaries for travel agents. GDSs earn their main source of revenue from booking fees which are negotiated with the travel supplier. For example, for airlines, it is based on segment booking fees and for hotels it is priced by transaction regardless of the length of stay. On average, for full-service carriers, there are 2.5 air segments per ticket. This is based on the hub economics of 30% local traffic and 70% connecting traffic and based on the assumption that one way and round trip bookings are equal. Travel agents that subscribe to a GDS earn incentives based on performance and in some situations the incentives could exceed 50% of the segment booking fees when performance thresholds are achieved.

4.5   GDS Discounts, Full Content, and SWABIZ Corporate Direct

Airlines receive trade discounts on segment booking fees through GDS programs in return for providing full content, including web-only fares, in the GDS. The GDSs ensured that the segment booking fee discounts were guaranteed, causing many carriers to abandon their channel shift strategy between 2000 and 2006. This action by the GDSs brought balance to the marketplace, causing airlines to move away from a channel shift strategy.

Undeterred by the GDS initiatives, Southwest Airlines continued to pursue a corporate direct booking strategy. Southwest airlines launched SWABIZ.com, a corporate booking portal designed for business and corporate travelers. It serves as a one stop shop for booking Southwest flights, hotels, and car rentals. By 2003, the SWABIZ.com direct connect model for corporate travelers had gained traction and Southwest expanded the booking portal. Through this portal corporate travel managers can book on behalf of employees, corporate travelers can get their corporate discounts, and the site permits ghost cards which allows corporate travelers to make bookings using a credit card without knowing the actual card details such as card number and security code. There are no fees to use the corporate direct portal. Today, 61% of Fortune 1000 companies are enrolled in SWABIZ.

4.5.1   Carlson Wagonlit Direct Connect Program

In a surprising move, Carlson Wagonlit Travel (CWT), a global travel agency, launched the direct connect capability with Navitaire's DirectNet on their Symphonie travel agency platform in August 2003.

Accenture owned Navitaire. DirectNet originated from Anderson Consulting, which was Accenture's predecessor. Anderson Consulting launched the Via World Network in 1996 as a low-cost channel for employee bookings, but it failed.

American Airlines, Continental Airlines, Delta Air Lines, and United Airlines joined the program, with US Airways joining later. DirectNet did not replace the existing GDS multi-carrier fare search capabilities on the Symphonie platform but instead complemented them. This allowed travel agencies and corporate customers to compare prices across airlines, giving them an advantage over airline websites. Accenture was the first corporate user. According to Navitaire, tickets issued through DirectNet cost an airline $3 per ticket compared to GDS tickets that cost $10 to $15.

The CWT direct-connect initiative was poorly timed since a content gap did not exist between the direct channel and the GDS platforms. While DirectNet accessed Web-only fares, these fares were already available on Sabre and the other GDSs followed a few months later.

Only a few dozen corporate accounts had adopted the direct connect technology a year after its launch. In 2005, CWT discontinued the direct-connect program that linked select clients to five airlines for bookings and substituted the GDS bypass with G2 Switchworks, a GDS new entrant (also known as a GNE). G2 Switchworks connected to six airlines, also utilizing Navitaire's direct-connect technology. The G2 software with added features was considered more robust than CWT's previous offering and could only be accessed by clients that use the CWT Symphonie platform.

Carlson Wagonlit Travel was officially renamed CWT in 2019.

4.6   Genies and the Threat to the GDS

In 2002, airlines spent an estimated $2.2 billion on GDS booking fees which average $4.36 per segment. From 1995 to 2002, the average GDS booking fees increased approximately 4.5% per year (Schaal, 2002).

The pressure on distribution costs imposed by the airlines on the GDSs gave rise to the so-called Global Distribution System New Entrants (GNE) that was pronounced "GeNiEs". They included Triton Distribution Systems, ITA Software, G2 Switchworks, and Farelogix.

The Genies entered the market in late 2004 and early 2005 with the compelling claim that they were developing alternatives to the traditional GDS model to reduce distribution costs. Further they claimed that their systems were developed from the ground up that allowed them to design flexible systems with a focus on customer-centric capabilities.

They were Internet-based access and distribution systems that did not require data to be stored internally. They search multiple supplier travel sites (airlines, hotels, rental car, cruise lines, tour operators) as well as OTA's (Expedia, Travelocity, Orbitz, and others) simultaneously to create a virtual dataset that is presented to agency customers based on parameters to filter the itineraries. Unlike the GDS, bookings can be made directly with the suppliers and a master itinerary obtained from multiple sites is produced for the customer.

The GNEs created a groundswell of support from airlines worldwide since they offered a glimmer of hope to drastically reduce the cost of air product distribution (Quinby, 2005). Triton and G2 Switchworks promised savings of 75% over GDS costs and ITA promised pricing at $0.40 per segment for its alternative GDS offering. Like the GDSs they each registered a global receiving id with IATA. ITA was 1U, Farelogix was F1, Triton was H1, and G2 Switchworks was A1. Yet, GNEs suffered from the technical and operational challenges of broad adoption and accounted for less than 0.5 per cent of the US domestic market by 2006 (Quinby, 2006).

Reduction in GDS booking fees with the full content contract renewals further mitigated the value proposition of the GNEs. G2 SwitchWorks was purchased by Travelport for its Kestrel agent desktop.

The GeNiEs also had a key marketing message that they were not shackled with legacy technology like the GDSs and were developing their products with new technology that will ensure customer centricity and better product and service levels. Some of the benefits included direct connection to carriers, unlimited capacity for new products and services such as private fares and preferred display of inventory to authorized agencies and interoperability with any back-office system. Yet, they were unable to penetrate and collectively represented less than 1% of the U.S. domestic market.

In less than 3 years since their inception, with no traction, the GeNiEs changed their strategic direction with enterprise solutions for airlines or travel agents. However, they served an important role for the airlines, which is providing leverage during contract negotiations with the GDSs after GDS deregulation. The GNEs helped the airlines lower the segment booking fees between 20% and 40%. It was also the first time that segment booking fees were going down. Unable to gain traction after three years, the GNEs changed their strategic direction since they realized that they were not likely to surpass or replace the entrenched GDS model.

4.6.1   Why Did the GNEs Fail?

Important messaging with new technology and reduction in the cost of travel distribution did not save the GNEs. Why?

The new technology was overhyped and the GNEs underestimated the complexity of GDS transaction flows. The GNEs promised direct connections to airlines which was largely unfulfilled. In 2005, only Orbitz had direct connections with a few major U.S. airlines to bypass the GDSs. A direct connect solution is less than optimal for TMCs that service corporate customers who have rigorous and complex requirements for fulfilment, exception handling for changes to the reservation, and back-office integration. These are features already supported by the GDS. Besides, there are complex requirements for interlining, airline alliances, complex itinerary management, price guarantees, customer service support, security for personal data, and back-office solutions for travel agencies that cannot be ignored.

A related factor is corporate travel where corporations use automated booking tools that interface with the GDS to manage an employee's travel needs. The switching cost to a GNE was not an option for travel agents.

The GDSs offer incentives to travel agencies for achieving a goal. The GNEs focused on the airlines and ignored the travel agencies. To reduce the cost of travel distribution, they did not offer incentives to travel agents. This was a fundamental issue since neither the smaller agencies nor the TMCs were amused. Incentives drive bookings through the GDS channel.

During the GNE era, GDSs introduced changes to their pricing model. They offered concessions by charging lower segment booking fees for home markets and charging more for away markets.

4.6.2 Lessons Learned

There are three takeaways from the GNE saga.

1. Airline GDS contract negotiations can be contentious, which could impact the travel agency with added fees imposed by airlines on the GDS during negotiations. In 2006, travel agencies realized the importance of pursuing a multi-GDS strategy to mitigate risk instead of using a single dedicated channel. The multi-GDS strategy shifts the balance of power from the GDS to the travel agency.
2. Airlines were able to negotiate deeply discounted segment booking fees in the 20%–40% range as a direct fallout from the promises made by the GNEs.
3. As a direct result, since 2006, travel agencies were charged 80 cents per segment by the GDSs for access to full content from participating carriers. This fee is deducted from the GDS incentives paid to travel agents.

In summary, the airlines were the primary beneficiaries of the entry of the GNEs into the market. By chance, their timing for entry could not be better, when airlines were getting ready to renegotiate GDS contracts after deregulation in 2004.

Of the original GNEs only Farelogix has a business model that is remotely similar to what they had originally planned. Acquired by Accelya in July 2020, Farelogix today offers an application layer (the FLX Platform) that enables travel agencies to source their content directly from airlines or from the GDSs. It is not viewed as an alternative to the GDS but is a bridge to enable travel agencies to manage inventory sourcing from multiple channels. FLX does offer direct connectivity to airlines, which makes it an alternative channel for suppliers to distribute their products to travel agencies outside of the GDS. During contract negotiations with the three GDSs, American made its fares and optional services (ancillaries) available through Farelogix in 2006, giving travel agencies an option to access all the fare content (Schaal, 2006b). Before the agreement with American, Farelogix had connectivity agreements with AirTran, Continental Airlines, Northwest Airlines, and Spirit Airlines.

4.7   GDS Value Pricing

After GDS deregulation, Amadeus was the first GDS to introduce two pricing categories for home markets and away markets which dictated the cost of the segment booking fees. The value-based pricing structure is based on the simple fact that not all air segments are created equal, and some markets are more valuable to airlines since it is out of reach of their direct channel. The value pricing model assumes that bookings in away markets are more valuable to an airline than a home market. In 2005, Amadeus further refined the value-based pricing model (Business Travel News, 2004) by adding two additional booking fee categories for short haul and long haul segments. The booking fee categories were called standard, standard plus, premium, and premium plus. However, the range for the segment booking fees for the four categories were between €2.67 and €4.90 and varied by airline.

After the introduction of value pricing by Amadeus, the other GDSs followed with a similar pricing model.

4.8   GDS Deregulation

In the late 1990s, the airlines started divesting their interests in the GDSs. Galileo acquired Apollo in 1992 and went public in 1997 as an independent company. Sabre separated from American in 2000 and became independent. Amadeus took back ownership from the founding airlines in the early 2000s. It is now listed on the Madrid stock exchange.

In July 2004, the U.S. Department of Transportation (DOT) deregulated the U.S. GDS industry and sunset all the CRS regulations in the United States. This included:

1. Display bias, which prohibited GDSs from preferential displays to favor any airline. This rule did not apply to hotels or rental car companies.
2. Parity clauses that required participating airlines to maintain the same level of service in all GDSs. This was significant since airlines no longer had to subscribe

to all the GDSs, but they could be selective, but this has not happened because every airline wants access to the premium corporate segment.
3. Full content agreements negotiated between a GDS and an airline. The GDSs previously would mandate full content agreements as a condition for participation to ensure access to all fares inclusive of web fares.

On January 31, 2004, the Department of Transportation (DOT) decided to phase out GDS regulation within six months by July 31. This was a significant change at that time since the GDS business had been heavily regulated by the federal government for two decades.

GDS deregulation also led to the onset of web-fares. Some airlines started offering cheaper fares on the airline direct channel that were not available in the GDS. To gain access to all the fares offered by an airline in the GDS channel, GDSs offered a discount for the segment booking fees.

Before the Internet era, the volume of bookings from the agency channel had increased steadily over the years, and the resultant mix was 80% of bookings came from the GDSs. The remaining bookings came from airline call centers, ATOs and CTOs. After GDS deregulation, full-service carriers have continued to participate in the GDSs.

4.8.1   What Does Deregulation of the GDS Mean to Suppliers, GDSs and Consumers?

The U.S. Department of Transportation (DOT) made the decision to deregulate the global distribution system business since the U.S. airlines no longer had an ownership stake in the GDSs. In 2004, Amadeus was the only GDS with airline ownership from Iberia, Air France, and Lufthansa along with public shareholders. An equally important reason for this decision is lessons learned from airline deregulation in 1978 that provided lower fares and more services for the traveling public. DOT anticipated that deregulation of the GDS business should also benefit customers since contract negotiations between airlines and GDSs were at a level playing field. GDSs could no longer require airlines to provide their web fares as a condition to participate in the GDS. With GDS deregulation, airlines could use their web fares as bargaining chips to negotiate better terms and discounts for booking fees for participation.

With deregulation, airlines were no longer mandated to participate in all GDSs. However, in the aftermath of deregulation, no airline has left a GDS due to the importance of corporate bookings, which on average have higher ticketed prices than leisure bookings. Hence, airlines have focused on actively negotiating and providing their Web Fares in exchange for discounts in booking fees. From a DOT perspective, they also do not have the authority to force airlines to put all their publicly available fares in every distribution channel. Despite deregulation, the GDSs are prohibited

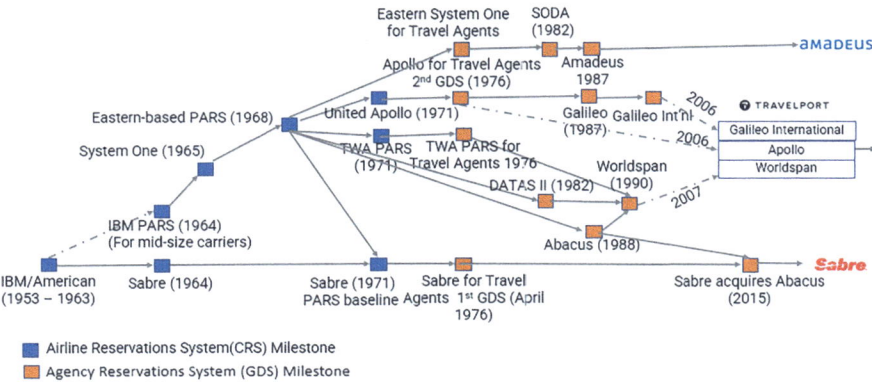

Fig. 4.3 Evolution of the GDS from reservations systems (1953–2007)

from practicing unfair methods of competition, commonly referred to as antitrust law.

Today the traditional GDSs Amadeus, Sabre, and Travelport operate at scale in a deregulated environment. The fourth, TravelSky, that operates in a regulated environment, enforced by the Civil Aviation Authority of China (CAAC). All domestic bookings in China and international itineraries that commence in China flow through the TravelSky system. However, traditional GDSs can book travel for customers who travel to China but commence the travel overseas.

Figure 4.3 illustrates the evolution of reservations systems and GDSs with IBM hardware and software as the initial baseline and the current state of the GDSs. After consolidation, by 2007, the oligopoly consisted of three GDS companies. Of the three GDSs only Amadeus has migrated away from IBM mainframe technology and TPF which took place in 2018. Travelport has the highest costs of operation since they operate three different GDSs under the Travelport brand (Apollo, Galileo International, and Worldspan). Of the three, Galileo has the largest subscriber base. In 2020, Travelport decided to converge on the Galileo International platform and sunset Apollo and Worldspan by migrating existing subscribers to Galileo. This is a slow process since besides the large travel agencies there is a long tail of smaller travel agencies that need to be converted. Conversion to the Galileo platform is expected in 2023.

4.8.2 The Travel Management Agreements (TMA)

After GDS deregulation, to protect the GDS business, GDSs like Sabre introduced the so-called Travel Management Agreements (TMA) for airline participation in the GDS as a replacement for the Participation Carrier Agreement (PCA). The purpose of the TMA contracts was to manage carrier business relationships in a deregulated environment, tighten language concerning carrier responsibility for content and treatment of GDS subscribers, extend the contracts for longer durations if possible and introduce the value-based pricing structure.

The elements of the new contract with airlines clearly stated the responsibilities of the GDS, participation option requirements (level of connectivity), payment requirements, term of the contract, and the responsibilities of the participating carrier.

By mid-2007, over 80% of the 440 airlines had migrated to the new TMA agreements. However, many carriers rejected some of the clauses in the new TMA contracts such as expected discounts on booking fees, surcharges, travel agency direct connections, and inclusion of ancillary products and services as part of the full content agreements.

The term TMA was short lived and was a defensive posture, reaction to deregulation by the GDSs, to protect themselves. Today, it continues to be called the PCA, though several terms and clauses from the TMA are now part of the PCA agreements.

4.8.3   Sabre, Amadeus Content Sharing Agreement

After GDS deregulation, as a protective measure against airline threats to withhold content for deep concessions in segment booking fees, Sabre and Amadeus signed a contract to ensure access to airline inventory content in 2006 (Schaal, 2006a). This unprecedented agreement was the first between the two companies. The initiative was called Project Cervantes for Sabre and Amadeus to remedy the potential lack of access to bookings by sharing content to complete bookings on airlines that do not participate in their respective systems.

For example, if United decided not to participate in Amadeus, then Amadeus would require Sabre to make United's flights available to Amadeus connected travel agencies and corporations through the Amadeus GDS. In this scenario, United would continue to pay the segment booking fees to Sabre as specified in the contract and a separate commercial agreement between Sabre and Amadeus governed the compensation paid by Sabre to Amadeus. The existing participating airline agreements (PCA) allowed Sabre and Amadeus to enter into a contract without authorization from the airlines. The agreement did not apply to airlines that only participated in one of the two GDSs due to the absence of an existing PCA with the non-participating carrier.

The content back-up agreement between the two GDSs did not apply to all countries. It also included a provision that allowed a GDS to reject an inventory request if the reservation system had an agreement with an airline preventing inventory-sharing.

4.9 The Regional GDSs

In December 2019, prior to the pandemic, the share of bookings worldwide through the indirect GDS channels (Amadeus, Sabre and Travelport (Apollo, Galileo International and Worldspan)) had slipped to the 50% range. This percentage excludes the LCCs who do not participate in GDSs. From an air market share perspective, Amadeus is the largest followed by Sabre and Travelport. TravelSky is a state-owned GDS in China and operates in a regulated environment.

There are also regional GDSs that operate in various regions of the world. Some of the well-known regional GDSs are Kiu System that operates in Latin America, Sirena-Travel in Russia, and Infini in Japan. Topas in South Korea, Axess in Japan, and SITA's GETS no longer operate.

4.9.1 Axess and INFINI

Japan Airlines (JAL) and All Nippon Airways (ANA) are carriers in Japan that carry large volumes of domestic passengers and have international flight operations. The split of passengers for domestic and international operations is roughly 80% and 20%, respectively. For many years, these two airlines used to operate two separate reservations systems for domestic operations and for international operations. The domestic reservations system used was the Unisys USAS reservations system and both airlines used variants of the British Airways Booking System (BABS) for international flights. BABS was an IBM mainframe product, running on IBM's proprietary TPF operating system. Internally, the JAL reservations and ticketing system were called JALCOM.

As the reservations landscape evolved, ANA migrated its existing reservations, ticketing, and check-in system from Able to the new Unisys AirCore reservations system in 2007 for domestic operations. For international operations, ANA initiated the migration from BABS to Amadeus' Altea Customer Management Suite in 2011 which was completed in 2014. In 2023, ANA decided to migrate to Altea for domestic operations to exploit the synergies of having the domestic and international transactions on the same platform. In 2014, Japan Airlines initiated the migration to the Amadeus Altea platform for both domestic and international operations, which was completed in 2017.

The Axess computer reservations system was established in 1991 by the Jalinfotech IT department of Japan Airlines. Established as a subsidiary of JAL, it was marketed to travel agencies in Japan. Sabre established a partnership with Axess to provide expanded booking and ticketing capabilities in the Japanese market. In 2012, Travelport signed a marketing agreement to host Axess out of its data center in Atlanta, Georgia as a partition of the Travelport GDS. Convinced of the halo effect of using Amadeus' Altea for reservations and inventory control, JAL switched to the Amadeus GDS in 2021 to be the carrier's recommended distribution partner in the Japanese market.

Infini was a computer reservations system that was created by All Nippon Airways and Abacus in 1990 to serve the Japanese market with its unique requirements. Today, Infini uses Sabre's technology for booking and ticketing capabilities to travel agencies.

4.9.2 TOPAS

Korean Air's subsidiary TOPAS is the regional GDS in South Korea. It started as a CRS in 1975 under the name KALCOS for the Korean market powered by a version of IBM's TPF. It was the main booking and ticketing system for travel agents in the Korean market.

In 2011, Korean Air decided to adopt the Amadeus Altea Customer Management Solution. Migration to Altea was completed in 2014. With that decision, Korean Air also selected the Amadeus Selling Platform, customized for the Korean market, to replace the legacy TOPAS travel agency system. The travel agency system in South Korea is called Topas SellConnect.

4.9.3 SIRENA

Sirena Travel is a regional GDS for the Russian market. Travel agencies access carriers in the Russian Federation and Commonwealth of Independent States (CIS) through the Sirena network.

It is an agency network of software and hardware connected to the inventory systems of airlines in Russia and the former republics of the Soviet Union (CIS) and provides a neutral display of flight schedules, flight availability, and shopping to travel agents. Sirena also has a PSS and check-in system that is used by small carriers.

Sabre was involved with the development of Sirena in the 1990s. Over 18,000 terminals of travel agencies and airlines are connected to Sirena through 7000 air ticket sales locations in Russia, CIS and overseas, supporting sales activity of 500 agencies.

4.9.4 Kiu System

Kiu System Solutions was created in 2003 to serve the Latin American market with a suite of integrated solutions. Kiu has a PSS and check-in system that is used by small carriers. The Kiu GDS is a reservations and ticketing platform that distributes Kiu PSS hosted airline content to travel agencies and OTAs. In 2022, Kiu established a partnership with TravelX for airlines to tokenize and distribute inventory as non-fungible tokens (NFTs) with unique digital identifiers through the blockchain ecosystem where they can be exchanged and resold. It is unlikely that network carriers will adopt this concept since it promotes revenue dilution and the middleman, the consumer, stands to profit from a change of ownership of the ticket.

4.9.5   SITA GETS

The SITA product for agencies was called Gabriel Extended Travel System (GETS) for travel agencies. It served as an alternative to the larger U.S. GDSs in markets where the flag carrier used its Gabriel reservations and inventory system. It served primarily in second and third world countries such as the former Soviet Union, Africa, India, Latin America, and Eastern Europe, GETS was short lived when Galileo International entered into a marketing agreement that added over 500 agencies in 19 countries.

4.10   The Origins of Preferred Channels and GDS Surcharges

The new round of negotiations between airlines and GDSs in 2006 introduced the concept of preferred distribution channels. It was an attempt by airlines such as Northwest Airlines and American Airlines to further reduce the cost of distribution.

However, even prior to the 2006 contract negotiations, Northwest Airlines introduced the Shared GDS Fee in August 2004 whereby the airline would charge travel agencies $3.75 for a one-way ticket and $7.50 for a round trip ticket. The surcharge was not well received by travel agents or the GDSs. With no support from rival carriers, Northwest abandoned the surcharge two weeks after launch.

In 2006, American was dealing with four contract renewals: Galileo, Worldspan, Amadeus, and Sabre. In July, American finalized agreements with Worldspan and Galileo. These agreements allowed for "additional services" to be offered to travel agents, specifically "Worldspan Super Access" and "Galileo Content Continuity." By opting for these new services, travel agents gained the ability to view and book all of American's fares. However, by entering into these agreements, agents were subject to a reduction or elimination of their incentive payment from the GDS. For example, a Galileo Content Continuity travel agent would give up $0.80 of its GDS incentive payment (Campbell, 2006). In contrast, Sabre and Amadeus did not budge and continued to charge American GDS fees that were in the range of $4 to $5 per segment.

American Airlines announced its Source Premium policy on July 12, 2006. With this new policy, travel agents in the U.S., Puerto Rico and U.S. Virgin Islands would be charged a $3.50 "*source premium*" fee per segment for GDS bookings starting September 1, unless American had designated it as a "competitive booking source." The fee applied to Sabre and Amadeus who were in contract negotiations. American had already negotiated new contracts with Worldspan (*Worldspan Super Access*) and Galileo (*Galileo Content Continuity*) who were exempt from the "source premium" surcharge. This was an opt-in program for travel agents. If the travel agent agreed to the new service, they would receive less or no incentive from the GDS, but the "source premium" fee would be waived.

Travel agents were not amused and thought that American's move was not justified since airlines had always paid a segment booking fee to distribute their product through the GDS and did not see a reason for change. Some were even bewildered that these surcharges were being imposed on Sabre, a GDS they had created and promoted for several years.

ASTA CEO Kathryn W. Sudeikis complained about the new policy (ASTA, 2006), *"American's announcement tells every travel agency in America: if you want to sell us, run your business the way we tell you or you'll be forced to pay us for the privilege of booking our services. This policy of shifting still more costs off of American's financial statements onto the backs of travel agents and their customers is unconscionable. American is trying to use its market power to impose its costs on other market players, as a condition to providing what travel agents clearly require to do business efficiently."*

Travel agents also objected to the surcharge since switching from Sabre or Amadeus to avoid the new fee was not an option since they had multiyear contracts with the GDSs. Further, they did not want to create a separate line item in the customer invoice to collect these new fees for flights booked on American.

American realized the veiled threat that loomed ahead. If the new policy went into effect on September 1 and competitors did not impose an identical surcharge, travel agencies had the power to shift traffic from American to its competitors in major markets and hubs at Chicago O'Hare and Dallas/Fort Worth. The veiled threat was unsettling, and it forced American to negotiate new contracts with Sabre and Amadeus that reduced distribution costs, but not to the extent they desired. The new contracts also preserved corporate travel agency relationships that produced premium revenue for the airline.

Ultimately the Source Premium policy was never implemented. On September 1, 2006, American and Sabre signed a new five-year agreement. It was called Sabre's *Efficient Access Solution,* which offered full AA content without a booking source premium charged by the airline for agencies participating in the program. Amadeus also signed a contract in October 2006 that erased the charges that were piling up starting September 1.

4.11   The Direct Connect Program from American Airlines

In 2011, American Airlines decided to single handedly take on the intermediaries to lower distribution costs. American launched "AA Direct Connect," a competing ticket delivery system in 2011 to decrease GDS booking fees (Bilotkach et al., 2014). The primary objective of the direct connect program was to encourage large travel agencies to adopt the program and bypass the GDSs.

Unfortunately, this was bad timing for two reasons: first, American Airlines was on the verge of filing for Chap. 11 bankruptcy protection, and second, it could not fight this battle alone without the support of other major carriers like United Airlines and Delta Air Lines who had contracts with Sabre and Travelport, and possibly Amadeus that extended through 2013.

Gerard Arpey, the CEO of American Airlines, remained unwavering in his refusal to consider bankruptcy as a solution to the company's financial problems, a route that many other U.S. airlines have taken. Despite facing pressure from the airline's board, Arpey prioritized his principles over the commonly accepted practice and made the decision to leave American Airlines, where he had worked for three decades.

In response to American's announcement, Sabre retaliated by displaying American Airlines flights unfavorably in city pair availability and shopping responses to its competitors. The TMCs supported Sabre and did not respond favorably to American's demands and eventually the direct connect program fizzled out. American filed a lawsuit against Sabre over alleged anti-competitive business practices.

Besides the Sabre lawsuit, American was also embroiled with Orbitz Worldwide and Travelport during the same period.

The OTA business model relies on utilizing web services provided by a GDS to incorporate responses into their workflow. Online bookings made through an OTA site are facilitated through the GDS. Most large OTAs have a multi-GDS strategy where they access more than one GDS and distribute the online bookings that is parameter driven or based on some predefined metrics such as conversion rates. The GDS charges the airline segment booking fees for each booking made, as agreed upon in the full content agreement. Additionally, the GDS provides performance-based incentives to OTAs based on achieving specific booking volume thresholds. These full content agreements between airlines and GDS are renegotiated every three to five years.

U.S. airlines have long held the view that when leisure customers book through an OTA site, they are influenced by the airline's brand rather than affinity to the OTA. They also question the value of low-priced bookings sold to leisure travelers, generated by the OTAs, which are lower than the higher-priced bookings generated by the offline travel agencies. For online customers, using an OTA to compare prices and book travel is convenient. If leisure travelers booked directly on the airline's website instead of the OTA, the airline would avoid paying the segment booking fees to the GDS.

After the IPO in 2007, Orbitz was Travelport's largest customer for GDS services with exclusive provisions. Travelport also owned 48% of Orbitz worldwide, the holding company behind the Orbitz brand, Ebookers, HotelClub, and Cheaptickets. In 2010, American and Travelport were in the midst of negotiating a new full content distribution agreement. In December 2010, American Airlines removed its content from Orbitz. This was a clear message to Orbitz and Travelport that American was unhappy with the distribution costs and how American's fares are displayed and marketed to travelers.

Both Orbitz and Expedia had refused to adopt American's direct connect technology which could influence how American's fares were displayed and purchased. Instead of displaying itineraries based on schedules and fares, the direct connect technology provided by Farelogix could display fares based on a traveler's preferences such as more legroom, priority seating, etc. that enables American to generate incremental revenues. In solidarity with Orbitz, Expedia did not renew its contract with American when it expired on December 31, 2010, and removed AA schedules and fares from its site in January 2011.

Expedia and American settled their dispute in April 2011. However, a court ordered American to resume providing content to Orbitz while Travelport continued to negotiate a new distribution agreement. American's schedules and fares were available on Orbitz.com in June 2011.

In April 2011, American filed an antitrust lawsuit against Orbitz and Travelport, stating that Orbitz made American's fares look higher than they were to consumers.

AMR Corp, the parent company of American Airlines, filed for bankruptcy protection under Chap. 11 in November 2011. The case was resolved out of court in October 2012. An undisclosed payment was made by Sabre to American Airlines, and it was approved by the U.S. bankruptcy court. As a result, the two companies signed a new distribution agreement (Reuters, 2012), and Sabre was no longer a defendant in American's federal lawsuit against Travelport and Orbitz Worldwide. The anti-trust lawsuit was settled with Travelport in March 2013 with a new distribution agreement, and the Orbitz suit was settled in April 2013.

American Airlines emerged from bankruptcy and merged with US Airways to become the world's largest airline. The parent company's name was changed from AMR Corp. to American Airlines Group Inc. Since the two companies could not reach a new agreement, American removed its fares from Orbitz in August 2014. American planned to withdraw US Airways flights on September 1. Customers who purchased tickets on Orbitz.com, cheaptickets.com, or ebookers.com could use them for travel, but any changes to the reservation had to be made by either American Airlines or US Airways reservations agents. This dispute was resolved on September 2, and American flights were restored. Corporate travelers who booked on Orbitz for Business, the corporate brand of Orbitz Worldwide, were not affected.

Travelport had divested a majority of its stake in Orbitz for $280 million, prior to its IPO in September 2014. Concerned about the growing dominance of Booking.com, Expedia acquired Travelocity from Sabre for $280 million in January 2015. In February 2015, Expedia acquired Orbitz Worldwide, a publicly traded company at that time, for $1.6 billion.

4.12   US Airways and "Monopoly Power"

US Airways brought an antitrust lawsuit against Sabre in 2011 in the United States District Court for the Southern District of New York. The lawsuit was filed under the Sherman Antitrust Act. US Airways alleged that Sabre used its "monopoly power" to impose a contract on US Airways, which included "full content and high fees". US Airways claimed that the provisions in the contract were not sustainable to stay competitive. If the contract had not been signed by the deadline, Sabre would have removed US Airways flights. Without access to corporate customers, the survival of US Airways would have been questionable.

In February 2013, a $11 billion all-stock deal was made between AMR Corp., the parent company of American Airlines and US Airways Group. The merger, initiated by US Airways CEO Doug Parker, gave control of the combined airline to the creditors of AMR Corp., which was bankrupt. As a result, Doug Parker became the CEO of the largest airline in the world. The merger was completed in April 2015 when the FAA granted a single operating certificate for both airlines, and the US Airways brand was phased out.

The jury determined that Sabre had violated antitrust laws by imposing unfavorable contract terms on US Airways, resulting in unfair restraint of trade. In December 2016, the Manhattan federal jury awarded $5.1 million in damages, which is tripled under federal antitrust law. The amount awarded was significantly less than the $72 million sought by American Airlines, representing US Airways. Additionally, the jury dismissed claims that Sabre conspired with its competitors to not compete with each other.

In 2019, a $15.3 million award was overturned by an appeals court. After 11 years of litigation, in May 2022, a federal jury in New York found Sabre guilty of exercising monopoly power in airline distribution, which confirmed the original claim made by US Airways in 2011. The victory for American Airlines did not come with financial gratification. The seven-person jury in the U.S. Southern District of New York awarded American only $1 as they could not prove that the Sabre-US Airways contract had restrained trade (Silk, 2022).

4.13 Merchandising

The retailing of ancillary products in the airline industry originated with the low-cost carriers (LCCs). In 1981, PeopleExpress, a low-cost carrier, introduced a novel approach of offering heavily discounted fares at the time of booking, while charging a fee of $3 for each checked bag. This strategy was a pioneering move aimed at generating additional revenue for the airline and to partially compensate for the deeply discounted fares.

The sale of ancillary services such as pre-reserved seats, checked bags, insurance, meals on board, etc. were introduced by the low-cost carriers (LCC) in the mid-1990s.

Most airlines, including both network carriers and LCCs, began charging for bags, overweight bags, and oversized bags twenty years later. Ryanair, an LCC, was one of the first airlines to implement this policy in 2006, charging travelers $10 for checked bags purchased online and $20 at the airport. In 2008, full-service carriers also started unbundling and charging for checked bags after experimenting with bundled fare families. Spirit Airlines received public criticism in 2010 for charging for carry-on bags, and other LCCs like Allegiant Airlines and Frontier Airlines followed suit. Additionally, other optional services related to bags such as priority bag drops, priority bag returns, and insurance were also introduced.

In 2007, Spirit Airlines adopted the ultra-low-cost carrier (ULCC) business model to provide basic air transportation at the lowest possible fare in the markets they served. To make the base fare as low as possible, Spirit unbundled the charge for checked bags and despite rising fuel costs, lowered the base fare to offset the unbundling of the product. In 2010, Spirit Airlines created an uproar when they started charging for carry-on bags that would not fit under the seat. To their credit, this charge did not apply to items such as medical equipment, baby strollers, and similar items. Allegiant Airlines and Frontier Airlines followed the Spirit model. LCCs specialized in charging a low base fare at the time of booking, and this was followed with the sale of extra services for an additional fee. An independent survey conducted by Leflein Associates in January 2006 (Alexander, 2006) showed that air travelers would pay for extra perks such as more frequent flyer miles, more overhead bin space, and the ability to sit in a child-free section of the aircraft.

Travelers did not respond well to initial attempts at merchandising in the airline industry, as they felt that features that were previously included in the ticket price were being taken away. American Airlines faced backlash when it introduced a fee for checked baggage in May 2008, as bags had previously been included in the ticket price. This was perceived by travelers as a takeaway, which is also known as the endowment effect in behavioral economics (Morello & Lopatko, 2012). The endowment effect suggests that changes framed as losses are more impactful than changes framed as gains. Baggage fees were introduced to offset rising fuel costs, but even though fuel costs have decreased, baggage fees have continued to increase over the years.

Airlines are increasingly prioritizing loyalty (Neff, 2017) and the customer experience as critical areas of investment to differentiate themselves from their competitors, safeguard, and grow their brand and the direct channel. With the adoption of NDC, airlines will have access to the customer data before they generate an offer consisting of the base fare for the itinerary and ancillary products as part of the bundle. This allows airlines to also focus on the indirect distribution channels such as brick-and-mortar TMCs and online travel agencies (OTAs) to actively promote their offers.

4.13.1   Branded Fare Families

In 2006, full-service network carriers Air New Zealand, Air Canada, and Qantas introduced merchandising with bundled fare products called *branded fares* in 2006. In the U.S. Frontier Airlines was the first in 2008. A branded fare is a bundle of popular ancillary products that are sold as an upsell during the sales process. It was a bold attempt to redesign the airline website to show availability by fare family bundles.

Product unbundling with the sale of à la carte ancillaries followed during the fuel crisis of 2008 and the recession of 2009. Airlines responded to the financial pressures by innovating their products and services to generate more revenue. This innovation began with airlines unbundling their base airfares and adopting a fee-based structure where customers only pay for the flight services they require. They introduced new fees for services that were previously included in the base airfare, such as charges for checked bags and seat assignments. Additionally, fees for existing services like exchanges and on-board pet checks were increased. These changes led to a significant increase in airline ancillary revenues. For example, as shown in Fig. 4.4, passenger baggage fees for U.S. airlines increased from $285,754,000 in 2004 to $3,395,471 in 2010 (BTS, 2012).

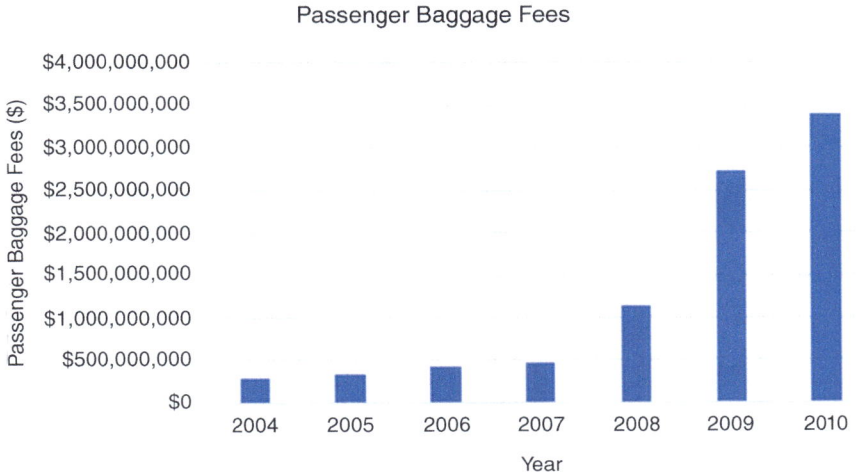

Fig. 4.4 Passenger Baggage Fees by Year. Source: Bureau of Transportation Statistics

To overcome the perception that an airline seat is a commodity, an approach is to introduce branded fare products that come with specific qualities. Airlines aim to provide transparency to customers by showcasing the value of their services and encourage the sale of additional services. Prior to branded fares, airline seats were always sold starting from the lowest eligible fare. However, with branded fare products, airlines can now sell seats at the middle or top range of the fare hierarchy, depending on a consumer's preference for the associated attributes of a branded fare product. The promotion of service differentiation with branded products and sale of ancillaries are collectively referred to as "merchandising" (Vinod & Moore, 2009). The introduction of branded fare products and unbundling has an impact on the entire travel value chain from shopping, booking, and fulfillment. Figure 4.5 illustrates the evolution of merchandising.

4.13.1.1   Branded Fares Example

Table 4.1 illustrates a hypothetical airline's branded fare products. There are four branded fare products called Basic, Standard, Flexible, and Premium with their associated attributes.

Branded fares pose several challenges that must be addressed from a pricing and revenue management perspective. They are described below.

4.13.1.2   Composition of the Branded Fare Products

The introduction of branded fare families begins with the design of the branded fare products. To ascertain the makeup of a branded fare offering, two assumptions must

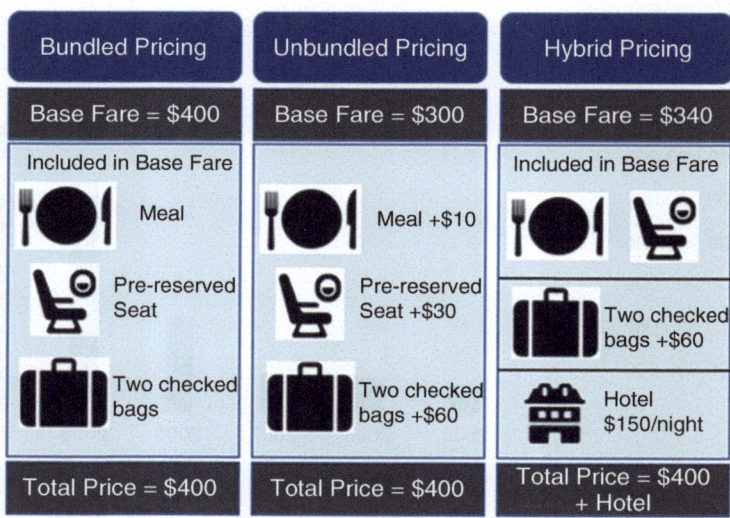

Fig. 4.5 The evolution of merchandising

Table 4.1 Branded fares example

Attribute	Basic	Standard	Flexible	Premium
Refundability	X No refund	X No refund	X No refund	✔ Yes
Seat Selection	$ For a fee	$ For a fee	X Standard	✔ Preferred
Checked Bags	$ For a fee	$ For a fee	✔ 1st bag free	✔ 2nd bag free
Changes	X No changes	$ For a fee	$ Fare difference	$ Fare difference
Loyalty Miles	25% accrual	50% accrual	100% accrual	125% accrual
Meals/Beverages	$ For a fee	$ For a fee	$ For a fee	✔ Included
Standby	X No	X No	✔ Select routes	✔ Yes
Fare	$450	$490	$550	$620

be established. First, customers will select a branded fare product that maximizes their satisfaction (utility). Second, it is critical to maintain consistent pricing, ensuring that the price of the higher-tier brand at a fare level is lower than that of the lower-tier brand at the next fare level (higher), considering the prices of additional items included in the higher-tier brand. This presents a difficult challenge as it necessitates identifying the optimal combination of ancillary services to include in each branded fare product to maximize profits.

4.13.1.3   Tariff Structure

Establishing the tariff structure according to the customer's willingness to pay (WTP) for attributes associated with a branded fare product is a challenge in the absence of experimentation. The tariff structure should maintain fare differentials between branded products to eliminate dilution and promote upsell.

4.13.1.4 Ancillary Services Pricing

Ancillary services, when they were introduced by network carriers, typically maintain consistent prices throughout the airline network. However, future pricing for ancillaries, such as pre-reserved seats, will vary depending on the market's competitive conditions and the length of the flight. There is ongoing research to quantify a customer's willingness to pay (WTP) for ancillary services to determine their value. Some established methods for this include multinomial choice analysis (Ben-Akiva & Lerman, 1985; Train, 2003; Balcombe et al., 2009), the van Westendorp price sensitivity model (van Westendorp, 1976; Hague, 2008), and conjoint analysis (Green et al., 2001; Hair et al., 1984), which assess the trade-offs between different combinations of price and product features. To provide empirical evidence of estimated valuations, experiments can be run based on stated preferences (Martin et al., 2008) to gauge traveler opinions on service attributes like comfort, food, ticket change fees, frequency, and reliability.

4.13.1.5 Discount Allocation Control

Discount allocations should rely on the value of the bundle, considering the projected ancillary revenues and the associated risk.

The industry relies on RBDs that are inadequate for branded fare families. Typically, an airline requires at least eleven fare levels. Hence, if an airline has three branded fares, it requires $11 \times 3 = 33$ booking classes, which is not feasible. To work with the constraint of 26 booking classes, airlines find innovative ways to assign booking classes to branded fare families, which will not guarantee that a given fare level is open or closed across all branded fares. Ensuring consistent availability across different branded fare products within the same fare level is a challenge.

NDC and continuous pricing (discussed in Chaps. 6 and 9) can address this problem since an airline can maintain fare levels for the lowest branded fare and derive the fares for all the other branded fare products based on a multiplicative factor of a fixed fare differential. Continuous pricing enables quoting a price for a branded fare between two successive fare levels based on supply and demand.

4.13.1.6 Branded Fares and Air Shopping

When branded fares were first introduced there was a fundamental problem with the air shopping algorithms which were designed to find the lowest economy fare and not the lowest fare by branded fare family. Air shopping did not recognize the higher valued branded fare families with their associated amenities.

When the shopping algorithms adapted to searching for fares by branded fare family, it was easier to implement in the airline consumer direct channel since every airline could uniquely define their branded fare families. The display and sale of branded fare families on a travel agency desktop is difficult to accomplish since the content FROM individual airlines are not homogenous. From the myriad of bundles created by airlines, the GDSs focus on the most relevant features across airlines to include and create a minimal set to display on the travel agency desktop.

4.13.1.7   Merchandising: The Onward Journey

Other network airlines followed, and the industry has never looked back. Airlines worldwide are selling ancillary air products such as checked bags, excess baggage, pre-reserved seats, meals on board, lounge access, priority check-in, premium seats, priority boarding, wireless Internet access, etc. to generate incremental revenues beyond the base fare. Branded fare families and the sale of à la carte ancillaries have generated billions of dollars for airlines in recent years. While the average base fare in the airline industry has declined by 0.9% per year over the past decade (IATA, 2018), ancillary sales have grown 40%.

The five largest U.S. carriers (American Airlines, Delta Air Lines, United Airlines, Southwest Airlines and Alaska Airlines) generated over $29 billion in ancillary sales in 2019. Global airline ancillary revenues in 2018 were $93 billion (Ideaworks and Cartrawler, 2018) and in 2019 prior to the COVID-19 pandemic, revenues exceeded $109.5 billion (Ideaworks and Cartrawler, 2019), which is a fivefold increase in ancillary revenues reported in 2010 of $22.6 billion. With the onset of the pandemic in 2020, Cartrawler projected a 47% decrease over 2019, to $58.2 billion dollars (Ideaworks and Cartrawler, 2020) and improved to $65.8 billion dollars worldwide in 2021 (Ideaworks and Cartrawler, 2021). After the decline during the pandemic, the total ancillary revenues were forecast to reach $117.9 billion in 2023 (Ideaworks and Cartrawler, 2023). However, as a percentage of global revenue, ancillaries continued to grow to 13.4% in 2020 during the pandemic compared to 12.2% in 2019. According to Cartrawler, in 2010, the ancillary revenue per passenger was $8.42, $12.13 in 2012, $28.97 in 2019, $31.39 in 2020, $33.45 in 2021, $42.11 in 2022, and $37.59 in 2023.

4.13.2   Ancillary Services and Fees Impact on Fare Management and Revenue Management

The airline practice of charging more for services that historically were included in the price of a ticket has created much customer angst and dissatisfaction (Reed, 2019). Airline executives view this revenue stream as a requirement for differentiation from the competition, survivability, and profitability. Ancillary revenue streams are here to stay and will continue to increase in the next decade, but adjustments may be made based on prevailing market conditions (Gottfredson, 2007; Straus, 2008). Examples of ancillary revenues are fuel and insurance surcharges, referred to as YQ and YR, which are settled during the airline ticket invoicing process. During the COVID-19 pandemic of 2020, U.S. majors stopped charging change fees due to the downturn in travel and the economic climate. Change fees were a large component of ancillary sales.

Beyond the sale of ancillary products, the last frontier for airlines, defined by the level of granularity in inventory control, is to price the air product at the seat level, either by seat type or the individual seat. The fundamental question is: *how much will customers be willing to pay for an aisle seat, a window seat, or an exit row seat?*

Like pricing and inventory control of individual seats in the airline industry, attribute-based room pricing and inventory control represents the last frontier for the lodging industry, with the goal of controlling inventory at an individual room level (Sorrells, 2018a; Vinod, 2019, 2022a; Vinod & Hobt, 2021). With attribute-based room pricing, a room with a collection of features can be confirmed at the time of booking. However, the actual assignment of a specific room is deferred until the check-in date to ensure that room inventory fragmentation is resolved to the extent possible (Vinod, 2022a), and rooms are optimally occupied.

Ancillary services introduce a new dimension to the fare management process. On a broader scale, ancillary revenues can be classified as ticket transaction fees, air extras, and travel extras. Ticket transaction fees pass on the merchant fee imposed by a credit card to the customer. Another example is channel fees for travel agents. These additional fees typically do not appear on the airline ticket but on a new passenger receipt which totals the airfare and additional fees.

Air extras are ancillary services that are consumed on board the aircraft. Examples are Internet service, pre-paid seat selection, meals, in-flight on demand entertainment, etc. This requires the issuance of a separate electronic miscellaneous document (EMD) and messaging infrastructure to support it.

Travel extras are travel related services that are consumed either before or after the flight. Examples are ground transportation, home to airport baggage forwarding services, lounge access fees, destination activities, etc.

With the renewed focus on ancillary products as a revenue stream that can augment the bottom line, airlines require the capability to sell, distribute and settle ancillary services across all channels of distribution. This implies that a capability is required to set the prices for ancillary services, distribute products with differentiated content, and conduct financial settlement across all channels. This has significant impacts on the capabilities of current airline reservations systems, GDSs, and revenue accounting.

ATPCO in conjunction with IATA provides an optional service fee solution. Table 4.2 illustrates the new types of fees being introduced.

The pricing for ancillary services will shift from a flat fee across the system to a market-based approach in the future. This is because customers perceive that a pre-reserved seat is more valuable on a long-haul flight than a short haul flight. However, without sufficient data at varying price points, determining a customer's willingness to pay for these services and establishing market prices will rely on survey data gathered during the booking process and while on the flight.

Table 4.2 Fee types

Fee type	Description
YQ/YR (S1, S2 records)	Carrier imposed fees. A standardized, automated collection, distribution, and pricing method that provides marketing carriers (carriers that appear on the flight coupon) the ability to control and collect fees at the sector (coupon), at the portion of travel (multiple sectors), or on the journey. ATPCO's application handles fuel, insurance, and carrier-imposed miscellaneous fees. • Fuel surcharges filed as YQF or YRF. • Insurance surcharges filed as YQI or YRI. • Not a tax, but a validating carrier specific fee. • Included in the total amount of the ticket, therefore no changes to ticketing or reporting are required. • Not applicable for interlines. • Not commissionable. • Filed by over 200 carriers worldwide.
OB (S4 record)	IATA defined code used for ticketing fees (optional, validating carrier only, not interline able). • IATA authorized the use of the OB tax code. • Distribution channel fees. • Carrier fees can be specified by call center, city/airport ticket office, department/station id. • Form of payment fees—One fee for all cards or a card specific fee based on BIN number. • Always part of the ticket transaction. • Calculated for the validating carrier like YQ/YR fees. • Available since 1Q 2007, but adoption has been limited pending resolution of several open issues.
OC (S5, S7 records)	IATA defined code used for fare related optional service or rule-based service fees (optional, validating carrier only, not applicable for interlining). • Passenger choices (seats, meals, movies, etc.). • May be linked to flights, not tickets. • The fees are *operating carrier based* not validating carrier as with YQ/YR and OB. • Issued as a separate EMD and can be issued simultaneously or after the ticket has been issued. • Involves inventory controls and messaging. Will involve inventory control and new messaging standards for pre-reserved seats, meals, headsets, etc. • Not applicable for interlines. Baggage fees • Excess baggage charges. • Overweight baggage fees.
OA	IATA defined code used for booking fees (optional, validating carrier only, not applicable for interlines.

4.13.2.1 Branded Fares Record (S-8)

ATPCO introduced the branded fares record, called the S-8 record, to standardize brand definitions and ancillary products offered by airlines. Specifically, this record provides the brand name definition, a mapping of RBDs, and/or fare basis codes to the airline branded fare product. ATPCO Table 166 provides a cross reference of

Optional Services to the branded fare product. ATPCO Table 189 serves as the fare identification table which cross references fares associated with a brand.

4.14   The Beginning of the End of Full Content Agreements and Most Favored Nation Clauses

The introduction of ancillary services by the network airlines that participated in the GDS was the first step toward the dissolution of the full content agreements, which was pivotal for the GDS to control the relationship with airlines. While the prevailing full content agreements were specific to the air fare, airlines were opposed to including airline ancillary products and fees as part of the full content agreements during contract negotiations. These were the first signs of a fractured relationship between airlines and GDSs during contract negotiations. Several leading network carriers even refused to file their ancillary products and services with fare aggregator ATPCO and demanded that if a GDS wanted access to ancillary products and services, the GDSs had to interface with the airline reservations systems with a direct connect link.

4.15   Total Itinerary Pricing with Ancillaries

GDSs are equipped to calculate the total cost of airfare and additional services, considering a traveler's preferences and frequent flyer status with the airline. This includes the base fare and ancillary services, such as checked baggage, which may be waived for travelers with elite status on a particular carrier. Other selected services, like lounge access, seat selection, Wi-Fi, and meals, will also be reflected in the total price based on frequent flyer status. Negotiated discounts or waivers for ancillary services may also be available for corporate customers. Comparable total itinerary pricing capabilities are also available on individual airline websites.

4.16   The GDS Travel Agency Desktop

The GDS travel agency desktop holds great significance as it serves as the customer facing primary product of the GDS. Despite common belief, the GDS is not just a singular product that aids in the shopping, booking, and ticketing of a passenger's itinerary. In fact, GDSs also provide web service APIs to their customers. As an illustration, an OTA may incorporate a GDS's shopping service into their own user interface workflow.

Raytheon, ICOT, Honeywell, and Videcom manufactured the original 'dumb" terminals that were deployed at travel agencies. In the late 1980s with the introduction of the personal computer GDSs transitioned from the traditional monochrome dumb "green screen" terminals to Windows based smart workstations with Intel-based 386 processors on a Local Area Network (LAN). The 386 processors were soon replaced with the more powerful Intel Pentium processors. While the backend GDS processing had not changed, the new desktops allowed the GDSs to provide user friendly, template-driven screens to travel agents to book travel at a faster rate, resulting in significant productivity gains.

Each GDS has its own travel agency desktop, which supports both the standard green screen and the graphical user interface. Sabre, Amadeus, and Travelport support Sabre Red Workspace (SRW), Amadeus Travel Platform (or Selling Platform), and Smartpoint Agency Desktop, respectively. SRW was rebranded as Sabre Red 360 in 2019. These are API-enabled and serve as access points for traditional GDS content (air shopping, hotels, etc.) and NDC content (air only). Sabre introduced Sabre Red Launchpad in March 2024, which serves as an extension of Sabre Red 360 for independent travel consultants.

According to industry data and feedback from travel agencies, Sabre's desktop product has consistently been rated as the most popular and considered the best for travel agency productivity. This implies that on high-volume booking days, agents can process more bookings using Sabre Red 360 desktop than with the Amadeus or Travelport versions of their desktops.

GDSs provide travel agencies with schedules, fares, and availability for both corporate and leisure travel. Sabre utilized the open source Eclipse Rich Client Platform in 2010 to introduce the industry's first agent workspace, which merges Eclipse's underlying capabilities with Sabre's technology. This allows users to create a custom work environment tailored to their business needs. Sabre also launched the Sabre Red App Centre in 2011, the world's first B2B app marketplace for the travel industry. This lets Sabre connected travel agencies access developers from all over the world to extend their agency workflow capabilities and create a custom workspace for their business needs.

Travelport bears the burden of maintaining and supporting three separate GDSs—Worldspan, Galileo International, and Apollo. In 2021, Travelport initiated Travelport Plus, to converge on the Galileo International GDS based on the strength of the agency community using the product. By the end of 2022, Travelport had migrated over 90% of agencies outside of North America to Galileo International. In North America, migration was slower with 40% of agencies converted to the common platform. The challenge for migration is addressing functional gaps in workflows, training to address differences in command line entries, and migrating the long tail of smaller agencies. The consolidation to Galileo International under the Travelport+ initiative leverages Smartpoint as the agency desktop. According to Travelport, all major agencies have already been converted to the Galileo International GDS in 2022. The smaller travel agencies that represent the long tail are expected to be converted in 2023 and 2024. When it is completed, all the agency subscribers of Travelport will operate on the Galileo 1G designator. Travelport+ also has airline retailing features such as the Content Curation Layer (CCL) to provide faster search responses and pertinent content such as fare comparison, offers, ancillaries, and personalization to travel agents.

4.17 IATA 2011 Initiative

In 2011, the IATA Board of Governors endorsed an initiative to facilitate the development of a New Distribution Capability. The development of the standards was the responsibility of the IATA Passenger Services Conference (PSC). IATA has historically played a role in standardizing communication codes and protocols for interline passenger communications.

Under governance of the PSC, the Passenger Distribution Group (PDG) was charged with developing the NDC vision to "increase competition, stimulate innovation, reduce distribution costs, and improve customer choice." While it may have been an intent, it was never stated that a goal was to break up the GDS oligopoly. The PDG consisted of 11 airlines: Air Canada, Alitalia, United Airlines Delta Air Lines, Air France/KLM, British Airways, Lufthansa, Swiss International Airlines, Korean Air, Cathay Pacific Airways, and Singapore Airlines. Interestingly, none of the GDSs, which excel in the development and deployment of connectivity products, were invited to be a part of the working team to paint a vision for NDC.

Chapter 5
Scale Matters: GDS Air Shopping

5.1 What Is Air Shopping?

When a customer plans a trip to a destination, they are interested in viewing the available options before booking travel. Shopping parameters in the request are the origin and destination, the departure date, return date (if applicable), preferred departure and return times (if applicable), and the number in party. This is called a shopping request. The shopping algorithm returns a set of priced itineraries from which the customer can select and book an itinerary. There are many types of trips such as one-way trip, round-trip, or multi-destination.

Air shopping is a complex problem, and it involves airline schedules and fares, itinerary selection, and pricing of the itinerary subject to booking class availability. It is the single largest product investment made by the Global Distribution Systems (GDSs). Enhancements, maintenance, and support costs are a multimillion dollar investment annually. Proprietary algorithms and patents are investments GDSs make for shopping efficacy to display a set of priced itineraries that maximize conversion rates (Vinod, 2015b; Vinod, 2021f). This is a balance of the lowest priced itineraries and itineraries that are convenient for the business traveler.

5.2 Historical Perspective

The largest application and infrastructure supported by a GDS is air shopping, which provides a range of priced itineraries for selection by a travel agent or end consumer (OTA customers) to make a booking. Interestingly, this application did not exist when GDSs were first introduced in 1976. At that time, travel agents would submit a command line entry request for availability of outbound flights specifying the city pair, date, and approximate departure time. The GDS would then display the booking classes available for sale, and the agent would select an itinerary by

B. Vinod, *Mastering the Travel Intermediaries*, Management for Professionals, https://doi.org/10.1007/978-3-031-51524-8_5

choosing a line from the city pair availability response. This process was repeated for the return flight. The agent would then submit a pricing entry to determine the total fare for the itinerary, which could be further refined by submitting a "new class pricing entry". This determined the lowest fare for the itinerary if a different combination of open booking classes on the outbound and inbound flights produced a lower fare subject to fare combinability rules. At that time, there was no mechanism to automatically search for alternate itineraries that may be more desirable or produce a lower price.

Basic air shopping features were not available on GDSs until the mid-1980s. Sabre took the lead in 1984 with the introduction of Bargain Finder^{SM}, the first automated low fare search capability that provided nine itineraries for a single shopping request. These shopping algorithms required travel agents to book an itinerary before displaying new prices and options. The entry used for this was called will price new itineraries (WPNI), and it quoted the lowest available fare for a booked itinerary, along with alternative itineraries with their associated fares, which could be with a different carrier or route. WPNI is an agency product.

Sabre's earlier versions of city-pair availability and Bargain Finder^{SM} lacked the capability to construct connecting itineraries dynamically. Instead, the connections were pre-built based on airline schedules and referred to as Sabre Sales Guide (SSG). The pre-built connections were refreshed daily to reflect schedule changes. The shopping algorithm was rudimentary, relying on simple heuristics to provide shopping responses. The shopping request included departure date for a one-way trip, or departure and return dates for a round trip. With the introduction of Bargain Finder^{SM}, all GDSs developed basic air shopping capabilities in the 1980s. For instance, Apollo/Galileo's flight search product was referred to as Best Buy Quote while Worldspan's was named Power Pricing (Worldspan Power Shopper®). Until the early 2000s, Sabre maintained a Fare Pricing Complex (FPC) running on TPF (IBM's transaction processing facility), which used expensive mainframes to support shopping and pricing.

Prior to the introduction of EDIFACT in 1988, the numerical availability associated with each booking class in the city pair availability displays and air shopping was reliant on teletype. With the implementation of EDIFACT and subject to an airline's level of participation in connectivity, the availability displayed could be based on the true, last seat availability retrieved from the airline's reservation system. This process is facilitated by availability request (PAOREQ) and availability response (PAORES) transactions. Additionally, EDIFACT allows for instantaneous confirmation of sell transactions, referred to as secure sell transactions, as availability is checked in the airline reservation system before the sale is finalized using sell request (ITAREQ) and sell response (ITARES) transactions.

Prior to the arrival of online travel agencies (OTAs) in 1996, shopping algorithms required a travel agent to book an itinerary before new itineraries were priced and displayed. In 1993, Bargain Finder Plus^{SM} was introduced as an enhanced version of the shopping algorithm that returned nineteen itineraries, but the itineraries displayed lacked adequate diversity. With the arrival of OTAs, demand for air shopping and related services increased dramatically. OTAs required a shopping service that could return a larger number of itineraries (200–1000+) for each shopping request without the need to book an itinerary first. Fare-led algorithms were augmented with schedule-led algorithms to ensure diversity of itineraries based on standard shopping parameters. A bundled sell entry was also created to provide the capability to sell multiple air segments as a single transaction. With the modifications to the shopping and selling model, the process begins with a shopping request which returns a list of possible itineraries and itinerary selection is followed with a bundled sell transaction to complete the booking.

To meet the growing demands from OTAs, Sabre's Chief Technology Officer, R. Craig Murphy, initiated an open systems version of shopping and pricing called the Air Travel Search Engine (ATSE) in 2001, which was later renamed to Air Travel Shopping Engine. ATSE was deployed in 2004. Dynamic Schedules (DSS), a component of ATSE, replaced SSG with a dynamic schedule builder. DSS did not rely on the pre-built static list of non-stop, single, and double-connect flights. Instead, they were created dynamically with no limits imposed on the multi-connect flights. This increased the sales opportunities by offering an expanded number of flights in the city pair availability and schedule displays. ATSE went through significant enhancements with the Estimated Seat Value (ESV) algorithm for Travelocity in 2008 (Benzinger et al., 2008) followed by the high-performance shopping engine (Steeb & Sohn, 2006), where data are organized by VTCR-Fare Class (Vendor, Tariff, Carrier, Rules, Fare Class).

ITA Software (de Marcken, 2003) launched QPX as a competitor to GDS air shopping and was deployed when Orbitz was launched and is used today by several airline websites and Google Flights. There is a key difference between ITA software and Google Flights. ITA Software does not rely on cached shopping responses, and all computations are made in real time. Google Flights uses a combination of ITA shopping responses supplemented with cached results for speed and to support inspirational shopping capabilities.

5.3   Aspects of Air Shopping Complexity

The primary inputs to the shopping algorithm for computing a shopping response are airline schedules, air fares, availability by booking class, surcharges, taxes, and exchange rates. Factors impacting air shopping are the volume of schedules and available fares, real-time volatility in fares, and booking class availability and the complexity of processing schedules and fares quickly to create a concise set of options that maximizes conversion rates.

5.3.1   The Search Space

To understand the enormity of the search space, consider the New York (JFK) to London Heathrow (LHR) market.

Figure 5.1 illustrates the search space that down-selects a few itineraries for display to a travel agent or a customer. The display typically has rich diversity from which an itinerary is selected, and travel is booked.

As observed in Fig. 5.1, there are 5 billion potential fares and routing possibilities to fly from JFK to LHR on a single day when airlines, cabin types, schedules, fares, connections, and interlines are considered. Of these options, over 3 billion might be available for purchase, while over 6 million represent reasonable price points and scheduling options. Fast sophisticated shopping, algorithms are required to minimize server response times and return a few itineraries (e.g., 50, 100, 200, 500, etc.) depending on the default travel agency parameter setting for itineraries to be displayed. OTAs typically display more itineraries than travel agency subscribers, usually between 500 and 1500.

The collections of itineraries returned for a shopping request depend on the market, carriers that serve the market, and the shopping parameter settings. Consider a collection of itineraries returned by the air shopping complex when a shopping request is made. Figure 5.2 illustrates the itineraries plotted with the price of the itinerary on the x-axis and the elapsed time on the y-axis. Ideally, a shopping response should have itineraries that belong to three out of the four quadrants, except for the top right quadrant. Sometimes, even the top right quadrant is important when diversity of air carriers is a parameter to generate the shopping response.

Potential ways to fly from New York to London Heathrow in a year	4,932,102,241,635
Itineraries for a specific day of travel	5,293,430, 799
Itineraries available for purchase	3,225,937,750
Itineraries likely to produce the best value	6,113,387
The best itineraries for display based on preferences and diversity[1]	200
The selected itinerary	1

1/Parameter setting in Air Shopping determines number of itineraries to display

Fig. 5.1 The search space for air shopping

Fig. 5.2 Types of itineraries in a shopping response

Fig. 5.3 Growth in air shopping requests

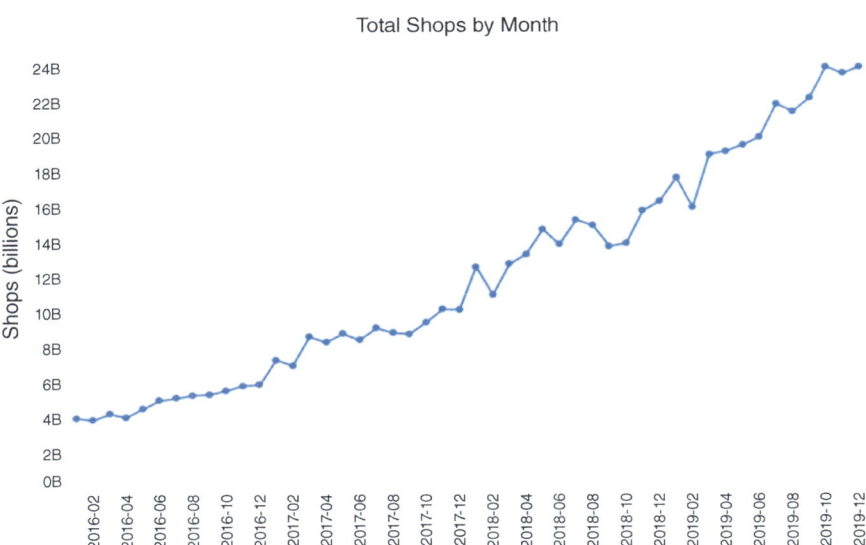

5.3.2 Growth in Shopping Requests

Another aspect is the growth in shopping volumes. Over the past decade until the pandemic of 2020, shopping requests have outpaced growth in bookings. Results shown are before the pandemic of 2020. Figure 5.3 illustrates the growth in shopping

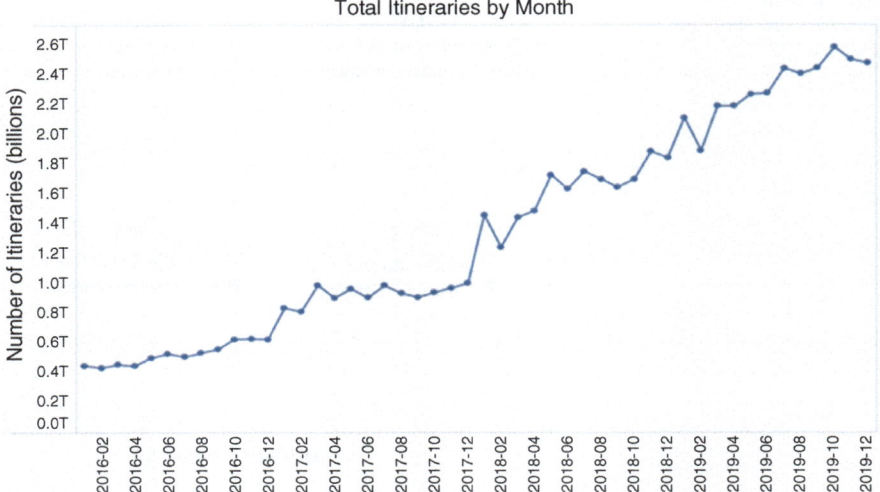

Fig. 5.4   Total itineraries returned by month

volumes at Sabre for the travel agency channel that was presented at NDTA-USTRANSCOM in Washington, D.C. (Vinod, 2020a).

The total shopping requests for 2019 exceeded 249 billion. The growth in shopping requests is primarily attributed to the OTAs who use the GDS provided web service APIs to submit shopping requests and receive shopping responses. Traditional brick-and-mortar travel agencies are far more efficient, and the look-to-book ratios are lower than the OTAs by an order of magnitude. Further, peak transactions per second for the agency channel surpassed 25,000 transactions per second (TPS).

Figure 5.4 shows a related statistic which is itineraries returned during air shopping which exceeded 2.6 trillion in October 2019 and remained at 2.5 trillion in Q4 2019 (Vinod, 2020a).

5.3.3   Volatility

Air fares, schedules, and seat availability are constantly changing. However, schedules have the lowest rate of change, followed by fares and seat availability. New fare transmissions are accepted hourly, schedules go through planned and ad hoc changes and seat availability by booking class and point of commencement changes dynamically based on bookings, cancellations, and airline revenue management flight and network re-optimization activity that updated inventory controls which impacts booking class availability. A shopping solution must absorb the changes in fares, schedules, and seat availability from airlines and their partners constantly throughout the day.

5.3.4 Response Times

Another layer of complexity is the response times for a shopping request—the response time is the time it takes to process the request on a server and the associated network latency. Server compute times are impacted by the inter-dependencies between schedules, fares, and seat availability. Network latency can be minimized with a distributed availability proxy solution (Vinod, 2007). A fast response time for each request is critical to avoid the risk of abandonment and to improve the conversion of shoppers to bookers. Internal studies have shown that a 33% reduction in response times leads to a 2.3% increase in conversion rates.[1]

5.3.5 Simple Versus Complex Shopping Requests

Shopping requests can be simple or complex. A simple shopping request is for a specific departure date and return date. A complex shopping request consumes more processing power and time to generate a response. Examples of complex shops from GDS perspective are as follows:

1. Multi-passenger requests
2. Single branded fare
3. Multiple branded fares
4. Alternate pseudo city codes (PCC) that an agency can access
5. Multi-ticket display
6. Alternate airports
7. Net fares
8. Private fares
9. Alternate origins
10. Alternate destinations
11. Alternate dates, typically a 7 × 7 calendar matrix

In addition, there is a plethora of corporate fares, which are essentially private fares, negotiated between an airline and a corporation, that are managed as a Category 25 Fare By Rule (FBR).

---

[1] Internal study at Travelocity, 2011.

5.3.6   Growth in Air Fares

There are two schools of thought on the growth of air fares in the future. With the adoption of NDC, airlines do not have to publish their fares to ATPCO, and they have an opportunity to simplify the complex fare rules, general rules, footnotes, and routing restrictions that exist today.

There is also the view that with NDC, the volume of air fares will increase with the introduction of date specific fares, time of day specific fares, and routing specific fares, increasing the complexity in data storage and processing costs for the airline (and not the GDS).

5.3.7   IATA NDC and the Computational Burden on Airlines

The adoption of IATA's new distribution capability (NDC) will significantly increase the computational burden on shopping algorithms managed by airlines. There are two reasons. First, airlines today only provide a shopping service to the direct channels (website and call centers). With NDC adoption they will have to respond to requests from travel agents since the GDS is no longer shopping and pricing itineraries in the GDS environment.

5.3.8   Doug Laney's Definition of Big Data

Volume, Velocity, and *Variety* are the three dimensions of Big Data as defined by the MetaGroup's Doug Laney (now part of Gartner) in a MetaGroup Research publication (Laney, 2001).

Volume refers to the amount of data which has been growing at an increasing rate. Travel shopping volumes have grown at an increasing rate over the past decade. For example, from consumers across the globe, Sabre processed around 249 billion air shopping queries in 2019 and over 25,000 shopping queries per second at peak times (Vinod, 2020a).

Velocity refers to the rate at which data is gathered and handled. Revenue management systems have traditionally operated in batches, processing data collected from the reservations system on a nightly basis since their establishment in the 1980s. However, in order to enhance the value derived from the revenue management process, these systems have evolved to handle streaming data, such as booking and inventory alert messages, in order to respond to real time changes in the market. This ensures that inventory controls are based on the most current information available. Another example of leveraging streaming data is dynamic intervention, where OTAs utilize it to promote offers based on the frequency of a customer's visits to their website.

Variety refers to the diverse forms of unstructured data, including text, audio, video, sensor data, documents, social media posts, web pages, geo-spatial data from satellites, as well as structured data that necessitates specialized techniques for processing. Technologies aimed at travelers, which encompass smartphone apps and wearable computing devices, have the potential to inundate the realm with vast quantities of comprehensive data sets comprising personalized information. Unlike unstructured data that is challenging to categorize, structured data can be arranged within a database and regularly stored and analyzed.

Storing and processing GDS shopping data for competitive advantage is discussed in Sect. 5.11.

5.4   Air Shopping Algorithms

Air shopping algorithms respond to each travel shopping request for travel options in real time. In the indirect space, to support brick-and-mortar travel agents and OTAs, the GDS shopping algorithms consider an enormous search space of fares, flights, and seat availability from over 400 carriers to find *both* low-priced and high-quality itineraries for each shopping request in a matter of seconds. There are two types of shopping algorithms—*fare-led* and *schedule-led* algorithms to respond to a shopping request from an end consumer or a travel agent (Vinod, 2015b, 2021f).

The search space for the fare-led algorithm is explored through fares. A key component of the fare-led approach is the construction of a lower bound table that captures the lowest fare in a market based on carriers and fare rules such as advance purchase, minimum stay requirements, and fare-by-rule for private fares. Once a cheap and valid fare combination is identified, flights are constructed for this fare combination and the availability of the booking classes is checked. Various techniques are used to reduce the response time during sold-out periods. As the algorithm is fare-led, the algorithm attempts to find the lowest fares in a market regardless of the quality of service. The fare-led algorithm, but its very nature, requests seat availability incrementally as it navigates through the search space and hence is more demanding than a schedule led algorithm where the availability requests are known in advance.

The schedule-led algorithm prioritizes and directs the search to determine desirable solutions. The method considers many outbound and inbound flights as input, for example a 1000 × 1000 matrix of outbound and inbound flights, which is a million itineraries. The outbound and inbound flights are selected intelligently based on several factors such as desirability and availability (usually AVS data), The flight options selected are subsequently priced. The algorithm explores a large search space of available flight options to determine a finite set of desirable itineraries based on the following criteria:

1. Low fare efficacy. This is the ability to find the *lowest fare* in the market regardless of how desirable the itinerary is.

2. Provide a rich *diversity of itineraries* that include online and interline non-stop flights and connecting flights. There must also be a minimum number of inbound itineraries for a selected outbound itinerary. This is required to maximize conversion rates, measured as the ratio of the number of bookings to shops. To achieve this goal, a rules-based approach can be deployed. Alternately, a calibrated consumer choice model that explicitly models a consumer's preferences such as schedule attributes (e.g., non-stop, departure time, elapsed time, carrier, etc.) and fare can be deployed. Calibration of the choice model is based on historical consumer behaviors including both shopping and booking, which facilitates the selection of the desirable itineraries based on schedule attributes and the applicable fare. This approach toward itinerary diversity enables the search for itineraries with the highest overall attractiveness, measured by customer utility and not just the lowest fare.
3. The itineraries displayed must be bookable. This is achieved with accurate booking class availability that ensures that there are no price jumps or sold out (UC, unable to confirm at sell) conditions. Accuracy of booking class availability is based on querying the airline's reservations inventory control system in real time before the priced itineraries are displayed.

The benefits of the schedule-led algorithm include fast predictable response time during sold-out periods, a price-service trade-off, and diversified set of itineraries.

5.5   Complexity of Airline Shopping

The complexity of airline shopping is directly attributed to airline pricing, which has grown in complexity following airline deregulation in 1978. The complexity stems from the growth in airline fares and itinerary pricing.

5.5.1   Airline Fare Management Process

Prior to airline deregulation, airfares were printed on the back of timetables. The airline fare management process for the past several decades has remained relatively unchanged except for the growing complexity in fares and fare rules (Vinod, 2010). This is illustrated in Fig. 5.5.

Since deregulation, an airline's main goal in controlling market pricing, has been to match a competitor's fare (Kretsch, 1995; Crandall, 1998). This is because adding an additional passenger has a relatively low marginal cost in a business that relies heavily on capital and assets. Hence, the focus is on protecting and retaining market share and paying down fixed costs. Consequently, an airline's fare actions are mostly reactive and not based on the quality of the schedule (the Quality of Service Index or QSI) to distribute a Quality of Service Adjusted Fare (QSAF). This behavior is

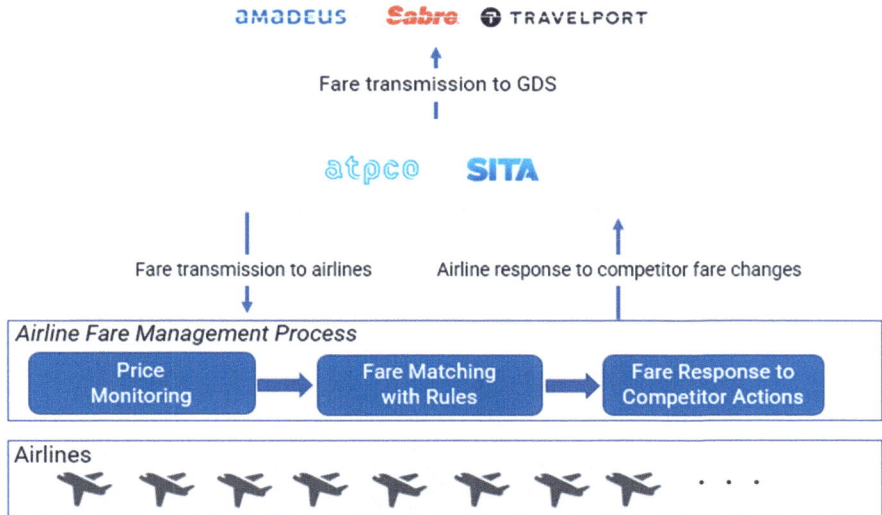

Fig. 5.5 Airline fare management process

consistent with the desire of airlines to maintain market share. Research dating back to the nineteenth century confirms that price matching in a competitive marketplace can be an optimal strategy (Curry, 1995). Advanced pricing decision support (Ratliff & Vinod, 2016) applications that set fares in a market based on quality of service and management objectives such as improve yields or market share have never gained traction in the airline industry for the fare management process.

Airlines file fares containing various purchase or travel restrictions with fare distributors such as Airline Tariff Publishing Company (ATPCO) and SITA. Today, 87% of the world's fare filings are made through ATPCO, and the fares database has more than 300 million fares that yields 14 million fare/rule changes daily.[2] It is owned by many of the major airlines and provides data collection and distribution clearing house services for more than 400 carriers worldwide. The fare aggregators consolidate fares received from various carriers and broadcast the fares to all participating airlines and subscribers such as the GDSs and third-party retailers. Hence, before the launch of IATA NDC, fare transparency was guaranteed.

There are six dimensions to a fare, as described in Table 5.1:

These six dimensions are physically separate. In other words, the fare basis code does not define a rule, but it refers to a rule that is defined elsewhere. The same is also true with footnotes, general rules, routings, and RBD validation.

---

[2] ATPCO website, 2023, https://www.atpco.net/about#:~:text=Today%2C%20our%20database%20holds%20more,million%20fare%2Frule%20changes%20daily

Table 5.1 Dimensions to a fare

Dimensions	Description
Fare record	Consists of the origin and destination, the fare amount, fare basis code, footnote, and the routing number. There are eight different record types that serve different functions in the processing of a fare.
Rules (fare categories)	Restrictions applicable to a fare.
Footnotes	Additional rule restrictions applicable to a fare. Commonly used to include travel restrictions or ticketing restrictions.
General rules	Additional rule restrictions that may be applicable to a fare.
Routing	The origin and destination and allowable cities in between
RBD validation	Validate availability of booking class (reservation booking designator)

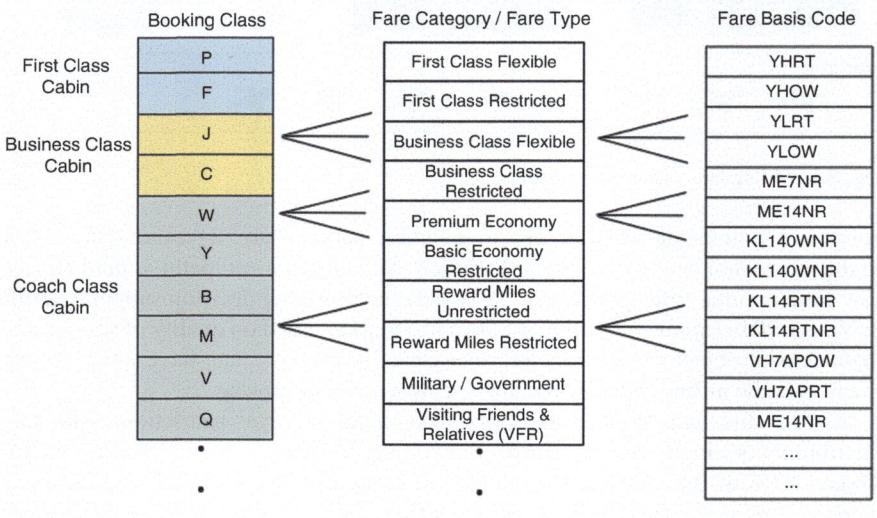

Fig. 5.6 Relationship between booking classes, fare categories, and fare basis codes

5.5.2 Booking Class, Fare Category and Fare Basis Code

Figure 5.6 illustrates the relationship between fare basis codes, fare categories (also known as fare types or fare groups), and booking classes.

Booking classes, also known as reservation booking designators, RBDs, are identified by single-letter codes (e.g., Y, B) and represent unique segments based on fare rules. Examples of segments are unrestricted full Y, restricted discount, promotional discount, visiting friends and relatives (VFR), etc. These booking class codes are not standard across airlines. Many airlines use all letters of the alphabet, except for I and O which are not distributed to intermediaries, to facilitate finer inventory controls. They are used by revenue management to forecast demand, set

Table 5.2 Interpreting the fare basis code

Examples of codes used in fare basis code	Interpretation
First letter of the fare basis code	Usually the RBD (booking class)
E	Excursion fare
OW	One-way fare
RT	Round trip fare
W	Weekend fare
X	Weekday fare
H	High season
L	Low season
CH	Child fare
IN	Infant fare
NR	Non-refundable
Numerals	Maximum stay, minimum stay restrictions
ID/AD	Industry discount/travel agency staff discount

inventory controls on the reservations system, and distribute airline seat availability to the GDSs. For example, if booking class B has a B5 display in city pair availability on a GDS it means that five seats can be booked in B booking class. Similarly, if the display is B0, it implies that B is not available.

The fare categories, also known as fare types, to avoid confusion with the ATPCO fare rule categories, define the market segments. The fare category is an internal classification by an airline to identify and group fares into predefined market segments. Fare basis codes, and booking classes are external classifications that are published to travel agencies, tour operators, and consumers.

The fare basis codes represent the fares with its associated rules, restrictions, routings, and footnotes represent the selling products that are ultimately distributed by an airline to the target channels. They serve as a unique identifier that is used to link fare rules and fare amounts. The first letter of a fare basis code is almost always the RBD. The fare basis is the code that appears on the ticket. It can include letters, numbers and up to two slashes (/). It is a compilation of the fare class or ticketing code and one or two ticketing designators.

A fare basis code is up to 15 characters long. It can include letters, numbers and up to two slashes, and is a shorthand method for describing pricing rules. It is a compilation of the fare class or ticketing code and one or two ticketing designators. The fare basis code is not published, it is constructed by the pricing process and is shown on the ticket. A full-service carrier (FSC) may have between 100,000 and 150,000 fare basis codes and many of them may be used for multiple markets.

Table 5.2 illustrates a sample of characters and numerals used in the definition of the fare basic code for a fare. In this table, it is assumed that the first character of the fare basis code is the RBD, and other codes follow.

For example, consider the fare basis code KL14RTNR. The letter K refers to the class of service, the RBD, L stands for low season, and it is a 14 day round trip refundable fare.

A many-to-one relationship exists between fare basis codes and fare categories. While ideally each fare category should map to a unique booking class, this is not the case, and a many-to-one relationship exists between fare categories and booking classes. For example, in the coach cabin, both Premium Economy and unrestricted frequent flyer redemptions may be mapped to the same booking class Y. Booking classes play an important role in both online fares and interline agreements. Incorrect mapping of a fare to a booking class or using the wrong designator on an interline agreement can result in loss of revenue. A booking class is a required element for auto pricing of passenger itineraries by GDSs. Booking classes are distributed globally through fare aggregators ATPCO and SITA. The airline reservations systems distribute booking class availability to the GDSs and third-party retailers.

Fare categories and fare basis codes are in the domain of airline pricing. Figure 5.6 illustrates the relationship between fare basis codes, fare categories (types), and booking classes.

5.5.3   Fare Classes and Booking Classes

A fare class specifies the rules of an airline's fare, specified in one to eight characters. Every fare has a fare class code which is used for pricing. Revenue management analysts incorrectly refer to booking classes (RBDs) and fare classes interchangeably, which should be avoided.

5.5.4   Classification of Fare Products

In airline pricing, there are three distinct fare products. They are public fares, private fares, and web fares.

5.5.4.1   Public Fares

Airlines create public fares and distribute them to fare distribution vendors who make them available globally through GDSs and other subscribers. All customer segments can access public fares, also known as published fares. However, the term "published fares" can be misleading because all fares, whether public or private, are published for access to certain entities with permission. Gross fares or IATA fares are other names for public fares.

5.5.4.2   Private Fares

An airline creates private fares using private tariffs. These fares are not widely distributed and are only available for discreet sales with specific permissions at the point of sale. Private fares use security in either ATPCO Category 15 (CAT 15) or Category 35 (CAT 35) to determine *who can sell* them and Category 1 (CAT 1) to specify *who can buy* them.

Private fares fall within the provisions of Category 15, Category 25, and Category 35 and have several variations.

A corporate fare is a private fare that is negotiated between an airline and a corporation. The creation of new fares can be enabled through Category 25 (CAT 25), which uses rules data to specify the market fares and amounts. These fares can either be calculated from existing fares and market rules or specified to create a new fare using the rule provisions in CAT 25. The fares generated by this process will not have a fare class application or be published fares in ATPCO systems. To ensure proper pricing, CAT 25 requires the Fare by Rule Index, also known as Record 8, to be coded for a given rule. By utilizing CAT 25, an airline can create new fares by filing discounts on a public fare (e.g., an ADT adult fare or a Passenger Type Code (PTC)), but the same rules will apply to the discounted fare. The public fare rules can be overridden or combined when creating the FBR fare, which is typically accessed at the point of sale. A discount applies to the amount of the public fare along with its associated rules and restrictions. Changes may also be permitted with a fee. For corporations, the CAT25 discount may not be applicable across all RBDs but is usually limited to the higher fare RBDs in the hierarchy. This can result in situations where the lowest public fare available is cheaper than the lowest corporate discounted fare since the RBD is higher in the fare hierarchy. In some cases, CAT 35 security may replace the CAT 15 security, but CAT 15 can still be used to enforce ticketing date and sales restrictions.

The CAT 35 *negotiated fare with markup* is a private fare with limited distribution. It is a type of negotiated fare that is designed to handle the requirements of net remit programs, IT fares, corporate fares, and other private fares. These fares can include multiple related fare amounts, special ticketing, fare markups, and enhanced security over existing sales restrictions (CAT 15). Negotiated fares are contracts between airlines and travel agencies or other entities such as consolidators that allow them to sell published fares for less or offer special fares authorized by the airline. These fares are always private data and are generally sold through wholesalers, tour operators, and travel agents. When selling a private negotiated fare, the travel agent agrees with the airline on the net fare amount for the tariff, marks up the fare, and sells it to a retail agent or passenger. Negotiated fares may consist of multiple fare levels, fare amounts, security, and ticketing data unique to the seller and use CAT 25 for fare generation and CAT 35 for defining the markup (or markdown) subject to the specified range.

When a customer purchases a private fare from a travel agent, the agent marks up the net fare negotiated with the airline and submits that amount to the airline through the BSP[3] for settlement. Private fares of this type are common in the Asia/Pacific region, Latin America, and the Middle East and present unique challenges to airlines, such as channel conflict, revenue dilution, and lack of control over the selling fare. Negotiated fares are also known as off-tariff fares, sanction fares, unpublished fares, market fares, confidential fares, net fares, secured fares, wholesaler fares, consolidator fares, bucket fares, and gray market fares. Managing the net/net fare (net of commissions and overrides) is crucial because gross fares do not accurately represent what the airline receives for these unpublished tariffs. In some cases, the net/net fare could be as low as 40% of the gross fare.

A commonly used private fare is a negotiated fare that follows CAT 35 rules for security ("who can sell" and ticketing requirements), as well as markup and commissions for travel agencies that sell the fare. Negotiated fares of this type can be created using a CAT 25 fare by rule (FBR), with or without a discount, to which a markup and/or commission can be specified. For example, a public fare of $1000 with an FBR 10% discount and a 5% markup would result in the airline receiving $900 (90% of $1000), the agency receiving $45 (5% markup on $900), and the customer paying $900 + $45 = $945. Through negotiated fares, airlines can promote the sale of specific high-value fares.

Travel agencies can receive a commission from airlines for promoting fare sales with negotiated fares. In countries like Brazil, where the agency/airline direct model is used, the fare is handled with CAT 35 over CAT 25. This means that fares are generated based on available public fares with CAT 25, viewed with CAT 35 security, and the net fare and FBR fare amounts are the same. Even though the airline sells the ticket, they can still specify a sales commission (e.g., 1%) for the agency.

To simplify the management of negotiated fare contracts with travel agencies, airlines can file fares directly with the GDS where they have a subscription. This direct filing process eliminates the ATPCO latency and appears quickly in the GDS for consumption. All GDSs offer a negotiated fares database and provide tools to create, manage, and distribute negotiated contracts.

5.5.4.3 Web Fares

The fares that are only available on an airline's website are called web fares. They were created in the early 2000s to help airlines increase brand recognition and lower distribution costs through direct sales to consumers. Airlines with full content agreements cannot offer lower fares on their website than what is available through the GDS channel. However, some airlines still publish web-only fares that can only be purchased and ticketed through their website. This is common in Latin America, where airline web fares may be cheaper than market fares.

---

[31] Billing and Settlement Plan (BSP) for IATA accredited passenger sales agents.

5.5.5 Fare Rule Categories

The complexity in pricing an itinerary is related to the different types of rules that may apply to a fare. These fare restrictions must always be met when pricing units are identified to price an itinerary. Table 5.3 summarizes the ATPCO fare categories that are in use today.

Routings are used in conjunction with both domestic and international fares. Routing data are also made available to host CRSs and GDSs with a subscription.

5.5.6 Journeys

There are three types of journeys that are shown on OTA and airline websites. They are one-way, roundtrip and multi-stop. A multi-stop trip can be a circle trip, open jaw (with surface sector), or around the world.

One-Way. The traveler goes from an origination airport to a destination airport. Examples are AUS-DFW and JFK-LON.

Roundtrip. The traveler goes from an origination airport to a destination airport and returns to the same origination airport where the trip began. Examples are BNA-JFK-CDG-JFK-BNA, and AUS-DFW-AUS.

Circle Trip. The traveler goes from an origination airport, goes to multiple locations, and returns to the same origination airport where the trip began. A circle trip has two or more stopovers. For examples, JFK-FCO-DEL-JFK is a circle trip with stopovers in FCO and DEL.

Round the World. The traveler goes from an origination airport in one direction (eastbound or westbound) around the world and crosses the international date line and returns to the same origination city where the trip began. The fares have restrictions such as the minimum and maximum number of flights, direction of travel, open jaws are not permitted, and a minimum number of stopovers. They are also limited to carriers within an alliance. For example, DFW-JFK-LHR = NRT-LAX = DFW is around the world trip.

Figure 5.7 illustrates various types of journeys.

Open Jaw. There are three variations.

The mileage for the open jaw (segment) must be equal to or less than the mileage of the shortest flown fare component. Otherwise, it will be treated as separate one-way fares.

Destination open jaw, where the traveler goes from an origination airport to the first destination airport, transfers to a different city using an alternate mode of transportation and takes a return flight back to the origination city where the trip began. Examples are JFK-LON (open jaw) CDG-JFK and JFK-AMS (open jaw) BRU-JFK.

Table 5.3 Summary of fare rule categories (ATPCO)

Rule category	Description
1	Eligibility, identification requirement and age range of a passenger type. E.g., ADT: Adult, CHD: Child, UNN: Unaccompanied child, INF: Infant without a seat, INS: Infant with a seat, BEV: Bereavement passenger, STU: Student, etc. There are over 150 passenger type codes (PTC).
2	Day/time application that defines the days and or times when travel is permitted.
3	Seasonal and promotional date restrictions.
4	Flight application indicates a fare is only valid on specific flight numbers.
5	Identifies advance reservation and advance purchase/ticketing requirements.
6	Minimum stay restriction, specifying when return travel may commence.
7	Maximum stay restriction, specifying the last day when travel may commence or may be completed.
8	Stopovers defines the number, locations, and charges of allowable stopovers within a fare component. A U.S. domestic flight is considered a stopover when the scheduled time on the ground is longer than 4 h. An international stopover exceeds 24 h. This category can override the default.
9	Transfers, conditions, and restrictions for transfers to occur.
10	Permitted combinations, process of using multiple fares or portions of multiple fares to arrive at an itinerary price for the passenger.
11	Blackout dates, used to define single dates or date ranges when travel is not permitted.
12	Surcharges, conditions when surcharges are applicable and the corresponding charge.
13	Accompanied travel when travel with one or more passengers is required to qualify for the fare.
14	Travel restrictions when travel dates are specifically stated in a rule.
15	Sales restrictions—where the fares can be sold and/or ticketed.
16	Penalties, if applicable for a fare and what charges will be assessed.
17	Higher intermediate point (HIP)/mileage exceptions.
18	Ticket endorsements requirements as specified in a rule.
19	Children discounts.
20	Tour conductor discounts.
21	Agent discounts.
22	All other discounts.
23	Miscellaneous provisions, states if fares should or should not be used in construction, proration and differential.
25	Fare by rule application to dynamically generate new fares by using a carrier's existing fares and rules as a base. Allows airlines to file in ATPCO increasingly complex corporate contract terms, thereby improving faring accuracy and reducing agency debit memos (ADM).
26	Groups, to define the requirements to qualify for a group fare. Not used in pricing but will display the rules text.
27	Tours, tour requirements for a fare.
28	Visit another country, requirements to qualify for a visit another country fare.
29	Deposits, defines deposit requirements to qualify for a fare, if any.

(continued)

Table 5.3   (continued)

Rule category	Description
31	Voluntary changes, process the reissue transactions programmatically.
33	Voluntary reroute and refund.
35	Negotiated/net fares. Enables travel agents to add their own markup to airline filed net fares, so that the agents' selling fare is displayed in the GDS. Includes security and agency commissions.
50	Rule title/application assumption contains the rule title, geographic application, type of journeys, and other conditions. This category exists for every rule and is used for rules text purposes only.

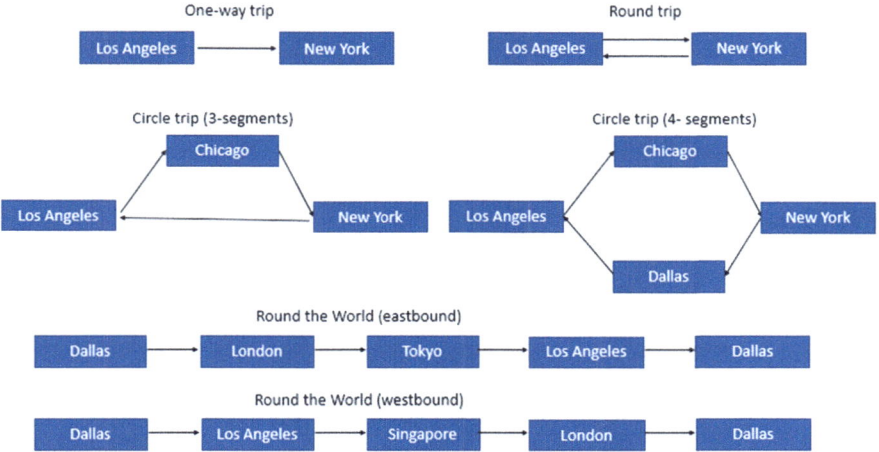

Fig. 5.7 Examples of journeys

Origin open jaw where the traveler departs from an origination airport to a destination airport and returns to a different airport/city. Examples are DFW-LON-SAT, DFW-FCO-AUS.

Double open jaw where two totally separate fares exist. It is a single return ticket where the O&D for the first flight and the second flight are different. An example is DFW-FRA for the first flight and VIE-AUS for the second flight.

Figure 5.8 illustrates various types of open jaw journeys.

Open jaws on an itinerary show up as ARNK, pronounced "ARUNK" means "arrivals unknown".

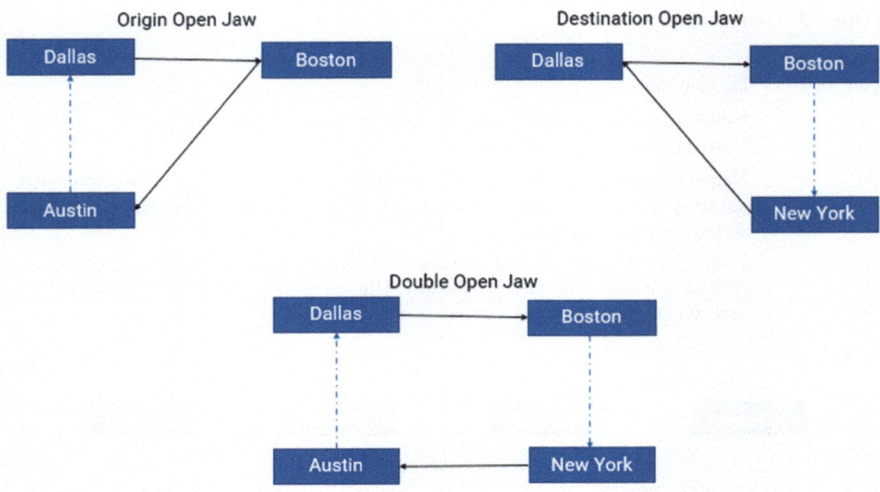

Fig. 5.8 Examples of open jaw journeys

5.6 Itinerary Pricing

The primary complexity in air shopping lies in pricing the itineraries. Itinerary pricing requires a deep understanding of fare components and priceable units (PU) or pricing units. The PU options are constrained by the customer's journey, and they constitute a significant source of complexity in pricing an air itinerary.

The fare component is the most basic unit of fare construction and represents a specific fare between two city pairs. A fare component is a section between two points along the travel path. One or more fare components make up a PU. One or more PU combinations produce a pricing solution for a trip. A PU is a domain where multiple fare rules may apply and consists of a group of fare components. Fares may require that other fares in the same PU are from the same airline.

Tickets are built from one or more PUs.

Flights can be broken into fare components and pricing units in many ways which increases the search space. PU's also introduce dependencies between different parts of a trip.

PUs can take several forms. The PU types are *one-way, roundtrip, circle trips, and open jaw*. This requires the itinerary to pass the fare rules such as fare combinability (Category 10). Pricing units are patterns and combination of PU types can form a pricing solution for a trip. Open jaw PU's are like circle trips with a missing fare component where the gap exists and normally the gap must be shorter than the distance flown in any of the fare components.

Flights can be broken into fare components and PUs in many ways. Consider the itinerary from LAX to LGA outbound connecting over ORD and LGA to LAX inbound connecting over DFW. Multiple combinations are possible for categorizing a particular group of flights into fares and PUs. In the given example of four flights,

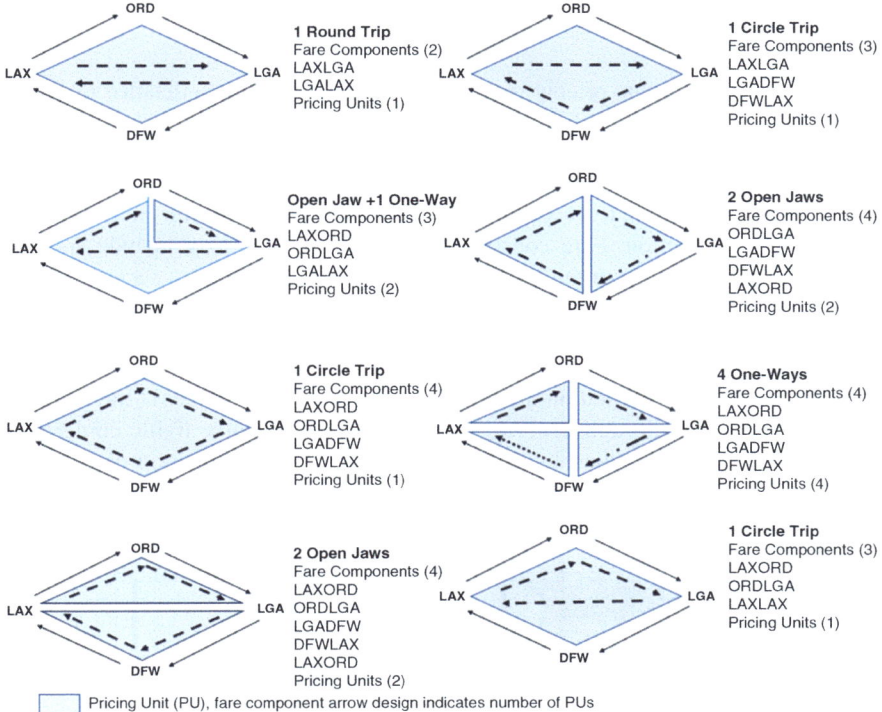

Fig. 5.9   Examples of possible fare components and pricing units

eight potential options are illustrated, although there are even more possibilities. Each fare component that is part of a priceable unit is represented uniquely by an arrow, while each priceable unit is depicted by a blue polygon. There are several combinations of fare components and possible PU patterns. A *few* are shown in Fig. 5.9.

   In summary, the process involves dividing an itinerary into fare components and combining them into one or more priceable units. Itinerary pricing searches for the most affordable solution that is both applicable and available, considering all taxes and fees, including government taxes imposed by the DOT such as passenger facility charges (PFC), Federal Excise Tax (FET), Segment Fee, September 11 Security Fee, U.S. or International Departure and Arrival charges.

5.7 IATA Traffic Conference Areas

Since deregulation airline pricing has grown in complexity and a detailed explanation of all the intricacies is beyond the scope of this book. From the public and private fares, a specific itinerary can be priced based on the fundamentals of *fare construction*. IATA fare construction is documented in the Maximum Permitted Mileage (MPM) Manual. It serves as the primary source for applying the fundamentals of fare construction. Fare construction for an itinerary may include specific routings, stop over charges, maximum permitted mileage, security fees, combinability of one-way pricing units, mixed classes of service, roundtrips, circle trips, open jaw, special mileage provisions, journeys with surface sectors, and ticket reissue or exchange.

Fare construction is based on the areas of the world the customer is traveling to, from, or connecting through. IATA divides the world into three traffic conference (TC) areas, which are further divided into sub areas. The composition of the three primary IATA Traffic Conference Areas is shown in Fig. 5.10 and Table 5.4.

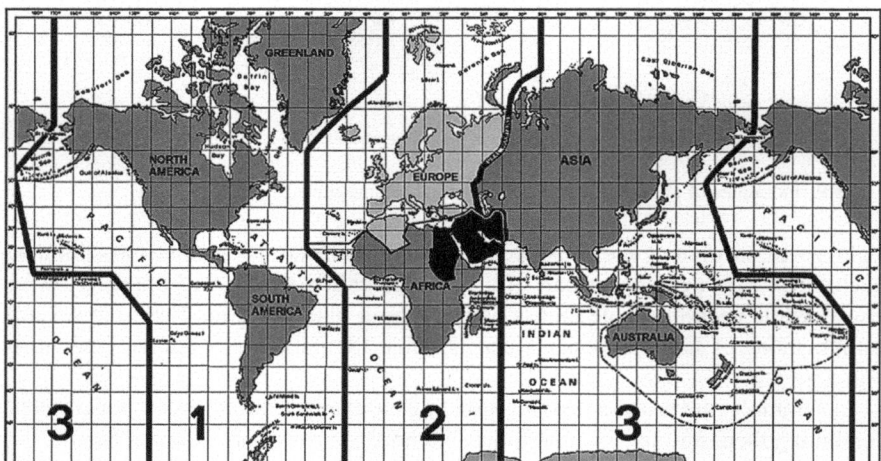

Fig. 5.10 IATA Traffic Conference (TC) geographic areas

Table 5.4 IATA traffic conference areas

IATA Area Identifier	Geographical locations
IATA area 1 (TC1)	North America, U.S. territories, Central America, Caribbean, South America.
IATA area 2 (TC2)	Europe, Russia (RU), Middle East, Africa, Indian Ocean Islands.
IATA area 3 (TC3)	Asia, Southeast Asia, Indian sub-continent, Japan/Korea, north/Central Pacific, Southwest Pacific, Oceania.

For example, a conference area of TC23 could incorporate fares between Europe (Area 2) and India (Area 3), while TC12 could incorporate fares between the United States and Germany. For example, TC3 could incorporate fares within the Pacific, between India and Fiji and between Hong Kong and Singapore.

In addition, The Ticketing Handbook (THB) serves as the official guide for airline passenger ticketing and covers ticketing entries for EMDs, additional optional services, YQ/YR surcharges, involuntary rerouting, fare selection criteria, baggage rules, fare calculation, and exchanges/refunds.

5.7.1 Constructed Fares

Constructed fares belong to a distinct category where the fare amount is determined only after undergoing a fare construction process. For instance, international carriers often use add-on fares, which allow them to establish fares between two cities without explicitly publishing each market they wish to serve. A new fare record is created by combining a published gateway fare with an add-on fare. As an illustration, Thai Airways may publish a gateway fare from their hub in Bangkok (BKK) to London, Heathrow (LHR). They may also publish an add-on fare from a domestic city in Thailand, like Phuket (HKT) to Bangkok (BKK). Consequently, the fare from Phuket (HKT) to London (LHR) can be constructed by utilizing the published gateway fare and the add-on fare.

5.8 American's Value Pricing Initiative

American implemented value pricing on April 9, 1992, under the leadership of CEO Robert Crandall. This was the airline industry's initial venture into value pricing (McDowell, 1992). American's objective was to transition to a more straightforward fare system that had become increasingly convoluted after airline deregulation (Michael & Silk, 1994).

Instead of offering seats at various price points, American introduced four types of fares at lower prices: first class, regular coach, and two discounted coach fares with 7-day and 21-day advance purchase restrictions. This simplified fare structure marked a significant departure from the intricate fare systems that featured discounts and restrictions. Crandall placed great emphasis on the new fare structure's "*simplicity, fairness, and value*," which made it easily comprehensible for all.

The updated fare structure includes four distinct fares for each market: first class, regular coach, discounted coach with a 7-day advance purchase requirement, and discounted coach with a 21-day advance purchase requirement. These fares are based on mileage and are cheaper than the prevailing fares in the market. First class fares have decreased by 20% to 50%. The regular coach fares, known as AAnytime Fares, have no restrictions and are at least 38% cheaper than the prevailing full coach fares. There are two PlanAAhead Fares available: one with a 21-day advance purchase and a Saturday night stay requirement, and the other with a 7-day advance purchase requirement. Both fares are significantly cheaper than the prevailing fares as well. In some cases, the lowest discounted fare may be more expensive. The lower discounted fares are nonrefundable but can be re-issued for a fee. Additionally, American Airlines eliminated all special discounts for corporate fares, conventions, and the military, resulting in an 86% reduction in the prevailing fares. The lower fare structure made flying accessible to a segment of the population that had never flown before.

American Airlines' value pricing initiative recognized as groundbreaking by industry experts, ultimately failed when its main competitors swiftly withdrew their initially matching tariff structure and introduced limited-time discounted fares. This move prompted American's CEO, Robert Crandall, to famously comment in an interview with Time magazine that *"this industry is always in the grip of its dumbest competitors"* (Castro & Crandall, 1992). During the annual employee President's Conference in Dallas, Crandall further emphasized the importance of outsmarting the competition and concluded his remarks on value pricing with the statement *"you are only as smart as your dumbest competitor."*

In response to American Airlines' revamped value pricing fare structure, Northwest Airlines and Continental Airlines filed a lawsuit alleging that American was attempting to monopolize the market. However, after a four-week trial in July/ August 1993, the jury swiftly concluded that American Airlines had not engaged in predatory pricing to eliminate weaker competitors the previous year with its value pricing strategy. Following the verdict (Jones, 1993; Crandall, 1993), Robert Crandall stated that his airline was simply aiming to compete and suggested that Northwest Airlines and Continental Airlines were *"hoping to accomplish in the courtroom what they couldn't accomplish in the marketplace"*.

5.9 Putting it all Together: Air Shopping Components

Air shopping constitutes a significant investment in advanced algorithms and hardware to respond to customers and travel agents when a request for itineraries available for sale in a market is made. Leveraging the cloud for deployment of the air shopping complex also has its unique benefits (Kavis, 2014).

The major entities that invest in air shopping are the three Global Distribution Systems (Sabre, Travelport, Amadeus), Google Flights, and Expedia (for U.-S. domestic shopping, not international). When a customer plans a trip, the first step is to explore itinerary options for travel to the desired destination and return if applicable. The itineraries displayed during shopping are the available itineraries that can be booked.

The major components of air shopping are shown below in Fig. 5.11. If a GDS also hosts carriers for reservations processing, then the shopping requests that come through the universal gateway are the direct and indirect (travel agencies. OTA's,

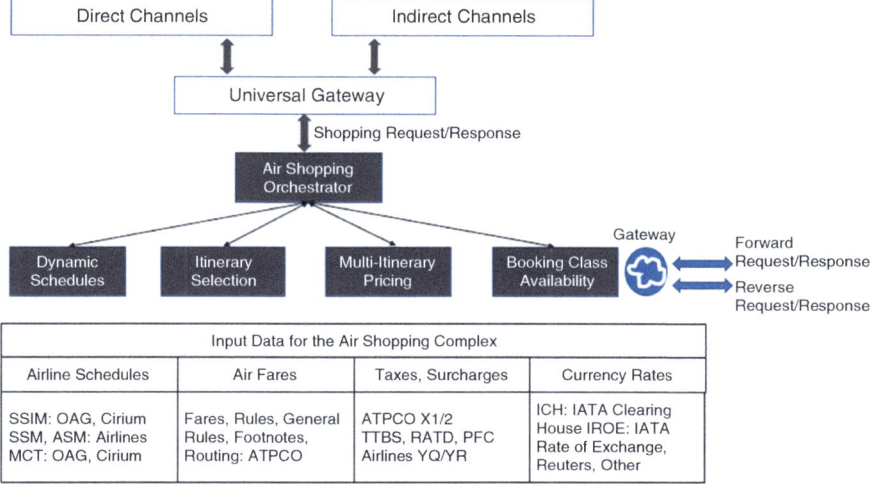

Fig. 5.11   Major components of air shopping

wholesalers) channels. The intelligent orchestrator, also called a broker, manages all service requests to downline shopping components to create a shopping response.

A brief description of the major air shopping components is summarized below.

5.9.1   Dynamic Schedules

The generation of outbound and inbound itineraries for round trip shopping requests is facilitated by dynamic schedules. These itineraries are subject to the minimum connect time (MCT) rules, as well as availability. It is crucial for the itinerary creation process to consider availability since if a flight has been sold out, it will not be considered.

5.9.2   Itinerary Selection

The itinerary selection process reduces the search space to determine candidate itineraries for the shopping response display for a travel agent or end consumer. Several passes are required to arrive at the selection of itineraries that are candidates. Diversity rules such as non-stops, single connects, double connects, carrier diversity, interline, and the minimum number of inbounds for an outbound are executed. During this process, the booking class availability will be sourced from the cache to ensure fast response times and minimizing network latency. In addition, rule validation shortcuts are used and typically only 90% of the rules, the most important, are validated.

5.9.3   Multi-Itinerary Pricing

The itineraries selected during the itinerary selection process are validated where all applicable pricing rules are executed to price the itineraries. In addition, forward DCA availability requests are sent to non-hosted airlines to ensure that true last seat (booking class) availability is considered prior to itinerary pricing. The list of itineraries is next sorted from the lowest fare to the highest. Diversity rules are considered before the collection of itineraries in the shopping response is finalized for display.

5.9.4   Booking Class Availability

A critical aspect of an air shopping algorithm is the usage of accurate flight availability to ensure that the itineraries displayed can be booked without UCs (Unable to Confirm at sell) and price jumps.

Figure 5.12 illustrates the logical architecture for a high performance booking class availability determination for the air shopping complex. The availability orchestrator determines the availability services that should be accessed to create a booking class availability response.

The major components for booking class availability are:

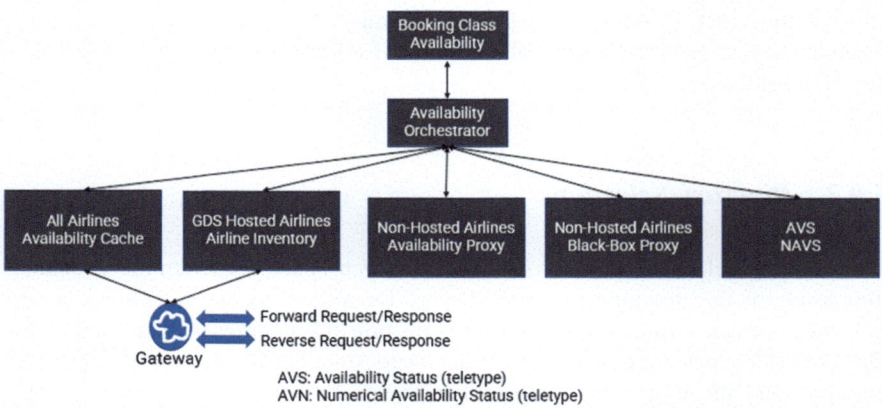

Fig. 5.12 Major components of booking class availability

5.9.4.1   Interactive Connectivity

GDSs have interactive connectivity agreements with many airlines worldwide. These are called the direct connect availability (DCA) and direct access interactive availability (DAI) with 100+ airlines worldwide providing the ability to query true last seat availability in real time from an airline's host CRS through the Gateway as shown in Figs. 5.11 and 5.12. In addition, there are direct access availability (DAA) with terminal emulation for many airlines. Queries for availability are sent to airlines through a gateway and are called forward requests. If a GDS also hosts airlines for reservations, the gateway will receive reverse requests for hosted airline availability from other GDSs, airlines and travel entities.

The volume of queries sent to airlines through the gateway for the transaction types are typically greater than 10 billion per month.

5.9.4.2   Availability Cache

The availability cache is an extensive in-memory database of booking class availability by carrier for markets that they serve. GDSs maintain an extensive availability cache for all airlines worldwide to ensure faster air shopping response times and simultaneously reduce the availability request load to airline reservations systems. The data stored in the availability service cache can be in a hierarchy such as point of sale, origin and destination, date, flight(s), and booking class availability. There are millions of unique items in this cache that are updated at the rate of 25–50 million/h. This availability cache serves as an alternative to real-time interactive connectivity to avoid excessive availability request loads to airline reservations systems. To keep the availability cache accurate, every item in the cache has an expiration date and the update process may deploy stochastic lifetime models for arrival rates modeled as a Compound Poisson process, Weibull, exponential power, or a discrete survival probability mass function. These types of models can predict arrival rates with an accuracy of 99.6%, which is used to determine the pro-active polling frequency to keep the availability cache accurate at all times. The availability cache is updated when responses to these *forward requests* are received.

5.9.4.3   Hosted Carrier Inventory

For airlines that are hosted by a GDS (this applies to Amadeus and Sabre), the hosted airline inventory system provides true last seat availability for both leg/segment and O&D revenue management based carriers. These are the reverse requests. The availability cache is updated when these *reverse requests* are received.

5.9.4.4 Availability Proxy

For a few select airlines, subject to collaboration, the GDSs also maintain an availability proxy, which replicates the availability processing logic resident in the airline's reservations system (Vinod, 2007). With the availability proxy, a participating carrier sends real time messages on changes in inventory. For an O&D revenue management carrier, this will be updated to the bid price curve. For a leg/segment revenue management carrier, it will be seats available by booking class. The availability proxy ensures that the air shopping complex has access to true last seat availability. The availability proxy is considered more accurate than the availability cache since it reflects the range of business rules resident in the airline host CRS for availability determination that is reflected in the availability proxy. For airlines with an availability proxy in the GDS air shopping complex, the availability is not stored in the availability cache.

Figure 5.13 illustrates the deployment of an Availability Proxy for air shopping.

5.9.4.5 Black Box Availability Proxy

Some airlines collaborate with the GDSs and deploy a "black box" Availability Proxy as an alternative to the Availability Proxy in the data center where the air shopping complex is maintained. This allows a partner carrier to deliver inventory data and rules 24/7 to achieve a higher level of accuracy in booking class availability. Partner carriers can also deliver software updates based on their software release schedule. It also allows a partner airline to not divulge details of their precise availability processing logic.

Both the Availability Proxy and the "black box" Availability Proxy provide a scalable solution to support the growing demands of air shopping.

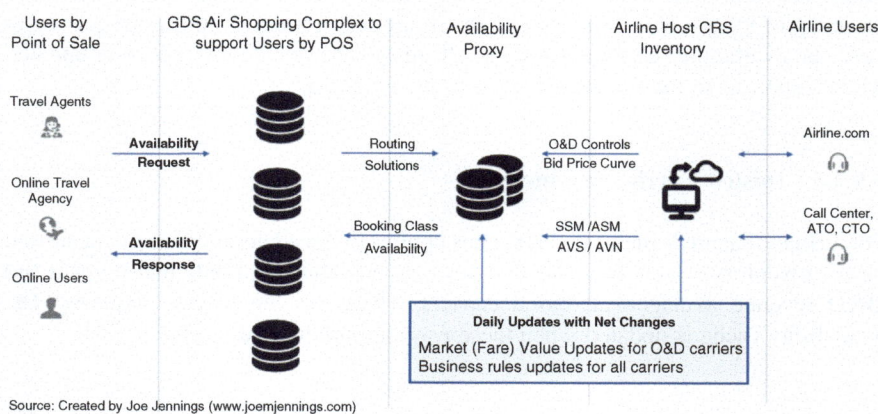

Source: Created by Joe Jennings (www.joemjennings.com)

Fig. 5.13 Availability Proxy deployment with the GDS air shopping complex

5.9.4.6 Teletype

Besides the variations of interactive availability, the availability cache and the two versions of the availability proxy, the air shopping complex of the GDS also receives teletype flight availability status messages (AVS) and numeric AVS (AVN) from airlines worldwide 24-h a day. These flight status messages are only used when interactive availability request is not sent to the carrier, or the availability cache does not have a flight status that has not expired. These teletype messages average several million per day.

5.10 Measuring Air Shopping Performance

Ultimately, the shopping results returned by the algorithm should resonate with customers to maximize conversion rates. The core measures of air shopping effectiveness are applicable for both round trip and one-way itineraries. The key metrics that a GDS monitors to measure the effectiveness of the air shopping algorithm are as follows.

5.10.1 Low Fare Efficacy

Low fare efficacy is the ability of a shopping algorithm to determine the absolute lowest priced itinerary for the market represented in the shopping request. Fare-led search algorithms perform much better than schedule-led algorithms in determining the lowest fare in the market. The quality of service for the lowest priced itinerary may be poor since it may involve undesirable connections. The reason why this fare is important is because it is used as a reference itinerary to the itinerary the customer eventually books.

5.10.2 Diversity

Corporate customers are schedule sensitive while leisure customers are price sensitive. Diversity of the itineraries in a shopping response is a key requirement since only a small percentage of corporate customers book the lowest fare. This is also true to a lesser extent with leisure travel among leisure agencies, OTAs, and supplier websites because customers value convenience based on their schedule and fare preferences.

So, what is shopping diversity and how is it measured? There are several dimensions to shopping diversity such as carriers, quality of the service expressed

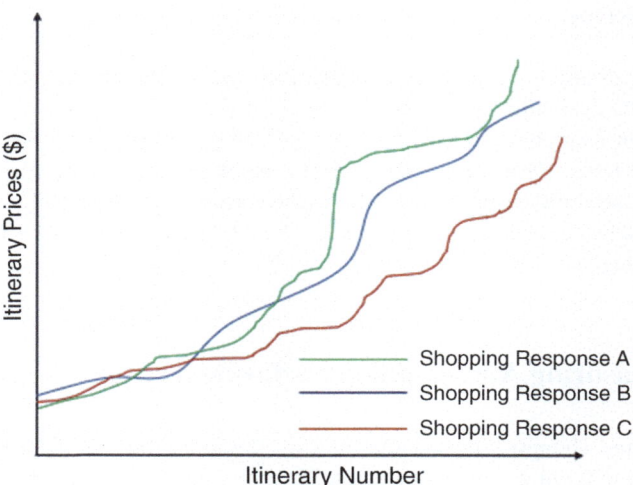

Fig. 5.14   Itinerary sets returned in shopping responses

by non-stops, single connect, double connect, interline, and inbound diversity. The inbound diversity is the minimum number of inbounds for a selected outbound itinerary. The diversity of itineraries displayed during shopping that will resonate with customers is of critical importance to maximize conversion rates.

Figure 5.14 illustrates three distinct shopping responses. Each shopping response is a collection of a fixed number of itineraries sorted from the lowest fare to the highest and the question is: *what collection of itineraries should be returned that maximizes conversion rates?* For example, if two itineraries with the same price are displayed one below the other where the quality of service (expressed as departure time and elapsed time) for the first itinerary is excellent compared to the second which has poor quality of service, there is a greater propensity that the customer will book the better itinerary.

An issue to be addressed with low fare efficacy is to provide low priced "value" itineraries instead of just 3 and 4 segment connections that is $10 cheaper.

Low fare efficacy can be measured against a competitor and results tallied as wins, losses, and ties for 1000s of markets and expressed in absolute value (wins and losses) and as a percentage.

Consumer choice models (CCM) can be calibrated (Ben-Akiva & Lerman, 1985) to determine the utility of schedule and fare attributes to determine the best set of diverse itineraries to display as part of a shopping response. Measuring the screen quality enables TMCs and OTAs to determine how effectively they convert shoppers to bookers. Utility is better computed as incremental, which implies that the itineraries in a shopping response are not considered independently. The utility of an itinerary depends in part on the other itineraries in the set. In addition, calibrating consumer choice models by market, and context for travel-based segmentation refines itinerary diversity in a shopping response.

Besides the individual itinerary level metrics for diversity, a composite measure called *entropy* is used to describe and compare screen diversity. So, what is entropy? Consider a variety pack of candy. The more different types of candy, the higher the entropy. If the frequency distribution is evenly distributed, the higher the entropy. This serves as a measurement tool to check the diversity of outbound flights, inbound flights, carriers, how many stops, etc.

Taking outbound flights as an example, entropy is designed to increase when more different outbound flights are available. On the other hand, the entropy value decreases if only a few outbound flights repeatedly appear in most of the itineraries, which gives customers poor flexibility. This characteristic makes entropy a good measure for screen shopping diversity. A word of caution is how code share flights are handled. Code share flights must be handled the same as an operating flight to avoid skewing the results with a high score in situations when there are many codeshare flights.

5.10.3 Book-Ability, Price Jumps, and Availability Errors

Book-ability is an important metric. There are two inverse measures of book-ability. They are unable to confirm at sell (UC) and price jumps.

There are two types of availability errors that occur when the availability cache does not reflect the true availability resident in the airline inventory system (host CRS).

The Type 1 error is interpreted as an "open-wrong" error. It occurs when the cached availability for a booking class is considered open while the booking class is truly closed in the airline reservations system (host CRS). This error can result in an inability to confirm at time of sell (UC). A Type 1 error can also result in the customer experiencing a price jump at the time of booking when the minimum available fare displayed is lower when only a higher fare is truly available. Type 1 errors can cause customer angst and influence shopping cart abandonment on OTA sites. It is less of an issue with TMCs. A travel agency's leisure customers may exhibit similar behavior to the OTA's customers. However, the travel agency's corporate customers are unlikely to leave since they are part of the corporate travel management program,

Type 2 error is interpreted as a "closed-wrong" error. It occurs when the cached availability for a booking class is considered closed while the booking class is truly open in the host CRS. Type 2 errors are more difficult to detect and require continuous monitoring with robotic simulations to identify situations when a "closed-wrong" occurs.

Despite the possibility of Type 1 and Type 2 errors, an availability cache is important during the discovery phase of shopping to guide the search. Once the final set of itineraries is determined, RBD availability can be validated against the airline's inventory system for the top N itineraries. This is frequently referred to as a *wash*. The *wash* ensures that the itineraries displayed are book-able. If the itinerary set changes because of booking class closures on an itinerary, it does not guarantee that the optimal set of itineraries was determined by the shopping algorithm.

Inaccuracies in availability can be mitigated in several ways—by accessing a local or distributed inventory solution in the cloud from an airline, improve the booking class availability cache aging algorithms to closely reflect market conditions (usage) and the deployment of an availability proxy solution from airlines in the same environment as the air shopping system.

5.10.4 Shopping Response Times

Response times need to be best-in-class worldwide to ensure customer retention (B2B and B2C) and maximize conversion rates. Response times are important to avoid the risk of abandonment and improve the conversion of shoppers to bookers.

The total response time is made up of two components: the server latency (time spent in determining the solution for the shopping request on the shopping servers) and network latency. Server latency can be addressed with algorithmic improvements to the shopping core so that applications that run on the core can execute faster. Application performance can also be improved. Network latency obeys the laws of physics and depends on the distance between the location of the server and the point of sale where the shopping request was executed. Network latency can be reduced by pursuing an edge deployment strategy. The shopping solution can be deployed in the cloud by region, instead of at a central location, so that they can be located physically closer to the customer requests (see Sect. 5.8). The biggest challenge for deployment of the shopping solution in the cloud by region is the availability cache, which is changing constantly and should be replicated in real time across all regions where the shopping solution is deployed.

5.10.5 Cost per Shop

The computational cost of shopping is an important metric, given the year over year exponential growth in shopping volumes leading up to the pandemic of 2020. The average cost per shop is the first metric that should be calculated across all shops regardless of the originating source. The total cost that is used in determining the cost per shop is the cost of the air shopping complex infrastructure and the cost of compute. The difficulty in estimating the cost of a shop (in cents) is because it should not be based on fully allocated costs across the server farm. It requires accurate attribution at a granular level since a server may process requests that are not directly shopping related.

An equally important measure that is tied to productivity is the revenue generated by point of sale to the number of shops in the channel. This data is important since it influences the cache settings. A channel that submits a lot of shopping requests and generates low revenue will be served predominantly from the shopping cache without resorting to a live shopping response, thereby reducing compute costs.

Knowledge of the cost per shop, the total volume of shopping requests by point of sale, and the revenue generated for the GDS based on segment booking fees can illustrate the worst offenders (usually OTAs) and the most productive agencies.

A laser focus on cost avoidance research for the shopping core is an absolute necessity to stay competitive. There must be a definitive plan to reduce CPU usage every year. Why is this important? Year over year growth in shopping volumes for GDSs have ranged between 50% and 120% between 2012 and 2019.

5.10.6   Look-to-Book Ratio

The look-to-book measure is meaningful for airlines in the context of city pair availability. City pair availability entries always rely on a live direct connect availability (DCA) call to query the airline CRS. In this scenario, it is the ratio of the number of live interactive availability calls to airline CRSs to the number of bookings.

The look-to-book measure for shopping transactions is less meaningful, but useful to understand growth in shopping volumes, the inefficiency of online channels, and planning for adequate capacity. In this scenario, the ratio is simply the number of shops divided by the number of bookings. Each shop, depending on the channel parameter setting, can return itineraries that range from a few itineraries to the hundreds. Look-to-book ratios are sometimes viewed in the context of conversion rates, but it is not very meaningful since look-to-book ratios vary widely by channel. Brick-and-mortar travel agencies typically experience a 10:1 to 20:1 look to book ratio while OTA's can range from 300:1 to 2000:1 in certain markets. From that perspective, OTAs are far less productive than TMCs.

When calculating the look-to-book metric, care should be exercised to exclude robotic shopping requests. Robotic transactions can be isolated with a machine learning pattern recognition model that determines if the shopping request is robotic. For example, the pattern recognition model can be trained to detect shopping request sequences such as consecutive length of stay requests for a given advance purchase, consecutive advance purchase requests for a fixed length of stay or combinations of a range of advance purchase and a range of length of stay. This approach, while effective, may not detect all the robotic shopping requests if the patterns change constantly. However, it serves as a lower bound on the total robotic transactions. This detection is done at the PCC level.

An alternative metric is the *modified* look-to-book ratio, that is based on the itineraries returned in a shop request that can vary based on channel and parameters specified by agencies at the time of the shopping request. The modified look-to-book ratio is simply the total number of itineraries returned during shopping divided by the number of bookings. Some GDSs monitor this metric since there is a wide disparity for look-to-book ratios between OTAs and TMCs.

5.10.7   Conversion Rates

The conversion rate is measured as the number of bookings divided by number of shops.

There are several ways to model and calibrate conversion rates. For instance, the conversion rate for shopping requests can be modeled as a simple logistic regression model or classification and regression trees (CART), wherein the dependent variable (the conversion rate) is based on the known values of the independent (causal) variables.

A logistic regression model is different from a linear regression model because the conditional distribution $Y|X$ is a Bernoulli distribution and not a Gaussian distribution. The predicted value, in this case the conversion rate, is restricted to $(0, 1)$ with the logistic distribution function. For example, the conversion rate can be calibrated as a function of the following independent variables ($x1$, $x2$, and $x3$), which are indicator variables (not values), where:

$x1$: Time of day (broken down into 24 buckets)
$x2$: Day of week (can be by individual day of week or mid-week vs. weekend)
$x3$: Context for travel-based segmentation.

Hence, in this scenario, the conversion rate is:

$$p = \frac{1}{1 + e^{-(\beta 0 + \beta 1 x1 + \beta 2 x2 + \beta 3 x3)}},$$

where $x1$, $x2$, and $x3$ represent the three independent variables. Based on the validated accuracy of this model, additional variables can be added if required. With an estimate of the conversion rate for each shopping request, decisions can be made on how to respond to the shopping request; whether it should be a response from a shopping cache or various levels of compute including a live shop to create a shopping response. Maximizing the efficiency of shopping responses is required to control the cost of compute against available server capacity.

5.10.7.1   Dual GDS Strategy, Conversion Rates and Low Fare Efficacy

Many large OTAs have a dual GDS strategy. What this means is they subscribe to web service APIs for shopping, booking, and ticketing with more than one GDS. These OTAs are increasingly measuring one GDS against the other with a single metric where "performance" equals conversion rates. In many ways, this is an unfair metric since several extraneous factors not under the direct control of the shopping algorithm go into this metric. OTAs that subscribe to multiple GDSs shift booking traffic from one GDS to the next based on this measure.

Some OTAs use low fare efficacy as the sole measure to shift bookings between GDSs. In this scenario, each shopping request is sent to the competing GDSs for a response and if one of them produces a lower fare, the booking goes to the GDS with the lower fare. In the case of a tie, the booking will be made by random assignment.

5.10.8 Missing Air Content

This is measured as the percentage of carriers whose content is available to a GDS by market. This is important regardless of a carrier's booking volume since absence of negotiated content can *directly impact* low fare efficacy. With the growth in low-cost carriers (LCC) that do not participate in the GDS, content from these carriers will require a direct connect to the airline CRS for a shopping response from the carrier for the GDS to create a blended response. The alternative to establishing a direct connect with each LCC is to negotiate an agreement with an LCC aggregator to access this content. Content gaps matter even if *no* GDS has full content since it triggers OTAs to establish direct connections with the nonparticipating carrier. This has similarities with what is happening with IATA NDC today.

There is also a secondary measure, "full fare content," which is increasingly an issue with LCC's, Value Focused Carriers (VFC), and Latin American carriers that participate in the GDS who frequently withhold their lowest fares and make them "web only fares". Even some network carriers like Air Canada withhold the lowest fares from the GDS when the point of sale is Canada. If a specific carrier withholds the lowest fares as "web only fares" from all GDS's, it is still an issue for two reasons: first, over time there will be an observable market share shift, and second, it may trigger OTAs to establish direct connections with the carriers to access the web-only fares. This has similarities with what is happening with IATA NDC today.

5.11 Monitoring GDS Shopping Performance

This may sound like a self-fulfilling prophecy, but GDSs periodically hire management consultants who specialize in airfare analysis to benchmark their shopping efficacy against competing GDSs. Examples of these benchmark specialists are TOPAZ, Dr. Fried, and Partner GmbH. These airfare analysis companies run controlled experiments comparing the shopping results for the host GDS (the one that has hired the benchmarking specialists) against the other GDSs. These studies typically use the top 50 or 100 markets as observed from MIDT data and multiple points of sale (POS) for the analysis. Using this input criteria, the benchmark specialists shopped different advance purchase and length of stay combinations from each GDS for the analysis.

The benchmark reports summarize a range of metrics such as low fare efficacy expressed as a win-loss measure (subject to a threshold of $3 or $5 per itinerary being termed a draw), expected savings that can be realized by using the host GDS, the number of itineraries with 10% of the lowest fare, usage of alternate airports and interline connections to find low fare itineraries, average response time, and many more metrics. Favorable results documented by a benchmark specialist are then used for press releases and to communicate to the travel agency and OTA communities who use the GDS shopping service. These reports are also used by GDS sales agents to drive agency conversions.

5.12   Network Latency and Air Shopping

Improving response times for air shopping is crucial to prevent abandonment and enhance the conversion rate from shoppers to bookers. A decrease of 33% in response times results in a 2.3% increase in the conversion rate.[4] The target audience for a GDS shopping service comprises agencies and OTAs who utilize web services for shopping. OTAs exhibit significantly high look to book ratios, typically ranging from 200:1 to 1000:1 or even higher. They account for over 70% of all shopping traffic in a GDS, placing even greater emphasis on the need for swift response times. Furthermore, larger OTAs adopt a dual GDS strategy and actively compare, contrast, and report on response times.

The total response time of an air shopping request includes the time it takes for a shopping algorithm to produce a response (server latency), as well as the network latency for both the incoming request and outgoing response. While computer scientists who specialize in algorithm development continuously make incremental advances to improve the performance of air shopping algorithms, reducing network latency can be achieved by deploying the air shopping infrastructure in a way that allows a user's shopping request to be fulfilled by a nearby shopping service, instead of sending it to the prime data center, which may be located on a different continent.

Network latency can be addressed by deploying the shopping service that functions in a primary computer center in a cloud environment by region, such as North America, South America, Europe, and Asia Pacific. This entails storing schedules, fares, and booking class availability in Couchbase or a similar technology, ensuring synchronization across all regional cloud deployments. As a result, shopping responses for the same shopping input parameters will be consistent across all regional cloud deployment locations when processing regional shopping requests. The only potential cause for any variations would be the synchronization of schedules, fares, and availability across different regions. Further, by deploying the shopping solution across regional cloud data centers shopping requests are distributed based on where the request originates, resulting in a scalable solution and faster response time.

---

[4]Internal study at Travelocity, 2012.

5.13   Storing and Processing GDS Shopping Data for Decision Support Applications

Big Data is composed of structured and unstructured data. Instances of structured data encompass booking, ticketing, and post-departure data. On the other hand, unstructured data encompasses user-generated content from travel reviews, social media posts, sensor data, audio, video, clickstreams, and log files. These data types are highly unstructured and cannot be stored or processed effectively in relational database management systems (RDBMS) like Oracle, DB/2, or Teradata. Analyzing these diverse data types together can provide valuable insights into consumer behavior, process efficiencies, and website design. Varian (2014) offers an overview of the analytical tools required by data scientists.

Big Data goes beyond managing the rapid increase in shopping data volumes. It encompasses the resources that enable efficient data processing, offer business insights, and enhance corporate agility. By storing shopping queries and responses in a Big Data environment, companies can create a cost-effective and scalable storage and processing solution for this data (Vinod, 2013a, 2016b).

Google developed a proprietary, distributed file system known as Google File System (GFS) in 2003 (Ghemawat, Gobioff and Leung). They also created a parallel programming technique and framework called MapReduce in 2004 (Dean and Ghemawat) for their web search purposes. These papers served as the foundation for the development of Hadoop, which consists of two core components: MapReduce and the Hadoop Distributed File System (HDFS). Google papers provided the inspiration for Doug Cutting (McKenna, 2017) to create Hadoop using Java, and he named it after his son's toy elephant. Hadoop, along with a few complementary open-source tools, enables easy access and analysis of shopping data using clusters of affordable commodity hardware. Apache Hadoop and Apache Spark, two popular open-source software solutions for Big Data, have become the industry standards for enterprise data lakes. A data lake refers to a vast repository for both structured and unstructured data. Over the years, Hadoop has evolved into an extensive ecosystem with various sub-projects such as Pig, Hive, HBase, and more.

With the traditional GDS model, shopping data from a GDS offers unlimited visibility into schedules and fares from all participating airlines. This data encompasses a consumer's interaction with a flight request, the corresponding shopping responses, and the subsequent booking made by the consumer. Traditionally, this information was stored in a data warehouse utilizing the Teradata DBMS. While these systems excel at handling structured tables and accommodating a high volume of users, they require data scientists to invest significant time in modeling the data before storing it as well as the cost of the machine for storing very large datasets. To optimize calibration, it is most effective to store the data in a raw file format instead of normalizing it into Teradata tables. This approach allows for direct inclusion and exclusion of attributes during the calibration process within a Big Data environment. By utilizing HDFS, shopping logs can be stored in their original format at a lower cost, while Map Reduce facilitates the calibration and development of advanced analytical models. The output generated by these models, which is considerably smaller than the raw data, can then be loaded into operational and warehousing systems for broader utilization.

Access to air shopping request and response data offers a competitive advantage to leverage large scale data analytics. Three examples are described below.

5.13.1 Demand Forecasting Based on Consumer Preferences

Traditional demand forecasting relies on time series models. Historical data is used to forecast demand using moving average and exponential smoothing models. An alternative is to utilize shopping data to predict consumer demand, which aligns with the natural demand process of browsing and choosing flights. Gaining access to shopping data is essential for comprehending consumer preferences and accurately modeling demand based on a consumer's first choice of air travel. Shopping data encompasses a consumer's interactions, including the requests made, the itineraries displayed, and the bookings made based on the consumer's observations.

Shopping data is required to calibrate a consumer the choice model for forecasting demand. This method involves sophisticated modeling of the consumer's selection process, considering schedule and fare attributes.

5.13.2 Competitive Revenue Management

Traditional revenue management relies on historical data to forecast demand and set inventory controls to maximize revenues. This is a myopic view since prevailing competitive marketplace conditions are not considered when inventory controls are established (Ratliff & Vinod, 2005). GDS shopping data can be used to enhance the traditional revenue management application into a competitive revenue management application that modifies inventory controls based on the selling fares of competitors in the market.

5.13.3 Optimizing Screen Display

Measuring the quality of screens displayed to online users allows an agency to assess their competitiveness in converting shoppers into bookers. The selection process for air itineraries can be overwhelming since, on average, shopping algorithms generate around 1000 outbound flight schedules and 1000 inbound flight schedules for each shopping request on an OTA. With a total of 1 million itinerary options (1000 × 1000), it is required to choose the optimal set of itineraries that offers the best alternatives for consumers. Customers do not automatically choose the lowest priced itinerary. Instead, the displayed itineraries for each shopping request should offer diversity in terms of service quality (such as nonstops, single connect, double connect, and interlines), fares, multiple carriers, and a minimum number of inbound itineraries for every outbound itinerary. To measure screen quality, a calibrated consumer choice model (CCM) can be used to determine the probability of a displayed itinerary being selected by the user. The choice model must consider various factors, including the selling fare and schedule attributes.

In addition, a variety of itineraries also improves conversion rates. A collection of highly rated CCM model itineraries may all be non-stops leaving in the morning, but some travelers might prefer cheaper connections in the afternoon.

By measuring screen quality, algorithms can be continuously improved to ensure that wrong or ineffective itineraries are not displayed, ultimately leading to higher conversion rates.

5.14   Potential Impacts of Air Shopping with IATA NDC

There are several changes to be anticipated with IATA NDC in the realm of air shopping and airline pricing. Air shopping capability will be resident in airlines systems with NDC. The GDSs can stop investing in air shopping since pricing power is gradually shifting from the GDS to the airline. The future of NDC shopping and how airlines can gain a competitive edge are discussed in Chap. 8.

The potential impacts of NDC on GDS air shopping in the hybrid state are summarized below:

5.14.1   Schedule Participation

When an airline migrates all content for all flights to NDC, it can be argued that they no longer have to publish their schedules to schedule aggregators like OAG and Cirium. However, this is unlikely to change since GDSs, to request offers from airlines, require access to schedules to determine the airlines who offer a service in a market requested by the customer.

5.14.2   ATPCO Participation

Full service airlines use fare aggregators to keep track of their competitors' prices and update their own prices accordingly. These changes are then sent to Global Distribution Systems (GDSs) and third-party retailers through the fare aggregators. This ensures that prices are transparent and easily accessible.

However, with the implementation of new distribution capability (NDC), airlines no longer need to share their market fares with fare aggregators. As a result, only the airline can determine the price of an itinerary. This has led some to argue that full-service airlines are no longer transparent in their pricing, resembling low-cost carriers in their lack of transparency.

5.14.3   Booking Classes aka Reservations Booking Designators

It is anticipated that over the next decade, with the adoption of NDC, booking classes will be phased out for various reasons.

Many airlines believe that the 26 letters in the alphabet that are used as booking classes do not offer sufficient flexibility. Eliminating the discrete price points and moving to a continuous scale can capture the gaps on the price demand curve.

After NDC adoption the city pair availability display will no longer be required since travel agents are no longer booking flights based on city pair availability in the GDS. Further, airlines will not have to distribute availability to GDSs with AVS/AVN messages or with a direct connect availability response. With the transition to offers and orders, many FSCs are adopting continuous pricing which is also known as classless revenue management (Isler, 2016). With classless revenue management, the end product of revenue management is not booking class availability, but the price of an itinerary.

5.14.4   Schedule and Fare Transparency

Schedule and fare transparency will no longer be the same when each airline distributes its offers on demand. While there will be some level of information, the bundled offers pose a problem since offers originating from different airlines are not homogenous. With classless revenue management, the end product of revenue management is not booking class availability, but the price of an itinerary.

5.14.5   Air Shopping with Non-standard Pricing Formats

Today all air shopping algorithms in the market from GDSs and third-party vendors abide by the ATPCO rules and restrictions to price an itinerary. Many airlines may decide to not publish their fares through ATPCO in the future and, instead, maintain the tariff structure in an internal format. Airline shopping engines licensed from third-party providers including the GDS will have to be adapted to work in this new custom pricing structure environment which can vary from one airline to the next.

5.14.6   Agency Debit Memos

After an itinerary is selected from the shopping response, a booking is made. Prior to ticketing, the itinerary goes through an itinerary pricing system that performs a comprehensive validation of all applicable fare rules to determine the final price. GDSs make a significant investment in the pricing system and ensure its accuracy. Unlike shopping, the final price inclusive of taxes and surcharges is determined for a single itinerary by the pricing system. The pricing system response prior to ticketing is accurate. However, situations could arise when the GDS has not received the latest fare changes, or the GDS auto-pricing did not consider all the subtle nuances of a complex international itinerary. If the airline determines that an itinerary is priced incorrectly as part of their quality check sampling process, a debit memo for the price difference is issued to the travel agency. A travel agency can dispute the debit memo or pay the invoice if it is an agency error. The GDS technically does not have to reimburse the travel agency for auto-pricing by default unless the reimbursement policy is explicitly stated in the contract between the travel agency and the GDS.

In an NDC world, debit memos will cease to exist since only the airlines will be able to price an itinerary.

Chapter 6
GDS: The Turbulent Years (2012–Present)

6.1 Introduction

Following the IATA Board of Governors endorsement in 2011, IATA and a small group of airlines attended the Passenger Services Conference (PSC) on October 18–19, 2012, to pass resolution 787 (Enhanced Airline Distribution). The resolution was designed to support the development of an open Extended Markup Language (XML) based data exchange standard to complement the existing teletype and EDIFACT data exchange standards already managed by IATA for the airline industry (IATA, 2012). Resolution 787, also known as the New Distribution Capability (NDC), had a stated objective to overhaul the existing legacy messaging exchange standard with a new XML standard between airlines and intermediaries. The underlying goal was to make it easier, faster, and cheaper to provide customers with comprehensive information about fare alternatives, airline ancillary products and services, onboard amenities, and rich graphics such as seat maps when booking through a travel agent.

Resolution 787 was adopted at the conference. Subsequently, it was sent to the U.S. Department of Transportation (DOT) for approval on March 11, 2013 [DOCKET OST-2013] with an intended effective date of June 1, 2013. DOT gave tentative approval in May 2014 and final approval in August 2014 for Resolution 787, paving the way for NDC adoption in the United States. The implementation of NDC was strictly on a voluntary basis, and the first set of NDC schemas were released on September 1, 2015.

6.1.1 Stakeholder Concerns

The evolution of airline product distribution has raised questions from stakeholders across the travel ecosystem. Marketplace dynamics for intelligent retailing are

B. Vinod, *Mastering the Travel Intermediaries*, Management for Professionals, https://doi.org/10.1007/978-3-031-51524-8_6

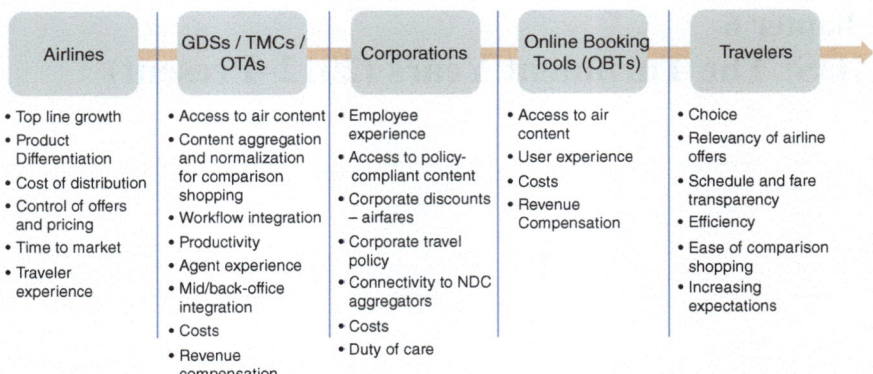

Fig. 6.1 Stakeholder concerns

Fig. 6.2 Primary GDS domains

driving the need for new capabilities in product distribution. Primary concerns across the travel value chain include workflow integration, process automation, agent productivity, access to content, revenue compensation for services rendered, duty of care, cost of IT investment, and many more. Figure 6.1 provides an overview of stakeholder concerns that will be elaborated throughout this chapter.

6.2   GDS Domains

The GDS has many domains, which are a subset of the domains in the airline reservations system. This is because loyalty, departure control, reservations inventory control, AVS (availability status) generation and recap, pre-reserved seats, weight and balance, etc. are specific to airline reservations processing. The primary domains of a GDS are shown in Fig. 6.2.

As seen in Fig. 6.2, many critical components are required to operate a GDS today when the GDS is responsible for offer creation and pricing. For airlines that participate in a GDS, when the industry transitions to NDC in its entirety, many of these core domains will no longer be required.

6.3 GDS Data Used in Decision Making

The GDS provides a wide variety of unique data that help travel management companies and travel suppliers such as airlines and hotels make informed decisions. In the case of airlines, GDS data types are bookings, tickets, and shopping. This data serves a variety of purposes for airlines and travel agencies.

Shopping data is technically classified as Big Data, which can be hundreds of terabytes or petabytes (10^{15}) in size. In the past decade, GDSs have seen an explosion in shopping volumes from a few terabytes per day to hundreds of terabytes per day, and this data is growing at an increasing rate. Air shopping volumes have outpaced bookings over the past decade.

Shopping data, along with booking and ticket data, supports a wide range of applications. Here are a few examples:

6.3.1 Airline Applications

There are several applications that benefit airlines, primarily in the revenue management and schedule profitability space. Here are a few examples.

6.3.1.1 Demand Forecasting Based on Consumer Preferences

Instead of relying on traditional time-series models that use historical data to forecast demand, another option is to analyze shopping data and use a top-down approach starting with an estimate of market size to estimate market share. This involves studying the consumer's preferences during the itinerary selection process, from the initial shopping request to the itineraries displayed and ultimately what is booked. By examining schedule and fare attributes, this method follows the actual demand process of selecting an itinerary and offers a more sophisticated approach to predicting demand.

6.3.1.2 Dynamic Availability

Dynamic availability is a competitive revenue management technique. It involves modifying existing inventory control recommendations by opening and closing booking classes based on prevailing competitive market conditions (Vinod, 2021a), particularly in key markets. GDS booking and ticket data are necessary to estimate an airline's market share and develop strategies to maintain it. Competitive selling fares from GDS shopping responses must be monitored to determine optimal inventory controls. To determine the attractiveness of each itinerary, a consumer choice model must be calibrated using data extracted from shopping requests and responses, including variables such as displacement time, elapsed time, fare, and schedule attributes.

While many airlines have benefited from dynamic availability over the past few years, it will no longer be tenable with IATA NDC since selling fares in the market may not be transparent and the offer creation process may not require booking classes.

6.3.1.3 Dynamic Pricing

Dynamic pricing is an integral component of an airline's offer that is distributed through the IATA NDC messaging standard. It is closely related to dynamic availability and involves determining a dynamic fare based on selling fares in the marketplace and the airline's remaining seat inventory in the market. A session-based fare optimizer determines the optimal price point for the host airline based on the competitive set and current selling fares of competing airlines in the marketplace. This approach will not work in the absence of fare transparency in a NDC world. However, the dynamic price can still be determined based on an airline's demand forecast and remaining capacity without access to competitive selling fare data. The dynamic price is used to approximate the ticketed price instead of converting it into an inventory control recommendation like dynamic availability.

Dynamic pricing is an active area of academic research (Gallego & van Ryzin, 1994; Christ, 2009; Nasiry & Popescu, 2011). While dynamic pricing can be deployed on a consumer direct airline website controlled by the airline, the travel agency channel that comprises a large percentage of bookings relies on pricing itineraries in their own environment based on the fares and rules filed by the airlines to fare aggregators. This will change with IATA new distribution capability (NDC) when the airline is the only entity that can price a dynamic offer.

The reservations booking designator (RBD) has played a crucial role in segmenting customers and distributing availability of an airline's product through intermediaries such as GDSs. However, this will change with IATA NDC, because the airline controls the creation of the dynamic price point across all channels of distribution, eliminating the need for a booking class. This concept is known as classless revenue management in the IATA NDC context.

While dynamic pricing of opaque travel products (Zouaoui & Rao, 2009) has minimal business process changes, the potential impacts of directly deploying dynamic pricing on third-party systems can be quite significant (Choubert et al., 2015).

6.3.1.4   The Value of Shopping Query Data

Academic journals and practitioners frequently discuss the value of shopping response data and how it can be leveraged to solve airline problems. However, shopping query data is rarely mentioned. Airline websites, OTAs, and GDSs receive millions of raw search requests daily (Vinod, 2011a). About 10% of all queries receive no shopping response for a range of reasons. For example, a customer on an airline website may request a destination that the airline does not serve. These queries can be captured and processed to develop metrics and analytics to gain additional insights into customer behaviors.

Typical shopping requests can be generic (origin, destination, departure date, and return date) or specific (origin, destination, departure date, return date, carrier, cabin, service, etc.). It can be argued that by developing the right performance metrics, this data can be used by an airline to influence key airline marketing planning functions such as flight scheduling, fare management, revenue management, marketing programs, and frequent flyer redemptions.

Shopping queries can be categorized as generic or specific. The categorized shopping queries can be aggregated and summarized by key input attributes of the shopping request as well as derived attributes.

Examples of key input attributes are destination city, origination city, number in party and distribution of number in party (adults, children), carrier preference, and service preference.

Examples of derived attributes are shopping queries by days to departure (e.g., three-day advance purchase, seven-day advance purchase, etc.) and minimum stay requirements (e.g., three days, six days, greater than seven days, etc.)

Sample measures that can be derived from shopping query data are:

1. Destinations ranked by popularity (queries from multiple origin points to the same destination). This index is across all carriers that serve the destination.
2. Destinations ranked by popularity from the most popular to the least popular (queries from multiple origin points to the same destination) by carrier. This data can be used by airlines to influence pricing actions and exploit pricing power in popular destinations.
3. Distribution of number in party by destination. This can determine the destinations for two-for-one promotions, child-travel-free promotions, etc.
4. Origin cities with the most to least requests by carrier. Origin cities with the least requests can be targeted by airlines for marketing spend in offline channels, e.g., radio, print magazines, etc.

6.3.2   Travel Agency Application

Travel agency applications are also frequently referred to as travel agency revenue management applications. These applications have an objective to either improve service levels for their corporate customers or enhance their revenues.

6.3.2.1   Forecasting Fare Volatility

The task of fare forecasting involves predicting future airline selling fares, which is a challenging problem due to various factors such as revenue management inventory control recommendations, competitor fare and fare changes, revenue management strategy, and promotional and seasonal fares. Fare forecasting cannot rely on ticketed data, as consumers may not have purchased the lowest priced fare. Instead, air shopping data is an ideal source for developing machine learning algorithms like Q-Learning (Watkins, 1989), which make recommendations to "buy" or "wait" based on expected fare fluctuations. A "buy" recommendation is made when fares are expected to increase and a "wait" recommendation is made when fares are expected to go down. Reinforcement learning techniques like Q-Learning are effective in prediction and involve an "agent" to classify the recommendations as right or wrong and the model learns from its recommendations and improves the accuracy of predictions over time.

Airlines dislike fare prediction models since it is an approach to reverse-engineer the revenue management process. Mining airfare data to minimize the cost of travel was and continues to be an active area of research in academia as well (Etzioni et al., 2003; Mantin & Rubin, 2016).

To optimize corporate travel budgets and reduce costs, travel agencies deploy automated re-shopping and re-booking of existing booked itineraries when the selling fare has gone down in price. To re-book, the savings must be greater than any change fees that may be incurred with a cancellation of the booked itinerary and re-booking at the lower fare. Instead of manually reviewing all booked itineraries, the fare prediction model can be used to evaluate only those itineraries that are expected to decrease in price.

6.3.2.2   Optimal Markup of Net Fares

Net fares are fares negotiated between an airline and a travel agency. What the agency owes to the airline is the negotiated net fare, and the agency can markup the fares for sale. While net fares are less prevalent in North America, they are more prevalent in Asia Pacific and the Middle East. GDS shopping response data is used to calibrate a choice model to determine the optimal markups to determine if the corporate fare can be substituted with a marked up net fare that generates revenue for the travel agency and reduces the cost of travel for the corporate or leisure traveler.

6.3.3 GDS Application

The primary GDS application improves the screen displays on the agency desktop to boost productivity.

6.3.3.1 Optimizing the Screen Display for the Travel Agency Desktop

By measuring screen quality, travel agencies can assess their competitiveness in converting shoppers into bookers. Selecting air itineraries can be overwhelming as shopping algorithms generate and evaluate thousands of options, and the optimal set must be chosen to offer the best alternatives to consumers. Corporate clients do not always choose the lowest priced itinerary, but rather the one that fits their schedule. Similarly, leisure travelers desire a diverse range of itineraries expressed by quality of service. For example, non-stops, single connect, double connect, interlines; fares and carriers on both the outbound and inbound schedules play a critical role in the selection process. A calibrated choice model can measure screen quality by determining the probability of selecting a displayed itinerary based on attributes such as selling fare and schedule. Measuring screen quality continuously improves shopping diversity algorithms by displaying suitable options, thus increasing the conversion rate. Itineraries can be ranked based on the utility score from the choice model to maximize the conversion rate.

6.3.4 Future State

With IATA NDC the existing GDS workflow of organically generating shopping responses from travel agencies shopping queries comes to an end. Shopping responses generated by a GDS based on their shopping algorithms will be replaced with airline specific content returned by each airline that was requested to provide offers. The airline *owns* this airline generated data consisting of bundled itineraries, and it is not clear if future GDS contracts will allow GDS entities to leverage this data for their airline partners to provide services such as forecasting demand based on schedule and fare attributes, and competitive revenue management that is based on influencing inventory controls based on the selling fares of competing airlines. It is a safe assumption that GDSs will be able to use the airline generated shopping data for their internal use such as improving travel agency displays, but not be allowed to bundle and sell enterprise software products to airlines.

6.4   Limitation of the GDS Platform and the NDC Initiative

Airlines often complain about the limitations of the GDS messaging through EDIFACT when it comes to generating offers for customers. Airlines and many in the travel industry classify the GDS as a legacy mainframe product that has not kept up with technological advances. With NDC, airlines hope to control and manage the indirect sales channel in a cost-effective manner with airline controlled offers, as they do with their consumer direct airline website.

The limitations of the GDS to serve the needs of airline retailing of the future are:

6.4.1   Open Marketplace

The traditional GDS model is in many ways a closed ecosystem. The successful marketplaces of the future will be open marketplaces. This requires the ability to allow business entities, buyers, and sellers of travel and non-travel related services, to register and connect to the marketplace. It should also promote the ability for third parties to build new products and services to complement the baseline and co-create value. This plug-and-play capability is a fundamental defining characteristic of a platform.

6.4.2   Access to the Customer

Carriers have for many years wanted to learn more about their customers to provide additional services such as flight insurance, baggage insurance, cabin upgrades, and meals on board. These optional services are called ancillaries, and they are a source of incremental revenues for the airlines. When a travel agent serves as an intermediary between the customer and the airline, the traditional GDS workflow to shop, book, and ticket a flight does not allow that since offers are created and presented by the GDS. The new standard proposed by IATA reverses the GDS workflow and travel agents must provide information about the customer before the offers are generated by the airline. In the context of transaction workflows for booking and ticketing customers, NDC is a seismic change.

6.5 Changes in the Distribution Landscape with IATA's New Distribution Capability

From an airline CFO's perspective, GDS costs are one of the last controllable expenses for an airline. It is not just GDS booking fees, but also credit cards and agency commissions. For over a decade (2005–present), the long-term strategy of network carriers has been to advance distribution from fare and schedule-led selling to merchandising to transform themselves from suppliers of a commodity, an airline seat, to product marketers of airline bundles of base fare and air ancillary products. Despite the value provided by the GDS in providing price transparency and delivering higher valued corporate customers through managed travel programs, this is seen as a shortcoming.

GDSs have been under the threat of disintermediation since the early 2000s. "Disintermediation" is technology and/or business process that will cut the GDS out of booking channels and fees. Full content agreements negotiated between airlines and GDSs do not apply for the sale of ancillary products sold by airlines. Further, travel agents and GDSs are not compensated for the sale of ancillary products such as bags, pre-reserved seats, lounge access, and meals at time of booking. Ancillary products are not part of the full content agreements, which implies that airlines do not have to distribute ancillary products to ATPCO and GDSs.

Low-cost carriers (LCC) are budget airlines that have seen double-digit growth over the past decade and most of them do not participate in the GDS to avoid the cost of indirect distribution. They account for a third of the air traffic volume worldwide. When a travel agent books an LCC itinerary for a customer, the LCC controls the offer creation process, and the GDS booking is a pass-through booking to the LCC's host CRS.

6.6 New Distribution Capability: A Decade in the Making

American physicist and Episcopal priest William Grosvenor Pollard once said:

> "Learning and innovation go hand in hand. The arrogance of success is to think that what you did yesterday will be sufficient for tomorrow".

This quote is apt in the world of airline product distribution where there are many forces at play that are defying the status quo as it has existed since 1976 to create a fundamentally different transaction workflow, business process, and revenue model.

IATA launched the new distribution capability (NDC) in 2012 based on feedback from the airline community. Lufthansa, British Airways, American Airlines, and Iberia were the early proponents of NDC. It is a data exchange format to create and distribute personalized offers to the customer across all channels of distribution, including the GDS. Order management processing was included in 2015.

In today's environment, a travel agent subscribing to a GDS can shop, book, price, and ticket an itinerary for a customer. With IATA's new distribution capability (NDC) that is currently being rolled out, pricing power shifts from the GDS to the airline. An agency must request itineraries and prices from an airline in an NDC world, and the task for the GDS is to normalize the nonhomogenous content across travel suppliers for display to a travel agent. Hence, creation and delivery of content with this approach will be within the domain of an airline's environment. The promise of NDC and the new XML-based messaging standard between airlines and GDSs are several.

6.6.1   Benefits of NDC

When NDC was launched, the key benefits of the new approach to air product distribution were identified to enhance the customer experience. While reducing the cost of distribution is the primary driver today, it was not mentioned when NDC was launched. They are:

6.6.1.1   Rich Content

The term rich content in the context of NDC has a very broad definition. It includes priced itineraries with ancillary products, value-added products and services, photos, videos and expanded text of facilities, policies, passenger reviews, etc. pertinent to the flight itinerary. All this is made possible with the new XML-based messaging standard instead of EDIFACT (Electronic Data Interchange for Administration, Commerce, and Transport), with its inherent limitations.

6.6.1.2   Product Differentiation

Enables an airline to differentiate and offer products and services to customers that differentiate it from a competitor. While the traditional GDS model only displays schedules and fares, the new messaging standard allows an airline to showcase their rich content of value-added services. While this is theoretically correct, there are practical limitations since the GDS had to efficiently manage the screen real estate with NDC channel data received from multiple airlines that are not homogenous.

6.6.1.3   Pricing Power

The traditional price management framework requires airlines to publish their fares through fare aggregators like ATPCO and SITA. This has the benefit of fare transparency across airlines that compete in the same market. Dynamic pricing is the latest generation of revenue management that bridges the chasm between fare management and revenue management by recommending a dynamic price at the time of shop based on prevailing, competitive market conditions.

There are two forms of dynamic pricing, laddered dynamic pricing and dynamic pricing on a continuous scale. Even though ATPCO standards were created by a working group consisting of airlines, GDSs and vendors for laddered pricing (Dezelak & Ratliff, 2018), laddered pricing has not gained traction as airlines focus on continuous dynamic pricing.

Further with NDC, pricing power shifts from the GDS to the airline. This requires the airline to price every itinerary for display in a GDS agent desktop. There is also the threat that fare aggregators ATPCO, and SITA may become irrelevant. Airlines are no longer required to publish fares to fare aggregators since the itinerary price is computed by the airline when a request is received and not by the GDS.

6.6.1.4 Offer Creation

The new offer creation process enables airlines to maximize network revenues with dynamic pricing of the itinerary and inclusion of dynamic ancillary bundles based on context for travel. There are two approaches to dynamic pricing for the base fare: laddered pricing and continuous pricing. Ancillary products and services in the offer can include seats, bags, in-flight entertainment, Wi-Fi, meals, airport lounge access, and many more to generate incremental revenues.

6.6.1.5 Pricing Structure and Improved Conversion Rates

With NDC, airlines are no longer constrained to a fixed number of booking classes which implies a fixed number of price levels. By quoting a continuous price on the price demand curve, NDC fares will be cheaper than traditional files fares through ATPCO, leading to improved conversion rates.

The new XML APIs enable airlines to display fare families at multiple price points with increasing value to promote upsell.

6.6.1.6 Personalization

A common theme for dynamic pricing, offer creation, and personalization is that the GDS no longer controls any of these processes and the airline is in control.

Under the traditional GDS model, both brick-and-mortar travel agents and online travel agencies do not provide any customer data to the airline. However, with NDC, the message format has a provision for the intermediaries to furnish airlines with customer data, which enables the airlines to customize the shopping experience by providing customized offers containing exclusive content not found within the GDS. By leveraging the customer data, airlines can create tailored offers with rich content that is not available in the GDS. These personalized offers can drive revenues at the time of booking by selling ancillary products and services like pre-reserved seats, baggage, Wi-Fi, lounge access, meals, and more. The base fare included in the offer can also be a dynamic price, otherwise referred to as continuous pricing.

6.6.1.6.1 Customer Relationship Management and Personalization

When loyalty programs were first introduced, the concept of customer relationship management (CRM) was not widely recognized. CRM allows airlines to target customers specifically through special campaigns and offers. It serves to generate new demand and convert that demand into loyal customers. CRM initiatives complement an airline's established loyalty programs. These initiatives rely on the customer database within the loyalty program as a starting point, enabling the development of a deep understanding of individual customers, capture their preferences, and estimate customer lifetime value (*CLV*) or the potential future value (*PFV*) which is synonymous with future customer value, remaining *CLV* or residual *CLV*. This knowledge allows for customer recognition at various touchpoints and provides valuable information on each customer's worth. Through advanced analytics, CRM also offers a framework to target customers with cross selling and upselling opportunities based on their profile, as well as providing enhanced service to each customer.

With the adoption of NDC, CRM tools to gain insight into the loyal customer base will increase in stature to generate personalized offers. Calculation of *CLV* and traditional CRM methods for customer segmentation are discussed in Chap. 9.

6.6.1.7 Payment Processing

Expenses for airlines related to payments in the industry come in second place only to the GDS segment booking fees, commissions, and incentives. Prior to NDC, travel agents could choose to receive payment from the customer and settle it through the IATA Billing and Settlement Plan (BSP) or use a GDS to authorize credit card payment to the airline's designated bank. However, with NDC, travel agents transition from a payment intermediary to a payment pass-through, and the airline is now responsible for processing the payment as the merchant of record. It is expected that NDC payments will decrease payment costs in terms of merchant fees and enhance the customer experience by providing more payment options in the future. For example, airlines will be able to use payment instruments such as PayPal, bank transfers, pay by installment, and frequent flyer miles to indirect distribution. Led by the Asian market, the list of payment instruments is growing rapidly and supporting new forms of payment will be required for an airline to be competitive and expand the customer base.

6.6.1.8 Cost of Distribution and NDC

It is important to analyze the distribution expenses from both the airline and GDS points of view.

The main factors that relate to the cost of distribution through the indirect channel are:

6.6.1.8.1 Cost of Compute for Offer Creation

Airlines adopting NDC now have greater control over pricing, which has significant implications for the cost of computing required to produce offers.

The current computing capacity of GDSs enables the creation of offers, with air shopping being the most substantial investment in their data processing environment. This investment costs millions of dollars annually and is used to generate schedules and priced itineraries for each shopping request. For instance, prior to the COVID-19 pandemic, Sabre GDS processed over 249 billion shopping requests from travel agents, peak shopping requests exceeding 28,000 per second, and itineraries returned during shopping exceeded 2.5 trillion per month in 2019 (Vinod, 2020a). In certain markets, the look-to-book ratios generated from OTAs and processed by the GDS can sometimes exceed 20,000:1. Additionally, shopping volumes have historically increased by 60% year over year. Amadeus and Travelport support comparable volumes.

In an NDC environment, the airline takes on the responsibility of generating priced itineraries, rendering the GDS compute power unnecessary. Consequently, airlines will require a larger investment over time to produce all offers, and the GDS is obligated to determine how to display the offers returned by airlines through the NDC XML messaging gateway on the travel agency desktops.

It is improbable that all airlines worldwide will adopt NDC for all bookings by the year 2035. This means that transaction processing for airline bookings will be in a *hybrid state*, with a combination of traditional GDS EDIFACT bookings and NDC content-based XML bookings. Both the GDS compute environment and airline computing environment for NDC offers will coexist within this hybrid state. However, during the transition to NDC, the GDS computing costs will decrease, while the airline computing costs will increase to compensate for the offers that the GDS is no longer generating. Airlines must prepare for the steep increase in look-to-book ratios, and the associated higher volume of queries that their systems will process.

6.6.1.8.2 GDS Booking Fees

During the transition to NDC, the GDS booking fees for traditional EDIFACT bookings are not expected to go down until a tipping point is reached for NDC bookings. This may not be true in all airline-GDS contracts. When the tipping point

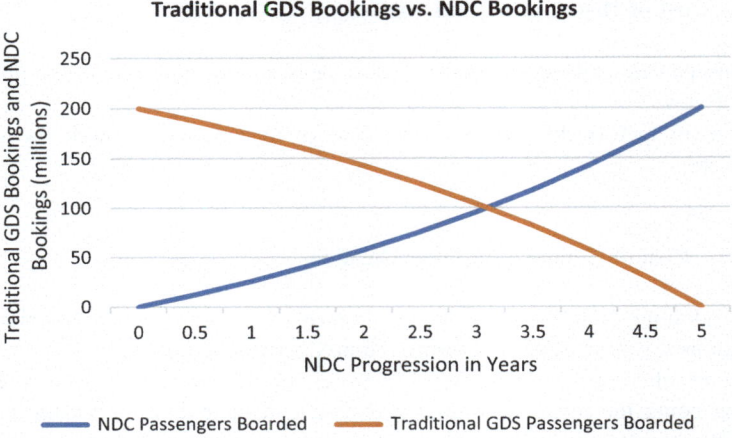

Fig. 6.3 An airline's transition to NDC bookings

is reached, there will be renewed pressure from the airlines on the GDSs to reduce the segment booking fees for traditional GDS EDIFACT bookings.

For NDC bookings made through the GDS, the pass-through bookings, the cost of distribution will be lower from the beginning. Airlines will continue to negotiate deeply discounted segment booking fees for NDC-content pass-through GDS bookings. To compensate for this loss in revenue, GDSs will not offer incentives to travel agents for NDC pass-through bookings. Further, they have two options to offset the revenue loss. First, since all NDC fares are private fares, a GDS can technically markup the NDC fares before they are presented to the travel agent. Second, they can add a booking fee surcharge like the booking fee charged by NDC aggregators. In either case, the total fare with either the mark up or the surcharge will be higher than the NDC fare provided by the airline as part of the offer. If the markup or surcharge is the same across all airlines, fare competitiveness is not an issue. However, if they are different based on GDS-airline contract economics, the risk associated with the markup or surcharge could render some airlines to be less competitive than others.

A data point that should be considered by GDSs is the cost of distribution through the airline preferred new-entrant NDC aggregator channels. NDC aggregators charge the seller (the travel agency) a booking fee to access and book NDC content directly with the airline. A booking made through an NDC aggregator is roughly 80 cents to $1.50 per booking, which is a fraction of the current distribution costs paid by the airlines to the GDSs. However, the key difference with using NDC aggregators is that the airlines never see this cost since the travel agency must pass this cost to customers as a customer service fee.

Figure 6.3 illustrates a hypothetical transition for an airline from traditional GDS bookings to NDC bookings regardless of the channel, the GDSs or NDC aggregators for an airline with 200 million passengers boarded per year. The transition in Fig. 6.3 is aggressive and assumes that 100% of bookings transition to NDC in five years.

Assuming the passenger boarded count stays the same over the five-year period, the graph shows the point when NDC bookings exceed the traditional GDS booking model. Based on the slope of the ascending and descending curves, the point at which the two curves intersect, the tipping point, is a little over three years. At the intersection point, NDC bookings and traditional GDS bookings are the same after which NDC bookings increase at a faster pace. It is likely that the pressure to further reduce traditional GDS (EDIFACT) booking fees will occur at the tipping point which will be in the vicinity of the intersection point. NDC bookings, however, will always be at a lower price point for GDS pass-through bookings.

6.6.1.9   Omni-Channel

There is an opportunity to promote an omni-channel strategy across all channels of distribution because only the airline has the responsibility to create and distribute offers. However, absence of discipline in not treating all channels equally will result in divergence from an omni-channel strategy.

6.7   The Transition to NDC

The planned migration to NDC is explained in three stages: the way GDSs have operated since 1976, the hybrid state with NDC and the end state when all airlines have migrated to NDC.

6.7.1   Standard GDS Booking Workflow

Figure 6.4 illustrates the standard GDS workflow for travel agent subscribers. Travel agents use GDSs to shop, price, generate offers (itineraries with ancillaries), book, ticket, and collect payment. Before NDC, the number of travel agencies and OTAs who sourced content directly from an airline were the exception rather than the rule.

Figure 6.4 illustrates how traditional GDS content is used to create a booking. Airlines create schedules and distribute them to schedule aggregators such as OAG or Cerium that in turn distribute the schedules to subscribers like the GDSs. The Schedule Change Information Manual (SSIM) is an IATA format by which airline schedules are distributed by the schedule aggregators weekly. SSIM is a standard schedule format that airlines use to exchange flight schedule data with GDSs, airports, and other partners. Changes to the SSIM are also submitted every other week. SSIM is used for bulk schedule data, like the complete schedule of an airline.

When schedule changes are published to the SSIM, there are two types of messages, called standard schedule message (SSM) and ad hoc schedule message (ASM) that are sent by the airline to the GDSs. An SSM transmission permanently

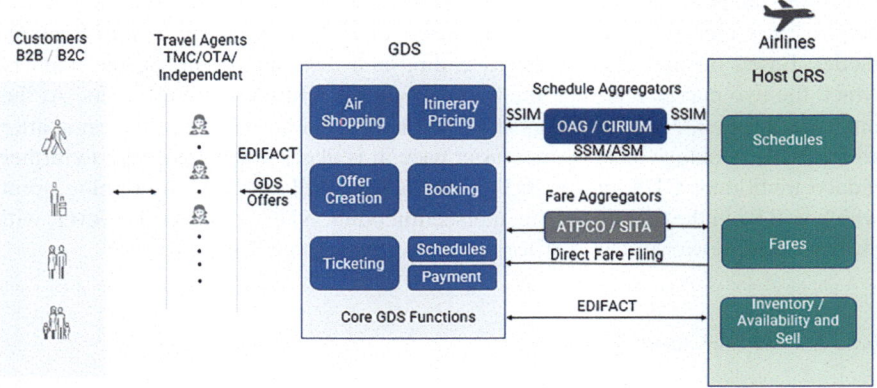

Fig. 6.4 Current GDS transaction workflow

changes an airline's regular operating schedule, while an ASM is an update to the regular operating schedule for single days, sent by the airline's operational control systems. SSM is usually used for schedule information outside the operational period and may include a collection of schedule changes. In contrast, ASM is normally used for the operational period for a single-dated schedule information. SSM and ASM are usually telex message formats.

Airlines have a fare management capability to determine if new fares need to be introduced into a market or if existing fares need to be updated based on competitor fare filings. Airlines initiate fare actions and submit their changes to fare aggregators like ATPCO, which in turn transmits the fares, routings, and automated rules to subscribers like GDSs and airlines. In the domestic U.S. market, due to the high volumes of fare changes, ATPCO transmits on weekdays (Monday through Friday) at 1000, 1300, 1600, and 2000 h. On weekends (Saturday, Sunday), the transmissions are once a day at 1700 h. International data transmissions are more frequent and are made every hour on the hour on weekdays and with a reduced frequency on weekends.

Figure 6.4 illustrates the current GDS transaction workflow.

In this environment, *pricing power is in the domain of the GDS,* which can generate shopping responses of priced itineraries and accurately price an itinerary for ticketing and payment. Air shopping is a multimillion dollar investment for a GDS and may consist of a combination of schedule-led and fare-led algorithms. In the GDS transaction processing world, *scale matters*, Air shopping volumes have grown rapidly, mostly fueled by shopping requests from the OTAs. For example, prior to the COVID-19 pandemic, the total volume of travel agency shops in 2019 was about 249 billion on the Sabre GDS which averages over 700 million shopping requests from agencies per day. Further, peak transactions per second in 2019 were approximately 25,487 (Vinod, 2020a). Displaying many itineraries for a shopping request is not the answer. It is important to display the best set of itineraries for a customer based on their schedule and fare preferences. All aspects of shopping, except for querying an airline host CRS for booking class availability, are performed internally in the GDS environment. Scale, from a GDS perspective, includes both business and technical considerations. While internal transaction processing needs to be efficient to deliver shopping responses after evaluation of the immense search space in under three seconds, business considerations are equally important to meet the needs of corporate travel booking tools, travel management companies, and online travel agencies who access the GDS.

Once the itinerary is selected, the GDS creates the PNR which belongs to the travel agency. The PNR is stored in the GDS and the airline owns a part of the PNR.

There are two alternatives for payment processing. If the agent is the merchant of record, payment is collected by the agent and settled with the airline at a later stage using BSP. Alternately the travel agent uses the GDS to provide customer card details. The agent does not collect payment. After the GDS payment authorization step, the payment is credited to the airline bank.

Once payment is completed, the agency requests the GDS to start the ticket issuance process with the airline. The GDS checks if the airline is approved to issue tickets through this agency, assigns a neutral ticket number, and forwards it to the airline for ticketing. The airline reviews the information provided and notifies the agency through the GDS once validation is complete.

In a future environment with NDC, what is most important is to address scalability, both technical and business, to deliver the same level of efficiency and service.

Note that all communications between the airline and the GDS are based on EDIFACT and in some cases teletype. For example, interactive availability requests (also known as seamless availability) from a GDS to an airline and its subsequent response are based on EDIFACT, while availability status messages (AVS) are based on teletype.

6.7.2 The Hybrid State with NDC

The hybrid state is what we have today. While some airlines have started to adopt NDC, many have not. Even for the ones that have adopted NDC, their booking volumes through the NDC channel are significantly lower than the traditional GDS channel. Hence, they must continue to distribute their schedules through OAG/Cirium and fares through ATPCO/SITA. However, the NDC fares may be different from what has been filed to the fare aggregators for traditional GDS bookings using EDIFACT.

Figure 6.5 illustrates the transaction workflow with IATA NDC and traditional GDS content co-existing and operating in a *hybrid state*. The key difference is that for *NDC content pricing, power shifts from the GDS to the airline* which generates offers and takes control of distribution of the airline product through the indirect channel.

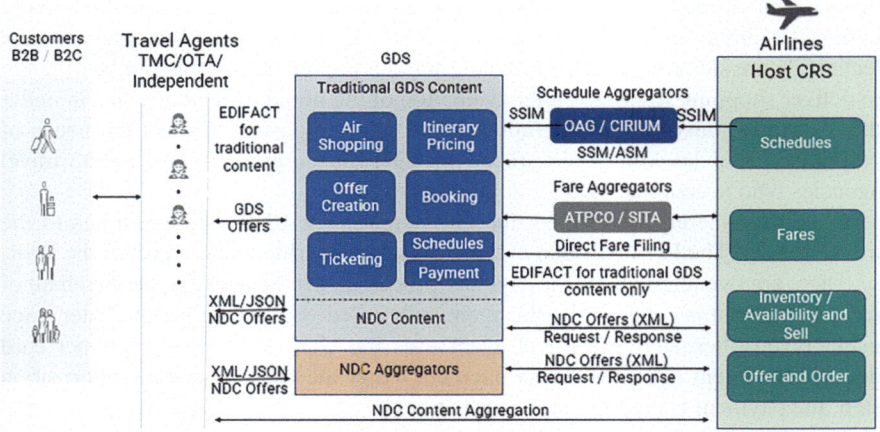

Fig. 6.5 Transaction workflow with NDC content and traditional GDS content (the hybrid state)

The process of migrating all airlines worldwide to NDC for 100% of bookings is no simple task. It is anticipated that the airline industry will be in a hybrid state for a minimum period of 10 years from 2024.

To receive NDC content, when a shopping request is made by a travel agent, the GDS sends the message to multiple airlines that serve that market for a response with NDC offers. The nonhomogenous airline responses are mapped to the predefined shelves on the travel agency desktop, and the content is normalized based on value before it is displayed to a travel agent. NDC content-based itineraries booked by a travel agent in the airline host CRS can be stored in the GDS as passive segments. The segments are passive, or informational segments, as they are also called. This is because any change to the itinerary must be requested by the travel agent and made by the airline. It also allows airlines to share rich content (e.g., images, text, video, etc.) and personalize the offer with dynamic pricing and ancillary bundles for product differentiation from competitors. NDC also gives flexibility to airlines to connect directly with OTAs, TMCs, and travel search engines without the need for a GDS.

For NDC content, once an offer is selected by the travel agent, additional ancillaries such as seats and bags can be added and repriced. The GDS or NDC aggregator sends a create order request to the airline who owns the order.

For NDC bookings, the GDS or NDC aggregator does not collect payment but sends the PCI compliant payment details to the airline. This triggers the creation of the PNR and E-TKT or the order. Like how payment is processed on the consumer direct channel, the airline can accept or reject the credit card that was submitted.

Once payment is accepted, ticketing can proceed and the airline issues the travel documents.

Note that the GDSs are by default aggregators of NDC content, but the GDS has significantly more capabilities than the new-entrant NDC aggregators. In this book, the term NDC aggregators refer to the new entrants in the marketplace.

The creation of the new NDC XML-based messaging standard has also given rise to many new entrants in the market called NDC aggregators. NDC aggregators offer connectivity to airlines, and they aggregate NDC content like a GDS. Sometimes, a large TMC may use a GDS and NDC aggregator for NDC content. This is because the NDC aggregator may have connected to airlines that the GDS still has on their roadmap for the future. This creates issues with mid-office and back-office reporting which is discussed in Sect. 6.23.2. When a travel agent makes a booking based on NDC content returned by an NDC aggregator, the booking bypasses the GDS and the NDC aggregator routes the booking directly to the airline. While disintermediation of the GDS exists in the hybrid state, it is still a very small percentage of indirect bookings in 2023.

Note that for NDC offers, the connectivity with the airline is based on the new NDC XML messaging standard and the GDS to travel agency connectivity is based on XML or JSON.

6.7.3 The End State with NDC

The end state assumes that all airlines worldwide, all 600 of them, have migrated to the NDC way of doing business.

Figure 6.6 illustrates the transaction workflow with IATA NDC in the end state. The key difference is that *pricing power has shifted entirely from the GDS to the*

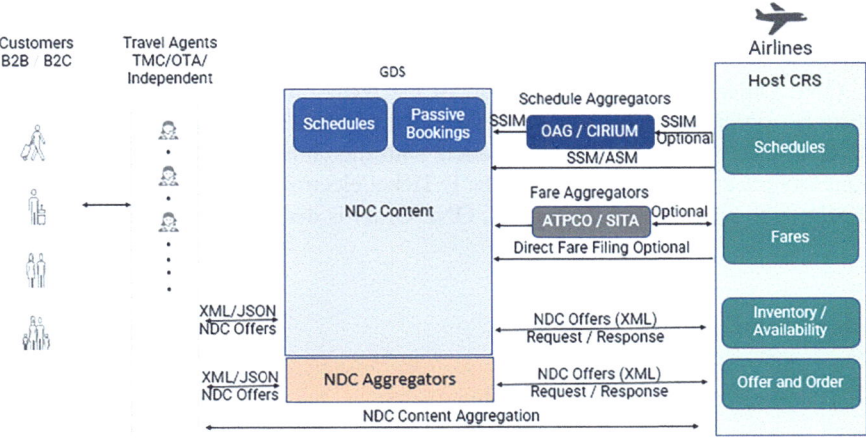

Fig. 6.6 Transaction workflow end state with NDC

airline which generates offers and takes complete control of distribution of the airline product through the indirect channel.

Airlines can use NDC to manage distribution through the indirect channel by sending customized rich content, priced itineraries, and air ancillary packages to travel agencies. NDC poses a threat to fare aggregators such as APTCO and SITA. They also pose a threat to schedule aggregators such as OAG and Cirium. Airlines may opt not to file their fares and schedules through these intermediaries in the future. However, the GDSs still require schedules to route requests from travel agents to the airlines that respond to NDC offer request, so airlines will continue to disseminate schedules to aggregators and GDSs.

On payment collection, prior to NDC, the travel agent would either collect the money from the customer and settle it through the IATA Billing and Settlement Plan (BSP) or use a GDS for payment authorization and credit card payment to the airline's designated bank. With NDC, the airline becomes the merchant that processes the payment. The airline is responsible for the payment transaction, which implies that the airline is responsible for any security breach at the seller when the seller is transmitting details of the client card to be used by the airline as a "merchant of record."

With a 100% NDC adoption, every travel agency booking will have to go to the airline for offers which is an itinerary with base fare and priced ancillaries. The GDS air shopping service used by travel agents and OTAs will no longer be required. While NDC may be in the early stages of adoption in 2023, it will eventually render the last remnants of full content agreements obsolete. This is because pricing power to create offers has shifted to the airline for indirect bookings by giving airlines the ability to merchandise their products and control the price quote. NDC also allows airlines to make the same offer across all channels of distribution or customize offers by channel or point of sale. There is a greater probability of diverging from an omni-channel strategy. It can be viewed as a financial restructuring of the airline industry for indirect bookings since the airline becomes the merchant that processes the payment.

ONE Order is also an integral part of the NDC program to introduce a single customer order record to capture all data elements that require to be fulfilled. This streamlines the sale of ancillary products with the elimination of multiple reservations records and the elimination of the E-Ticket/electronic miscellaneous document (EMD) with a single reference order. ONE Order is discussed in Chap. 9.

6.8   The Direct Channel

NDC impacts the indirect channel while the direct channel does not change. Online users access an airline's schedules and fares through the airline website. The response displayed on the airline website will be very different from one airline to the next because branded fare family definitions are unique to an airline. Airlines may also use an external shopping service that is typically provided by a GDS or

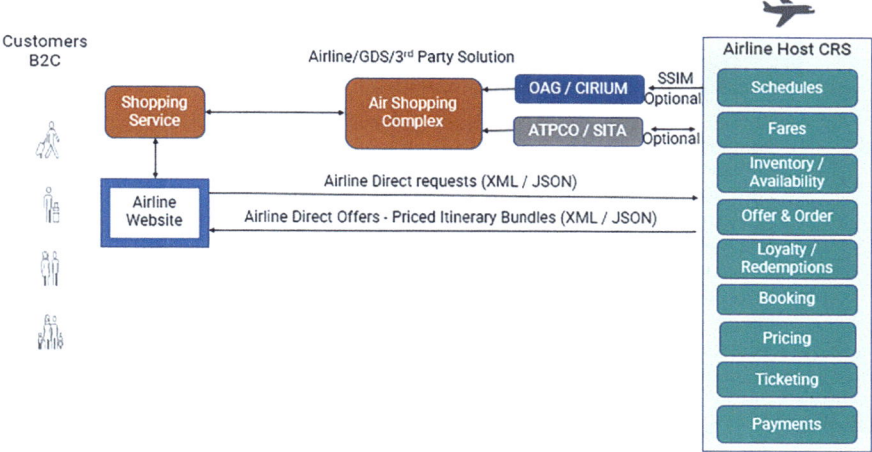

Fig. 6.7 The direct airline channel

Google/ITA. With NDC adoption, filing schedules and fares with aggregators is optional. This is illustrated in Fig. 6.7.

With NDC, many airlines aspire to extend leisure bookings to include corporate bookings. By providing a unique corporate id on the website, the pre-negotiated discounted NDC corporate fares can be displayed to a corporate user. The issue is that travel agencies do not like the idea of having to use alternate entry points for corporate bookings. Second, mid-office and back-office reporting will be an issue for TMCs that want to see consolidated agent sales and commissions reports.

6.9   Connectivity with NDC

Unlike connectivity described in Chap. 3 (Sect. 3.6), connectivity between an intermediary and an airline is very different to support NDC workflows since the offers are created and controlled by the airline. Traditional GDS commands like city pair availability and shopping is no longer required. Hence, the GDS no longer has to request booking class availability from an airline's reservations system to price an itinerary during shopping and ticketing.

The NDC workflows are offer and order management centric.

The creation of an offer begins with a travel agency's NDC shopping request. The offer management system has an offer store that allows an airline to distribute its range of products and promote additional services using visually engaging content such as pictures, videos, and virtual reality. This process can be anonymous or personalized depending on whether the travel agency had provided information

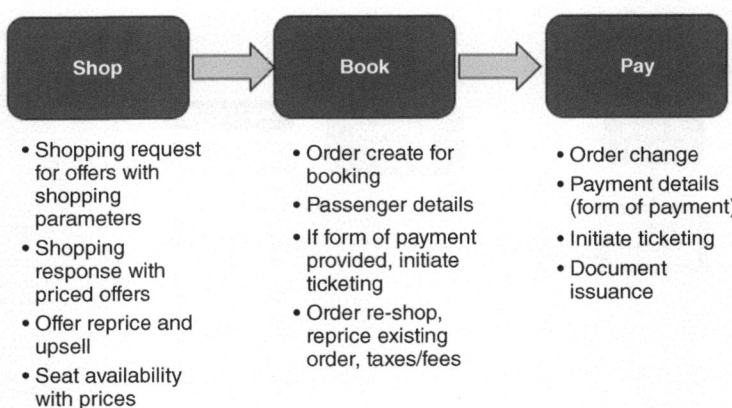

Fig. 6.8 The NDC booking workflow

about the traveler. The pricing of the offer can be based on traditional methods or continuous pricing.

Order management involves the creation, storage, and administration of orders. An order provides a view of the different products and services that have been requested. A unique identifier, called on Order Id, is used to reference the passenger name record (PNR), electronic ticket (ETKT), and electronic miscellaneous documents (EMDs) associated with an order.

An NDC booking and ticketing workflow consists of three steps, outlined in Fig. 6.8.

6.10   The NDC Certification Process and ARM

The success of the NDC program will require uniform airline adoption and demonstrated scalability of transaction volumes that are comparable to the GDS model, which is still largely unproven after a decade.

Multiple versions of the XML-based messaging standard exist and cannot be classified as an industry standard. Governance is required to unify the NDC standard into a current version and a limited number of prior versions. This does not exist today.

Airlines that utilize an NDC API along with sellers and aggregators that consume an airline NDC API can apply for the NDC certified designation from IATA. The designation is specific to one or more available versions of the NDC APIs at a specific level.

Sabre and Amadeus, in their role as IT providers, provide hosted reservation services for airlines on the NDC standard, and sellers/aggregators can also apply for the NDC capable status. This designation is available for a specific version, or multiple available versions, of NDC schemas, at Level 2, 3 or 4. At each designation level, different combinations of NDC message types are assessed based on the relevant schemas. This evaluation applies to both NDC certified and NDC capable designations.

Broadly speaking, the certification involves offers and orders. The shopping schemas enable airlines to distribute their complete range of products that may consist of a dynamic fare and ancillaries. These ancillaries can include baggage, seat selection, Wi-Fi, airport lounge access, and a range of ancillary products and services with engaging content either anonymously or tailored to the individual traveler. The order management schema empowers airlines to oversee NDC-driven orders at every stage, from booking to fulfillment.

All entities applying for a NDC Certified designation must have a live deployment with messages in a production environment through a seller. IT providers that apply for a NDC capable designation can demonstrate in a live deployment or a test environment. Progress toward NDC enablement is maintained in a NDC registry.

Global distribution systems (GDSs) and online travel agencies (OTAs) have been investing millions of dollars in IATA's new distribution capability (NDC) for airline content that was initiated in 2012 (IATA, 2012). This new open XML (eXtensible markup language) and JSON (Java Script Object Notation) messaging standard is a requirement for intermediaries to have access to air content from suppliers, consisting of schedules, fares, and air ancillaries. This is a significant change in the airline industry. Before NDC, shopping for flights and pricing of itineraries had always been in the domain of the GDS for the travel agency indirect channel. With the NDC model pricing power shifts from the GDS to the airline which is now responsible for responding to shopping requests from travel agents with priced airline itineraries and ancillary products (Vinod, 2021a).

6.10.1 NDC Certification Levels and ARM

Initially there were five levels, and this information is maintained in an NDC registry.

1. Level 1: Post Booking Ancillaries. The most basic capability relies on current or past NDC schemas for the sale of ancillaries after the booking has been made.
2. Level 2: Offer Management. This is the deployment of the shopping request and shopping response API.
3. Level 3: Offer and Order Management. In addition to Level 2, this includes the deployment of the order create request, the order view response, and the order change request APIs. This is deployment of offer and order management where the airline takes full control of shopping, booking, ticketing, and payment.
4. Level 4: Full Offer and Order Management. In addition to Level 3 capabilities, it includes a limited range of post-booking activity such as changes and cancellations. This includes the order re-shop request, order re-shop response, and order change notification APIs.

5. NDC@Scale. Specific to airlines, this level covers evaluation along four dimensions: technical setup, organization setup, user cases, and capabilities. It does not specify minimum transaction processing requirements during peak periods.

Level 1 was considered elementary and removed from the certification process in 2019. In October 2021 at the Digital, Data, and Retailing Symposium (DDRS) in Madrid, Spain, IATA decided to replace the NDC registry and Level 1 through Level 4 certification with the new multilayered program called the Airline Retailing Maturity (ARM) index.

The objective of ARM was to foster better retailing practices across the industry and for airlines to assess where they were on their retailing journey against their peer group.

The new index registry details specific NDC capabilities as before and adds the ONE Order fulfillment capabilities which were separate, but it does not have the certification level designations. The certification steps were required to eventually demonstrate scalability. Initially, ARM met with skepticism from industry analysts (Silk, 2021). However, many airlines, intermediaries, and other travel entities have transitioned to the ARM index.

The new ARM index registry is one of the three pillars of the broader ARM index. While it still details the IATA-verified NDC capabilities of each travel entity seeking certification, it does so without using those certification levels.

While the ARM index registry is public, the remaining two pillars are private. The second ARM index pillar is called partnerships deployment and is intended to give airlines and intermediaries insights into how well they are performing in scaling their NDC-enabled capabilities. IATA collects data on the partnership between airlines and intermediaries to produce a maturity report that illustrates how a deployment compares to peer groups.

The third ARM index pillar, called Value Capture Compass, is private to the airline community. It enables airlines to evaluate their performance in various categories such as bundled offers, content, pricing, revenue management, payments, and customer engagement against a peer group in deploying modern retailing strategies. The maturity report serves as a self-assessment on how an airline is performing against its peer group in those categories.

6.11   GDS Investment in NDC and Legacy Infrastructure

While the NDC messaging standard provides clear benefits for airlines, the benefits to OTAs, wholesalers, and TMCs are less clear. It is viewed as a mandate: when an airline is NDC-ready, they may choose to use the NDC messaging standard as the only conduit for OTAs and TMCs to communicate with them. The GDSs have multimillion dollar investments to handle the new communication protocol since they realize that without the investment, they may no longer have content.

NDC is also a radical transformation of capabilities required in a GDS in the future. For example, air shopping constitutes the single largest recurring annual investment in advanced algorithms and infrastructure for a GDS. For example, at Sabre, these volumes averaged over 675 million shops a day in 2019 (Vinod, 2020a). When all airlines are using the NDC messaging standard, pricing power will shift from the GDS to the airline, and this investment is no longer required since it is now the responsibility of the airline to provide priced itineraries. The same is true for itinerary pricing, which requires investment to maintain its accuracy. By the end of 2020, there were approximately forty aggregators that had achieved NDC certification level 3 or greater. Several nimble aggregators have entered the market with modern technology and APIs that provide airline connectivity to travel agencies out of the box that are NDC compliant and at a price point that is acceptable to airlines. What they lack is expertise in agency workflow management and security. To eliminate the legacy costly infrastructure and capabilities that will not be required in the future, GDSs should consider acquiring or partnering with a new entrant that uses modern technology and augment their weak spots of agency workflow management and security. This enables a GDS to migrate away from aging technology rapidly. For GDSs operating in a legacy TPF environment, it can provide a faster and more cost-effective path forward than an expensive TPF offload to open systems. It can also accelerate reducing the cost of distribution, which is an airline priority.

6.12   Deployment Scenarios of NDC for TMCs and OTAs

There are several deployment scenarios of NDC with TMCs and OTAs. The most likely scenarios are illustrated in Fig. 6.9.

By default, GDSs are NDC aggregators. Option A is the default option. It is the most common, low-risk alternative, with minimal changes to travel agency

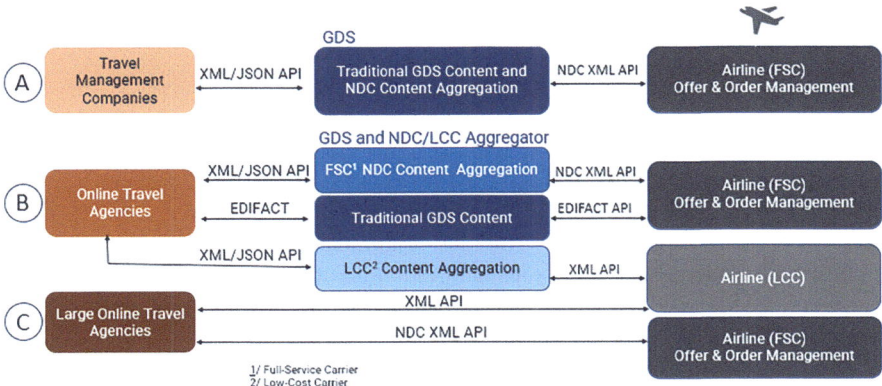

Fig. 6.9 Likely deployment scenarios for NDC

workflows, where the GDS provides traditional GDS content and NDC content. However, there could be situations when a TMC supplements NDC content from an NDC aggregator.

With Option B, the travel agency engages multiple entities to guarantee content coverage that can include GDSs, NDC aggregators, and LCC aggregators.

Option C is the most expense and only very large OTAs and a few large TMCs can consider this option to source content directly from the airlines.

In this book, the term NDC aggregators refer to the new entrants in the marketplace.

6.13   NDC Aggregators and the Threat of Disintermediation

IATA uses the term "aggregators" instead of GDSs within the context of NDC. They encourage GDSs to become aggregators, but they are indifferent about which entity fills that role. GDSs by default are NDC aggregators and incorporate NDC, LCC, and classical ATPCO content (traditional GDS content booked via EDIFACT). The issue is that coverage of NDC and LCC content may be limited to a handful of airlines.

Creation of the XML messaging standard with NDC has also spawned a very large number of new entrants in the marketplace called NDC aggregators. In a nutshell, they offer connectivity to airlines and aggregate NDC responses from airlines. NDC aggregators distribute a shopping inquiry from a travel agent to various connected airlines that serve the market, collecting and consolidating the airlines' responses. Unlike a GDS, they do not have an internal capability for shopping or pricing. They are focused on ensuring that they can handle the NDC message volumes, manage rich content when applicable, and add connectivity to more airlines to aggregate NDC content.

6.13.1   GDS Bypass

When a travel agent receives NDC content from an NDC aggregator (not GDS) and a booking is made, the transaction bypasses the GDS and goes directly into the airline's reservations system. When a travel agent receives NDC content from an NDC aggregator and a booking is made, these bookings are not integrated with the mid-office/back-office systems. This has significant impacts on daily agency reporting on bookings, commissions, and individual travel agent productivity metrics since the transactions that bypassed the GDS will not be reflected in the daily reports.

A GDS bypass refers to a distribution channel that completely avoids using GDSs. This means that an NDC aggregator directly integrates with an airline's NDC API, bypassing GDSs entirely. When NDC was launched in 2012, GDSs were resistant to adopting NDC because they feared it would enable widespread GDS bypassing and reduce GDS usage. However, by 2016, when it became clear that NDC would become prevalent, GDSs began to adapt. The travel agency community also favored the traditional distribution channel for two reasons. First, because they received a revenue stream from GDSs called incentives, and second, integration to the mid-office and back-office systems.

6.13.2 GDS Pass-Through

A GDS pass-through occurs when GDSs integrate with airline NDC APIs and deliver content to the point of sale. In this case, the GDSs act as intermediaries, passing along messages between travelers/agents and the airline's NDC system as a message router since the airline manages the offer creation and order management processes. The primary role of GDSs is to ensure a consistent presentation of offers from different airlines on the agency desktop. Now that all GDSs have committed to NDC, it is likely that if the non-GDS NDC aggregators exit the market, the term "GDS past-through" will eventually fade away.

In summary, when a GDS aggregates NDC content, and a travel agent makes a booking request, it is a GDS pass-through transaction. However, in the NDC framework, airlines have more control over offer creation and order management, whereas with traditional indirect distribution, the GDS is responsible for generating the offer, creating, and retaining the passenger name record (PNR), and only sharing a copy of the PNR with the airline after the transaction has been completed..

6.13.3 NDC Aggregator Economics

NDC aggregators do not provide the NDC content to the travel agents for free. Instead, they charge a booking fee that in some cases can vary by market, volume, airline, and the type of customer. Travel agencies that use NDC aggregators to access NDC content typically must pay a booking fee. In most cases, this fee is less than $1.50 per booking. Travel agencies pass this booking fee to customers as part of a service fee.

6.13.4 Transition to NDC and the Role of the GDS

GDSs continue to play a significant role in the travel industry, but their role is evolving and potentially diminishing. To support the future of distribution and adoption of NDC, they are focused on broadening the content offerings with multi-source, source agnostic content. It is not clear how their role is going to change. They may evolve into a tool for direct corporate bookings and direct leisure travel. They are challenged with the changes in the travel industry such as growth of direct bookings, incentives offered by airlines to travel agencies to book directly and avoid GDS booking fees, surcharges imposed by some airlines for tickets booked through a GDS, and the NDC messaging standard for data transmission initiated by IATA for airlines to communicate with intermediaries such as the GDS.

6.13.5   Customers and NDC

A frequently asked question is: *What does the customer lose with NDC?* Customers lose schedule and fare transparency as we know it today. They may also not have access to the lowest fares in a market.

In the current environment, OTA and GDS customers have access to multi-supplier content in an instant when they conduct a low fare search. With NDC, booking an airline ticket will be like buying a new car. To buy a car, a customer goes from dealership to dealership comparing prices, features, and packages offered without instant visibility to all car dealerships and their offers. With air bundles offered by airlines using the NDC messaging standard, the content across airlines is not homogeneous. Hence, a customer must make a judgment call on which airline has a better offer since the ancillary content is never the same across airlines. The GDSs will have to normalize the content returned from multiple airlines from an offer request, thereby making it easier for travel agencies to compare offers and make a booking. The science of normalizing nonhomogenous content for display is far from perfect and is discussed in Chaps. 9 and 12. This increases the likelihood of a customer paying more for a trip than they really should.

6.14   Schedules, Fares, and NDC

GDSs have always relied on two key data sources to facilitate the booking process. They are schedule aggregators and fare aggregators.

NDC allows airlines to take control of the indirect channel and send customized content, dynamically priced itineraries, and air ancillary bundles to a travel agency customer. NDC poses a threat to the future of fare aggregators like ATPCO and SITA since the airline may not file fares. It is also a threat to schedule aggregators like OAG and Cirium, as airlines may choose not to publish schedules through them, though this is less likely. This would require all travel agency bookings to be made directly with the airline. NDC is currently in the early adoption stages and will eventually make full content agreements obsolete, as pricing power gradually shifts from GDS to the airline for indirect bookings. ONE Order is an integral part of the NDC program, introducing a single customer order record to capture all necessary data elements that require to be fulfilled. This simplifies the sale of ancillary products, eliminating multiple reservation records and the E-Ticket/electronic miscellaneous document (EMD) with a single reference order.

6.14.1 What Happens to Schedules and Fares?

Airline content consists of schedules, fares, and availability. Seat availability is always resident in an airline's reservations system. Every GDS must query the airline's reservations system to determine true last seat availability. However, prior to NDC, the GDS subscribed to schedule aggregators to access airline schedules. Similarly, they subscribed to fare aggregators like ATPCO and store all future fares in the GDS.

The question remains, what will happen to schedules and fares in a NDC world?

6.14.1.1 Current State of Schedules (Before NDC)

Airlines create and publish their schedules to schedule aggregators who in turn disseminate the data to the GDSs. The Schedule Change Information Manual (SSIM) is an IATA format by which airline schedules are distributed by the schedule aggregators weekly. SSIM is a standard schedule format that airlines use to exchange flight schedule data with GDSs, airports, and other partners. Changes to the SSIM are also submitted every other week.

6.14.1.2 Future State of Schedules (After NDC)

While airlines do not technically have to publish their schedules to schedule aggregators like OAG and Cirium, they will continue to do so. The NDC model requires a GDS to send requests for offers to multiple airlines that serve a market requested by a customer. GDSs need to have access to schedules to orchestrate the shopping request to multiple airlines to return offers (base fare and ancillaries for an itinerary).

6.14.1.3 Current State of Fares (Before NDC)

There are three different types of fare products in airline pricing: public fares, private fares, and web fares. Public fares are available for purchase by anyone. They are created and filed by an airline, and then distributed by fare distribution vendors such as ATPCO and SITA to all global distribution systems (GDSs) for global accessibility. Public fares can be accessed by all customer segments and are automatically subject to travel agency commissions, which vary depending on the fare and route. The applicability of a travel agency commission is determined by the booking class associated with the fare. Typically, the cheapest fares, lower in the RBD hierarchy, do not offer commissions. When a travel agent sells a public fare to a passenger, the gross fare is written on the ticket, and the agent submits the net amount to the airline after deducting the commission. Public fares are sometimes known as gross fares or IATA fares.

Private fares are fares that are created and regulated under private tariffs. These fares have limited distribution and are used as a discreet sales outlet. There are various ways to create private fares. These fares utilize security in ATPCO Category 15 (CAT 15) or Category 35 (CAT 35) to determine who can sell them, and the Category 1 (CAT 1) rule to specify who can purchase them. Corporate fares are generated using a Category 25 (CAT 25) fare-by-rule (FBR). CAT 25 fares are negotiated discounts between an airline and a corporation, offering lower prices compared to the public fares. CAT 25 enables the creation of new fares by utilizing rules data to determine the fares and their amounts. These fares can be calculated from existing public fares and rules in the market or specified to create a new fare using the provisions in CAT 25. However, it is important to note that the CAT25 discount for corporations is typically applicable only to the higher fare RBDs in the hierarchy, rather than across all RBDs. In certain cases, airlines may also directly distribute private fares to the GDS.

Web fares are essentially restricted to the airline's website and are often considered private fares. These fares emerged in the early 2000s as a response to the increasing significance of the consumer direct channel, which aimed to enhance brand visibility and minimize distribution expenses. Despite the existence of full content agreements, web fares continue to exist in various regions worldwide, such as Latin America. Although the GDS has historically had issues with airline web fares, it accommodates them by offering discounted segment booking fees to ensure their availability in the GDS.

6.14.1.4   Future State of Fares (After NDC)

With NDC, pricing power shifts from the GDS to the airline for bookings that originate in the GDS.

If all bookings for an airline are NDC bookings, they do not technically have to publish their fares to fare aggregators like ATPCO. However, this will impact the GDS fare quote (FQ) display that shows an airline's reference fares in a market with associates restrictions to travel agents.

If every airline in the world were on NDC, GDSs would no longer have to support the extensive computing environment for air shopping and itinerary pricing. This is because with NDC the GDS functions as a message broker and the airline generates the offers with priced itineraries.

A hybrid state will persist for a long time perhaps into the 2030s. This is because there are 600 airlines worldwide and only a few have committed to investing in NDC, which means the GDSs will have to retain and continue to enhance their air shopping and pricing capability to support traditional GDS EDIFACT bookings.

With NDC, public fares will disappear from the market and all fares—public, private, and web will be treated as private fares that are offered using the NDC messaging standard as part of an airline's bundled offer. Likely these priced itineraries, derived from reference fares, will be calculated by the method of dynamic continuous pricing.

6.15 ONE Order

The commercial airline industry (FAR 121 operators) is going through a multimillion dollar information technology investment from airlines and intermediaries to transform the legacy passenger name record (PNR) to what has been called as ONE Order. IATA introduced ONE Order at its World Passenger Symposium in 2015, describing it at the time as "*a single customer order record holding all data elements required for order fulfillment across the travel cycle*". ONE Order is an integral part of the NDC program.

The fundamental objective of the ONE Order initiative is to extend the PNR by consolidating a passenger's personal information and purchases such as an airline ticket and air ancillaries within a single record. ONE Order phases out PNRs, E-Tickets, and electronic miscellaneous documents (EMD) by combining them into a single customer focused order called the Order Id. It aims to simplify the traveler experience by eliminating the need to keep track of different reference numbers and documents when checking in or making changes to an existing itinerary. Given the legacy environment that commercial aviation has been operating in, ONE Order is a major multimillion dollar investment from airlines and intermediaries to transform the legacy passenger name record (PNR) to what has been called ONE Order. It has taken several years to develop since the introduction of the "new" concept in 2015.

While this concept is new to the commercial airline industry, it is in active use in other industries such as manufacturing and retail. Manufacturers and retailers use order management systems (called OMS) to get a single view of the customer order regardless of where the individual components are sourced.

6.15.1 What ONE Order Means to the Airline Industry

To understand what ONE Order means for the future of transaction processing, we need to first review the current transaction processing environment.

To complete a booking, multiple documents are required for passengers to book, pay, board, and ultimately fly. Many of these documents are remnants of legacy processes that are still necessary for functions that are no longer current. For instance, between 1990 and 2000, the standard travel procedure involved making a reservation by contacting the airline or a travel agent, receiving a paper ticket upon payment, reconfirming itinerary details by phoning the airline or travel agency a few days before traveling, and obtaining a boarding pass at the airport by exchanging the flight coupon with an agent.

The utilization of ticket numbers is another example of legacy procedures. For each passenger itinerary, a single coupon is assigned to a specific flight. However, if an itinerary consists of more than four flight segments, a conjunction ticket is necessary. A conjunction ticket is a ticket issued to a passenger which also contains another ticket for passage which together constitute a single contract of carriage. This results in two distinct ticket numbers. These two ticket numbers, such as 125-1234567890 and 125-1234567891, signify the use of two ticket stock from the airline. In this example, the first three numbers (125) represent the unique airline number, in this case 125 is British Airways, followed by the 10-digit ticket number. The reason behind this is due to the original paper tickets, specified by IATA in 1930, only having four flight coupons. In addition to these four coupons, there was a coupon for airline accounting purposes and a final coupon for the passenger's receipt. When the industry transitioned to E-Tickets (ETKT) in 2008, the coupon limits carried over from the legacy environment.

The documents associated with a passenger's journey are:

1. Passenger name record, or PNR, that includes details about the itinerary and personal information of the passenger, such as their name and contact information.
2. The traditional paper ticket was replaced by the E-Ticket, in 2008. This electronic ticket is now mandatory for all flights.
3. The EMD, or electronic miscellaneous document, is a non-flight record that can be used to pay for ancillaries, services, and fees. This document has replaced some usage of the miscellaneous charge orders, or MCO.
4. To board a flight, a passenger must have a boarding pass. While the traditional paper version is becoming outdated, electronic boarding passes can be accessed through online check-ins and smartphone applications.

The objective of ONE Order is to simplify the fulfillment, servicing, delivery, and accounting procedures pertaining to airline goods and services.

The concept of ONE Order refers to a unique reference number that includes everything the traveler requires. The PNR number, passenger personal data, passenger contact information, itinerary details, payment status, ticket information, and ancillary information are all combined into a single representation called ONE Order.

The sharing of information between the parties (such as airline partners or even airline and hotel providers) will be made easier with the ONE Order unique reference number. This will occur regardless of where, how, or from which supplier the passenger purchased his services or seats.

The financial and accounting operations will be made easier by eliminating the necessity for the reconciliation between tickets and reservations.

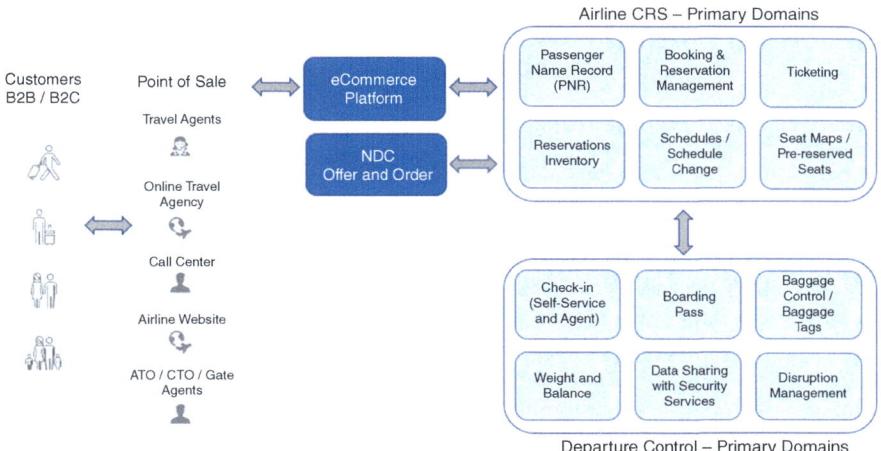

Fig. 6.10 Airline reservations processing with offer and order

6.16 Airline Reservations with Offer and Order

Offer and order management are integral components of airline reservations with NDC. Offer and order management capabilities are extensions of the airline host CRS (PSS) that are either developed by the reservation system provider or by technology providers such as Farelogix. Figure 6.10 illustrates the airline host CRS with offer and order management.

6.17 Passive Bookings

Booked airline segments in a GDS are active segments. Passive bookings existed in the GDS environment before NDC. A passive booking is a flight segment that is stored in the GDS that does not result in a ticket being issued. Passive segments are utilized by travel agents to create itineraries, take notes, and give details to the mid/back-office and expense management systems. When a booking is not made within the GDS, passive segments are generated to ensure a comprehensive perspective of all bookings made by the travel management program. In such cases, the GDS functions as a repository. GDSs provide a restricted number of passive segments at no charge and apply fees when the threshold is surpassed. For example, after Lufthansa imposed surcharges for Sabre bookings in 2015, travel agents booked directly with Lufthansa to avoid the surcharge and posted these bookings on Sabre as passive bookings for which Sabre charged Lufthansa a *courtesy fee*. A courtesy fee is not part of the standard segment booking fee in the contract between an airline and GDS. Passive bookings provide itinerary data to the back-office and expense management systems. This supports TMCs with consolidated reporting on agency sales, commissions, print itineraries, generate invoices, and track records.

There are three scenarios in the traditional GDS model where passive segments are booked. They are.

1. Basic passive segment.
2. Passive segments with due or paid fare or taxes from a refund or exchange.
3. Passive segment as a retention segment.

NDC air segments in a PNR are not valid for manual pricing since all the fare data originates from the airline. If non-NDC content is included in the PNR, then the manual pricing can only apply to the non-NDC air content.

In the GDS, NDC air segments will be reflected as confirmed with the pertinent status code in both the ONE Order and the PNR. However, in the PNR, these segments are not active air segments but serve as informational segments only. Changes to an NDC segment requires the airline to effect the change through the order management API's. The GDS Offer and Order API's for NDC bookings are available as REST API's.

In summary, NDC content bookings made by the GDS, NDC aggregators, or any other channel are booked and ticketed by the airline. A copy of the booking can be stored by the intermediary, but the bookings are passive and not active like the traditional GDS bookings. The agent cannot make changes to the reservation such as adding a segment to the itinerary, rebook an earlier, or later flight since the segments are in the control of the airline. This has a significant impact on agent productivity.

6.18   Codeshare and Interline

Interline traffic, which makes up 8% to 12% of global airline traffic, plays a crucial role in the industry. Over 30% of leisure bookings on OTA sites are interline bookings in the pursuit of a lower fare. Interline connections are displayed when the carriers have an interline ticketing agreement.

Codeshare flights, which are widely used today, are a vital element in airline alliances. Coordinated pricing and revenue management decisions drive incremental revenues for the alliance (Boyd, 1998; Vinod, 1999, 2005b; Wright et al., 2010). NDC also addresses the management of codeshare and interline flights.

6.18.1   Origins of Codeshare Flights

The concept of codeshare can be traced back to Richard Adams Henson, a test pilot and operator of a flight school, who established the idea of commuter airlines. Henson launched the Hagerstown Commuter airline, offering commuter services to Washington, D.C. In 1967, Henson formed a partnership with Allegheny Airlines, marking the first instance of a codeshare agreement with a major airline, albeit before the term "codeshare" was coined. The term "codeshare" was officially introduced in 1989 by American Airlines and Qantas Airways when they established a collective codeshare agreement connecting various cities in the United States to cities in Australia. Codeshare agreements gained significant traction in the early 1990s.

6.18.2 Codeshare Flight Example

Code sharing plays a vital role in the development of airline alliances. It involves the operation of flight segments by one carrier, but their marketing and sale are carried out by one or more different carriers. Codeshare flights predate the establishment of alliances. However, within an alliance, the scale of code sharing is significantly greater. By engaging in code sharing, an airline can expand its range of services beyond what is possible through bilateral agreements alone. Various amenities and services, including mileage accrual, seamless check-in (also known as through check-in), and access to premium status, are facilitated through code sharing arrangements. This practice allows the operating carrier to extend its reach without the need for additional investments in aircraft, crew, or ground resources.

Consider the simple interline network shown in Fig. 6.11. American Airlines (AA) is the operating carrier from LAX-DFW and DFW-LHR, while British Airways (BA) is the operating carrier from LHR to CDG. AA is the marketing carrier from LHR to CDG with flight number AA 6556. Similarly, BA is the marketing carrier from LAX to DFW (BA 6503) and from DFW to LHR (BA 1580). Hence, AA is sharing its operating flight with BA from LAX to DFW and DFW to LHR, which allows BA to put its own flight number on the same physical aircraft and sell its seats. Similarly, BA is sharing its operating flight with AA from LHR to CDG, which allows AA to put its own flight number on the same physical aircraft and sell its seats.

Government regulations mandate the disclosure of the operating airline. Additionally, an operating flight may have various marketing flights linked to it. Another benefit of codeshare flights is the ability for an airline to present connecting services as online rather than interline. Passengers prefer online connections, which is also the favored option in most GDS and online displays.

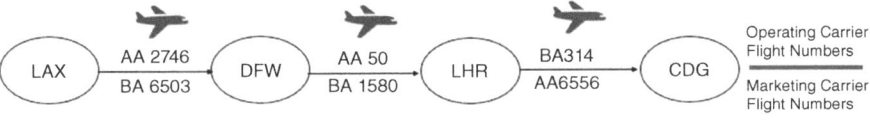

LAX: Los Angeles International Airport
DFW: Dallas/Fort Worth International Airport
LHR: London Heathrow International Airport
CDG: Paris Charles de Gaulle International Airport

Fig. 6.11 Codeshare flights

DEI, an abbreviation for "Data Element Identifier," is a widely used term in the Standard Schedule Information Manual (SSIM). Within the context of SSIM, specific data elements are denoted as DEI 9 or DEI 101. The DEI serves as a repository for mapping operating flights to marketing flights.

6.18.3   Types of Codeshares

The three main codeshare agreements that exist between an operating carrier and marketing carriers are hard blocks, soft blocks, and free sale.

In the case of hard blocks, the operating carrier sets aside a fixed number of seats on a flight specifically for the marketing carrier that has purchased these seats. Any unsold seats cannot be returned to the operating carrier. The marketing carrier is responsible for managing and filling these purchased seats. From the perspective of the operating carrier, the hard block of seats is treated as if they have already been sold in their inventory system.

As for soft blocks, the operating carrier provides a block of seats to the marketing carrier, with a specific release date for any unsold seats. Typically, this release date is set to occur 7–14 days before the scheduled departure. This arrangement exposes the operating carrier to some level of risk, as the marketing carrier may release seats back to the operating carrier that are left unsold at departure.

Free sale is the most used codeshare agreement. It operates under a "sell and report" system, where each codeshare partner is responsible for managing and controlling the inventory of the flights they operate.

6.18.4   Codeshare Availability for Free Sale

For free sale codeshare, there are various options for codeshare availability.

Standard Codeshare Availability is determined by AVS. The operating carrier notifies the marketing carrier to stop selling with a stop sell message. The marketing carrier can keep selling the booking class until they receive the stop sell message.

The transmission of AVS stop sell teletype messages from the operating carrier to the marketing carrier are unreliable, often resulting in delays before they are implemented on the marketing carrier's reservations system. This delay in availability presents opportunities for knowledgeable travelers to secure lower fares when booking the marketing flight. The potential savings when booking marketing flights on international itineraries can be substantial. This is because the booking class is closed on the operating carrier's host CRS but is still open on the marketing carrier's host CRS. It is important to note that this practice is entirely legal. To mitigate this issue, airlines can opt for cascading codeshare availability or utilize the bid price exchange (available for O&D carriers only) to ensure true last seat availability.

True Availability refers to the circumstance in which both carriers are hosted on the same reservations system. In this scenario, the availability request can be directed to each inventory partition of the operating carriers to ensure true last seat availability.

Cascading Availability is the procedure in which the availability request is forwarded (cascaded) on to the operating carrier to ensure true last seat availability. This is also known as seamless codeshare availability. Cascading availability is dependent on the DEI record, which offers the flight number mapping between the operating carrier and the marketing carrier.

Bid Price Exchange enables the operating carrier to instantly share the current bid price for a specific flight leg and date with the marketing carrier. This is essential for carriers utilizing O&D control. Bid price exchange ensures metal-neutral availability. It is preferred over cascading availability because it allows for evaluation of an O&D request with codeshare flights based on the total reservation's value and the bid prices of the corresponding legs of the itinerary.

6.18.5   Codeshare and NDC

In an NDC world, existing code share agreements will remain in place, but bilateral prorate agreements that determine revenue sharing between airlines may no longer exist. This change simplifies the financial settlement process between airlines. The transaction flow for code share bookings will still be similar, as the airline creating the offer will request a bundled offer instead of booking class availability from the codeshare partner.

In the case of hard blocks and soft blocks, the airline that receives the offer request will respond for all flight segments. These include the airline's flight segments and the hard blocks or soft blocks of the partner airline, if applicable.

In the case of free sale codeshare, when an airline receives a request for an offer, it may not be able to cover the entire service on its own. As a result, two or more airlines are involved in making the offer. The offer responsible airline (ORA) sends the request to the offer partner (participating) airline (OPA), which responds with a price for the bundle and any associated restrictions. This process resembles cascading availability, except that the marketing carrier returns a bundled offer. The second similarity is with a special type of bilateral prorate agreements that exist, where the marketing codeshare partner requests payment for a fixed amount regardless of the total fare for the itinerary. It is important to note that with NDC, code share partners do not align their booking classes, and the airlines involved in the offer creation process may not be using booking classes at all.

The absence of AVS-based availability for codeshare and interline flights with NDC results in price consistency in the marketplace since the total fare will be the same whether the marketing carrier or operating carrier flights are booked. With AVS, booking a marketing flight may produce a lower priced itinerary than booking an operating flight because the booking class may be closed on the operating flight, but the stop sell message has not been received by the marketing carrier due to delayed AVS message updates. This eliminates revenue leakage for the partner airlines caused by delayed updates of AVS messages from the operating carrier to the marketing carrier.

6.18.5.1 Financial Settlement Process

The financial settlement process consists of two parts. The first part involves the settlement between the travel agency and the airline, while the second part involves the settlement between the airline and its interline and code share partners.

The ORA collects payment for the entire journey and submits a sales report to the IATA BSP (Billing and Settlement Plan). This report includes a list of all the tickets issued the previous day. The BSP then calculates commissions and taxes on commissions and sends the data back to the airline. In the case of credit card payments, the BSP generates remittance files for the acquiring banks of the airlines. These banks will then pay the airline directly for all the credit card transactions.

If the customer paid for the ticket in cash through the travel agent, the airline must collect the payment from the travel agent. To streamline this process, the BSP sends a billing statement to the travel agent at the end of each period. This statement includes a list of tickets, commissions earned, and cash collected on behalf of the airline. The travel agent must submit the payment to the BSP by the designated date, at which point the BSP will pass the funds to the airline.

An airline can also utilize NDC to transmit an NDC order receipt to the BSP for real-time sales calculation. The airline revenue accounting system updates sales information based on the created order values. Once the flight has flown, the data is sent to the airline revenue accounting system to reconcile the flown passenger with the received payment.

For airlines to reach a settlement, the ORA is required to pay the OPA. The OPA can access ticketing information and transfer sales data to the revenue accounting system. This data includes the accepted order values from the ORA at the time the offer was made. Once the flight has been completed, the OPA will be provided with a settlement authorization code by the ORA. This code is then submitted to the IATA Clearing House (ICH) as a claim for payment from the ORA. Upon receiving the claim, the ORA will proceed to make payment to the interline or codeshare partner through the ICH.

6.19   Technical Barriers to Adoption of NDC

Since its launch in 2012, several challenges need to be overcome for the rapid adoption of NDC. They are:

6.19.1   Is NDC a Standard?

IATA coined the term "New Distribution Capability". In the digital era where technological advances are moving at a breathtaking speed, the term "New" in NDC requires a reset. Adoption of NDC by even the most vocal airline proponents of NDC has been limited, measured by NDC booking volumes So, the debate continues: *Is NDC a failure? What is the future of NDC? How long will it take to achieve mass adoption? Despite these concerns, the future of NDC is secure, driven by the commitment for adoption by the major network carriers.*

The NDC messaging standard introduced by IATA is legacy technology, burdened with a verbose XML standard that is slow and inefficient. After over a decade since its launch, the standard continues to be refined and is far from where it needs to be for mass adoption. Unlike its predecessor EDIFACT, which was a rigid standard, NDC is anything but standard. Attempts to standardize NDC have failed since the implementation is dependent on the IT providers of each airline that is adopting NDC. What this really means is that there is no actual standard to base it on, making it difficult if not impossible for GDSs and OTAs to integrate the technology. It also has resulted in significant delays toward time to market.

Since its introduction in 2012, only a relatively small number of airlines have adopted NDC with limited booking volumes. There are multiple versions in use today which indicates a lack of standardization even among the airlines adopting NDC. After a decade, the latest version of the XML schema in 2021 was 21.3, which does not support personalization and is still very far away from being a standard. At issue is that airlines are not implementing NDC in a standardized way. Sharply different versions of NDC create additional implementation cost and complexity while reducing efficiency (Silk, 2023b).

Besides scalability, this impacts adoption of NDC.

A larger issue with NDC is uniform airline adoption and demonstrated scalability on transaction volumes comparable to the GDS model, which is also largely unproven. Multiple versions exist, and it is not a standard for the airline industry.

To gain traction, IATA has replaced the NDC registry and the Level 1 through Level 4 certification process with the so-called Airline Retailing Maturity (ARM) model. ARM is ambiguous since the new index registry details specific NDC capabilities as before and adds the one order fulfillment capabilities which were separate before but does not have the certification level designations. The certification steps were required to eventually demonstrate scalability. The ARM index is expected to measure progress in air travel retailing and is structured around three pillars: verification of capabilities, partnership deployment, and a value capture compass. ARM has met with skepticism from industry analysts (Silk, 2021).

While airlines wrestle with the transition to ARM and focus on scalability, the hotel transaction flow model between the GDS agency channel and a hotel CRS is similar to the IATA NDC model. Pricing power has always been with the hotels. For example, when a travel agent that subscribes to a GDS submits a request for room rates with a GDS entry, the transaction is not processed internally by the GDS.

Instead, the transaction is sent to the hotel's CRS which returns available selling rates and applicable taxes.

6.19.2   Investment in NDC by Intermediaries

The GDSs were slow at accepting the new reality of NDC when it was first launched in 2012. They even hoped that it was a knee jerk reaction by airlines and IATA that would go away. With the realization that it was here to stay, GDSs have been investing heavily in NDC since around 2015 to the tune of $15–$35 million dollars a year. They deem the investment as important to ensure that they have access to content and in the GDS industry content is king. In the absence of airline content— schedules, fares, ancillaries, and availability, a GDS's revenues will dwindle over time.

6.19.3   Total Cost of Ownership for the Airline to Control the Offer

A topic that has rarely been discussed in industry forums is the total cost of ownership (TOC) for the airline to control the offer across all channels of distribution. As pricing power shifts from the GDS to the airline, the airline IT costs will increase by an order of magnitude to respond to all requests for offers.

Distribution costs will go down with NDC adoption because of the two-tier revenue model for booking fees. While EDIFACT booking fees may remain unchained, booking fees for NDC bookings will be vastly lower (see Sect. 7.4).

Airlines need to have clear financial controls and activity-based costing to estimate what the true costs are to control the offer in an NDC environment.

6.19.4   Scalability

A larger issue with NDC is uniform airline adoption and demonstrated scalability on transaction volumes comparable to the GDS model, which is still largely unproven after a decade. Multiple versions of the XML messaging "standard" exist and hence it cannot be certified as a standard for the airline industry.

Scalability is a major concern, and it remains to be seen if NDC can scale to current GDS volumes for transaction processing without caching the customer offers (Vinod & Huff, 2019). Caching offers has a drawback. Offers can be stored by trip-purpose segment but storing offers by individual customers may not be viable.

In the absence of caching offers, scalability in the NDC context consists of two components. First is the airline's fast response system (FRS) to generate a sub second bundled offer response for every request from an intermediary, airline website, or any other channel. This requires an ultrafast air shopping engine and offer creation engine to keep pact with booking requests. FRS is discussed in Chap. 8 for shopping and offer management to generate the NDC bundled offers. Second is interoperability of transactions between airlines and intermediaries, which needs to be highly efficient. As airlines take on the workload of generating offers for all channels of distribution, a far greater investment is required by airlines in software and computing power to keep pace with the expected volumes.

6.19.4.1   XML and Scalability

There are two primary approaches to APIs: RESTful architecture, which is typically used for low to medium complexity B2C APIs, such as those utilized by Google or Facebook, and SOAP (Simple Object Access Protocol) web services, which are commonly employed for complex, large-scale B2B systems in enterprise-level implementations. SOAP is considered complex to build, use, and debug. RESTful APIs, proposed by Roy Fielding in his dissertation in 2000 (Fielding, 2000), are easier to implement and gain rapid adoption. However, SOAP is preferred by many for its robustness, established reputation, and strong security features.

SOAP necessitates the use of XML; however, other formats such as JSON (Java Script Object Notation) can be utilized in NDC's RESTful API implementations. Despite NDC being an XML messaging standard, it provides support for JavaScript developers by presenting JSON-to-XML-to-JSON transformation tools. These tools can be employed in front of existing NDC XML APIs to handle the transformation seamlessly without making changes to NDC APIs. These tools are openly accessible for download on IATA's AIR Tech Zone's GitHub repository. Those companies that opt for JSON as a data transport medium will still be required to comply with NDC's XML-based standards, which can be achieved with the seamless transformation between XML and JSON.

Representing information is not an efficient method with XML. This is because XML relies on metadata to describe the meaning and context of the message, in addition to its content. While the use of a text-based, human-readable, and metadata-encoded markup language has various benefits, it can also result in the size of the XML document being much larger than that of an equivalent binary representation of the same information, sometimes by ten to twenty times.

Moreover, XML's processing overhead, storage requirements, and bandwidth consumption become problematic when transaction volumes are high, despite its suitability for moderate transaction volumes. As a result, some companies resort to tactics that can be potentially hazardous, such as compressing XML, ignoring XML validity, and altering the parsing rules for XML to optimize its performance. These tactics to optimize XML performance are generally not recommended.

The conventional XML format is known to be an ineffective method for representing data. This issue is further exacerbated with additional layers of complexity that come with Web Services. However, trying to improve XML performance can have a negative impact on interoperability and standards-based computing. A compressed, non-validating "XML-like" format essentially becomes a proprietary data format, leaving behind standardization. Further, optimization tactics require agreement between both ends of a communication path. This can transform the loosely coupled, implementation-agnostic XML format into a tightly coupled, proprietary implementation, erasing most of the benefits of using XML for system-to-system communication.

So, the question remains, is there an alternative to the verbose XML representation of data to improve performance? Two alternatives to consider for the future are GraphQL and gRPC.

In 2012, Facebook developed GraphQL, which offers a novel approach to interacting with external services. It allows developers to control the structure of the dataset being returned. GraphQL comprises both a query language and a runtime for fulfilling those queries. This technology is considered by many to be a superior solution to SOAP and RESTful APIs for building modern software in an efficient manner. By making software easier to modify and adaptable to changing requirements, GraphQL is a preferred choice by many developers for contemporary front ends with more flexible data structures. Furthermore, it delivers better performance and efficiency. IATA has assessed the technical feasibility of using GraphQL in the context of NDC, but it found that the GraphQL Schema Definition Language was not descriptive enough to represent the full complexity of XML schema definitions (XSD) utilized in NDC and ONE Order.

During the 1970s, Remote Procedure Call (RPC) was created as a method to initiate a function on a remote server. Unlike newer APIs, it requires a specific format and expects the same format in response. In 2015, Google developed gRPC as a variant of the RPC architecture to facilitate faster data transmission between microservices and other interacting systems. This API architecture diverges from previous APIs in several aspects. First, gRPC utilizes a distinctive format known as protocol buffers ("protobuf"), which are a language-neutral and platform-neutral mechanism for serializing structured data developed by Google. Unlike XML and JSON, it uses a binary format that is machine readable but not human readable. It is more efficient than XML and JSON. Second, it uses HTTP2 instead of the original HTTP1, which significantly increases its speed compared to RPC. This feature makes it simpler to implement since it can transmit messages up to 10 times faster than previous versions. For larger applications, this enables better management of communication processes, although it is slower than REST when implementing the API. The primary challenges with adoption are the inability to validate the content of messages against the IATA standards unlike the XML messages and the absence of support with an active community.

6.19.5   Transition from NDC Level Certification to the ARM Model

IATA replaced the NDC registry and certification levels with the Airline Retailing Maturity (ARM) model (refer to Sect. 6.9) in 2021. Upon its introduction, ARM was met with ambiguity by many users. The new index registry still provides details on specific NDC capabilities, as well as the addition of One Order fulfillment capabilities. However, it no longer includes certification level designations. These certification steps were previously necessary to demonstrate scalability. During its debut at the Digital, Data and Retailing Symposium (DDRS) in 2021, ARM faced skepticism from industry analysts (Silk, 2021), which has since been overcome.

6.20   The Business Limitations of NDC

There are some fundamental issues with the rollout of the new IATA NDC messaging standard. They include the scope and depth of personalized offers, shopping transparency to opaque bundles, absence of agency incentives for NDC content and airlines losing control of their tariffs because of potential markups to NDC fares or surcharges imposed by GDSs and travel agencies to compensate for the absence or steep reduction of GDS segment booking fees.

6.20.1   Personalization: Myth or Reality?

Personalization is a term that can be applied at different levels of aggregation. At the heart of NDC, the intent is to personalize the offers for an individual customer. Is personalization for a segment of ONE, myth or reality?

In travel, the context for making a trip influences a customer's preferences and propensity to buy. Customer segmentation based on context is also called trip-purpose segmentation (TPS). The TPS can be augmented with additional attributes such as the primary purpose for the trip such as weekend getaway for two, attending a wedding, and college reunion to make the offer even more precise. Creating personalized offers begins with context for travel-based segmentation of the customer base. This is a core input into a recommendation engine to generate offers that resonate with customers. Most online users are anonymous when they begin their search for a flight or a hotel room, yet the offers generated should be relevant. Once the person is declared, historical bookings associated with the customer that are stored on the customer profile can be used to improve the recommendation and even offer a discount for the total offer based on a customer's projected lifetime value.

The context for travel-based segmentation to cluster customers together with similar preferences and generate an aggregate offer is foundational to airline offer management. It is the single most important concept for effective offer management (Sorrells, 2018b). However, how feasible is personalization of the offer for a segment of ONE? For example, consider full-service carriers (FSC) in the United States like American and United who have stated that over 50% of an airline's revenue came from 85% to 87% of its customers who flew only once. This also means that for FSCs, true 1:1 personalization is limited to loyal business flyers and high-end leisure travelers that are part of the 13%–15% of customers who generate half the airline's revenues. This information was stated by Scott Kirby when he was first president at American Airlines. On an October 25, 2015, conference call with stock market analysts, Kirby stated that half of American's revenue in the previous year came from 87% of its customers, who only flew the airlines once. He went on to state that 50% of the company's revenue comes from passengers "for whom air travel is largely a commodity" (Yanofsky, 2015). A year later, Kirby was President at United Airlines when he confirmed that 85% of United's customers fly less than once a year, and like American, account for close to 50% of revenue (Zhang, 2016). The importance of retaining customers who spend the most money cannot be understated, while the airline maintains a positive relationship with the rest of the customer base.

Another reason is the enactment of General Data Protection Regulation (GDPR) in 2018 which gives EU citizens control over their data, what is collected and how it is used. With GDPR, models that work with aggregated data instead of individual personal data are more relevant which implies that personalization may be limited to creating offers at the trip-purpose segment level, instead of the individual. Like GDPR, the California Consumer Privacy Act (CCPA) of 2018 is a state statute to give residents in California more control over their personal data collected by businesses. Drivers of change in digital advertising and personalization are legislative actions to protect consumer privacy. This brings us to personal identity and how it is managed.

6.20.2 Air Shopping Transparency

NDC significantly alters the shopping experience for travelers. While the shopping response content may be rich and diverse, it creates added complexity since the shopping responses from individual airlines are not homogenous, making comparison shopping difficult. Second, with the demise of full content agreements, air shopping transparency is negatively impacted. In theory, a dynamic price based on continuous pricing, which an airline can promote through the NDC channel for a specific request, is a calculated fare based on various factors such as the price demand curve, the demand booking rate, and competitor selling fares is unique to the request, adding further confusion on the minds of travelers when they shop for flights are fares.

Further, corporate cost control policy is at risk with NDC along two dimensions, first is the ability to compare a corporate fare against a public fare, and second is the effectiveness of comparison shopping since NDC content returned from airlines are not homogenous.

6.20.3   Absence of Transparency Leads to Airlines Losing Control of Airline Tariffs

In the traditional GDS transaction flow model, airlines pay a segment booking fee to the GDSs but *do not manipulate* the published tariffs filed by airlines to the fare aggregators like ATPCO and SITA. In the traditional scenario, based on the agreement between airlines and GDSs, GDSs receive the segment booking fees from the airline for each itinerary booked and, in 2022, they pay incentives to the travel agents based on volume performance which could be as high as 53% of the segment booking fees. GDSs do not normally alter the public fares published by airlines but limit it to net fares that are designed for markups.

The introduction of NDC content has changed the level playing field to the extent that airlines may decide not to pay segment booking fees to GDSs or NDC aggregators for booking NDC content. Further, all NDC fares can be termed as private fares. GDSs have the capability to markup the airline fare before they are presented to the travel agencies, or they can impose a surcharge. The NDC content delivered by the GDSs to the travel agents is also void of any travel agency incentives paid by the GDSs, which in turn will force the travel agencies to markup the fare delivered by the GDS or NDC aggregator or charge a service fee to the end consumer. As a direct result of these incremental fares or service fee actions by GDSs and travel agencies, airlines can potentially lose control of NDC tariffs. There is an added risk if the markup amount or surcharge is not the same across all airlines but a function of GDS-airline contract economics. If the incremental charges are not uniform across airlines, it could render some airlines to become uncompetitive based on the total fare which may include a fixed markup or surcharge. *This is probably the biggest shortcoming of the NDC initiative,* and no one has thought through this outcome. It can damage an airline's reputation and result in financial loss that is beyond their control. Rather than eliminating segment booking fees paid by airlines to GDSs, paying a segment booking fee is a winning strategy since it can prevent intermediaries from manipulating airline fares or imposing surcharges.

Another aspect of NDC is the loss of travel agency incentives paid by the GDSs. Since travel agencies no longer collect incentives from the GDSs for NDC bookings, their long-term affinity to a GDS also becomes questionable.

6.20.4   Steep Unplanned Investment in IT by Airlines

To support NDC, airlines need deep pockets to invest and support all the related NDC workflows such as shopping, continuous dynamic pricing, booking, ticketing, refunds and exchanges, repricing of itineraries, ONE order, and many more.

6.20.5   Affordability of NDC for the Smaller Airlines

Airlines that operate with fewer than 10 million passengers boarded per year are considered small and may not have the resources and capital to invest in NDC. So, the largest question is: who will foot the bill for the smaller carriers? Or is the industry destined to operate in a hybrid mode with traditional GDS content and NDC content for air bookings?

6.20.6   Travel Agency Incentives

New contracts between GDSs and travel agencies limit or eliminate incentives for NDC bookings. Ironically, the objective of NDC is to offer more content to travelers than non-NDC bookings. GDSs introduce these clauses in the contract for two reasons. First, the segment booking fees negotiated with the airlines are lower for NDC bookings, and this is a way to protect GDS margins. Second, it allows GDSs to recover some of the costs incurred in aggregating NDC content from suppliers and distributing it to travel agencies.

6.20.7   Leveraging NDC Aggregators for Online Bookings

Leading online booking tools will struggle with assimilating NDC content when airlines set deadlines for discontinuing full content availability through traditional GDS EDIFACT channels. This forces online booking tools to rely on third-party NDC aggregators (e.g., Travelfusion, Duffel, HitchHiker, etc.) to have access to full content, which presents several challenges. These challenges include:

1. Manual servicing of bookings by the TMC or direct assistance from the airline.
2. Ensuring NDC aggregator content includes *all* content during a specific carrier's search rather than being limited to NDC content.
3. Surcharges to access and book NDC inventory from NDC aggregators, which TMCs must pass on to the customer (corporation).
4. Unavailability of pre-trip approvals for business travel.
5. Manual updates with passive segments in the GDS for TMCs to provide duty of care services that they are contractually bound by to corporations.

6. A disconnect with mid-office and back-office daily reconciliation and reporting for a range of metrics.

6.20.8   Overcoming Business Limitations

Regardless of all these pitfalls, there are multimillion dollar investments from suppliers and intermediaries to make NDC work. It will require many more years to achieve mass adoption.

6.21   NDC, Direct Connect and GDS Market Share Erosion

When NDC was launched, the defensible position taken by stakeholders was around personalized offers and rich content that the airlines wanted to push to intermediaries. A subtle undercurrent of NDC is the push toward direct connect from a point of sale to the airline, bypassing the GDS. Prior to NDC, the connectivity was based on EDIFACT, standards were maintained and enhanced by PADIS/IATA, and the connectivity was controlled and managed by the GDSs for shopping, availability, sell, and ticketing transactions.

Compared to EDIFACT, the newer NDC is an XML-based message that provided the airline with more flexibility for pushing bundled offers and rich content to the GDSs. The threat to the GDS is that there are many new entrants who invest and adopt this new messaging framework for NDC content aggregation. Hence, an airline has the choice of using the traditional pipe supported by the GDS (EDIFACT) for the traditional GDS booking process, or the new messaging framework (XML) of the GDS or other third-party aggregators such as Accelya (Farelogix), TP Connects, Travelfusion, Duffel, HitchHiker, AirGateway, and many more. In this context, Farelogix is only a partial aggregator of NDC content and serves primarily as a fast response function for offer requests from sales channels. However, the choice of EDIFACT versus XML does not really exist since over time traditional GDS content as we know it will cease to exist. In addition, in 2022, many airlines like Finnair, SAS, and American Airlines have publicly announced that they do not plan to support EDIFACT in the future.

When NDC was first launched, the GDSs were dismissive of this initiative and were slow to embrace the alternative booking workflows. For the first few years, they were even resistant to investing in NDC. For a GDS to survive, content is king and gradually the GDSs grudgingly decided to invest in NDC to gain access to the NDC content that would not be available through the traditional GDS model. A key question is whether the GDSs will lose transaction volumes over time to the new entrant NDC content aggregators who bypass the GDS. With several airlines investing in corporate portals with the aid of NDC aggregators for travel agencies to book travel, the answer is a "maybe". This is because the GDSs can accelerate NDC content aggregation from airlines and displace the NDC aggregators.

Bookings made through an NDC aggregator bypass the traditional GDS channel. This leads to a loss in market share. The fundamental question is whether airline NDC content is being transacted on the GDS in a pass-through model or are they being transacted at scale on the new entrant NDC aggregator platforms. It is highly unlikely that NDC aggregators will ever scale to GDS volumes. So, how significant is the loss of market share anticipated to be? Prior to the pandemic, excluding LCCs who do not participate in a GDS, in 2019, GDS bookings represented approximately 52% of the total. It is likely that the market share for GDS bookings, represented by traditional EDIFACT (on the decline) and NDC XML enabled (on the rise) will go down and be in the 30%–35% range in a few years. However, a negative impact on total GDS booking volumes may be minimal due to the expected organic growth in air traffic.

Air travel is growing at twice the rate of GDP. The GDS situation may be acceptable since volume of bookings may not be impacted based on the argument that organic growth in travel over the last two decades has averaged above 5% per year. Leading up to the pandemic of 2020, there are only two data points in recent memory, the September 11 terrorist attacks in 2001 and the financial meltdown from 2009 when there was negative growth in air traffic.

6.21.1 American's Ticketing Restriction for Legacy Channels

In December 2022, the world's largest carrier American Airlines alerted travel agencies that while NDC fares would be visible through the legacy EDIFACT-based connections of a GDS, they would not be bookable starting in April 1, 2023 (Airoldi, 2022). American urged travel agencies to connect to their NDC technology to have access to comprehensive rich content. While the fares would continue to be filed via ATPCO and visible in booking systems, new fare rules would prevent ticketing these fares through the legacy GDS EDIFACT channel. The new fare rule is based on the eighth character of the fare basis code. If it is numeric, it can be booked via EDIFACT. American's plan was to withhold approximately 40% of the fares available via EDIFACT starting in April 2023. There is no innovation on the part of American to differentiate content, but merely a ticketing restriction for the legacy EDIFACT channel. This exemplifies the chaotic disconnect between suppliers and intermediaries in the context of NDC that IATA is powerless to manage with prudent governance. Pricing and ticketing in third-party systems like GDSs never query a specific field in a fare basis code to achieve a specific outcome. It is usually limited to use by internal airline reservations systems for point-of-sale identification or mapping of fares to branded fare families on airline websites.

All three GDSs were able to assimilate American's NDC content by the deadline. However, leading corporate booking tools like SAP Concur required additional testing beyond the target date. This resulted in customers having to pay higher fares. To ensure access to non GDS content, SAP Concur leveraged new-entrant NDC aggregator Travelfusion as an alternative to the GDS NDC APIs. This allows

Concur to display non GDS content in the booking tool alongside traditional GDS airline content.

6.21.1.1   ASTA Complaint to DOT

This move by American is very far from the purported omni-channel strategy advocated by all the major airlines. Ultimately customers end up paying more for a ticket than what is available in the marketplace, which destroys trust between the airline and the customer.

On July 31, 2023, American Society of Travel Advisors (ASTA) filed a complaint with the Department of Transportation (DOT) (Davis, 2023; McCarthy, 2023), alleging that American Airlines has caused significant harm to consumers, travel agencies, and TMCs with its implementation of NDC technology that purposely held back 40% of the fares through the traditional GDS EDIFACT booking channel.

ASTA released the following statement to the media:

> That action has resulted in substantially higher air ticket prices for consumers and frustrated TMCs and their clients in fulfilling the duty of care owed to business travelers.

ASTA called upon DOT to talk immediate action and mandate American to restore all fares though the GDS EDIFACT channel. Joining ASTA in the filing were the Travel Management Coalition, the Business Travel Association (U.K.), and Folatur.

The ASTA formal complaint was docketed, forcing American to submit a written response by November 21 (Biesiada, 2023b). Today, the back-and-forth complaints and responses between ASTA and American continue without any measurable progress toward a resolution (Silk, 2024a).

6.21.1.2   GDS Surcharges from Singapore Airlines

Singapore Airlines introduced a GDS surcharge of $12 for GDS bookings starting January 4, 2021. Effective June 1, 2023, Singapore Airlines raised the existing surcharge to $20 for GDS bookings. This applies to every EDIFACT booking through the GDS as a YR tax (see Table 4.1) that is included in the fare. In addition, the airline doubled its fare discount for NDC bookings from 3% to 6% below GDS fare levels. For corporate fares the differential was 7% (Silk, 2023a). Following American's lead, Singapore Airlines withdrew K and V booking classes from the legacy GDS on August 1. This automatically implies that full content will no longer be available via the traditional GDS EDIFACT channel.

6.21.1.3   Future of Full Content Fees

There is another aspect to American Airlines removing 40% of fares from the traditional GDS channel and only making them available through NDC connections (Pestronk, 2023). Does this mean that agencies no longer must pay the 80-cent "full content fee" that they have been paying since 2006 to ensure access to full content? For example, Sabre charges 80 cents for each Efficient Access Solution (EAS) segment. Similarly, Travelport charges the 80-cent segment fee for full content access as part of the Content Continuity Program and Amadeus charges carriers 80 cents per segment for participation in the Amadeus Content Plus program. There is ambiguity in these contracts between GDSs and TMCs on the *availability of content* versus availability of full content, which is a significant difference. What this ambiguity in contracts means is that if a travel agency is not charged a surcharge, the GDSs can continue to charge the 80-cent fee for all segments on participating carriers. However, this may change, since NDC content has a lower price point for segment booking fees and there are no travel agency incentives. Likely, the 80-cent fee to travel agencies for access to full content is not sustainable.

6.21.2   An Uncertain Future

A key question is whether the GDSs will lose transaction volumes over time to the new entrant NDC content aggregators who bypass the GDS. With several airlines investing in corporate portals with the aid of NDC aggregators for travel agencies to book travel, the answer is a definitive "yes".

For GDSs, NDC continues to be the single largest information technology investment year after year to ensure that they have access to airline content of base air fares and ancillaries. However, despite this investment to ensure access to GDS content, the question remains: Is airline NDC content being transacted on the GDS or is being transacted at scale on the new entrant NDC aggregator platforms? If the new entrant NDC aggregators gain traction, it will result in a loss of GDS market share. Further, segment booking fees if they do exist with some airlines will continue to go down. Some airlines are already planning on a fee to charge GDSs when they book NDC content through the GDS channel. All of this indicates that the trend is toward a channel shift away from the GDS.

6.21.3   Future Impact of Airline Loyalty Points Accrual

In the world of hotel bookings, all revenue is not created equal. Two key metrics are *RevPar* and *Net RevPar*.

$$Revenue\ per\ Available\ Room\ (RevPAR) = \frac{Total\ Room\ Revenue}{Total\ Rooms\ Available}$$

Besides *RevPAR*, the net revenue per available room (*Net RevPar*) is the ratio of net revenue to total available rooms. The net revenue is the room revenue minus direct variable costs associated with an occupied room such as distribution costs, room service and cleaning costs, etc. The cost of distribution associated with hotel bookings can range from 15% to 40% with the merchant model and pay at checkout (PAC) model. In the early 2000s, the rise of online travel agencies led to a significant increase in bookings made by intermediaries rather than direct bookings. OTA hotel impacts have a deep impact on hotel earnings. To counter this threat, hotels launched various initiatives such as promoting the benefits of direct bookings, member-only rates, and stopped awarding loyalty points for bookings made through OTAs.

Two decades after the hotel initiative against the OTAs, a similar move is afoot in the airline industry. Airlines have always awarded loyalty points to customers through any channel including GDSs due to the power of airline loyalty programs to generate incremental air bookings. This feature changed in 2024, when American Airlines announced that they would stop awarding loyalty points for bookings made through travel agencies that were not certified as "qualified" by the airline. Leisure agencies that do not achieve a percentage of bookings via NDC will be the most impacted. Corporate travelers will continue to accrue loyalty points along with the AAdvantage business program for unmanaged travel of small and medium sized businesses. The corporate travel exception will be in place regardless of whether the associated TMC is part of the preferred agency program.

With its unstated goal of being able to sell 100% of its tickets and ancillary products through the direct and NDC channels, this is a move to accelerate a channel shift. Further, American plans to discontinue loyalty point accruals for the basic economy tickets for all travel agency bookings. The new loyalty points accrual policy goes into effect on May 1, 2024 (Silk, 2024b). It remains to be seen if other larger network carriers will follow American's lead.

6.22   Information Technology Services to the Rescue

The fundamental changes in how airlines want to distribute their product through GDSs and other lower cost channels will have a negative impact on the GDSs. Airlines want to replace the traditional EDIFACT connectivity owned by the GDSs with the new NDC XML format that is more flexible to connect with new points of sale that did not exist before. The ultimate objective of the endeavor is to make both direct and indirect distribution cheaper and easier over the long haul. The only consolation for the GDSs is that adoption of NDC has been slow. Except for a few large carriers like Lufthansa Group, Finnair, American, Air France KLM, and others, many of the airlines have not committed to invest in NDC, slowing the penetration of NDC in the years ahead.

While changes in the distribution landscape are inevitable, how can GDSs address the projected revenue shortfall in the years ahead due to declining market share for the indirect channel, though impacts on booking volumes may be minimal based on the historic growth of airline traffic which has averaged over 5% for the past two decades. NDC bookings are being executed on both GDSs and NDC aggregator platforms that have support from the airlines. This is disintermediation of the GDS leading to loss of market share to the GDS channel.

Amadeus and Sabre have strong IT services divisions that sell hosted reservations services and airline products. These divisions can benefit from the migration to NDC. Losses in the traditional distribution division can be partially compensated with incremental revenues in the IT services division. The growth in IT services to airlines is related to the passenger service systems. Key incremental services that can generate new revenue streams are NDC connectivity to intermediaries and third parties, offer management with creation of bundled offers and dynamic pricing, and order management capabilities. The strength of the reservations hosting business will benefit Amadeus more than Sabre by virtue of its size. Travelport is at a disadvantage since it does not provide reservations services to any airlines.

6.23   Travel Management Companies and the Content Gap

Since the introduction of airline ancillary products, there has been a growing gap in content that is available on individual branded airline websites but is not available in corporate booking tools or the GDS. With the advent of NDC, airline content is seeping outside the managed channel at an accelerating pace. This gap in content has widened with NDC since private fares offered by an airline may only be available through the NDC channel. This is a less than perfect solution since access to comprehensive content is a fundamental requirement for a TMC to be competitive and successful. Larger TMCs are addressing the content gap problem by independently sourcing content not available in the GDS through aggregators like Travelfusion, TP Connects, Duffel, and others or from airline direct connect. Travel agency productivity is negatively impacted when shopping, booking, refunds, exchanges, and cancellations are not handled within a single workflow but managed with different workflows based on how the content is sourced. Besides, these investments can be expensive and ultimately the customer will have to pay a service fee to the TMC to defray ongoing development and support costs associated with acquired content not available in the GDS.

6.23.1   Traveler Profile System and NDC

The GDSs have always maintained a traveler profile system that encapsulates key personal details and preferences of individual travelers. Over the past decades, many of the larger TMCs such as American Express GBT, BCD Travel, CWT, and others have invested in their own profile systems. This is also true with smaller tech-savvy agencies like Navan (formerly TripActions), AmTrav, and WTMC.

With NDC, many of the smaller agencies will also be forced to develop their own traveler profile systems to give them flexibility to make bookings outside the GDS environment. This is because, in a NDC world, a travel agency can make a booking through a GDS, book directly with the airline through an airline provided dedicated NDC portal or use a third-party NDC content aggregator to make a booking.

In summary, the proliferation of TMC-built profile systems rather than reliance on the GDS profile systems will accelerate in the NDC era. It provides convenience and flexibility for the travel agency in addition to increasing the robustness of customer profiles with data gathered from third-party providers.

6.23.2 Mid-Office and Back-Office Systems

Travel management companies will continue to invest in mid-office and back-office systems to have a single view of all bookings, including those made through channels outside the GDS. This is required for aggregate sales reporting and to ensure corporate policy compliance is in place across all booking channels. Besides agent sales reports, this function produces agency commissions reports and agent productivity reports. The function also automates the process of comparing the traveler profile data with corporate policies and ensures compliance without human oversight. For example, an employee's position in the corporate hierarchy can be used to ensure that an air booking is made with the price of the itinerary that is within the established policy for that employee level.

NDC poses some unique issues with mid-office and back-office reporting. To ensure availability of content, a TMC may contract with a NDC aggregator who has NDC connectivity with one or more airlines not supported by the GDS to which they subscribe. In this scenario, if a TMC contracts with an NDC aggregator in addition to the incumbent GDS, consolidated reporting will be an issue. While GDSs already have integration to mid-office and back-office systems, NDC aggregators often lack out-of-the-box integration with these systems, which is critical to support the daily operations of a TMC. There is no guarantee that an NDC aggregator has mid-office and back-office integration capabilities, and further investment may be required on the part of the TMC to consolidate data from the NDC aggregator into the mid-office and back-office systems. This is illustrated in Fig. 6.12 which assumes that the industry is operating in a hybrid mode where some GDS transactions are based on EDIFACT for traditional content and NDC content is based on XML. This is an important reason why corporate travel stays with a GDS. Changing the TMCs mid-office and back-office systems for NDC content delivered by NDC aggregators is a major investment.

Mid-office automation also provides other capabilities such as re-shopping booked itineraries within the booking window to realize a lower fare. This cancel and rebook feature is demanded by many corporations to reduce the cost of travel. Other important mid-office features are generation of virtual cards for one-time use secure travel payments and markups of net fares negotiated by the TMC with the airline that results in a fare that is lower than the discounted corporate fare.

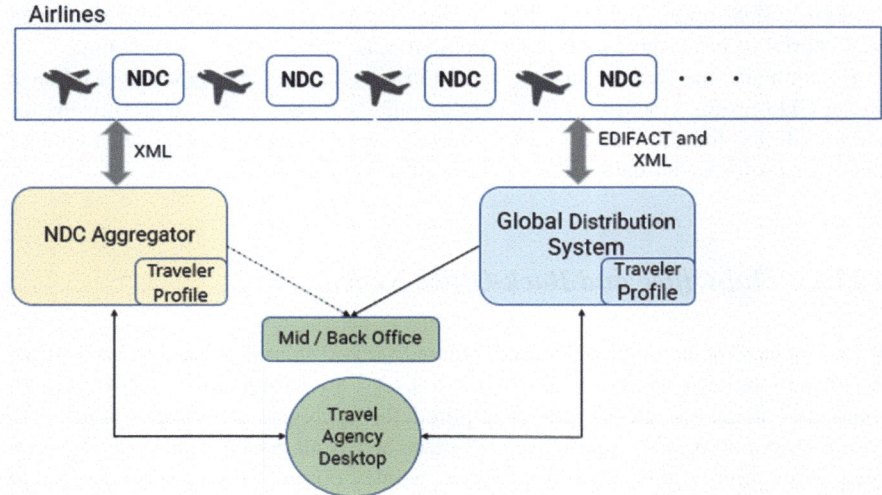

Fig. 6.12 NDC and consolidated reporting

6.24   The Mid-Market and NDC

In the past, GDSs did not prioritize serving the midmarket, the small and medium size businesses (SMBs) that make up the long tail. Due to long time-to-market and high cost of sale, GDSs lacked a clear strategy to cater to smaller companies seeking to control the cost of managed travel.

The mid-market for corporate travel could be transformed by NDC, with airlines and TMCs leading the way by providing booking tools for small- and medium-sized businesses through airline preferred NDC channels. Airline corporate booking portals with NDC content in partnership with NDC aggregators are evolving. Some TMCs are focusing on the mid-market that showcases NDC content. This presents an opportunity for airlines and TMCs to work with NDC aggregators and tap into a previously overlooked market segment. The opportunity is significant, as smaller accounts can add up to a sizable volume of bookings. Additionally, it allows for a balanced approach to both expense and employee satisfaction.

There is also the threat of disintermediation of TMCs and GDSs if full-service carriers insist that to receive the discounted NDC fares, the bookings made by SMBs must be supplier direct and not by a TMC that accesses NDC fares through a GDS.

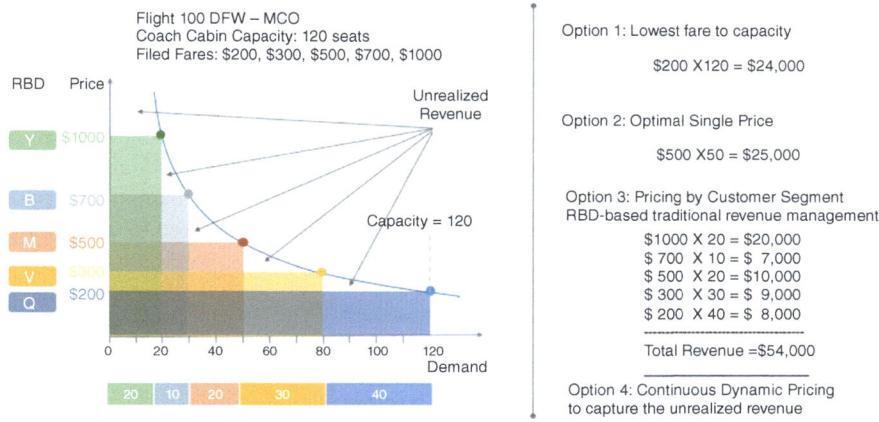

Fig. 6.13 Price demand curve and unrealized revenue

6.25   Dynamic Pricing and NDC

While much of the focus of NDC has been on offer and order, the biggest value proposition is the dynamic pricing of itineraries that is part of the offer with air ancillaries.

Traditional revenue management has relied on booking classes (also known as revenue booking designators or RBDs) for inventory control in the airline host CRS to maximize revenues. The RBDs are also used by the airlines to distribute availability of the booking classes to the GDSs. In an NDC world, airlines no longer have to communicate booking class availability to intermediaries. Instead, every customer request is processed by the airline for a response on the offer bundle that consists of an itinerary price and ancillaries.

Figure 6.13 illustrates the value proposition of dynamic pricing in the context of the traditional price-demand curve and shows the realized and unrealized revenues under four scenarios: the lowest fare to aircraft capacity, the optimal single fare, fares by customer segment defined by the RBDs, and continuous pricing. For illustrative purposes, 5 RBDs are shown.

6.25.1   Price Differentiation

Price discrimination produces increased efficiency. After airline deregulation customer dislike for differentiated pricing has grown into grudging acceptance. The justification for price differentiation based on a morning flight that is more desirable than a mid-afternoon flight, advance purchase requirement, minimum stay requirement, and many more fare rules.

Price discrimination in the airline industry is governed by three conditions. First, there must be customers with varying levels of willingness to pay. For example, schedule sensitive business travelers have a greater willingness to pay than price-sensitive leisure customers. Second is the absence of arbitrage. A customer cannot sell a lower priced ticket at a higher price. Name changes are prohibited on a passenger name record (PNR) primarily for security reasons. The third contributing factor is the low marginal cost associated with transporting each incremental passenger. The airline industry is asset intensive, and most costs are fixed to operate the flight schedule, providing flexibility for differential pricing based on customer preferences rather than cost considerations.

6.25.2   The Case for Continuous Pricing

Traditional revenue management forecasts demand by booking class and determines allocations by booking class to maximize revenues for leg/segment carriers and uses bid price controls to control availability by O&D. The RBDs are used to open and close booking classes to get the optimal mix of passengers that maximizes total revenue. Airline reservation systems have evolved over the past few decades to expand the number of RBDs from seven booking classes to 13 and now to 24, since two booking classes (I, O) are not distributed externally but retained for internal use such as staff travel. With 24 RBDs, there is still unrealized revenue that can be captured with continuous dynamic pricing. In this scenario, an airline may file reference fares, but the price quoted is a continuous dynamic price. In the traditional approach when the $700 fare (booking class B) is closed, the next available fare is $1000. However, with continuous pricing, a derived fare can be quoted that lies in between $700 and $1000. Note that continuous pricing works with the airline's published and contracted fare products. Fare combinability will be enforced based on fare rules when the itinerary is priced.

A requirement for the implementation of continuous pricing for corporate customers is an audit trail that provides a trace of how the corporate fare was quoted. Since every continuous pricing dynamic fare is a derived fare, corporations will require an audit trail and reporting to ensure that they are indeed receiving their corporate discounts.

6.26   Corporate Travel and the GDS

The U.S. Travel Association estimates that domestic business travel will not recover to 2019 pre-pandemic levels until 2024. It is expected to reach 76% of its 2019 levels in 2023. International business travel lags domestic business travel by a year. International business travel is expected to reach 72% of its pre-pandemic levels in 2023 and full recovery by 2025 (French, 2023).

Corporate travel volumes will continue to rebound, but at a slower rate than the growth in leisure traffic (Curley et al., 2020). Before the pandemic, in 2019, corporate travel by volume represented 35% of the total. Growth in leisure traffic is expected to grow at a faster rate than recovery in corporate bookings. Corporate bookings will represent a smaller share of the market, closer to 23%–25% by 2025.

Corporations are scaling back on employee travel for visits to corporate offices and industry conferences. Sales travel will see a minimal impact due to its sheer necessity to close business deals.

Corporate travel managers will continue to work with TMCs to deploy fare approval and rebooking solutions to reduce the cost of managed travel.

A major issue for corporate travelers is due to the hybrid state where some bookings are NDC bookings, and the rest are the traditional EDIFACT bookings. For example, a traditional ticket cannot be exchanged for new NDC tickets. This is an issue for TMCs customers who may have many unused tickets that were booked by the traditional EDIFACT based on GDS content. Since these tickets cannot be exchanged for cheaper NDC tickets in an automated way, it puts corporate travelers at a disadvantage. Typically, 10% of a TMC's sales include an unused ticket exchange. In June 2023, Accelya announced an automated capability to exchange unused American Airlines tickets purchased through the GDS connections (EDIFACT) for those issued through NDC connections (XML) (Baker, 2023). This automation will be made available to other NDC-enabled carriers who can provide access to EDIFACT tickets from their host CRS.

6.27 Dynamics of GDS Surcharges, NDC Aggregators, and the GDS Revenue Model

Travel Management Companies (TMCs) rely on a GDS to shop and book airline tickets for their corporate customers. The GDSs in turn offer incentives to travel agencies for booking through their GDS. While the incentives are performance based, they range from 40% to 55% of the segment booking fees the GDS collects from the airline. The incentives vary by TMC based on size and productivity. The traditional GDS-airline business model is based on the fundamental premise of access to full content from the airline in return for a discount on GDS segment booking fees. The GDS-airline full content agreements stipulate the discount with the caveat that if the airline withholds content, the GDS had the right to revert to the standard segment booking fees.

With the advent of NDC in 2012, there is growing uncertainty on how TMCs will service corporate customers in the future. The complexity stems from the various moving parts initiated by airlines such as surcharges for traditional GDS bookings, private channel, surcharges for NDC content booked through a GDS, and agency created portals for access to shop and book NDC content.

GDSs invest in the NDC messaging framework by developing the NDC content aggregation connector to ensure that the travel agency booking goes through the GDS to protect their segment booking fee revenues. A threat to GDS market share is the proliferation of third-party NDC content aggregators. There are many such aggregators in the market such as TPConnects, Duffel, and others that connect travel agencies to airlines with their NDC connector hub. This results in disintermediation of the GDS. Some airlines use these third-party NDC aggregators to deliver NDC content to agencies through a proprietary agency portal for agencies to avoid the GDS surcharges. This is a problem for GDSs that will never see these bookings in their system and result in lost revenues due to disintermediation.

An evolving trend has been the introduction of surcharges by airlines for indirect agency bookings. Lufthansa introduced surcharges for travel agency bookings in 2015. This was followed by the International Airline Group (IAG) members British Airways and Iberia in 2017 and Air France KLM in 2018 (Vellapalath, 2018). Singapore Airlines introduce a surcharge of $12 for GDS bookings starting in 2021 (Parsons, 2020).

A surcharge for GDS bookings was first introduced by Lufthansa in September 2015, called the distribution cost charge (DCC). The surcharge of €16 per ticket was nonrefundable and non-commissionable. It created a significant backlash from travel agents and GDSs. With the surcharge, Lufthansa opted out of the full-service agreement and had to pay the standard segment booking fees without the discount. Lufthansa defended the surcharge as a move to "secure freedom in distribution". Further, the Lufthansa Group offered €1 for every booking made through their direct NDC channel.

International Airline Group's British Airways and Iberia introduced their introductory version of the GDS surcharge in November 2015 of $10 at a fare component level. The surcharge for a nonstop outbound and return would be $20. If the outbound and return involves a connection and the total fare that was calculated had three or four components, the surcharge would be even higher. Air France-KLM followed in April 2018 with a one-way surcharge of 11 euros. A few years after the introduction of the surcharge, Lufthansa amended the fee to vary by GDS to reflect the cost of the GDS channel. Airlines that impose a surcharge and abandon the full content agreements always pay the higher, non-discounted booking fee to the GDSs. Airlines that opt out of the full content agreements realize that it is a strategic move to reduce long-term distribution costs. When an airline opts out of the full content agreements, it results in an increase in costs and fewer bookings for the airline over the short term. The increase in costs is partially offset by the surcharge. When a travel agent initiates a low fare search for flights in a market, the default sorting of itineraries to be displayed to a travel agent is based on the lowest total fare which means airlines that introduce surcharges will not be displayed on the first page because the total fare is higher. The airline's intent is to break even in the second or third year by promoting their NDC portal to travel agents and other channels of distribution that are cheaper.

While the surcharges were initially used to negotiate better discounts for segment booking fees until NDC reaches maturity, it is also being used today by many airlines to accelerate adoption of NDC content through intermediaries. Airlines like Finnair, Emirates, and many more have introduced surcharges for GDS bookings that give them leverage over the GDSs to accelerate the adoption of NDC content through the agency channel.

These actions by airlines to opt out of the full content agreements led travel agencies to invest in a direct connection infrastructure. Farelogix, an IT provider, contracted with many airlines to use the NDC API and support its adoption in the travel agency community.

When surcharges were introduced, they were waived for some TMCs that booked through the GDS. These TMCs were considered part of the private channel club. Airlines that promoted the private channel realized that it created a perception in the marketplace among travel agencies that the primary focus of its NDC strategy was controlling costs rather than the benefits touted by IATA such as enhanced rich content, loyalty status, and personalization of offers to the corporate customer. Airlines imposing the surcharges have phased out transitional "private channel" agreements that have shielded select travel agencies from the surcharge fee.

To offset segment booking fees, some airlines like British Airways and Iberia have imposed a smaller fee per passenger booked with NDC content through the GDS instead of the original DCC charge for traditional GDS bookings. This new fee will be ultimately paid by the customer.

How can an agency avoid the surcharges for their customers? Many of these airlines that have imposed surcharges have developed portals for corporate clients and TMCs to access NDC content of base fare and ancillary products to directly book, change, and cancel tickets with an airline. This is disintermediation of the GDS since bookings made through the portal bypass the GDS resulting is lost revenues. The airline portals also support corporate compliance policy rules by corporation. However, the implementation of corporate compliance policy is limited to the single airline instead of the GDS marketplace. For example, a common corporate compliance policy is that the price of the selected itinerary cannot exceed X% of the lowest fare across all airline itineraries presented in the response to the shopping request.

So, the question remains: *what is the theoretical end state of the GDS revenue model?* While it is difficult to predict the revenue model a decade from now on how the GDSs will generate cash flow, this much is clear.

1. The traditional GDS bookings model as we know it where GDS prices an itinerary will cease to exist. The traditional EDIFACT-based messaging for traditional messaging will be eventually sunset.
2. Air content consisting of base fare and ancillaries will be promoted by an airline to a GDS through the NDC XML messaging path.
3. Traditional GDS booking surcharges imposed by some airlines will cease to exist as the industry transitions to the NDC channel and content.
4. Even with an NDC connection, GDS bookings made by a travel agent will incur a surcharge from the GDS for the NDC content, though it will be a fraction of the GDS booking surcharge imposed by some airlines.

5. GDS segment fees paid by the airlines to book NDC content will be much lower than traditional EDIFACT bookings.
6. Travel agencies will have to collect the NDC booking surcharges imposed by the GDS from the customer (for business travelers, it will be the corporation) as a service fee.

All this indicates that the future of the GDS is precarious. With the airline adoption of NDC, disintermediation of the GDS is a foregone conclusion as they introduce their own direct portal for NDC content. This implies that the market share of GDS bookings will decline over time. In 1994, before the Internet, offline indirect GDS bookings constituted 85% of the bookings, and the balance was made up of offline direct bookings from call centers and wholesalers. In 2019 before the pandemic, the GDS share of total bookings was approximately 50% (this excludes the LCCs that do not participate in GDSs) and is expected to decline over the next decade.

Part of the problem is the flawed thinking from GDS executives that the traditional revenue model that has existed for five decades will prevail. It is highly unlikely. With NDC, GDS segment booking fees will continue to decline and agency incentives from the GDSs will vanish. Comparing the GDS business to Uber and Airbnb (Sheivachman, 2017) is also incorrect, The GDS business does not look anything like Uber or Airbnb. The GDS business fundamentally has two B2B contracts. They are:

1. The contract between a GDS and a supplier. This is a multi-year contract, typically for three years, which covers the distribution of airline content through the GDS. In the United States, they are called direct connect availability contracts that guarantee the GDS access to all published fares, promotions, and schedules of a participating airline at all points of the sale of the GDS, both online and offline. The contract also guarantees the highest level of connectivity for city pair availability request where the airline CRS will be queried using an EDIFACT message for true last seat availability. In exchange, the GDS provides a discounted booking fee for the duration of the contract.
2. The contract between a GDS and the travel management company that gives them access to the GDS content either with the GDS agency desktop or Web service APIs if the agency had their own custom desktop application. The contract also stipulates the performance-based incentives that GDS would pay the travel agency.

In the case of Uber, it is a consumer-to-consumer (C2C) business with an intermediary where neither the driver of the transport vehicle nor the customer who orders a ride has the bargaining power to negotiate with Uber. It is very different from B2B where size and scale matters at the bargaining table. For a GDS to survive, customers do matter. Today, the GDS is strictly an intermediary, and the customers belong to the travel management companies. This much is clear from the initiatives of airlines that have opted out of full content agreements: *they do not want to pay segment booking fees.* Period. This is an unstated goal of NDC, to eliminate booking fees paid to intermediaries. As the revenue model evolves, GDS executives must realize that ownership of the TMC and the end consumers are important because ultimately they can charge a service fee to the customer to compensate for the loss in segment booking fees. In December 2021, when Sabre was cash strapped during the pandemic, they made a multimillion dollar investment in the American Express Global Business Travel SPAC deal (West, 2021). It is likely that Sabre now realizes the value of access to the end customer as the revenue model changes as we know it today. Another example is Travelport, who acquired Deem, a corporate travel management platform to provide access to the point of sale (Travelport, 2023). Travelport's investment is a direct acknowledgement of the importance of owning the point of sale.

6.28   The ATPCO NDC Exchange

The NDC exchange is an aggregation service launched by ATPCO in collaboration with SITA in 2019. The objective was to eliminate the complexity of point-to-point connections and multiple versions of the XML message supported by the airlines with a single application program interface (API) on each side of the hub for airlines and sellers to exchange orders and offers. The aggregation service has standard requests and responses and had internal orchestration to support the nuances of each airline (which is why NDC is not really a standard). The NDC exchange is like NDC aggregators supported by the GDSs and third-party aggregators. The aggregation service allows sellers to submit a single shopping request and receive responses, including ancillaries and rich content, from multiple airlines which are aggregated by the exchange for a response to the seller. Figure 6.14 illustrates the ATPCO NDC exchange.

However, in July 2022, ATPCO chose to sunset the NDC exchange by the end of the year because of ATPCO's limited ability to drive change, faced with competition from existing NDC aggregation service providers.

6.29   ATPCO Acquires Routehappy

Routehappy, a metasearch company with flight and amenity rating system, was launched in 2011. However, it later shifted its focus to providing comprehensive content, including images and information about in-flight amenities such as Wi-Fi availability, seat pitch, power outlets, and additional services for travel agencies

Suppliers   Sellers

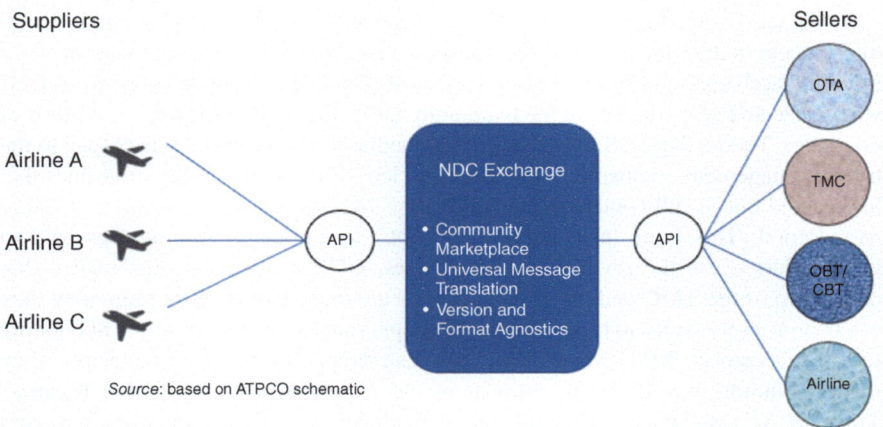

Fig. 6.14 ATPCO NDC exchange

looking to book flights. To adapt to the ever-changing digital landscape, ATPCO acquired Routehappy in February 2018, marking its first acquisition in over five decades. By joining forces, Routehappy enhances ATPCO's ability to provide accurate and extensive airfare information. However, the future of ATPCO's role in fare filing and distribution in the NDC era remains uncertain.

As a wholly owned subsidiary of ATPCO, Routehappy acts as a centralized source for airline-approved merchandising content, offering shoppers a visually immersive shopping experience that presents a comprehensive overview of the various offerings from different airlines.

There are three content types supported by Routehappy for airline offers. They are as follows:

1. Universal Product Attributes (UPA). This feature brings unique airline fares, products, and services with images, messaging, videos, cabin tours, and enhanced description of airline services.
2. Amenities. This feature enables airlines to promote features of the onboard experience such as seat pitch, Wi-Fi, power plugs, entertainment, hygiene, and safety protocols. This feature enables a travel agent to compare features between seat types within an airline or across airlines.
3. Universal Ticket Attributes (UTA). This feature, sourced from ATPCO Fares, Branded Fares, and Optional Services, explains the fare rules and restrictions in simple language.

As of August 2023, airline participation in Routehappy content is summarized in Table 6.1.

With the acquisition of Routehappy, ATPCO can create and distribute all content types for carriers to all third-party distribution channels such as GDSs and OTAs that subscribe to Routehappy.

Table 6.1 Airline participation in ATPCO Routehappy content[a]	Routehappy content	Number of airlines
	UPAs	473
	Amenities	473
	UTAs	423

[a]Source: ATPCO, August 2023

Traditional GDS shopping services for travel agents also support various aspects of Routehappy such as product attributes and ticket attributes.

Airlines are currently transitioning to NDC, and as part of this, Routehappy content has been integrated into the NDC exchange.

Regardless of what happens with the future of the NDC exchange, airlines can distribute Routehappy UPAs to various third-party sites of their choosing in multiple languages. This content can also be distributed to preferred channels using the NDC messaging standard.

6.30   The Role of Revenue Management for Channel Optimization

Airlines that opt out of full content agreements and MFN with GDSs pay a higher booking fee. This also provides an opportunity to optimize the various channels of distribution based on their underlying costs to the airline. Channel optimization is the process of promoting the right collection of selling air fares through the right channel to maximize airline revenues net of total distribution costs through the channel. The key performance indicator is the net contribution for each channel. The net contribution is simply the difference between the cost of the airline ticket before taxes minus all the distribution costs through the channel such as segment booking fees, credit card fees, reservations handling fees and any commissions, if applicable.

6.31   Travel Agencies: The Road Ahead

Travel agents or travel advisors have expertise in destination markets and fares, which is a compelling reason to use them. A customer can always make a booking online for a domestic or international trip and the price quoted by a travel agent may be the same or can be lower than the price quote determined online by the customer. Several factors contribute to the savings such as complex itineraries, travel agency access to net fares or wholesaler fares, and creative routing with a competitive elapsed time. Other factors that contribute to the value of a travel advisor are knowledge about a destination, creating customer itineraries at destinations and finding the best deals.

Table 6.2 Travel management company consolidation

Travel management company	Travel agency acquired	Date
American Express GBT	Hoff Robinson Group (HRG)	February 2018
	DER Business Travel	September 2019
	30SecondsToFly	October 2020
	Ovation Travel Group	January 2021
	Egencia	May 2021
BCD Travel	Adelman Travel Group	February 2020
	Ventura Spa	August 2019
	Air Club Travel	October 2017
	Ticket Biz	December 2015
	World Travel Service	May 2015
Carlson Wagonlit Travel (CWT)	WorldMate	October 2017
	Kaleva Travel	February 2011
	Viajes Lepanto	September 2008
	Piedmont Travel	July 2008
	Ark Travel	December 2007

Travel agencies have witnessed a sea of volatility and churn since Delta Air Lines first imposed limits on front-end commissions in 1996.

An ASTA survey study with Sandals Resorts in 2021 concluded that consumers are increasingly turning to travel advisors to assist them with their travel needs. The study found that 27% of travelers used travel agents before the pandemic and 44% of respondents confirmed that they were more likely to use a travel agent to assist with travel planning (Biesiada, 2021).

Like the airline industry, travel agencies also have gone through consolidation cycles. Table 6.2 illustrates the wave of acquisitions of the smaller agencies by the larger more established top three TMCs. On March 25, 2024, American Express GBT announced that it has reached a definitive agreement to purchase CWT. The acquisition, expected in the second half of 2024, is valued at around $570 million and is subject to regulatory approvals (Catron, 2024)

6.31.1 Access to Airline Content

With their commitment to NDC, many leading airlines have worked closely with technology partners to provide travel agencies access to NDC content that may not be available in the traditional GDS. For travel agencies who prefer an off-the-shelf solution, access to an NDC aggregator or GDS can deliver the content to the agency desktop. These tickets are issued on neutral ticket stock.

Larger travel agencies that can invest in IT resources would typically pursue the direct connect path, which allows agencies to connect directly to the airline's API. In this scenario, the agency should be an ARC/IATA accredited agency for ticketing. Travel agency direct connect tickets are issued on neutral ticket stock.

Figure 6.15 illustrates the options available to travel agencies in the NDC world.

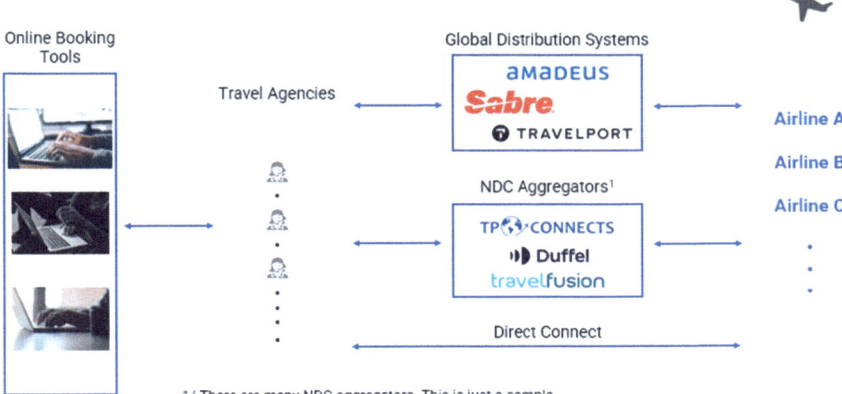

Fig. 6.15 Travel agency options for access to airline NDC content

6.32 Low-Cost Carriers and their Impact on Product Distribution

Since the mid-1990s, low-cost carriers (LCC) have transformed the air travel market, disrupted the world of full-service carriers (FSC), and captured a significant share of the market. They entered the market with aggressive pricing and marketing tactics that resulted in renewed competition and threatened the very existence of major U.S. airlines, global pure play, and international flag carriers (McDonald, 2006). The largest LCCs are Ryanair, easyJet, Air Asia Group, Lion Air Group, and GOL Transportes Aéreos. Spirit Airlines is the largest LCC in North America.

A subcategory of the LCCs is the so-called ultralow-cost carriers (ULCCs) who offer fewer amenities and hence have a greater range of add-on ancillaries for a fee. The ULCCs are typically short haul segments, have less frills, and are cheaper to operate. Today, LCCs and ULCCs have viable and sustainable business models, accounting for more than 30 percent of industry capacity (Fig. 6.16) since 2017.

While traditional network carriers have reduced in size, LCCs have grown at the rate of 10 percent to 20 percent per year. With their rapid growth, they now compete directly with traditional U.S. carriers on more than 90 percent of their route networks. Today, LCCs and ULCCs control 33% of the airline travel market.

LCCs are known for having a simple, no-frills product offering for the price conscious customer—high frequency, point-to-point operation to secondary airports, low operating costs with a homogeneous fleet, high fleet utilization, and consumer direct distribution. With their distinct lower-cost-per-seat-mile advantage, airlines such as AirAsia, Spirit Airlines, GOL Transportes Aéreos, easyJet, Frontier Airlines, JetBlue, Ryanair, and Southwest Airlines, are proactively setting the fare structure in markets where they have a strong presence and rapidly altering customers' valuation of air travel. The growth in LCC airlines and traffic started in the 1990s and

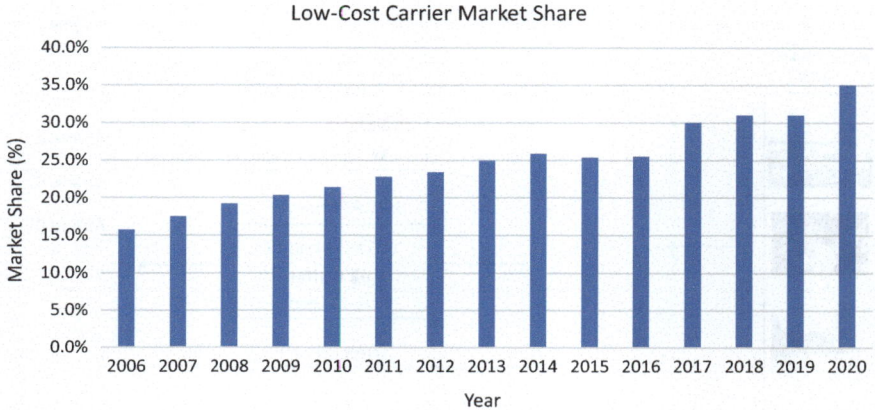

Fig. 6.16 Growth in Worldwide Market Share of LCCs. Source: www.statista.com

continues to exercise its dominant presence in the marketplace. Today, LCCs control approximately 33% of airline traffic worldwide.

Another important parallel is IATA NDC. Direct access to airline content was initially created for LCCs since they did not distribute their product through the GDSs. The IATA NDC paradigm looks very similar to accessing NDC content from full-service carriers (FSC).

6.32.1   Low-Cost Carrier Dynamics

The LCC operating model is very different from that of full-service carriers. Key differences are as follows:

1. One-way fares, fewer fare rules, and restrictions. Typical restrictions imposed are refund penalties and ticketing constraints (instant purchase or 72-h time limit). The fare is the primary determinant of the customer segment.
2. In a restriction free or lightly restricted tariff world, the assumption that demand for each booking class is independent is no longer true. This has an impact on demand forecasting and discount allocation models.
3. Overbooking is very conservative. No-show rates are lower because of ticketing requirements and most fares are nonrefundable. This is augmented by the absence of interline agreements to book denied boarding passengers on alternate carriers.
4. There are fewer booking classes and smaller fare differences between booking classes.
5. Most LCCs do not participate in a GDS. All bookings are consumer direct, on the airline website, thereby avoiding GDS fees.

In addition, there is also little or no barrier to entry for a new LCC, which can change the landscape of competitive fares overnight.

Full-service carriers (FSC) participate in the GDS since they want access to the lucrative corporate bookings. GDS content is fragmented due to the lack of participation of LCCs in the GDS. LCC content is significant since they control approximately one-third of all bookings worldwide.

Metasearch sites like Momondo.com. Skyscanner, and Kiwi.com display schedules for some LCCs and provide a direct link to the LCC reservations systems to make a booking. Travel data aggregators like Travelfusion, a direct distribution platform that is IATA NDC certified, also displays LCC content and facilitates direct bookings.

LCCs have four distinct alternatives to distribute their content:

1. On the consumer direct channel, bookings can be made on the airline website.
2. Global distribution systems. Some LCCs participate in GDSs if the cost economics of indirect distribution is acceptable. A second reason is to attract corporate customers who predominantly book through a TMC that subscribes to a GDS.
3. IATA NDC channel, the new alternative that either bypasses the GDS or serves as a GDS pass-through. The NDC channel used by travel agents typically uses an NDC aggregator or GDS to access NDC content from airlines. A growing number of LCCs now participate in the IATA NDC channel.
4. There are several LCC aggregators (e.g., Ypsilon.Net AG, DoHop.com, Fly.com, Mobissimo.com, Travelfusion, Hitchhiker, etc.) that can be used by a GDS or a TMC to acquire LCC schedules and fares. However, the bookings cannot be made with the GDS and require a direct connection to the LCC's reservations system.

6.33 Sabre and Farelogix

Farelogix was one of the original GDS new entrants (GNRs) that went through a transformation to build and deliver NDC-aligned airline controlled offer and order management distribution technology. Technically, Farelogix is not a GDS since it is a one-sided platform while the GDS is a two-sided platform. Farelogix has no travel agency customers and no commercial relationships with travel agencies.

Farelogix developed a data transmission based on the NDC standard which allows airlines to bypass a GDS and connect directly to travel agencies to make bookings using the Farelogix Open Connect (FLX-OC) solution. Airlines have the option to utilize FLX-OC in the airline direct channel to distribute their services directly to online customers. Additionally, airlines can use FLX-OC in the indirect channel through direct connections with travel agencies and non-GDS aggregators, which serve as an alternative to the traditional GDS. The solution also allows for "GDS passthrough," which involves sharing NDC content with travel agencies through the GDS to create pass-through bookings.

Before the pandemic, Sabre's revenues in 2018 were about $3.9 billion, and the Sabre GDS contributed the lion' share of the revenue, about 75%. Farelogix had revenues of $42 million and half the revenues were derived from FLX-OC. Many airlines like Lufthansa, American, United, Emirates, Air Canada, and others use Farelogix for airline controlled offer and order management.

Sabre first proposed buying Farelogix in 2018 valued at approximately $360 million. The transaction required both parties to file and observe a premerger mandatory waiting period under the Hart-Scott-Rodino Act.

Based on complaints from airlines about the merger, the U.S. Department of Justice (DOJ) following a preliminary inquiry, issued a "Second Request" for data related to the discovery process. The U.K. unlike the U.S. has a voluntary process where the parties are not required to notify the antitrust authority for merger clearance. Regardless, triggered by complaints from airlines about the merger, the U.K.'s Competition and Market Authority (CMA) opened a merger investigation in June 2019. In August 2019, at the end of the Phase 1 review, the CMA identified substantive concerns with the transaction.

Later that month, the DOJ filed suit under the U.S. merger laws (Clayton Act, Section 7) in the U.S. District Court for the District of Delaware to block the parties from closing.

In September 2019, the CMA opened an in-depth Phase 2 review, which is the UK equivalent of the U.S. Second Request process. Phase 2 referral triggered U.K. law prohibiting the parties from closing the deal before obtaining clearance from the CMA.

In the United States, the U.S. District Court had ruled unfavorably for the Department of Justice (DOJ). On April 7, 2020, the court rejected the DOJ's plea for an injunction to prevent the transaction from proceeding. The court's decision was primarily based on its belief that the DOJ did not adequately define a relevant market in which the parties involved were in competition. The court asserted that Sabre functioned as a two-sided platform, whereas Farelogix operated solely on the airline side of Sabre's platform as a one-sided platform.

In April 2020, following a U.S. federal court's decision to not impede the impending merger, the UK's Competition and Markets Authority (CMA) issued a ruling against the deal, citing concerns of anti-competitiveness.

The approval process for Sabre's acquisition of Farelogix was delayed due to actions taken by antitrust regulators in the United Kingdom. and the united States. Despite the absence of any evidence, these regulators had concerns that Sabre might increase prices and limit the availability of FLX-OC to airlines. Sabre's financial situation was severely affected by the COVID-19 pandemic in 2020, leading them to withdraw their bid for Farelogix. It is widely known that airline customers of Farelogix quietly supported the rulings made by the CMA and Sabre's subsequent decision.

Ironically, after the U.K.'s CMA blocked the Sabre-Farelogix deal in April 2020 alleging antitrust concerns, Farelogix was acquired by Accelya in July 2020. Accelya is a financial back-office settlement company. The acquisition was backed by Vista, the private equity company, that had acquired Accelya in November 2019.

Chapter 7
The Impact of Evolving Business Models on Global Distribution Systems

7.1 Overview

The commercial revenue model is all about the cash flow between the supplier (airlines), global distribution systems (GDS), travel management company (TMC), and Online travel agencies (OTAs). With the evolution in the distribution landscape, the revenue model, which dictates how the intermediary is paid, has evolved as well. In this chapter, we start with the GDS—airline revenue model and its impacts on TMC's and OTAs. The commercial revenue model for hotel distribution will be covered in Chapter 11.

7.2 The Changing Landscape of the GDS Revenue Model

The traditional GDS revenue model that has been in existence since 1976 is in the process of being reinvented as a byproduct of the transition to NDC content and NDC bookings. The fundamental issue with the GDS model is that it was viewed by its inventors as a technology play of orchestrating middleware between a point of sale (travel agent) and a supplier (airline). To truly understand customer preferences, changes in customer behavior patterns and customer segmentation requires access to the point of sale, which is something the GDS never had. From that perspective, the brick-and-mortar travel agencies (TMCs) and the online travel agencies (OTAs) are one step ahead; they leverage the GDS to facilitate transactions between suppliers and customers, but they have complete access to and control the point of sale.

The major milestones are outlined in Figs. 7.1 and 7.2.

B. Vinod, *Mastering the Travel Intermediaries*, Management for Professionals, https://doi.org/10.1007/978-3-031-51524-8_7

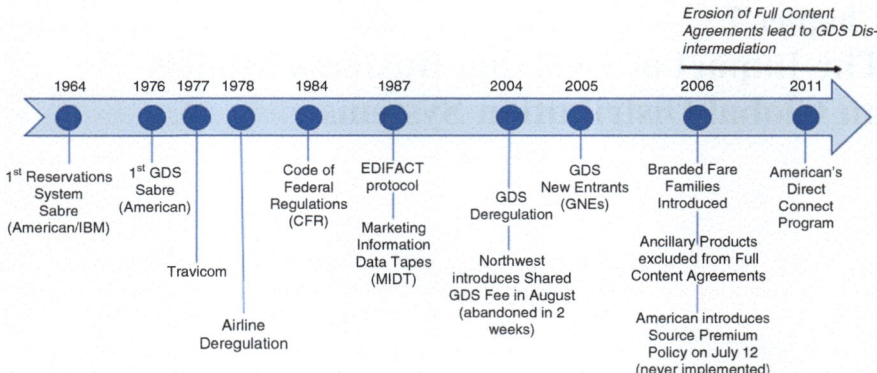

Fig. 7.1 Major milestones and path to disintermediation of the GDS (1964–2022)

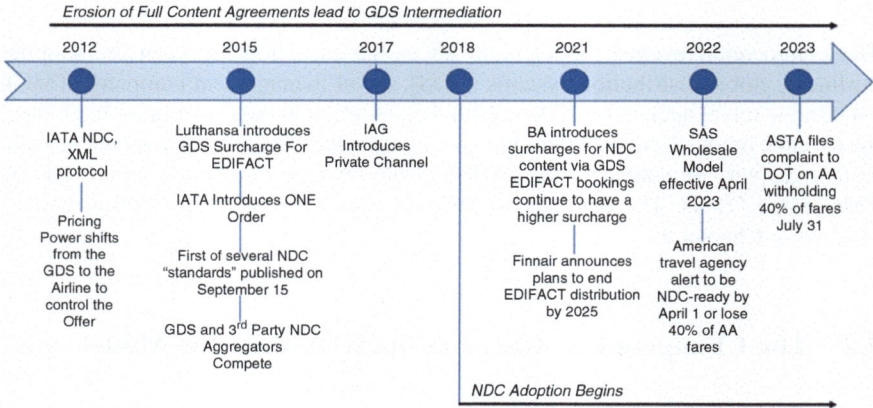

Fig. 7.2 Major milestones and path to disintermediation of the GDS (1964–2023)

7.2.1 The Traditional GDS Revenue Model

The traditional GDS revenue model is straightforward. For every booking made through the GDS, the airline pays the discounted segment booking fee that was negotiated as part of the full content agreement (FCA). In this situation, the customer pays the price of the ticket and any service fees if applicable. The booking fees are net of cancellations. The fee for net bookings is at a segment level. Figure 7.3 illustrates the traditional GDS revenue model.

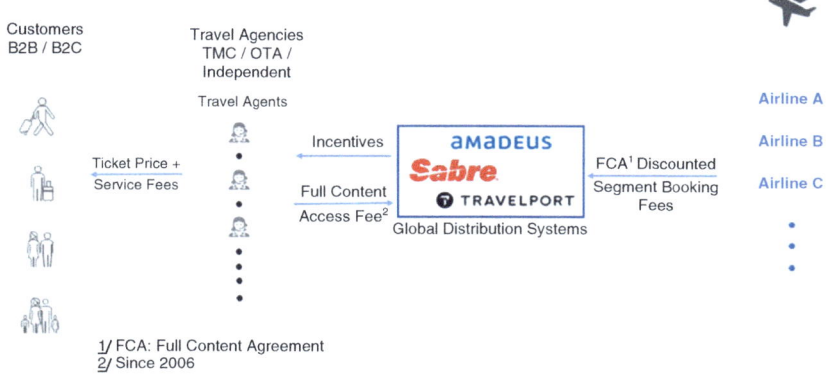

Fig. 7.3 The traditional GDS revenue model

7.2.1.1 The Strategic Importance of Owning the Point of Sale

The biggest issue with the GDS model when it was first conceived in 1976 was that the GDS did not own the point of sale. The focus of the GDS was on technology power to connect buyers and sellers and not to provide customer centric power, where they had direct access to travelers. Travel agents were independent entities that subscribed to a GDS, and they managed their own profit and loss performance. They also have access to the end consumer. If the GDS model had owned the point of sale, many of the issues related to the pressure on segment booking fees and disintermediation would not exist today because the product distribution costs currently borne by the airlines will be passed on to the end consumer.

Travelport recently acquired Deem, a corporate travel management platform to provide access to the point of sale (Travelport, 2023). It is a direct acknowledgement of the importance of owning the point of sale.

The importance of owning the point of sale is illustrated by drawing parallels from other industries.

7.2.1.1.1 Movie Theatre Ticket Purchase Example

Take for example the purchase of tickets to watch a movie in a theatre.

If a customer goes directly to the theatre and purchases a movie ticket, they only pay the price of the movie ticket. The theatre maintains a reservations system that keeps a count of sold seats and the associated seat map for each showing.

Alternately, a customer can purchase the tickets at Fandango Media, LLC, a ticketing company that sells movie tickets via their website as well as through their mobile app. Fandango uses an API to access the theatre's reservation system to display the seat map to a customer. Customers who book tickets through Fandango can view the seat map, select seats, and generate the tickets. Fandango earns revenue through a per-ticket convenience fee from the customer. They also earn revenues from digital advertising on its site and mobile platforms, gift card sales, and promotional sales. In 2022, Fandango charged a convenience fee of about $2 per ticket, and for a $10 ticket, this translates into a 20% fee, attributed to the cost of customer acquisition that the customer, not the theatre, must pay. This is a very significant charge for the cost of distribution, compared to the airline industry, which is approximately 2.2% of the value of the ticket.

7.2.1.1.2   Professional Sporting Event Performance Ticket Purchase Example

The second example is event ticketing by Ticketmaster. If you want to see your favorite NBA team in action, you can purchase the ticket online from Ticketmaster. The total price of a ticket is made up of several components. The face value price of the ticket is set by the Ticketmaster client, which can be the venue, sports team, event promoter or reseller. The service fee, also known as a convenience fee, is added per ticket to the face value price of the ticket. The service fee varies by event based on their agreement with each individual client. Next is the order processing fee, which is added to each order (not per ticket) to offset the cost of ticket handling, shipping, and support costs. The order fee is typically waived for in-person box office purchases. The convenience fee and order processing fees are shared with the Ticketmaster client. Next is the ticket delivery fee, which can vary by option selected such as Mobile Tickets, Will Call pickup, Print-at-Home, U.S. Mail, or UPS. The ticket delivery fee is higher than the true cost, and Ticketmaster realizes a profit for each transaction. In some cases, the venue operator may add a Facility Charge to upgrade and maintain the venue. This is a pass-through revenue stream from Ticketmaster to the venue operator. The final component is city, state, and local taxes, which are listed as a separate charge. So, in this scenario, the customer pays the value of the ticket and all the add-on fees and charges.

7.3   Changes to the Revenue Model

Changes to the traditional GDS revenue model were inevitable, caused by aggressive supplier initiatives to reduce the cost of air distribution, a changing marketplace, and changes in the technology landscape.

7.3.1   The GDS Content Access Fee

The primary beneficiary of the entry of the GDS new entrants (GNEs) in the marketplace in the early 2000s were the airlines. When the GNEs disappeared from the market, the airlines improved their position with deeply discounted booking fees that forced the GDSs to charge travel agencies a fee for access to full content from participating carriers. This fee is deducted from the GDS incentives paid to

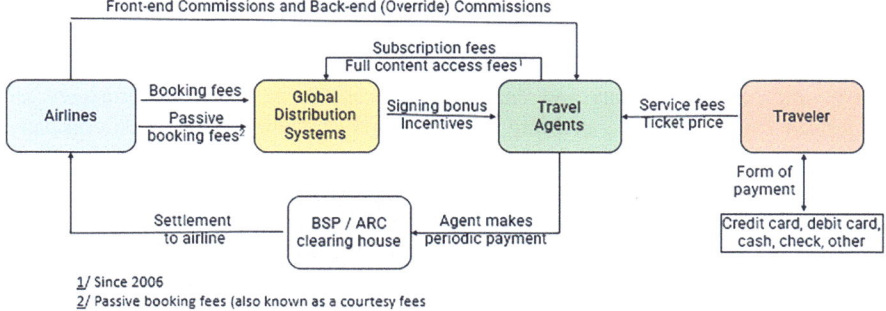

Fig. 7.4 Modified flow of payments in the travel distribution market

travel agents. Each GDS has a unique name for this new fee imposed on travel agents since 2006. They are:

1. Amadeus: Amadeus Content Plus
2. Sabre: Efficient Access Solution
3. Travelport: Content Continuity Program

These are travel agency opt-in programs that are preferred channels by the airlines. If an agency decides to opt-out, agencies may not receive all published, private, and promotional fares from a participating carrier. Further, they would be subject to a booking fee from the airline. The opt-in programs guaranteed access to all content and protection from booking fees from airlines in return for a reduction in agency incentives of 80 cents per segment.

The language used in contracts with reference to the full content access fee between GDSs and travel agencies is vague (Pestronk, 2018). GDSs simply state that content from a participating carrier will be accessible through their platform. However, if the participating carrier decides to withhold certain fares, the GDS cannot be held responsible for incomplete content. These contracts require travel agencies to pay a fee of 80 cents per segment, regardless of whether they have access to the full content from participating carriers.

Figure 7.4 illustrates the changes to the agency revenue model. This is a variant of Figure 3-9 from Chapter 3 with new revenue streams for the GDS from travel agents (full content access fees) and from airlines (passive booking fees).

When the opt-in program was rolled out on September 1, 2006, by the GDSs, travel agencies such as CWT decided to charge $2 per airline ticket to offset the costs generated by the opt-in programs and protect their agency incentives.

7.3.1.1   American Express and TravelBahn Distribution Solution

In 2003, American Express introduced the TravelBahn Distribution Solution (TravelBahn DS) with great excitement. The goal was to decrease reliance on the GDS and lower distribution expenses for airlines like American and United.

However, the financial details of the TravelBahn initiative remained undisclosed, shrouded in secrecy.

TravelBahn aimed to minimize dependence on the GDS network by establishing a network communications and data infrastructure. This initiative connected all American Express offices, enabling the management of traveler profiles, company policies, and personal travel information. The system featured a super PNR, which allowed the agency to share limited data with the GDS for air and hotel bookings. By storing a super-PNR in a central repository, travel agents gained a comprehensive view of customer itineraries, including restaurant bookings, tee times, event tickets, and more. It also claimed to provide access to Internet fares and direct connections with airlines.

By virtue of their size and stature, American Express was not aligned with the opt-in strategy followed by major and mid-size travel agencies because of its TravelBahn DS model.

In 2007 (The Beat, 2007), it was revealed that the smoke and mirrors strategy to reduce distribution costs did not involve direct connections to airline reservation systems or bypassing the GDS. Instead, American Express agreed to a "long-term financial commitment to these airlines" in exchange for access to all fares, including web fares, from airlines like American and United.

Competitors BCD Travel and HRG announced similar plans to launch their separate super-PNR capabilities. A GDS bypass was not part of the plan, but the focus was on improved reporting and data accuracy (Harris, 2007).

7.3.2   The Surcharge Model

When an airline charges a surcharge for bookings made through the GDS channel, they are in violation of the full content agreement and have to pay the standard booking fee, which is higher than the segmented booking fee. Over the short-term, airlines that charge a surcharge will see a drop in bookings, since they are no longer competitive on GDS low fare search shopping displays and an increase in GDS segment booking fees. In this situation, the customer pays the price of the ticket, the GDS surcharge, and any service fees if applicable. As before, the fee for net bookings is at a segment level. Figure 7.5 illustrates the GDS surcharge revenue model.

7.3.2.1   Lufthansa and Sabre Litigation

In 2015, Lufthansa implemented a €16 surcharge for bookings made through Sabre, Travelport, and Amadeus. Lufthansa defended this surcharge based on the expectation that the new XML messaging standard would enable more sophisticated methods of selling airline ancillary products and services and generate additional

Fig. 7.5 The GDS surcharge revenue model

revenue through continuous dynamic pricing of the itinerary and promotion of rich content through the indirect GDS channel.

When Lufthansa imposed the surcharge on Sabre bookings in 2015, travel agents that were Sabre subscribers booked directly with Lufthansa, for which Lufthansa paid them €1 per booking and posted the passive bookings on Sabre, for which Sabre charged Lufthansa a courtesy fee for the passive bookings. A courtesy fee is not part of the standard segment fee contract between an airline and GDS. Passive bookings provide itinerary data to the back-office and expense management systems. This supports TMCs with consolidated reporting on agency sales, commissions, print itineraries, generate invoices, and track records. Lufthansa challenged the courtesy fee, and the following year Lufthansa took legal action in Texas to clarify its ticket distribution contract with Sabre. The lawsuit also sought damages for alleged breaches of contract and interference with agency contracts. In response, Sabre counterclaimed, arguing that Lufthansa's surcharge violated their agreement. Sabre requested a court order to enforce Lufthansa's alleged contractual obligations.

After seven years of legal and commercial disputes, the Lufthansa Group and Sabre resolved their differences and negotiated a new distribution agreement in December 2020. The new distribution agreement introduced distinct commercial models for traditional GDS content and NDC content. Lufthansa Group is a strong advocate for continuous pricing of NDC itineraries. The new commercial model includes itinerary fares based on derived continuous dynamic pricing through the NDC channel. It also required agencies to reach bilateral agreements with the airline regarding NDC implementation.

In August 2023 the Lufthansa Group (Lufthansa, Swiss, Austrian, Brussels, and Air Dolomiti) introduced a distribution cost charge for tickets sold through GDS NDC aggregators. While the fee is lower than the EDIFACT surcharge, the €8 per ticket surcharge applies to NDC-based tickets booked through all three GDSs and issued by travel agencies that did not have a bilateral agreement with Lufthansa Group.

7.3.3   The Private Channel Model

The so-called private channel concept was introduced by airlines such as British Airways and Iberia in 2017. The private channel relationship involves four entities: the airline, GDS, TMC, and the corporation. It is a variant of the surcharge model with a twist. It creates a bifurcated channel within a GDS. With this model, an agency that reaches a private channel agreement with an airline is not subject to the GDS surcharges imposed by the airline. Through the private channel, the airline can also offer differentiated content that is not available in the traditional GDS channel. Bookings made through the private channel will be subject to lower segment booking fees paid by the airline to the GDS. Since the GDSs earns less per booking, they do not pay incentives to travel agencies that book through the private channel (Biesiada, 2017). The private channel concept exists in NDC channels, eliminating incentives to travel agents that are tied to NDC bookings.

The private channel arrangements break the standard full content agreements between GDSs and an airline. It is also the first time that two types of bookings could be made through a GDS with different cost economics. Agencies that book through the private channel for better content typically make up for their loss in GDS incentives by charging a service fee to customers. For the travel agencies, having access to full content is a determining factor in using the private channel. Airlines that promoted the private-channel concept had focused on the large TMCs like Amex GBT, BCD Travel, and Carlson Wagonlit. When the private channel was launched, ASTA had raised concerns since it could never reach the smaller (weaker) travel agencies. The long tail of smaller travel agencies has little or no knowledge of the private channel agreements, and thus they have not become an industry norm.

The private channel was a stop-gap measure by a few airlines prior to the adoption of NDC to bypass the GDS. It was the first concrete step toward disintermediation of the GDS.

7.3.4   The Wholesale Model

The so called "wholesale" revenue model for the GDS was a radical departure since the flow of revenues in the form of booking fees and incentives changes. The flow of revenue is reversed with the wholesale model. Originally proposed by Scandinavian Airlines System (SAS), where the airline would pay agencies tiered upfront commissions for sales and leaves it to the agency to select a technology partner such as a GDS, NDC aggregator, or corporate booking portal to facilitate the transaction. While the SAS wholesale model was never deployed, it is an interesting case study.

The wholesale model based on NDC, first proposed by SAS, was expected to be effective on March 1, 2023 (Frary, 2022; Boehmer, 2022a) in Denmark, Norway,. And Sweden. With this new model, SAS plans to pay a commission of €1 per fare component for Go Smart Fares, €2.50 for Plus/Business fares, and no commissions for the cheapest fares; the Go Light fares. Bookings made with the airline's NDC API or booking portal will not be charged. However, the airline plans to recover up to €12 per segment from the travel agency for GDS charges incurred by the airline for bookings made outside of the preferred channels with airline debit memos (ADM).

The model is unique in the sense that the airline only pays a commission to the travel agency. The travel agency is free to choose their booking channel, which could be a GDS, NDC aggregator or corporate booking portal that uses a NDC aggregator for content. If the agency uses a GDS to transact the booking, then the agency pays the GDS a fee directly for each booking or indirectly through the airline. In this situation, the customer pays the price of the ticket and any travel agency service fees if applicable. The booking fees are net of cancellations. The fee for net bookings is at a segment level. These bookings are for NDC content only, not the traditional GDS bookings using EDIFACT that typically has a surcharge from many European carriers.

Deployment of the SAS solution has its own unique challenges. They are:

1. The agency commissions are paid by fare component and vary by fare type, which requires auditable, automated reporting from SAS to the travel agencies.
2. The airline debit memo is an outdated model. The sheer administrative burden of managing ADMs impacts productivity of both the airline and the travel agency.
3. If SAS charges the distribution cost of their GDS channel to the travel agency, they are in effect sharing their GDS costs with the travel agency, which is a breach in confidentiality between SAS and the GDS.
4. SAS will have to update its existing contracts with travel agencies.
5. SAS also encourages travel agencies to negotiate with their GDS of choice. Few agents have the resources, time, and money to negotiate with the GDSs.
6. The SAS NDC offering to travel agencies in the Scandinavian countries is limited to one-ways and round trips. Besides, ancillaries are limited to seats and bags, which is less than what the GDSs offer.

This is yet another illustration of channel fragmentation where GDSs and NDC aggregators compete for bookings. Figure 7.6 illustrates the revenue dynamics of the wholesale distribution model.

The SAS model was not a sustainable model and was limited to the Scandinavian countries where SAS could exert some influence with travel agents.

Figure 7.7 contrasts three distinct scenarios: the wholesale model, the airline direct booking portal, and agencies that avoid the wholesale model. With the wholesale mode, an airline like SAS provides access to all its fares including the lowest fares via NDC certified channels and approved technology partners like aggregators and potentially GDSs.

The travel agency also has the option to do nothing and avoid the wholesale model mandate. This is a bad option, since the agency will pay the highest cost for minimal content. The cost is the highest because the traditional high segment booking fees charged by the GDS will trigger the airline to recover the full GDS

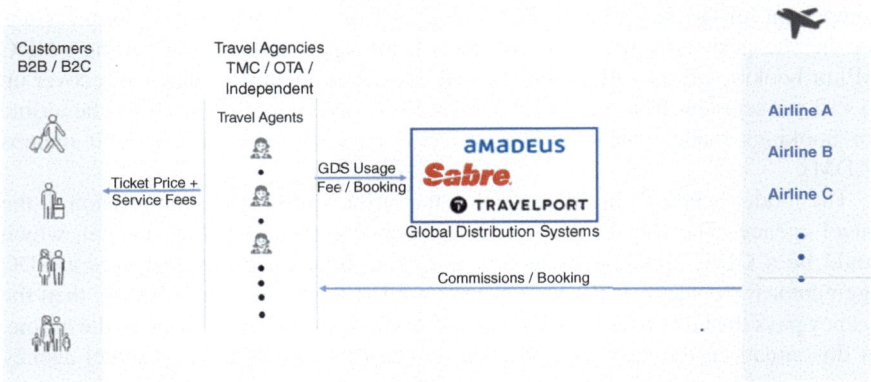

Fig. 7.6 Revenue flow with the wholesale model

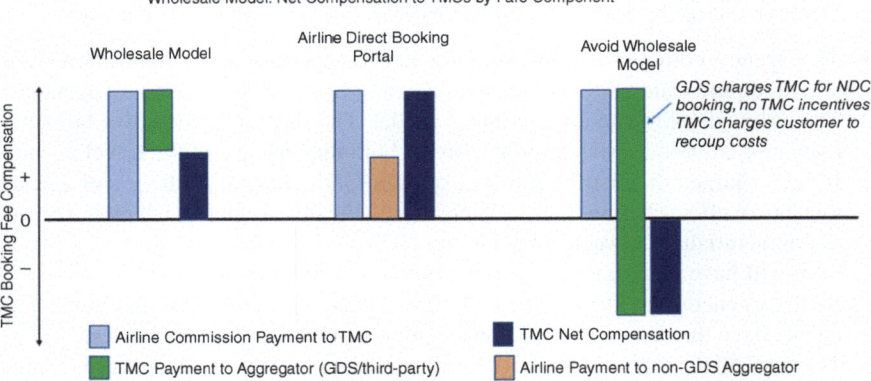

Fig. 7.7 Comparison of alternatives with the wholesale model

booking fees net of any financial incentives paid by the GDS to the agency from the travel agency with agency debit memos (ADM). Netting out the agency incentives is difficult, since the airline has no knowledge of the actual incentives paid by the GDS to the travel agency.

These alternatives also demonstrate that NDC is far from mainstream adoption. While it may represent the future, NDC will be a step backward for a decade (2023–2033) for travel agency productivity. Polarizing marketing one liners from IATA, and the airlines has not accelerated adoption of NDC even though it has been gaining momentum since 2023.

The wholesale model is also a burden for travel agencies that need to navigate and determine the best path forward. For example, if an agency is locked into a GDS contract and SAS bookings are significant for the agency, they will have no choice but to negotiate a new wholesale rate for these bookings. A typical market rate charged by competing aggregators is between \$0.50 and \$1 a segment. This drastically changes the GDS economics who are used to segment booking fees in the \$4–\$7 range in Europe.

Alternatively, if the travel agency has flexibility in their GDS contracts, they can force the GDS and NDC aggregators to compete for the airline's bookings based on unit economics and access to all the content offered by the airline through each channel.

In January 2023, Scandinavian Airlines alerted travel agencies that they planned to delay the launch of the controversial distribution model by six months to September 1 *"in order to allow all partners enough time to prepare."*

In March 2023, SAS abandoned the wholesale model after they reached a new agreement with Amadeus, the dominant GDS in the region (Hoskins, 2023). Effective October 2023, SAS decided to support both NDC content and EDIFACT for travel agencies, but there are commercial differences. SAS NDC content is available through Amadeus and Sabre, and Travelport rollout was planned for 2024. Through the GDS EDIFACT channel, public fares, except for Go Light fares on short haul flights and flights with point of commencement (POC) Japan were available However, negotiated fares and special promotional fares were withheld from the GDS EDIFACT channel. For EDIFACT bookings, SAS also imposed surcharges of €5.50 per fare component for all GDSs except Sabre, whose surcharge was €8.50 per fare component. While public fares, negotiated fares, and promotional fares were available through the GDS NDC channel, SAS imposed a surcharge of €1.50 per fare component for Go Light fares.

7.3.5   NDC Aggregators, GDS Booking Fees and Agency Incentives

The new revenue models with NDC aggregators continue to evolve, and it also impacts the GDS revenue model. NDC aggregators charge booking portals for the NDC content that they aggregate. The booking fee varies between 80 cents and \$1.50 per booking. The airlines compensate the NDC aggregators with performance-based commissions. These are like the back-end (or override) commissions that airlines pay the travel agents.

In contract negotiations with the airlines, GDSs are faced with lower segment booking fees for NDC bookings. While the segment booking fees for the traditional EDIFACT bookings may be the same, the lower segment booking fee NDC bookings are increasing while EDIFACT bookings are declining. With the lower segment booking fees for NDC bookings, GDSs do not pay incentives to travel agencies. Incentives are limited to the traditional EDIFACT bookings that are on the decline. Further, GDSs have informed travel agencies that they will have to pay a fee for the NDC content to compensate for their multimillion dollar annual investment in NDC and the associated lower booking fees that comes with the NDC channel. The multimillion dollar investment by the GDSs in NDC will continue to adapt and modify the agency workflows to ensure backward compatibility with the same versions of the NDC XML message. Airlines will continue to pay both front-end commissions in markets, where they are warranted and back-end performance-based commissions to travel agents. When this model shakes out, it is unlikely that the travel agency revenue stream will be comparable to the legacy EDIFACT model. To compensate for the loss in agency incentives, agencies will have to tag a customer service fee on all invoices. Ultimately, with the NDC model the customer is impacted with new service fees that they did not pay before.

7.4   A Bifurcated Revenue Model in the Hybrid Era

During the hybrid era, as NDC adoption increases, there is a one-for-one replacement of traditional GDS EDIFACT bookings with NDC bookings assuming that the passenger boarded count is static year over year (see Fig. 6.3). With NDC, the airline is the only entity that is responsible for shopping, creating the offer, booking, and ticketing. As airline computing costs increase with NDC adoption, they expect significant cost reductions from GDSs for all bookings, the traditional EDIFACT bookings, and the NDC bookings. For GDS pass-through bookings from travel agents for NDC content, a two-tier revenue model has evolved. While airlines continue to pay the standard segment booking fee for standard GDS EDIFACT bookings, the booking fee paid for NDC content is significantly lower.

NDC aggregators have an alternate revenue model where they charge the travel agencies for access to NDC content, usually between $0.75 and $1.50 per booking, and they do not collect a segment booking fee from the airlines whose NDC content they aggregate and deliver to the point of sale.

Chapter 8
A Fast Response System: Airline Shopping in an NDC World

8.1 The Transition from GDS Shopping to Airline Shopping for the Indirect Channel

Global distribution systems (GDS) shopping, discussed in Chap. 5, involves various complexities like schedules from multiple airlines, interlines, multiple pseudo city codes (PCCs), branded fares across several airlines, and dimensions in itinerary diversity, whereas airline shopping is less complex. It is usually limited to the host airline and their code share partners. In a new distribution capability (NDC) world, what matters is airline shopping and not GDS shopping. As the adoption of NDC by airlines increases, the value of GDS air shopping diminishes over time. This is because pricing power is gradually shifting from the GDSs to the airline. With more NDC bookings being made, airlines have the formidable task of executing shopping requests and selecting the priced itineraries to which an ancillary bundle is attached. These shopping request responses, inclusive of network latency, must be under 3 s to ensure scalability. What this entails is an air shopping and offer management response system that is optimized for accuracy and speed.

If all airlines worldwide were entirely on NDC, then GDS air shopping can be completely sunset. However, the airline industry will be in a hybrid state for at least another decade or more (2035). So, what is the hybrid state? Traditional GDS content and NDC content will co-exist. In a hybrid environment, the value of GDS air shopping will continue to diminish from the point of view of total shopping transactions processed until it is no longer needed. During the hybrid phase, some early adopters may move to NDC shopping and offers for *all* requests, while others will transition over time.

© The Author(s), under exclusive license to Springer Nature Switzerland AG 2024
B. Vinod, *Mastering the Travel Intermediaries*, Management for Professionals,
https://doi.org/10.1007/978-3-031-51524-8_8

8.2   Attribute-Based Shopping, Pricing, and Booking Is the New Reality

Attribute-based shopping, pricing, and booking is the new reality in air, hotels, rental car, and cruise lines. This is the new reality for airline-led shopping to generate offers based on specific attributes captured in the shopping request.

When attributes for a trip are specified by a customer, the search space needs to be refined to return content with the selected attributes. For example, customers like to select itineraries based on seat availability—premium (more leg room) seats, aisle seats, window seats, exit row seats, seats together for a family, etc. free bags, pre-reserved seats, schedule, fare preferences, etc. In this paradigm, the attributes specified by a customer come first before itineraries (e.g., flights) are returned as part of the air shopping response.

Similarly for a memorable hotel stay, users may specify property attributes (e.g., gym, pool, location, etc.) or room level attributes (ocean view, room type, connecting rooms, etc.) as part of the search request. Hotel rates have historically been bundled that has afforded little flexibility. The advent of attribute-based pricing for hotel stays offers flexibility for hoteliers with unbundled rates.

8.3   Segmented Shopping

NDC provides a unique opportunity for airlines to reinvent and fine tune air shopping. GDS shopping algorithms generate bookable itineraries based on various factors such as low fare efficacy and diversity, but do not focus on segmented shopping since *customer segments are managed and controlled in the airline domain.* Unique ways of segmenting customers based on the parameters received in the NDC air shopping API request by an airline and are totally within the control of the airline. This assumes that airlines are looking beyond traditional booking classes for customer segmentation. The context for travel of a customer can be determined from attributes of the input shopping data. These customer segments based on the context for travel are known as trip-purpose segments (discussed in detail in Chap. 9). Every airline can define their own unique trip-purpose segments based on data attributes received in the NDC schema that can then be used for segmented air shopping and offer management (see Chap. 9) with or without 1:1 personalization.

8.3.1   Trip-Purpose Segmentation

Airlines have utilized the reservations booking designator (RBD), also known as a booking class code, since the 1960s with the introduction of airline reservations systems (host CRS) and global distribution systems (GDS). The RBD serves as a substitute for customer segmentation in various airline processes such as fare management, seat availability and inventory control, shopping, booking, pricing, and ticketing. A single RBD is associated with multiple fare basis codes and their respective fare rules. Additionally, the RBD is the primary means for product distribution through marketplaces such as Amadeus, Sabre, and Travelport that determine seat availability. While RBDs are useful in displaying seat availability, airlines are currently seeking alternatives to traditional RBDs for greater flexibility in customer segmentation. An option with NDC is to go classless since distribution of booking class availability to the GDS is no longer required. This is also influenced by purchase behavior patterns that cannot be captured by an RBD.

Unlike a survey, deducing the customer segment based on revealed preferences identified in the shopping request is an approach that is closely aligned with revenue management. This approach is helpful in achieving personalized shopping even when the identity of a customer is not known. Trip-purpose segmentation recognizes that a typical customer has numerous profiles that are based on the context of their travel. Examples of context are business, romantic getaway, family vacation, extended trip to visit friends and relatives, etc. customers who shop on a website may not have declared their identity and may be anonymous. Nonetheless, the shopping response should be sensitive to the context of the trip to ensure that the itineraries displayed are meaningful. Trip-purpose segmentation is useful in identifying customers based on the trip characteristics provided in the shopping request and implicitly determines the customer segment.

Trip-purpose segments are mutually exclusive and collectively exhaustive. What this means is that shopping attributes captured in a shopping request can only be associated wih a unique customer segment. Table 8.1 illustrates an example of trip-purpose segments. These customer segments can be global or defined by market entity (collection of markets) or market. Trip-purpose segments are unique *personas* of the population.

Itineraries generated from a shopping request are based on shopping parameters. To provide the most appropriate results, shopping responses should be optimized according to the dominant values of the attributes that make up the trip-purpose segments. For instance, business travelers may prefer short connections while families with children may require longer ones for the same shopping requests defined by market, departure, and return dates. The first step toward narrowing down the search and return relevant options involves defining trip-purpose segments. In corporate travel, corporate compliance can also help filter the search by augmenting the trip-purpose-segment shopping parameters.

Segmenting customers based on the context for travel is a learning process that needs to be fine-tuned with periodic updates to reflect changes in market conditions and customer behaviors over time. Targeting customers with specific offers and generating incremental revenues begins with attribute-based shopping. This new reality of airline product unbundling demands attribute-based shopping, pricing, and booking. When users specify trip attributes, the search space must be refined to return relevant content. For instance, air travelers prefer itineraries based on seat availability, premium (more legroom) seats, aisle seats, window seats, exit-row

Table 8.1 Air trip-purpose segments

Air customer segment id	Type	Business/leisure	Advance purchase	Length of stay
A1	Individual	A typical business traveler	0–6 days	0–1 days
A2	Individual	Business traveler with a longer LOS	0–6 days	2+ days
A3	Individual	A business or leisure traveler	7–13 days	0–3 days
A4	Individual	A leisure traveler with a longer LOS	7–13 days	4+ days
A5	Individual	A leisure traveler that plans ahead	14–20 days	Any
A6	Individual	A leisure traveler that plans ahead	21+ days	0–3 days
A7	Individual	Leisure traveler, longer LOS that plans ahead	21+ days	4+ days
A8	Couple (2)	Couple on leisure trip	0–20 days	Any
A9	Couple (2)	Couple that plans ahead on leisure trip	21+ days	Any
A10	Family (>2)	Family vacation	Any	Any
A11	Individual	One way business travel	0–6 days	One way
A12	Individual	One way, business or leisure traveler	7–13 days	One way
A13	Individual	One way, business/leisure traveler that plans ahead	14–20 days	One way
A14	Individual	One way, leisure traveler that plans ahead	21+ days	One way
A15	Couple (2)	One way, couple on leisure trip	Any	One way
A16	Family (>2)	One way, family vacation	Any	One way

seats, seats together for a family, free bags, pre-reserved seats, schedule and fare preferences, loyalty tier, and other attributes. The new paradigm mandates user-specified attributes over cheaper itineraries. Shopping for air travel has become more complex due to the unbundling of fares to promote the sale of ancillaries. Unlike online retail stores, air travelers usually remain anonymous while shopping and reveal their identity later in the sales process. Additionally, the context of the trip, such as leisure, business, or visiting friends and relatives, plays a significant role in determining the itinerary that would interest the customer.

8.3.2   A Common Definition for Air Shopping and Offer Creation

Every airline can uniquely define their trip-purpose segments based on their viewpoint of their business model and customers they serve. Customer segmentation is an art and not a science. While analytical techniques such as hierarchical clustering

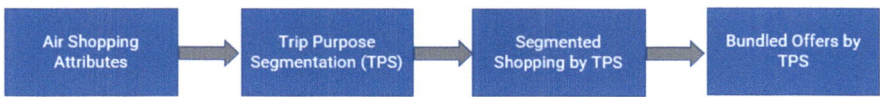

Fig. 8.1 Trip-purpose segment influenced shopping and offers

models can be used to identify trip-purpose segments, the final definition of trip-purpose segments requires manual intervention from experts that truly understand their customers. Once defined, trip-purpose segment definitions should be reviewed periodically and adjusted as necessary.

Trip-purpose segmentation serves a dual purpose: to fine tune the itineraries that are returned in a shopping response and to recommend offers (bundle of ancillary product and services along with the itineraries).

Figure 8.1 illustrates the common definition of trip-purpose segmentation that is required for air shopping and offers (discussed in Chap. 9).

8.3.3   Data Analytics for Segmented Shopping

Identifying a customer's trip-purpose segment from the input data in the shopping request is the first step toward personalized air shopping displays. For example, consider the following parameters that may be important for a customer to book a flight. Historical booking and ticketed data can be used to identify dominance for each of these attributes shown below by trip-purpose segment.

1. Nonstop vs connection
2. Outbound departure time window (day parting)
3. Outbound arrival time window (day parting)
4. Inbound departure time window (day parting)
5. Inbound arrival time window (day parting)
6. Elapsed time (shortest, shortest +30%, greater than shortest +30%)
7. Fares: refundable, partially refundable, nonrefundable
8. Preferred connection airports
9. Preferred connection times (e.g., less than an hour, 1–2 h, greater than 2 h)

For example, consider outbound departure times. Historical booking and/or ticketed data can be used to determine the time slot that is dominant for each trip-purpose segment. These attributes can be calibrated by market, market entity or systemwide, though market or market entity are recommended. For example, Fig. 8.2 illustrates preferences for outbound departure times by trip-purpose segment (based on 7 segments).

For example, if the customer belongs to trip-purpose segment A, the departure time slot between 8:00 am and 10:00 am is dominant, as observed with 35% of bookings. The air shopping algorithm should consider this data to ensure that the

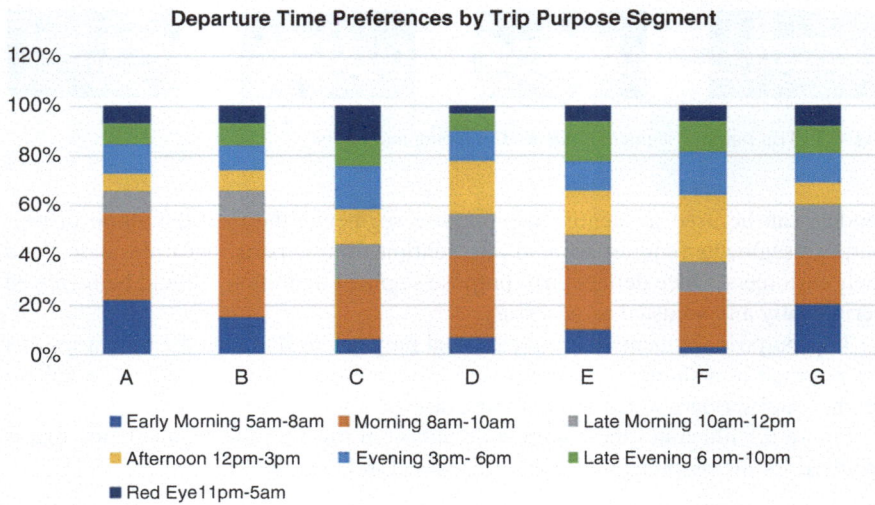

Fig. 8.2   Outbound departure time window preferences by trip-purpose segment

itineraries returned as part of the shopping response are biased toward this time range.

8.4   Mastering Scalability with NDC Airline Shopping and Offers

Scalability is a primary concern with orchestrating NDC transactions. If caching past shopping responses created by airlines for previous requests is not an option since the responses are personalized, the process of generating priced itineraries (air shopping with dynamic prices) and bundles (offer management) must be executed with sub second response times. This fast response capability is required every time a GDS submits a request to an airline for offers (including base fares and ancillaries).

Airlines can create new fares and store them internally on a database and optionally publish these fares to fare aggregators. If the airline is using continuous dynamic pricing for determining the price of an itinerary, the dynamic fares that are created are from reference fares. This is because continuous dynamic pricing is a perturbation off the reference fare, and the quoted fare will be the ticketed fare. Continuous dynamic pricing will generate itinerary prices that are typically lower than the itinerary pricing based on reference fares since discrete jumps are avoided. For example, if a K fare at $100 is closed and the B fare at $120 is open, continuous pricing will determine a price point between $100 and $120.

Generating quick responses to NDC shopping requests from GDSs is critical to address scalability. GDS shopping involves complexities like schedules from multiple airlines, interlines, and multi-PCCs, whereas airline shopping is less complex and usually limited to the host airline and their code share partners. This allows for faster offer generation in less than a second.

Some approaches to expedite NDC offer requests with significantly lower CPU usage and faster response times for the airline shopping component are summarized below.

1. Pre-built flights can be stored in cache, including nonstop, single connect, and double connect options. If the number of schedule options exceeds a certain threshold, flights can be generated dynamically. Airlines can generate flights hourly to accommodate both standard and ad hoc schedule changes.
2. Binding of fares and rules ensures instant access to fare rules associated with a fare. Fare-rule binding can be conducted hourly to address an airline's fare changes. The reference fares can be changed based on market conditions and stored internally. Publishing the reference fares to fare aggregators is optional.
3. Pre-calculate market values to establish lower bounds to reduce the search space.
4. Fare market values are also required for O&D carriers that use continuous nesting or bid price controls.
5. Date-only rule validation that can eliminate fares based on date(s) only without specific flights.
6. Rule summaries, which enable rapid rule validation to remove invalid fares quickly in the early stages of the search process. Directionality for rule validation is reflected in the summary.
7. A "rule set" of a fare is the set of rules used for a specific fare, determined by fare-rule binding. Further, the application of rule summaries can be at the rule-set level.
8. Seat led shopping, wherein a request for an aisle or window seat can pre-screen flight schedules with a seat map cache.
9. The algorithm must be focused on calendar shopping with fixed-date requests.

These powerful concepts are captured by summarizing and combining rules in date-calendar form. These highly summarized rule data allow rapid and effective elimination of inapplicable fares in the early stages of the search process. Modifying the shopping algorithm along these lines will result in faster response times to create the NDC offer response.

8.5 Elimination of the Market Class Fare Value Process

A by-product of developing a fast response shopping system is the elimination of the market class fare value (*MCFV*) table (see Sect. 3.5.2.1) in the airline reservations inventory system.

The market class fare values (*MCFV*) used by O&D carriers for availability determination are surrogates for the actual fare in the market and is an approximation. *MCFV* are used in conjunction with bid prices to determine available seats by RBD. In an NDC world, some airlines may continue to use bid price controls while others will opt for continuous dynamic pricing. In either case, the *MCFV* data is required to determine internal booking class availability or calculate the lower bound and net contribution threshold for continuous pricing. The steps of the fare conditioning process to create the *MCFV* table include:

1. Revenue accounting data is used to average the historical fares with future fares by matching the first four characters of the fare basis code. Data aggregation is by point of commencement of travel inclusive of fare qualification rules (Vinod, 2021a) that are known when a shopping request is made. This fare conditioning process produces the *MCFV* data.
2. Creation of the (*MCFV*) data is typically a batch process and loaded into the inventory component of the host CRS daily.

The size of the *MCFV* table is dependent on the carrier. Before NDC, the size of the *MCFV* table for a large carrier with over 100 million passengers boarded per year can exceed 60 million rows.

With the rollout of NDC, it is anticipated that there will be a significant increase in the number of rows in the *MCFV* table because of filed fares, internally or externally through ATPCO, along the following dimensions.

- Time of day specific fares.
- Date specific fares.
- Routing specific fares.

This introduces a failure point to creating and loading a large dataset into the inventory system in the host CRS each day. This could be as high as a billion rows. A second issue with the *MCFV* table is that the market values used for availability determination for O&D carriers are always going to be different from the actual ticketed fare because the actual fare is not used to compare against the bid prices to determine availability.

In addition, this does not consider dynamic pricing and APTCO Tariff 037, the dynamic pricing engine tariff used for laddered dynamic pricing.

The *MCFV* table can be eliminated altogether by calculating the fare value in real time by utilizing the pricing component of the ultrafast shopping algorithm to generate the fare values dynamically for each service in an availability request. For each shopping request, the fares and rules data can be used to compute the cheapest fare for each booking class. The objective is to validate date related rules and flight related rules by the fare flight binding logic to deliver sub second responses.

8.6   Incremental Benefits for the Airline Direct Channel

Another by-product of developing a fast response sysetm for air shopping and offers to distribute NDC content through the indirect channel (GDS) is airline shopping for the direct channel. A faster shopping engine can enable various aspects of inspirational shopping capabilities with "live" shops that does not rely on a shopping cache for the airline direct website.

For the direct channel, there are varying degrees of personalization that start with trip-purpose segmentation that ultimately leads to personalization of shopping results for a segment of ONE. Personalization at the individual level can elevate airline shopping by utilizing trade-off analytics applied interactively to determine the optimal fare. Many full-service carriers (FSC) that are investing in NDC anticipate corporations will book directly on their website to access lower fares that may not be offered through the legacy GDS channel. This presents an opportunity to improve the shopping experience for both leisure and business travel on the airline website.

In the process of air shopping, customers prioritize finding the lowest fare available in the market as a reference point for their desired fare. Diversity of itineraries by persona maximizes conversion rates. To maximize conversion rates, the air shopping algorithm can return a range of itineraries by persona, using dominant shopping parameters identified from historical booking and/or ticketed data, such as departure time window, nonstop vs. connection, return time window, and elapsed time for outbound and return itineraries, that is relevant for the persona. For instance, on a business trip, travelers would prefer nonstop flights even if the fare is more expensive. However, the collection of itineraries returned, while consistent with the persona, may not necessarily produce the best fare for a specific customer. Itineraries returned from segmented shopping can be evaluated by multi-criteria decision-making models that work with interval data such as TOPSIS (Technique for Ordering Preferences by Similarity to Ideal Solution) and VIKOR (VlseKriterijumska Optimiza-cija I Kompromisno Resenje) to rank itineraries based on a traveler's relative trade-off between schedule and fare attributes (Opricovic & Tzeng, 2004).

8.6.1   Segmented Shopping to Personalization of Shopping for a Segment of ONE

The attributes that define the importance of a persona's trip are determined by typical trip characteristics, expressed as a percentage of degree of importance and value (if applicable). These characteristics include outbound and inbound departure or arrival time, travel time, connecting airports, connection time, and refundable/partially refundable/nonrefundable fare.

The default percentages will form the baseline for a customer to adjust the relative importance of these parameters to create a 1:1 personalized response with the best fare. The default weights can be created by normalizing the dominant value of the attribute determined from the booking and/or ticketed data for each attribute that is deemed important by the customer to be included in the trade-off analytics. These default percentages can be adjusted by the customer to personalize their response with the best fare based on their unique preferences. Any modifications to the

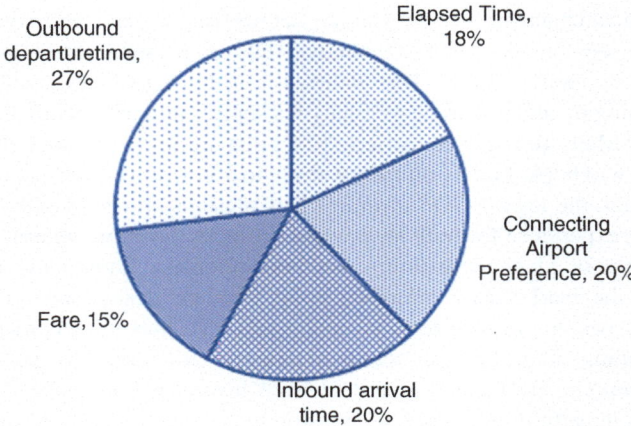

Fig. 8.3 Default preferential weights by trip-purpose segment

weights made by the customer should be stored on their profile against the trip-purpose segmentation id, to allow them to resume their search later without having to start over.

The default weights by trip-purpose segment, which can be represented as a pie chart, are used to initiate the shopping process. This is because, tyically customers are anonymous when they start shopping and the default personas from a baseline that will be updated as the user interacts with the default weights to personalize the itineraries displayed.

An example of default and modified preferential weights by a customer can be seen in Figs. 8.3 and 8.4, respectively. A preference driven air shopping display algorithm (Vinod et al., 2015; Vinod, 2016b) based on trade-off analytics is superior to traditional travel website filters, since a filter would exclude an itinerary based on one attribute, even though it would have been outweighed by the goodness in the other attributes.

8.6.2   *Inspirational Shopping Features for the Airline Direct Channel*

The audience for the airline consumer direct channel is primarily leisure travelers, yet the booking workflow is business travel centric since the flight search always begins with a defined departure date and return date for a market. While many leisure travelers may want to travel to a destination on a fixed departure date, most have flexibility if the price is right.

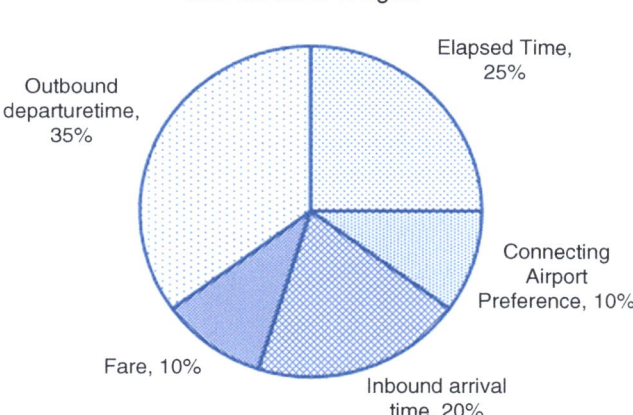

Fig. 8.4 Customer modified preferential weights

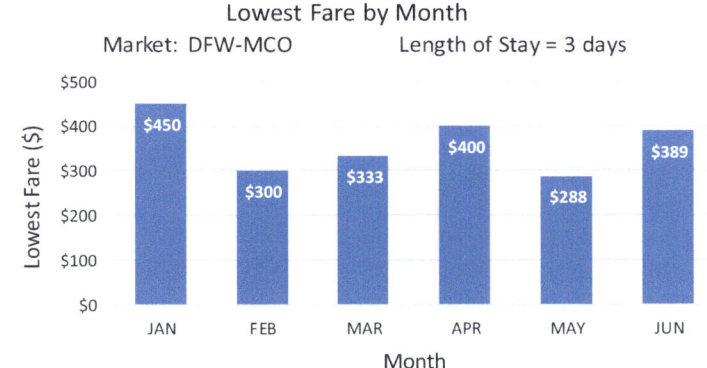

Fig. 8.5 Lowest price by month

Today, many airline websites address this flexibility with a ± 3 day matrix of outbound × inbound flights with a lead price for each applicable cell in the matrix. This is a 49 × 49 matrix of lead prices. Looking beyond ALT dates ± days, the extended calendar is very appealing to leisure travelers.

If the shopping algorithm has been optimized for speed, then the default display should be an extended calendar for a 6 month or 12 month period with the lowest price by month for a fixed length of stay that was captured in the customer input when the flight search was initiated. This is shown in Fig. 8.5 for the Dallas Fort Worth (DFW) to Orlando (MCO) market.

Next, the customer can select a month that will then return the lowest price for each departure date in the selected month. This is shown in Fig. 8.6.

MAY 2023
Lowest Fare by Day
Market: DFW-MCO, Length of Stay: 3 days

SUN	MON	TUE	WED	THU	FRI	SAT
	1	2	3	4	5	6
	$345.09	$411.01	$355.09	$291.35	$377.11	$381.14
7	8	9	10	11	12	13
$311.35	$355.09	$32135	$355.09	$351.66	$375.57	$299.38
14	15	16	17	18	19	20
$399.35	$401.35	$331.35	$355.09	$291.35	$296.65	$325.35
21	22	23	24	25	26	27
$345.54	$451.39	$361.35	$288.00	$333.30	$389.35	$466.25
28	29	30	31			
$387.12	$296.15	$291.45	$355.09			

Fig. 8.6 Lowest fare by departure date in May

Once a specific departure date is clicked, then the standard display of itineraries for that departure date is shown.

Building upon this capability, additional capabilities to drive traffic to the airline website with an ultra-fast shopping engine that produces results that are not dependent on a shopping cache are described below.

8.6.2.1   Theme-Based Shopping

Many leisure passengers do not care about a specific destination, but are more interested in experiencing a theme such as skiing, gambling, sightseeing, romantic getaway, wine tasting, history and culture, nature tour, beach, safari, adventure, and hiking. A theme-based shopping workflow on airline websites allows customers to evaluate a range of destinations that are specific to a theme.

Theme-based shopping also allows an airline to redirect demand from high-demand destinations to low-demand destinations that support the same theme by ranking the destinations in the search results based on available unused capacity (unsold seats) or ranked in ascending order of the total bid price for the various trips.

Theme-based shopping can also be on a budget. A traveler's budget constraint can be added to limit the itineraries shown that fulfill a budget.

8.6.2.2 Frequent Flyer Redemption Shopping

The single biggest problem with frequent flyer redemptions for air travel are the high booked load factors with revenue passengers. As redemption opportunities decline with higher load factors, the new generation shopping framework must enable frequent flyer redemption shopping for markets over a date range from a month to six months. The shopping request to a traveler's redemption request should clearly indicate the minimum loyalty points required by departure date for a fixed length of stay by market on the calendar.

8.6.3 Seat-Led Shopping

An important consideration during flight search is seat led shopping. With this capability, a customer can state their seat preferences as part of the shopping request to avoid having to go through multiple shopping iterations when the desired seats are not available on the seat maps displayed for the itinerary. Seat-led shopping must only display itineraries that can fulfill the seat request criteria, and the total itinerary price must reflect the price range if seats are individually priced.

8.6.3.1 Why Is It Important?

Here are a few scenarios for seat led shopping.

> Show me flights from Los Angeles to New York departing on June 4 before noon and returning June 8 in the evening for a family of 4 and we want to sit together.

> I want to go to London for 1 week, departing June 1 from New York. Only show me flights where an aisle seat in Business Class is available.

> I want to go to New York from Phoenix on June 1, returning on June 6. Show me flights where exit row seats are available. In addition, I want to avoid regional jets.

> I want to go to London for 1 week, departing July 1 with my spouse. Only show me flights where seats 5A and 5B are available in Business Class on a Boeing 777.

The workflow for seat selection today in both the agency channel and the consumer direct channel occurs after the itinerary has been selected by the customer. This is a significant shortcoming since the seat map may display available seats that are not acceptable to a traveler, and they must start over.

8.6.3.2 Seat Map Cache

For airline shopping, post processing of itineraries returned from shopping to select itineraries that fulfill the customer seat request is not an elegant solution since none of the itineraries returned may fulfill the seat request constraint. Seat-led shopping as a post process of the shopping process has its limitations, since there is no guarantee that the itineraries returned during shopping fulfills the seat request. Ideally, seat-led shopping should be *in path* in the shopping algorithm and not a post-process. This implies that the seat selection constraint imposed by the customer is considered a priori when schedules are generated by the shopping algorithm. For example, if a customer wants four seats together, the only itineraries that should be displayed are those which have four seats together that can be selected. With seat-led shopping, the prices for the itineraries returned by shopping may be higher with the seat type constraint. Seat-led shopping advances the user experience to select itineraries that guarantee the requested seat type request at time of shop. Querying the pre-reserved seats function on the host CRS in real time to support shopping is an expensive proposition, besides increasing latency in response times. Access to a seat map cache in real time during shopping addresses this gap to support seat-led shopping to fulfil the seat requirement.

Individual seat pricing is the last frontier in airline revenue management. An individual seat is the lowest atomic unit of inventory. To address customer preferences for a seat type, shopping algorithms should be able to only display itineraries that fulfill this contraint. Adding a seat type constraint in air shopping when schedules are generated dynamically also implies that the itineraries displayed may not be the lowest fare in the market. For efficiency, seat-led shopping also requires a seat map cache that stores various counts of seat types in memory in real time that will be accessed when dynamic schedules are generated. Besides, the airline website may also want to display seat map counts by seat type based on customer preferences to enhance the display of itineraries returned as part of the shopping response. To avoid the transactional and computational burden of querying seat counts and seat maps on the host CRS in real time, a viable approach is to deploy a seat map cache. A seat map cache addresses the issue of an airline's pre-reserved seats system resident on the host CRS from being overwhelmed with several hundred seat map requests following a shopping request from an OTA. Every time a seat is consumed from any channel, agency or direct, the seat map cache can be updated in real time, and shopping can access this cache in real time to support seat lead shopping. The seat map cache counts of inventory will have the same level of accuracy as RBD availability. The seat map cache is also required for individual seat pricing.

The seat map cache should maintain counts of sold and available seats along the following dimensions in real time: total seats, aisle/window/center seats, exit row seats, premium seats, preferred seats, no charge seats, bulkhead seats, pay-for seats, minimum noise zone, extra leg room, uninterrupted view, seats together for a specific party size (two seats together, three seats together, etc.), and details for a specific seat (e.g., seat 12 A—window, premium, etc.). For hosted airline inventory, the various counts by seat type are maintained current in real time for every sell and cancel of a seat assignment, regardless of where the request came from. For code share and alliance partners, a seat map cache controller determines when an item in cache needs to be refreshed. The refresh frequency can be based on static predeparture reading day concept from revenue management or dynamic based on actual activity. The cache should also be able to display the physical seat map.

Deployment of a seat map cache should support all the Passenger and Airport Data Interchange Standards (PADIS) seat characteristics an airline chooses to send. There are over 100 characteristics, which can be grouped into categories such as location of the seat (e.g., front of cabin, upper deck, adjacent closet, etc.), missing seats (e.g., no seats because of exit door, no seat because of upper stairs, etc.), seat characteristics (window, aisle, etc.), seat occupation details (e.g., occupied, advanced boarding pass issued, etc.), and seat blocking details (e.g., blocked for airport, blocked for through passenger, etc.).

There are two *necessary and sufficient conditions* that should always be satisfied to ensure consistency and accuracy of seat map counts. Consider a flight that goes from A to B to C.

Condition 1: If a seat is consumed on a through flight (e.g., A to C), then the same seat should be consumed on the legs that make up the through flight (A to B and B to C).

Condition 2: If a seat is consumed on a leg, it should not be available on a through flight that includes this leg.

Hence, all seat maps are stored at a flight leg level and not at a flight segment level since all flight segments are made up of the underlying legs. When a seat map request is made for a specific flight, the cache examines the line of flight for this flight date to determine which leg maps to combine to create the requested segment map.

8.6.3.3 Pricing of Seats

Airlines permit customers to select seats on the aircraft while selecting an itinerary and making a booking, apart from exit row seats, blocked seats for premium passengers, bulkhead seats for families with small children or weight and balance restrictions. Customers are charged a static amount for paid seats with extra leg room, preferred seats, etc. However, those who do not pay for seats can follow the standard seat selection process for unoccupied seats. The policy varies from 24 h to 2 days before flight departure of the first segment. The same policy is applicable for unpaid seats during return trips.

Throughout the process of selecting seats, each customer holds a certain perception of the inherent value associated with the chosen seat. This perception ultimately influences the rate at which seats are purchased, measured by the ratio of seat purchases to the total number of seats available for a given seat type. The value of a seat can be capitalized, particularly for prolonged trips, by implementing dynamic pricing strategies based on various flight and customer attributes. Examples of flight attributes include the market being traveled, type of aircraft, travel distance (short, medium, long, and ultra-long haul), departure and arrival times, season, and day of the week. Examples of customer attributes include frequent flyer tier, gender, advance purchase, departure day of week, length of stay, number of people in the party, sales channel, point of commencement, point of sale (country and city), currency of purchase, and form of payment.

Seat prices can be determined through either rule-based methods or advanced decision support systems that calculate prices for ancillary services based by flight leg and date. Personalization based on rules can become cumbersome when there are too many pre-established rules that require periodic updates. To simplify maintenance, updates, and deletions, the rules must be categorized and stored in a catalog. A common issue that arises when a catalog is absent is the accumulation of inactive rules that must be periodically removed.

Predictive personalization is an advanced approach that does not necessitate the creation and management of rules. Recommendations are based on a combination of purchases made by customers in the same segment and actual purchases from past behaviors. More complex techniques can also be utilized to establish seat pricing. Machine learning methods, including logistic regression, gradient boosting machine (GBM), random forest, and deep learning, can be calibrated to gauge precision, sensitivity (recall or true positive rate), precision, and false-positive rates. The calibrated models can forecast the likelihood of a purchase at a specific price point. However, these tactics typically require extensive calibration before the predictive model can be launched, lowering the scalability of the solution. Additionally, to account for market conditions and competitor responses, these models must be frequently re-calibrated, adding to the scalability burden. The monetization of seats with a dynamic seat price is an active area of research.

It is recommended to use a reinforcement learning-based test and learn experimental approach instead of traditional predictive model calibration to support continuous calibration. With this approach, changes in customer preferences and willingness to pay behavior over time can be addressed in real time without periodic recalibration. This method also continuously adapts to the environment based on market conditions.

If airlines use a leg-based seat pricing model, they can incorporate a variable pricing concept like a bid price curve. This makes the seat price a function of the seats available for a specific type of seat. The seat counts by seat type can be accessed from the seat map cache to determine the price for a seat.

Figure 8.7 illustrates variable pricing for a seat as a function of remaining aisle seats available for sale. Based on competitive market factors, these prices can be by flight leg.

The pricing of ancillary seats is based on the remaining seats available for each seat type. An ATPCO OC (optional service fees) filing for an aisle seat will only have a reference fare for informational purposes. The ancillary price for a seat will depend on the bookings by seat type and potential type of customer, identified by the trip-purpose segment and frequent flyer status. The fewer seats available for a particular seat type, the higher the price. All indirect sales channels must go to the airline to price ancillary services, which is also the NDC paradigm for itinerary pricing.

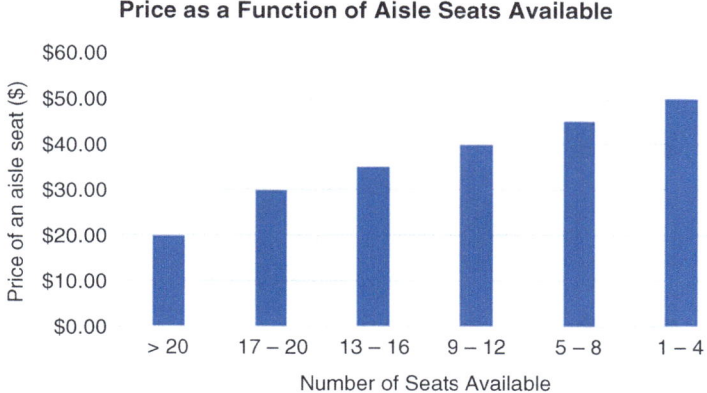

Fig. 8.7 Variable price of an aisle seat

8.7   Enhanced Shopping and Website Redesign

With NDC, many airlines aspire to go beyond leisure traffic with corporate booking portals that are linked to their website, like Southwest Airlines has been doing since 2003. If an airline decides to go in this direction, ideally the website should support distinct workflows for corporate business travel and leisure travel. Perhaps airlines should incorporate a third workflow for bleisure, if bleisure traffic is sustainable in a post-COVID world.

Leisure traffic is expected to outpace corporate traffic. The audience for the airline consumer direct channel today is primarily leisure travelers, yet the booking workflow is business travel centric since the air search always begins with a city pair, departure date, and return date (if applicable). While many leisure travelers may want to go to a specific destination, many do not and are interested in experiencing a theme such as skiing, gambling, sightseeing, romantic getaway, wine tasting, history and culture, nature tour, beach, safari, adventure, and hiking. With an ultrafast air shopping algorithm, airline websites should support inspirational shopping capabilities such as extended calendars for a fixed length of stay, theme-based shopping, and budget shopping. This allows customers to explore the dream and plan phase of travel on the airline website to evaluate multiple destinations that are specific to a theme. Theme-based shopping also allows an airline to redirect demand from high-demand destinations to low-demand destinations of the same theme by ranking the destinations in the search results based on available unused capacity (unsold seats) or ranked in ascending order of the total bid price.

Figure 8.8 illustrates an air shopping workflow for the direct channel that is expressly targeted for leisure traffic.

The first step is to determine the trip-purpose segment based on the shopping parameters input by the user. For calendar shopping with an anonymous use, perhaps the only input is the duration (length of stay) and month. In the single day shopping

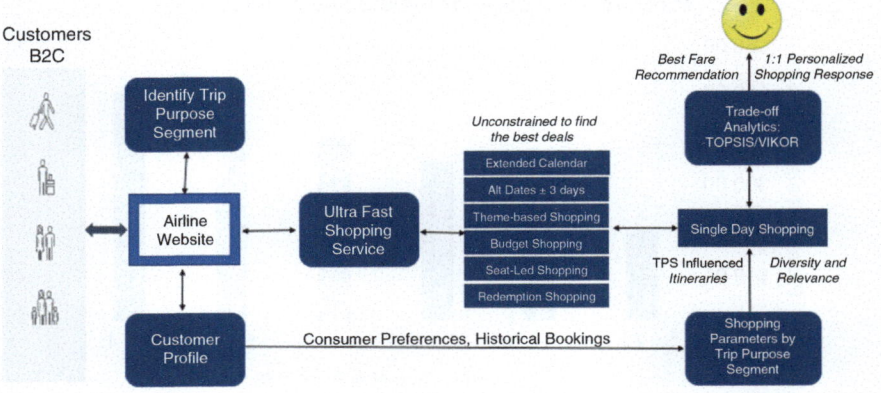

Fig. 8.8 Personalized air shopping for the airline direct channel

stage, the trip purpose is refined based on expected departure date, return date, and number in party. Itineraries returned will be influenced by the default shopping parameters by trip-purpose segment from historical aggregate booking data. The itinerary selection is then refined based on preferences with the trade-off analytics algorithm, which is a recommendation of the *best fare* for the customer.

Chapter 9
Offer Management, Dynamic Pricing, and Order Management

9.1 The Transition of Revenue Management to Offer Management

For three decades after airline deregulation, the focus of revenue management in the airline industry has focused on maximizing revenues by selling the right product to the right customer at the right price at the right time. It was one-dimensional in the sense that the selling fare was all that mattered. Revenue management advanced from leg/segment inventory controls to origin and destination inventory controls. With leg/segment controls, seat availability was based on either net nesting or threshold nesting. To control inventory by O&D, availability is based on the continuous nesting, also known as bid price controls. Restriction-free pricing (RFP), a feature introduced by LCCs, was first adopted to some degree by the FSCs when they had to compete directly with the LCCs on key routes. In the absence of fare restrictions, when price was the only determinant of the customer segment, demand for booking classes is no longer independent. The revenue management models had to go through significant enhancements to support lightly restricted and unrestricted tariffs.

With schedule and fare transparency during shopping, competitive revenue management gained rapid adoption (Ratliff and Vinod, 2005), whereby the selling fares in the market offered by competitors was considered as an input before an airline determined the optimal inventory controls subject to prevailing competitive market conditions. The latest advances in revenue management of the base fare are in the areas of dynamic availability and dynamic pricing. Dynamic availability works with the existing airline tariff structure that has been published by the airlines to the fare aggregators. Dynamic availability factors the competitive marketplace to dynamically open and close discrete inventory controls in the airline host CRS to maximize revenues. Dynamic pricing takes it a step further to generate a dynamic price that may not be a fare that was filed by the airline to fare aggregators. The

B. Vinod, *Mastering the Travel Intermediaries*, Management for Professionals,
https://doi.org/10.1007/978-3-031-51524-8_9

advantage of dynamic pricing is that it fills the missed opportunities (gaps) on the price demand curve by recommending a continuous dynamic price.

With the growing importance of the sale of ancillary products and services bundled with the base fare to improve the bottom line, this transition of revenue management of the base fare to offer management was spearheaded by the LCCs. With FSC's following the LCC lead by selling ancillary products and services, the science of revenue management has transitioned from *inventory control tactics to sell the base fare* to *selling a bundle consisting of the base fare with ancillary products and services*. With this change every dimension of the offer—the base fare and ancillaries included in the offer with their respective prices are important to maximize an airline's network revenues and profitability.

The offers recommended to customers can be by customer segment based on the context for travel. This is a recommendation by trip-purpose segment. Alternately, if the customer has purchase history on their profile, this information can be leveraged to *personalize* the offer at the individual level for a segment of ONE.

With the IATA NDC messaging standard, the key difference is that all offers are created and distributed by the airlines (Touraine, 2021).

9.1.1   How Revenue Management Is Evolving in a Post Pandemic World

Besides the fundamental transition from revenue management of the base fare to revenue management of bundles, revenue management is evolving in a post-pandemic world.

Historical demand patterns may not accurately predict future demand, which was observed after the onset of the COVID-19 pandemic in 2020. Adaptive revenue management is a reliable framework that allows businesses to quickly adapt to changing market conditions with ad hoc schedule changes by incorporating key concepts from sales and operations planning (Vinod, 2022b). During the pandemic, there was a significant decline in customer demand, and airline revenue management systems struggled to accurately forecast flight demand (Vinod, 2020c, 2021b). To effectively manage the business without relying on accurate forecasts, it is necessary to shift toward an adaptive robust revenue management framework. It is an operationally driven approach that continuously manages demand. This approach, known as continuous demand management (Vinod, 2021a, b), involves actively monitoring key performance indicators and proactively taking corrective action to ensure that revenue targets are met. By embracing continuous demand management, businesses can effectively respond to demand fluctuations and achieve the goals outlined in the revenue plan. This approach bears similarities to the sales and operations (S&OP) workflow (Palmatier & Crum, 2003), which involves synchronizing demand, available capacity, and resources to optimize performance, particularly during times of high uncertainty and volatility.

To promote bundles, calibration based on historical data must be avoided at all costs. First, the historical purchase history data may not be sufficient or nonexistent when new ancillary products are introduced. Second, periodic calibration to revise the estimates of the parameters of the offer model is a time-consuming task and will *always lag* evolving consumer preferences over time. The more adaptable approach is to conduct *test-and-learn experimentation*, using a reinforcement learning technique commonly known as the multiarmed bandit (MAB), to determine the composition of the offer bundles. This approach learns continuously based on changing consumer behaviors and prevailing competitive market conditions. The implementation of the test and learning experimentation model should be context sensitive (Byrd & Darrow, 2021). Contextual reinforcement learning tracks customer responses by trip-purpose segment.

Personalization of the offer based on the context for travel is important to improve conversion rates. Offers can be recommended by trip-purpose segment, and in situations where the customer travels frequently, the offer can be personalized based on past purchase history and preferences.

9.1.2   The Transformation with NDC

With the traditional GDS transaction flow model, the GDS calculates the price of an itinerary based on availability of booking classes determined by the airline's revenue management process. Many airlines publish their ancillary services (ATPCo records S5, S7, S8), and GDSs also create offers consisting of a base fare and ancillary products and services.

As airlines transition from revenue management of the base fare to offer management consisting of bundles, they want to have complete control of the offer creation process given its strategic importance to maximize revenues. There are two fundamental changes, and they are related to itinerary pricing and the ancillary bundle.

1. The itinerary price is not based on static filed fares, but a dynamic fare that is derived from reference fares based on market conditions. When a fare is derived based on business rules or advanced statistical and machine learning models, the GDSs can no longer price an itinerary. The creation of an itinerary price on demand for each customer request is called dynamic pricing and no longer requires RBDs since availability by booking class is no longer distributed to the GDSs.

2. Creation of the bundled offer at the time of request is a fundamental change. This requires advanced analytics to determine the context for travel to generate a recommendation consisting of an airline bundle of ancillary products and services that are pertinent to the customer request to enhance conversion rates. Creation of the bundle by an airline will be based on simple business rules or customer data

complemented with advanced statistical and machine learning models to generate the offer. With this new process, the GDSs can no longer create bundled offers.

9.2   Is Personalization For a Segment of ONE Achievable?

Personalized offers consisting of the dynamic price for an itinerary bundled with air ancillaries is one of the key value propositions of NDC. One of the key drivers of IATA NDC is personalization for a segment of ONE, also known as 1:1 personalization, which is difficult if not impossible to achieve at scale due to the nature and frequency of customer bookings. Most customers who travel make a trip every 12–18 months and are not part of a frequent flyer program. FSC carriers like American Airlines and United Airlines have confirmed that over 50% of annual airline revenues came from 85% to 87% of its customers who flew only once. This implies that for FSCs, true 1:1 personalization is limited to loyal business flyers and high end leisure travelers that are part of the 13% to 15% of customers that generate half the airline's revenues. This was confirmed by Scott Kirby when he was first president at American Airlines. On an October 25, 2015, conference call with stock market analysts, Kirby stated that half of American's revenue in the previous year came from 87% of its customers, who only flew the airlines once. He went on to state that 50% of the company's revenue comes from passengers "for whom air travel is largely a commodity" (Yanofsky, 2015). A year later, Kirby was President at United Airlines when he confirmed that 85% of United's customers fly less than once a year, and like American, account for close to 50% of revenue (Zhang, 2016).

The absence of historical purchase behavior data for a vast majority of travelers is a major impediment to pursuing 1:1 personalization of the airline offer. The bundle consists of a base fare which could be a dynamic fare and a collection of ancillaries that is expected to resonate with the customer. While dynamic pricing can still be pursued to maximize revenues based on historical price-demand curves, the composition of the ancillary products included in a bundle is a challenge. Perhaps the answer lies in a hybrid version where the vast majority of offers generated by an airline *will not be at the individual level* but based on the *context for travel for a collection of travelers*, also known as trip-purpose segmentation. True 1:1 personalization will be limited to customers who make multiple trips a year, and their historical purchases can be leveraged to fine tune and personalize the offer.

There is an expectation that advances in artificial intelligence, machine learning, advanced statistical modeling, and NDC preferred channels will revolutionize travel retailing. With the presence of an airline's NDC preferred channels competing directly with the GDS channel, for the first time, travel agencies will be exposed to the true cost of the channel they select. Further, if the GDSs start charging agencies for an airline's NDC content to offset lower segment booking fees for NDC bookings from airlines and to recoup their investment in multiple versions of NDC and their workflows, travel agencies could easily shift their allegiance to a lower cost channel to preserve their margins.

With NDC preferred channels, an airline's goal of an omni channel strategy of uniformly promoting their products across all channels of distribution is also at risk. NDC preferred channels will lead to a bifurcation of the omni channel strategy to create a seamless customer experience through all channels with identical offers. For an airline's customers to experience the airline brand through all channels; direct and indirect, may be an aspirational goal and for the most part, out of reach.

The larger question that remains to be answered is the degree to which personalized offers can be achieved.

9.3   Customer Lifetime Value and Traditional CRM Segmentation

If personalization is important in the NDC era, it must begin with frequent flyers who have a rich history of flown bookings.

Investing in customer loyalty and enhancing the customer experience are critical areas for airlines to distinguish themselves from competitors and safeguard their brand and direct channel. Personalizing the customer experience for frequent flyers is a key component of achieving this. The foundation of personalization lies in cultivating a loyal customer base. A common approach to creating a personalized experience involves segmenting loyal customers according to their purchasing behavior and overall expenditure.

Determining the customer lifetime value (CLV) is an effective measure of a customer's worth, which can be based on their past purchases, potential future revenue, or a combination of both. The ultimate objective of loyalty customer analytics is to estimate the customer's share of wallet. However, this can be challenging due to the lack of access to sales data from competitors on an individual customer level.

The formula for customer lifetime value is:

$$CLV = Customer\ Revenue - Cost\ of\ Customer\ Acquisition\ and\ Service.$$

Estimating CLV in the airline sector does not revolve around granting preferential seat availability to higher-tier frequent flyers. Instead, its significance lies in optimizing the costs of acquiring customers and segmenting frequent flyers in innovative ways to bolster promotional initiatives and engagement. A more effective approach to calculating CLV involves analyzing both past historical performance and anticipated future performance. This method is widely recognized as the standard CLV formula

Table 9.1 RFMVT segmentation definition

Measure	Description of measure
Recency (R)	The length of time in days since the customer made a purchase.
Frequency (F)	The number of times the customer has visited the storefront and made a purchase.
Monetary (M)	The total dollars spent by the customer on all visits combined.
Variety (V)	The number of distinct purchases by the customer in all visits.
Tenure (T)	Number of days since the first purchase.

$$\text{Customer Lifetime Value (CLV)} = -\text{CCA} + \text{VTD} + \sum_{t=1}^{T} \frac{(R_t - C_t)RR^t}{(1+i)^t}.$$

In this formula, CCA represents the cost associated with acquiring a customer, VTD represents the cumulative value from the beginning until the present time, R_t is the forecast revenue in future period t derived from predictive analytics, C_t represents the variable cost in period t, RR denotes the rate of customer retention, i represents the cost of capital, and T represents the timeframe for the calculation. The residual CLV, or potential future value (PFV), which is represented by the term in the summation, holds greater significance than the CLV itself in situations when past revenue performance does not align with expected future revenue performance. The formula has limitations since the retention rate is not consistent over time but tends to improve (increase) as customers become more familiar and confident with the brand. To create customer segments and formulate strategies for customer acquisition, estimates of CLV from existing frequent flyers can be utilized. These estimates aid in persona identification. Loyalty segmentation tiers can be constructed based on the CLV estimates. The CLV tiers can serve as a foundation for developing guidelines and strategies to attract and retain customers.

An alternative to CLV is the conventional segmentation method employed in retail by Customer Relationship Management (CRM) software, known as RFMVT. The analytics for RFMVT focuses on five essential factors—Recency (R), Frequency (F), Monetary value (M), Variety (V), and Tenure (T). All RFMVT metrics are compared to a specific customer and measurement timeframe (week, month, quarter, etc.).

Table 9.1 describes the RFMVT measures.

Adapting the RFMVT approach in the airline industry to account for variety (V) involves considering the number of different trip-purpose segments that a customer was assigned to for past purchases. RFM segmentation is the most used level of segmentation. This type of segmentation can be utilized for targeting promotional offers, analyzing market penetration, and evaluating profitability. To enhance the RFM segmentation and effectively summarize customer data, it is important to address the issue of "clumpiness," which arises from intermittent purchases made by customers. Zhang et al. (2015) argue that although statistical models based on RFM summaries may perform well at an aggregate level, they can lead to significant prediction errors when ranking customers at a micro-level unless the concept of "clumpiness" is properly accounted for.

There is a large body of research on CLV and RFM and a detailed treatment is beyond the scope of this book.

An advanced method to calculate the expected lifetime value, E(CLV), was developed by Fader and Hardie (2009) using probability modeling.

The E(CLV) for a yet to be acquired customer is stated as follows:

$$E[\text{CLV}] = \int_0^\infty E[v(t)]S(t)d(t)dt,$$

where $E[(v(t)]$ is the expected value of the active customer at time t, $S(t)$ is the probability that the customer has remained active to at least time t, and $d(t)$ is the discount factor that reflects the present value of money received at time t.

The expected residual CLV at time T for an existing customer is given by

$$E[\text{Residual CLV}] = \int_T^\infty E[v(t)]S(t|t > T)d(t - T)dt.$$

Schmittlein et al. (1987) introduced the Pareto/NBD model to forecast future customer purchases based on booking history, purchase frequency, and recency. A simplified version of this model, known as the beta-geometric/NBD model (Fader et al., 2005a; Fader & Hardie, 2009), offers easier parameter estimation and implementation. By incorporating RFM inputs for CLV projections, the concept of "isovalue" curves is utilized to group customers with distinct historical behaviors but comparable future values (Fader et al., 2005b).

9.4 Branded Fare Families and a la carte Pricing of Ancillaries

Branded fares, also known as fare families, were initially introduced by Air New Zealand, Air Canada, and Qantas in 2006. In 2008, Frontier Airlines became the first U.S. carrier to implement branded fares. The concept of branded fares involves offering additional optional products and services alongside the base fare, which are grouped together in each fare family. The prices for these bundled offerings are predetermined. Today, most airlines follow a hybrid approach that combines branded fares with à la carte pricing for ancillary products. This strategy aims to maximize revenue by encouraging customers to upsell from the base fare and increase the likelihood of purchasing additional ancillary products separately. There is an ongoing debate about the extent to which ancillary products influence the decision to upsell from a fare family, compared to purchasing them à la carte.

The sale of air ancillary products bundled with the base fare is called offer management. Airlines generate significant revenues through the sale of ancillaries such as baggage fees, pre-reserved seats, seat type, and access to the frequent flyer lounge. Customers purchase these ancillaries as a prerequisite to completing their travel plans.

For air, offer management is the process of selling the right bundle of base airfare and air ancillaries to the right customer at the right price at the right time. The science of revenue management of the base fare has evolved to the sale of a bundled offer (Vinod, 2017, 2021c, e; Vinod et al., 2018).

9.5   Ancillary Sales

Airlines worldwide are selling air ancillary products to generate incremental revenues beyond the base fare. Branded fare families and the sale of à la carte ancillaries have generated billions of dollars for airlines in recent years. While the average base fare in the airline industry has declined by 0.9% per year over the past decade (IATA, 2018), ancillary sales have grown 40%.

The question is not whether customers will pay for air ancillary products. An independent survey conducted by Leflein Associates in January 2006 (Alexander, 2006) well before ancillary product sales gained momentum showed that travelers will pay for extra perks such as more frequent flyer miles, more overhead bin space, and the ability to sit in a child-free section of the aircraft. The question is the degree to which offers can be made that resonate with customers and drive conversions.

In 2019, the five largest U.S. carriers (American Airlines, Delta Air Lines, United Airlines, Southwest Airlines, and Alaska Airlines) generated more than $29 billion in additional sales. According to Ideaworks and Cartrawler (2018), global airline additional revenues reached $93 billion in 2018. Furthermore, before the COVID-19 pandemic in 2019, these revenues surpassed $109.5 billion (Ideaworks and Cartrawler, 2019), marking a significant increase of five times the $22.6 billion reported in 2010. After a decline during the pandemic, the total ancillary revenues were forecast to reach $117.9 billion in 2023 (Ideaworks and Cartrawler, 2023).

Cartrawler predicted a significant decline of 47% in 2020 due to the pandemic, amounting to $58.2 billion dollars (Ideaworks and Cartrawler, 2020). However, in 2021, there was an improvement, with global revenue reaching $65.8 billion dollars (Ideaworks and Cartrawler, 2021). Despite the challenges posed by the pandemic, ancillaries saw growth in their share of global revenue, increasing from 12.2% in 2019, to 13.4% in 2020, 14.4% in 2021 and 15% in 2022. In 2023, there was a slight decline to 14.7%. Cartrawler's data reveals that in 2010, ancillary revenue per passenger was $8.42, which increased to $12.13 in 2012, $28.97 in 2019, $31.39 in 2020, $33.45 in 2021, $42.11 in 2022, and $37.59 in 2023.

The airline practice of charging more for services that historically were included in the price of a ticket has created much customer angst and dissatisfaction (Reed, 2019). But airline executives view this revenue stream as a fundamental requirement for survival, incremental revenues, and profitability. Ancillary revenue streams are here to stay and will continue to increase in the next decade, but adjustments are frequently made based on market conditions. For example, during the COVID-19 pandemic of 2020, U.S. majors stopped charging change fees due to the downturn in travel and the economic climate. Change fees were a large component of ancillary sales.

9.6 Airline Offer Management

Offer management requires a capability for customer segmentation, a recommendation engine to recommend bundles, and an offer engine to fine tune the offer based on a customer's unique preferences to generate offers that maximize conversion rates. Customer segmentation determines the type of customer and associated preferences, the recommendation engine determines likely bundles that are pertinent for the customer segment or persona, and the offer engine determines the specific personalized offer, for a segment of ONE. An important consideration during online interactions with customers is that they could be anonymous or registered, and recommendation engines and offer engines need to be relevant and targeted in both scenarios.

At the heart of IATA NDC, contrary to statements made by various entities in the travel value chain, only the supplier (airlines) will be able to generate an offer to a customer, intermediaries, while they may want to participate in the offer creation process, will be left out. The GDS storefront has an important role to play with product normalization of nonhomogenous NDC content from airlines (see Sect. 9.8) to manage the screen real estate, but not in the offer creation process. Visualization of the airline generated offers on the GDS storefront is a barrier to NDC adoption since offer responses from one airline to the next are vastly different.

While the airline is well positioned by IATA NDC to generate offers to customers, the larger question is whether airlines can become retailers? Will they be effective in generating offers that improve conversion rates? At a minimum intelligent retailing is an aspirational goal that requires changes to the revenue management business process, and it remains to be seen how successful airlines will be with offer management. Can they truly generate significant incremental revenue beyond the baseline ancillary revenues generated in 2019 that was widely reported by Ideaworks and Cartrawler (2019)?

The task of offer management for airlines can be overwhelming. A comprehensive program to advertise and sell additional products requires a deep understanding of the customer base to appropriately price, advertise, sell, and fulfill the product. This, in turn, necessitates coordination from various divisions within an airline such as sales, marketing, pricing, revenue management, E-commerce, in-flight services, airports, loyalty programs, call centers, ticketing, finance, mobile applications, and revenue accounting. Customer insights are vital in increasing the likelihood that customers will purchase the ancillaries in the bundle. Each airline must reach a consensus internally on the level of detail at which offers will be generated to match customer expectations. This can be at a context for travel-based segmentation level or personalized offers for a segment of ONE based on the understanding that most customers travel every 12–18 months and may not be part of the loyalty program. A hybrid model may work where one-to-one personalization is confined to frequent travelers with history, and the rest are based on context for travel-based segmentation.

As the airline industry transitions from selling airfares to intelligent retailing, airlines want to know the ancillaries customers want and what they are willing to pay. In turn, travelers are empowered to control their travel experience. Despite the technical hurdles associated with offer management, there is an opportunity for travel retailing to be a positive experience for both airlines and customers.

9.6.1 Customer Segmentation Based on Context for Travel

Airlines have for many years wanted to segment customers beyond the traditional booking class that is used for inventory control and distribution of availability through the global distribution systems (Vinod, 2008). Customer segmentation outside of travel may focus on demographics such as age, income, and resident postal code to understand purchase habits and propensity to buy a specific product.

Airline customer segmentation has less to do with demographic data and is based on behavior patterns and reason for taking a trip. For example, business travelers book closer to departure and require flexibility with flight exchanges and refundability, while leisure passengers are budget conscious and can accept more stringent fare rules associated with cheaper fares.

Before personalized travel became a priority, traditional customer segmentation was based on the class of service: first class, business class and economy. Within each class of service, customer segmentation was based on RBD. Frequent flyer segmentation was based on tiers, and qualification was based on the amount of travel spend.

Corporate employees are required to adhere to different levels of corporate travel policies, known as unmanaged, lightly managed, managed, and highly managed. Unmanaged travel is generally applicable to small companies with less than 500 employees. Within the corporate realm, there are three main customer segments: road warriors, who frequently travel for work (e.g., sales executives); corporate masses, who engage in a considerable amount of travel; and infrequent travelers, such as employees attending conferences or visiting a regional corporate office. Traditional corporate travel is closely governed by travel policies, whereas executive travel falls under the lightly managed traveler category. Bundled offers for corporate employees must conform to corporate policy.

The act of categorizing customers during shopping by using only a few details, like origin and destination, how long they plan to stay, the number of people in their group, and the day of the week they arrive or depart, allows for the creation of tailored shopping responses that cater specifically to the persona, all within the constraints of the information available at that point. This categorization of customers is known as trip-purpose segmentation (TPS), which is a way of grouping customers based on the purpose (context) of their travel. It is the single most critical concept that should be implemented to generate customized offers (Sorrells, 2018a) for each persona, which can lead to personalization for a segment of ONE.

Customer segmentation is an essential initial step in the creation of offers (Kothari et al., 2016). Various attributes of users, such as travel preferences, schedule and fare preferences, site activity, affinities, and travel history, can be utilized to create customer segments. It is important to note that customer segmentation is an ongoing process that should be periodically adjusted based on historical purchase data and evolving customer preferences. However, it is crucial to strive for accuracy with a small margin of error. A flawed customer segmentation model can result in offers that fail to convert into bookings during the sales process.

In the realm of travel, the significance of considering context for segmenting travelers cannot be overstated. Segmenting customers based on context for travel creates personas that combine the traits of groups of travelers who share similar preferences. However, in the travel industry, these personas come with a unique twist. This is because a customer's preferences can vary depending on the context of their travel, such as a business trip, a weekend getaway, visiting friends and relatives, a family vacation, or a "bleisure" trip that combines business and leisure activities. To define these different segments based on trip purpose, we need to ensure that they are *mutually exclusive and collectively exhaustive*. This practical step allows us to group customers who exhibit similar purchasing behaviors. There are two reasons why this is important. First, not all customers are registered users, and many remain anonymous until they make a booking. This is especially true when using online travel agencies that rely on browser cookies. Second, the typical traveler may have multiple profiles influenced by the purpose of their trip. Understanding the context of a trip, as determined by shopping parameters, greatly impacts customer preferences and price sensitivity. To further refine the offer recommendations based on trip-purpose segmentation, customer-specific data can be considered from the customer profile, such as name, credit card information, frequent flyer status, and past trips when the customer is declared. By taking all these factors into account, the offer can be refined to meet the needs and desires of each customer.

The segmentation of travelers in the airline industry has historically relied on airline pricing and is determined by the fare rules and restrictions outlined in the reservations booking designator (RBD). These RBDs or booking classes are also utilized to distribute availability to global distribution systems (GDS) using AVS, direct access, or seamless availability.

Price discrimination based on customer segmentation or individual customer behavior was not effective for mass product items like DVDs sold by online giant Amazon in 2000 (Turow et al., 2005). However, travel suppliers do not expect a

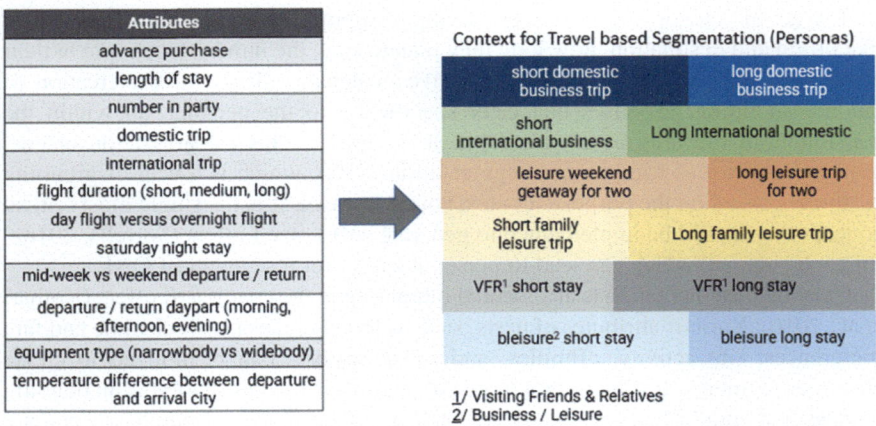

Fig. 9.1 Sample attributes for customer segmentation based on context

negative reaction from customers when it comes to price differentiation. Today, travel customers are more accepting of different price points for perishable products. The issue is further mitigated by bundled offers responses from airlines through the NDC gateway, since the content is not homogenous.

Every traveler has a single profile, but multiple personas based on the context for travel. Flight and bundle (air and non-air) vary based on the type of trip, business trip and its variants, and leisure trip and its variants. Given the importance of the context for travel which can lead to different offers for the same customer, Fig. 9.1 provides a sample set of attributes for customer segmentation. Unsupervised or supervised (if a marketing survey customer surveys past customers and labels their segment) learning techniques can be used to generate the various personas, which have been identified with a label as shown in Fig. 9.1.

Providers of machine learning toolkits often promote the use of black box techniques for customer segmentation. The management of inventory-controlled ancillaries falls under the purview of airline revenue management analysts. It is crucial to have visibility and interpretation of customer segments to respond to changes in custom offers and inventory controls. Besides, the trip-purpose segments must be mutually exclusive and collectively exhaustive. This implies that during a low fare search, a customer cannot be identified as belonging to two or more trip-purpose segments but must belong to a single unique segment. Therefore, a black box approach is not advisable due to the existing revenue management business processes and flight control methods. The argument holds true for dynamic segmentation as well.

Note that the trip-purpose segments for the coach cabin shown in Table 9.2 are mutually exclusive and collectively exhaustive. Hence, in an online session, a customer can only belong to one persona which defines the traveler's context for travel.

Table 9.2 Air trip-purpose segmentation example

Trip-purpose segment	Customer type	Business or leisure trip	Advance purchase	Length of stay
A1	Individual	Business	0–6 days	0–1 days
A2	Individual	Business/leisure	0–6 days	2+ days
A3	Individual	Business/leisure	7–13 days	0–5 days
A4	Individual	Leisure	7–13 days	5+ days
A5	Individual	Business/leisure	14–20 days	Any
A6	Individual	Business	21+ days	0–3 days
A7	Individual	Leisure	21+ days	4+ days
A8	Couple (2)	Leisure	0–20 days	Any
A9	Couple (2)	Leisure	21+ days	Any
A10	Family (>2)	Leisure	Any	Any
A11	Individual	One-way, business	0–6 days	One-way
A12	Individual	One-way, business or leisure	7–13 days	One-way
A13	Individual	One-way, business or leisure traveler that plans ahead	14–20 days	One-way
A14	Individual	One-way, leisure traveler that plans ahead	21+ days	One-way
A15	Couple (2)	One-way, couple on leisure trip	Any	One-way
A16	Family (>2)	One-way, family vacation	Any	One-way

Further, booking classes are not required with NDC for classless revenue management. For revenue management purposes, demand can be forecast by trip-purpose segment to determine the optimal dynamic pricing (continuous) of itineraries.

The trip-purpose segment definitions can be further refined based on marketing objectives. Micro segmentation is the process of further breaking down the trip-purpose segments into a more granular level of detail to create micro segments by adding dimensions to clustering customers such as *CLV* tiers and additional customer preferences. Micro segmentation adds precision to the recommendation engine to recommend offers.

9.6.1.1 Segmentation Based on Context for Other Travel Verticals

Personas can be created for other lines of business such as hotels, rental car, cruise lines, and ferry lines and rail in a similar way. This is a necessary step to generate custom offers by these suppliers and promote the offers through the various channels of distribution including the GDSs. A few examples follow.

Table 9.3 Hotel trip-purpose segmentation example

Trip-purpose segment	Customer type	Business/ leisure	Midweek/weekend arrival	Advance purchase	Length of stay
H1	Individual	Business	Midweek	0–6 days	1–5 days
H2	Individual	Business	Midweek	7–21 days	1–5 days
H3	Individual	Business	Midweek	>21 days	1–5 days
H4	Individual	Business/ leisure	Midweek	>21 days	>5 days
H5	Individual	Leisure	Weekend	0–13 days	1–2 days
H6	Individual	Leisure	Weekend	>13 days	1–2 days
H7	Individual	Business/ leisure	Weekend	0–13 days	2+ days
H8	Individual	Business/ leisure	Weekend	>13 days	2+ days
H9	Couple (2)	Leisure	Anytime	0–20 days	Any
H10	Couple (2)	Leisure	Anytime	>20 days	Any
H11	Family (>2)	Leisure	Anytime	Any	Any

A hotel example for the run-of-house product category is shown in Table 9.3.

A rental car example is shown in Table 9.4. These trip-purpose segments must be defined by rental car product category such as Economy (ECAR), Intermediate (ICAR), Full Size (FCAR), Luxury (LCAR), Premiums and Convertibles (PCAR), and Minivans (MVAN).

A cruise line example is shown in Table 9.5. For illustrative purposes, it is assumed that the cruise line has six major cabin types.

Why is context for travel-based segmentation important? Since only a small fraction of an airline's customers make several trips a year, personalizing an offer for a segment of ONE for all customers is not a viable alternative, regardless of what the proponents of IATA NDC may want. Trip-purpose segmentation is the default for creating offers, which can be modified to create a 1:1 offer for a customer based on the number of trips the customer has taken. An equally important reason why trip-purpose segmentation will prevail is because of the constraints imposed by the General Data Protection Regulation (GDPR) that is in place today. With GDPR, calibration must be done at a higher level of aggregation such as market or trip-purpose segmentation data. Shao and Kauermann (2020) recommend calibration price elasticity at the market level to segment markets and determine what offers to recommend to customers.

9.6.2   Branded Fares and Offer Management

Branded fare families are an important consideration in creating the final offer. When a bundled offer is determined, it should be mapped to the attributes included in each

Table 9.4 Rental car trip-purpose segmentation example

Rental car segment id	Type	Business/leisure	Midweek[a]/weekend pick-up	Advance purchase	Length of rental
R1	Individual	Business	Midweek	0–6 days	0–5 days
R2	Individual	Business	Midweek	7–21 days	0–5 days
R3	Individual	Business	Midweek	> 21 days	0–5 days
R4	Individual	Business/leisure	Midweek	> 21 days	> 5 days
R5	Individual	Leisure	Weekend	0–13 days	1–2 days
R6	Individual	Leisure	Weekend	> 13 days	1–2 days
R7	Individual	Business/leisure	Weekend	0–13 days	2+ days
R8	Individual	Business/leisure	Weekend	> 13 days	2+ days
R9	Couple (2)	Leisure	Anytime	0–20 days	Any
R10	Couple (2)	Leisure	Anytime	>20 days	Any
R11	Family (>2)	Leisure	Anytime	Any	Any
R12	Individual	B/L/one-way	Anytime	Any	Any
R13	Group ≥2	B/L/one-way	Anytime	Any	Any

[a]Midweek is pick-up Sunday through Thursday. Weekend is pick-up Friday or Saturday

Table 9.5 Cruise line leisure trip-purpose segmentation example

Cruise segment id	Type	Type of cabin[a]	Duration
C1, C2, C3, C4, C5, C6	Individual	ST, IC, OV, BA, SU, PH	3–4 days
C7–C12	Individual	ST, IC, OV, BA, SU, PH	1 week
C13–C18	Individual	ST, IC, OV, BA, SU, PH	>1 week
C19–C24	Couple	ST, IC, OV, BA, SU, PH	3–4 days
C25–C30	Couple	ST, IC, OV, BA, SU, PH	1 week
C31–C36	Couple	ST, IC, OV, BA, SU, PH	>1 week
C37–C42	Family (>2)	ST, IC, OV, BA, SU, PH	3–4 days
C43–C48	Family (>2)	ST, IC, OV, BA, SU, PH	1 week
C49–C54	Family (>2)	ST, IC, OV, BA, SU, PH	>1 week

[a]ST = Studio, IC = Inside Cabin, OV = Ocean View, BA = Balcony, SU = Suites, PH = Penthouses

branded fare product to ensure the lowest cost option is offered to the customer. Hence, for example, it may be cheaper to offer a customer a cheaper branded fare product and an ancillary priced separately than offering the higher valued branded fare product with the ancillary included. Hence, offer management provides a range of upsell opportunities tied to the various branded fare products that are available for sale, and the recommendation is based on the least total cost to the customer.

9.6.3   Recommendation Engine and Offer Engine

Recommendation systems play a key role in offer management, which is the capability to recommend relevant offers to customers (Jiang et al., 2014; Dadoun et al., 2021).

An offer management system requires a recommendation engine to generate a bundled offer to customers based on their trip-purpose segment and individual preferences. The bundle consists of the dynamic fare and air ancillaries. Advanced data analytics powered recommendation engines for bundles process historical purchase data and require periodic recalibration to detect and adapt to changes in customer behavior patterns and preferences over time. Using historical purchase data by trip-purpose segment, recommendation engines can deliver bundled offers for the persona (customer segment).

Techniques may consist of either traditional statistical predictive models or machine learning models (Fox, 2019). Examples of models that can be used include techniques such as collaborative filtering, classification and regression trees (CART), and gradient boosting machines (GBM).

An offer is a bundle that is displayed to a customer during shopping and can be refined from the point of booking to the checkout process based on their specific preferences (Vinod, 2020a). Customers who visit an airline website can be anonymous or declared. Customers are usually anonymous early in the sales process. A challenge is to display relevant offers to customers when they are anonymous. Based on shopping parameters input by the user, the context for travel can be identified with the associated trip-purpose segment (see Table 9.2). From historical customer bookings, the trip-purpose segment can be identified to determine what customers have purchased in the past. This data can then be used to infer the composition of the most appropriate bundles for each trip-purpose segment.

Figure 9.2 illustrates a variant of a persona-based recommendation engine (Vinod, 2020a) to maximize conversion rates of the air bundle.

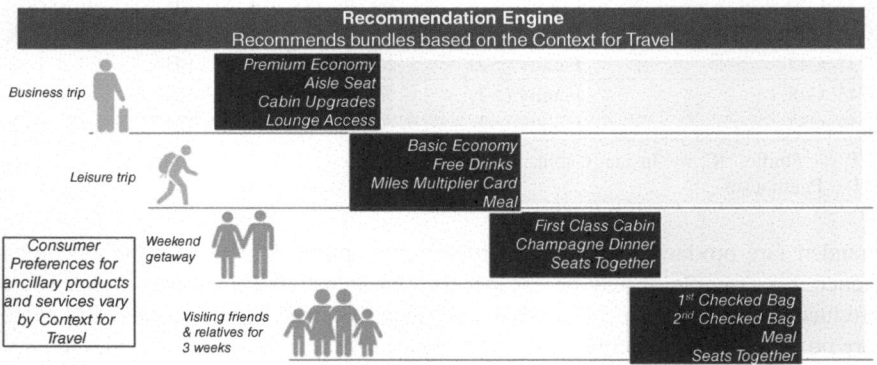

Fig. 9.2 Recommending offers based on context for travel

   This approach to recommending bundles requires periodic calibration to ensure that changes in consumer behavior over time are considered. Periodic calibrations can be time consuming and expensive and will always lag the market. Ideally, this approach to determining bundles by trip-purpose segment should serve as a warm start for the multiarmed test and learn experimentation model, a reinforcement learning technique (Robbins, 1952; Gittins et al., 2011; White, 2013) where experimentation is continuous, and recommendations are always current and reflect market conditions (described in Sect. 9.6.4).

9.6.3.1   Recommending Offers on the Airline Website

Determining the appropriate offers to present to online customers and the order in which they should be shown is an important design consideration for the airline website. Research has shown that customers tend to lose interest when presented with more than three (or a maximum of four) options to consider. This is because the brain's capacity to hold information in the conscious mind is limited to three or four items at a time (Moskowitz, 2008). As a result, it is advisable to display three alternatives.

9.6.3.1.1   The Best Offer

The best offer is based on the inferred trip-purpose segment identified from the shopping parameters. Historical sales data for the customer segment is used in its entirety to determine the most likely bundle that the customer will purchase.

9.6.3.1.2   The Decoy Offer

The decoy offer is a less desirable package that is presented alongside the best offer, which enhances the likelihood of the customer choosing the superior offer (Ariely, 2010; Segmentify, 2021). It is also known as the decoy effect, or asymmetric dominance. The decoy offer may have a price in between or be the costliest option. It serves as a reference point for customers to make their selection of the best offer.

9.6.3.1.3   Popular Attributes Across All Customer Segments

The third choice does not consider the travel context, but instead relies on the combined sales data of the most purchased ancillary products and services across all customers. The goal of this ancillary bundle is to guarantee that even if the assumed customer segmentation is incorrect and the customer falls outside of this group, the customer will still buy individual popular ancillary products and services as a bundle.

9.6.3.2   Finalizing the Offer

When the customer is declared (identity is known), subsequent refinements to the bundle can be made based on past purchase history, consumer intent captured during the sales interaction, frequent flyer tier, and preferences stored on the customer profile can be used to personalize the offer.

Optionally, to spur sales of ancillaries, a dynamic quantity discount model, calibrated with airline input, can be used to customize the air bundle (Vinod, 2017, 2021e; Vinod et al., 2018). The greater the total value of the bundle purchased by a customer, the greater the discount, subject to minimums and maximums. For example, the quantity discount model can be a hyperbolic discounting model to calculate the discount amount. The model can be calibrated to define the maximum discount regardless of the total value of ancillaries purchased. Figure 9.3 illustrates the maximum discount percentage of 10% applied as a function of total ancillary sales. With airlines migrating to NDC, the discount can be applied to the total price of the ticket inclusive of the base fare, since the financial settlement is direct with the airline.

The equation to generate the dynamic discount based on total spend for a booking is as follows.

$$\text{Discount}(\%) = \frac{\text{Maximum Discount } (\%) \times \text{Incremental Spend}}{\text{Constant} + \text{Incremental Spend}}.$$

The value of the constant can be evaluated by a revenue management analyst to determine the rate at which the discount should be applied based on the incremental spend. In Fig. 9.3, the constant was set to 100. When the constant is lower, the discounts are higher for lower incremental spend.

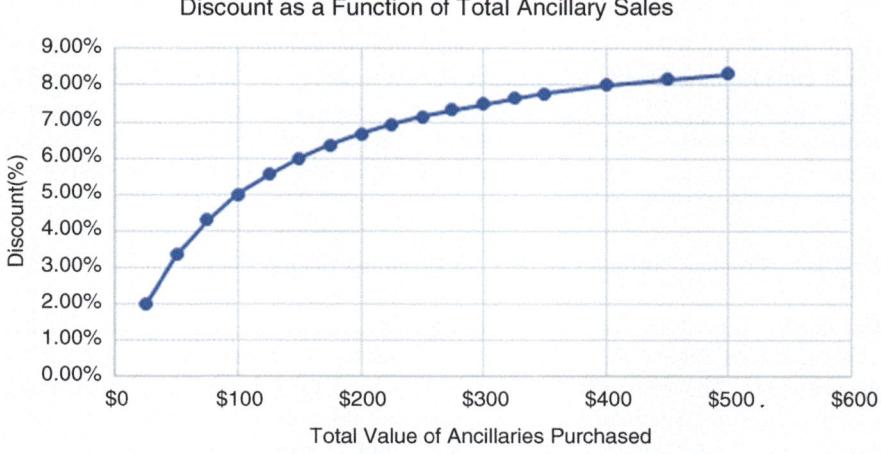

Fig. 9.3   Example of a dynamic discounting model based on total ancillary spend

This type of equation was first proposed by biochemist Leonor Michaelis and Canadian physician Maud Menten in the study of enzymes (Michaelis & Menten, 1913) and called the Michaelis–Menten equation.

9.6.4 Test and Learn Experimentation

In travel, where consumer preferences and attitudes are dynamic and are influenced by their interactions outside of the travel domain, test and learn experimentation must be an integral component of any recommendation engine. Experimentation provides a direct approach to running controlled test experiments to learn customer behavior. Unlike model-based decision support tools, business experimentation with the multiarmed bandit, a reinforcement learning application, in many respects is simpler to use and implement. The idea is to test new prices directly. Product designs or systems/processes are compared to current ones to see if there is a performance improvement. Simple binary tests (e.g., old versus new) are commonly referred to as A/B testing. A multiarmed bandit implementation is more general, adapts to the current environment, learns over time, and can support running multiple experiments simultaneously. For example, test and learn experimentation can help answer common business questions such as

1. *What are the price points for air ancillaries?*
2. *What is the best price point for the bundle to maximize revenue?*
3. *Which algorithm or business approach works best?*
4. *How are my revenues and conversion rates impacted by changes to my screen display rules?*
5. *Which bundled offer maximizes revenue?*

When the number of ancillary products offered by an airline increases, the number of potential unique bundles that can be offered increases by an order of magnitude. The number of possible bundles when there are n ancillary products is $2^n - 1$. For example, with 10 ancillary products, the number of unique bundles is 1023. This is illustrated in Fig. 9.4.

The demand for the ancillary products is not uniform, but a few such as pre-reserved seats, checked bags, and upgrades are popular and dominate the composition of bundles. However, the long tail of less popular ancillary products should not be ignored but sampled periodically by the multi-armed bandit model. The objective is to find the right bundles that maximize customer conversion rates.

The multiarmed bandit problem can be implemented with different sampling methods such as epsilon-greedy, upper confidence bound (UCB), and Thompson Sampling to improve future actions that will maximize the reward potential.

The concept of contextual bandits expands upon the multiarmed bandit test and learn experimentation method (Byrd & Darrow, 2021). By introducing the current state as a factor in decision making, the contextual bandit model goes beyond reliance on historical data. Instead, personalized decisions can be made based on

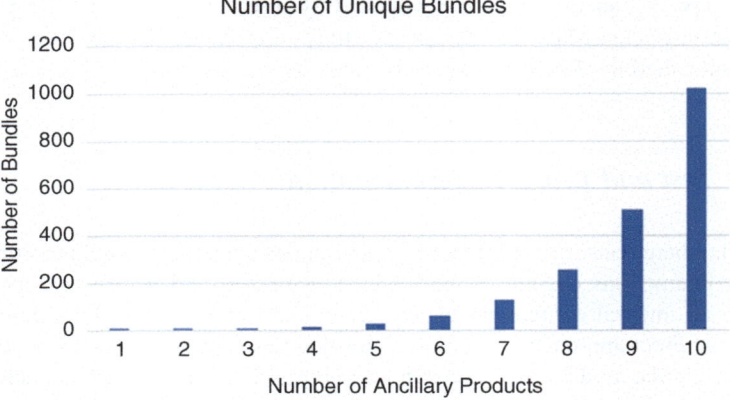

Fig. 9.4 Unique bundles based on number of unique ancillary products

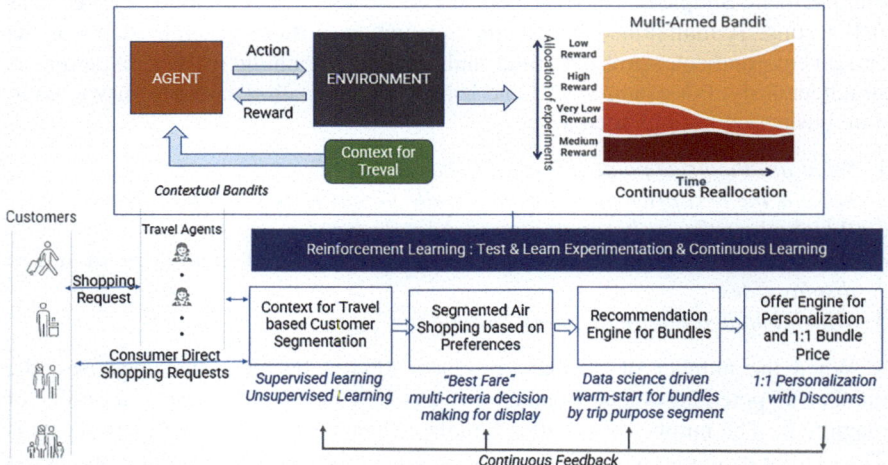

Fig. 9.5 Offer creation process with continuous learning

contextual information about the user, such as their origin, previous site activity, and device usage. Microsoft research has created a testing service called Vowpal Wabbit, which utilizes contextual bandits through open-source online and offline training algorithms (Dudik et al., 2011). The model was originally developed by researchers Miroslav Dudik, John Langford, and Lihong Li at Yahoo! Research in 2007 before being implemented by Microsoft. In the airline retailing industry, the trip-purpose segment serves as the primary contextual factor, and additional micro-segmentation can be appended as needed.

Figure 9.5 illustrates an approach for generating offers to customers with the multiarmed bandit test and learn experimentation model serving as a horizontal enabler during the offer creation process.

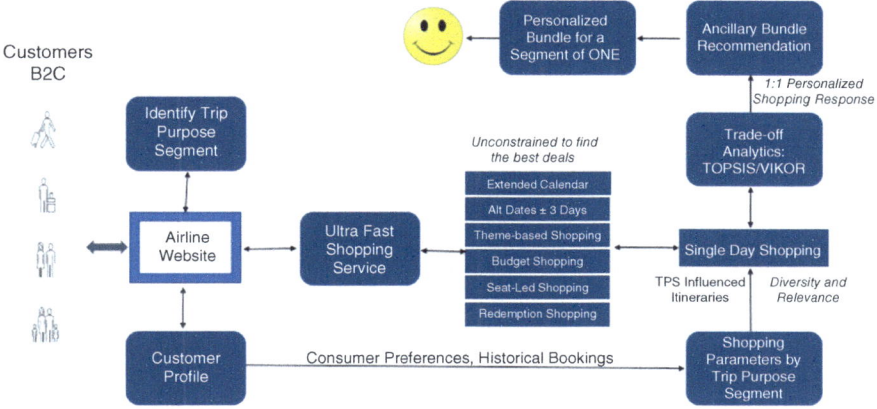

Fig. 9.6 Air shopping and offer management for the direct channel

9.7 Air Shopping and Offer Management

As shown in Fig. 9.6, the first step is to determine the trip-purpose segment based on the shopping parameters input by the user. For calendar shopping with an anonymous use perhaps the only input is the duration (length of stay) and month. In the single day shopping stage, the trip purpose is refined based on expected departure date, return date, and number in party. Itineraries returned will be influenced by the default shopping parameters by trip-purpose segment from historical aggregate booking data. The itinerary selection is then refined based on preferences with the trade-off analytics algorithm, which is a recommendation of the *best fare* for the customer. This is followed by a recommended offer (the bundle) by trip-purpose segment, which is then personalized with the offer engine for a segment of ONE inclusive of any dynamic discounts.

Figure 9.6 illustrates the consolidated workflow for air shopping and offer management for the direct channel.

9.8 Personalization and GDPR

The European Parliament and Council of the European Union (EU) initiated discussions on online privacy rights in 2012, leading to the adoption of the General Data Protection Regulation (GDPR) on April 14, 2016. The GDPR became enforceable as a law on May 25, 2018, and applies to all companies worldwide that engage in business with the EU. It affects companies that handle personal data of EU individuals, regardless of their geographical location. The GDPR empowers EU citizens to exercise control over their data, including the types of data collected and its usage.

Any violation of the GDPR can result in penalties of up to 4% of global turnover or 20 million euros, whichever is higher.

GDPR has not directly prohibited personalization but has instead implemented a system that requires permission for the collection of data by travel intermediaries and suppliers. What does this legislation mean in a NDC world? The individual or user is known as the *data subject*, while the entity that stores and handles the data, such as an OTA or travel agency is referred to as the *controller* and the *data processor* is the supplier that performs the data processing on behalf of the controller.

For example, to process NDC requests, a TMC may initiate a shopping request on behalf of a traveler, they will provide the necessary information for the airline to respond with relevant offers. This request may also include additional details about the traveler to create a personalized response. This is the stage where personal data is processed according to GDPR regulations. Before implementing NDC, all parties involved in the distribution chain, such as airlines, GDSs, TMCs, and NDC aggregators, must evaluate their personal data processing and understand their responsibilities under applicable laws. The scope of obligations and privacy protection requirements may vary between the travel entities and jurisdictions.

With the implementation of GDPR, personal data usage for personalization is challenged by data privacy rules. Approaches that utilize aggregated data instead of individual personal data have become more relevant (Shao & Kauermann, 2020; Millet, 2023). This implies that personalization may be limited to the creation of offers based on broader categories, such as trip-purpose segments, rather than targeting individuals specifically.

The impact of GDPR on personalization of NDC offers is a topic of frequent debate. It poses a challenge to the current system in which travel companies gather and retain customer information. However, GDPR can be navigated by giving customers ownership of their profile data and allowing them to decide who can access it for specific purposes. Under this approach, travel companies do not possess customer data; instead, customers grant access to their profile data on a one-time basis, as needed. To facilitate this process, a universal customer profile and a universal data exchange platform are necessary to transmit information to retail entities and fulfill service requests.

The California Consumer Privacy Act (CCPA) of 2018, like GDPR, is a state law aimed at granting California residents increased authority over their personal data, which is collected by businesses. This legislation safeguards additional privacy rights for consumers in California.

9.9 Dynamic Pricing and NDC

Dynamic pricing is an integral component of the offer management process with NDC. The key difference from a pricing and revenue management perspective is that the price quoted to a customer is a derived fare and not a fare that was published through ATPCO. The derived fare can be based on competitor selling fares (competitive revenue management) or based on traditional revenue management techniques.

Dynamic pricing has been hailed as the next breakthrough in revenue management (Choubert et al., 2015; Fiig et al., 2016). Numerous airlines are attracted to the idea of dynamic pricing and recognize its potential as a cornerstone in the context of NDC for dynamically priced offers (Fiig et al., 2018). It enables airlines to transcend traditional pre-filed fares and booking classes (RBDs), to generate personalized fares in real time, considering competitive market dynamics. This approach is often referred to as continuous pricing or continuous dynamic pricing.

Dynamic pricing of the base fare is closely related to dynamic availability. Both techniques can leverage competitive selling fares to arrive at dynamic price or inventory control recommendation. Instead of converting the optimal price point from dynamic availability to an open/close inventory control recommendation, the dynamic price that is derived is directly used to approximate the ticketed price.

The willingness to pay is the maximum price that a customer will pay for an itinerary assuming there are no other options available in the choice set. However, when there are multiple alternatives, the customer's conditional willingness to pay is the highest price the customer will pay for a specific itinerary from the set of alternative itineraries in the choice set. The customer choice model (CCM) may be used to rank each itinerary and determine the markup or markdown required to optimize the net contribution. Table 9.6 provides an illustration of this concept. When there is only one option (Choice Set A), a customer may be willing to pay more than the selling fare. However, when there are multiple choices, the maximum willingness to pay decreases to $700.

A calibrated choice model (Ben-Akiva & Lerman, 1985) can estimate the probability an itinerary will be selected in a shopping session based on host and competitor itineraries. If an airline increases its price for an itinerary, the probability will drop and vice versa. The price optimization model understands these trade-offs and can search to determine the set of itinerary prices that maximize the expected contribution, which is the difference between the expected revenue for the host airline's itinerary and the total bid price. A heuristic based on these factors was proposed by Gallego and Hu (2014), which identifies price decreases in uncompetitive situations and price increases when the host airline is over competitive.

Table 9.6 Choice sets and customer willingness to pay. Market: DFW-SFO, November 29

Choice sets	Available itineraries	Fare	Willingness to pay
Set A	Departure time 8:00 am, nonstop, Flight #100.	$800	Up to $1000 for the schedule/fare attributes.
Set B	Departure time 8:00 am, nonstop, Flight #100. Departure time 8:00 am, nonstop, Flight #200.	$800 $700	Not more than $700 for the schedule/ fare attributes.

The net contribution calculation is summarized below:

$$\text{Net Contribution}_i \mid \text{Sale}_i = \left[\text{Fare}_i - \sum_{i \in L}\text{Bid Price}\right] * \text{Probability of Selection}_i,$$

where i is an itinerary in a specific line in the shopping response and L is the set of legs in the airline network. The form of the equation displayed above is conditional on a sale occurring (i.e., someone makes a purchase in the session). It can be extended to estimate net contribution per session by multiplying by the sales conversion rate per session.

In the competitive revenue management scenario, dynamic pricing models can leverage customer choice models to compare various itineraries offered by different airlines. These choice models consider the appeal of available itineraries to customers, considering factors such as departure time, travel duration, number of stops, airline brand, aircraft type, cabin class, interline and codeshare options, and, of course, price. By estimating the likelihood of a customer selecting a specific itinerary, these choice models can be further calibrated according to different trip-purpose segments to account for variations in preferences between business and leisure travelers. The price optimization model for dynamic pricing then determines the most optimal prices for the host airline's itineraries by balancing the trade-offs between the probability of selection from the choice model and maximizing yield. Through a search across possible price points, this price optimization process identifies a set of itineraries that maximizes expected profit while taking into consideration the airline's schedule and fare attributes.

The likelihood of choosing an itinerary, the selection probability, can vary up or down depending on the price in the net contribution calculation. Therefore, the concept of dynamic pricing optimization aims to directly adjust the price to achieve the desired increase or decrease. By utilizing a session-based fare optimizer, the host airline can determine the best price for its services by considering the selling fares offered by competitors in the same market. Instead of converting the recommended price into an inventory control suggestion, the dynamic price is used as an estimate for the actual ticket price.

An alternative is to not use competitor selling fares as input, especially if the data is not available in the NDC era. In this scenario, traditional revenue management techniques with a calibrated price demand curve can be used to determine the dynamic price using reference fare levels as input.

9.9.1 Laddered Pricing and Continuous Pricing

There are two primary variations for deploying dynamic pricing: laddered dynamic pricing, and continuous dynamic pricing.

A pan-industry Dynamic Pricing Working Group was established in 2016 with ATPCO as the sponsor to develop specifications for laddered pricing. With this approach intermediate private fares are filed between public fares that can only be activated by an airline's dynamic pricing engine (DPE). Airlines, GDSs, vendors, and related entities participated in these working group sessions to finalize a specification and messaging standard for dynamically priced fares for distribution to airline websites and the indirect channel. The ATPCO JSON/XML schemas for connectivity were intended to provide an *interim step* toward NDC (Dezelak & Ratliff, 2018). The ATPCO specification allowed airlines to connect their dynamic pricing engines to traditional distribution prior to NDC cutover.

The second approach is continuous pricing, where the dynamic price generated by the dynamic pricing engine is used as the selling fare. Continuous pricing is the most popular dynamic pricing approach. Of the two approaches, continuous pricing has the most traction, and several airlines have implemented first-generation versions on their websites for selected markets and with NDC to promote continuous pricing to the indirect channels. With this approach, all gaps on the price demand curve caused by discrete fare filings are theoretically eliminated with a continuous price. The absence of discrete price jumps that exist with the discrete fare filings allows for lower prices to be quoted to customers. Without the existence of full content agreements with NDC, the ultimate objective of continuous pricing is to determine the optimal price point for a specific request, by customer segment for a specific channel.

The laddered pricing approach never gained traction with airlines for two reasons. First, it relied on existing booking classes to create intermediate private fares between public fare levels associated with RBDs and the future of RBDs is uncertain at best. Second, the continuous dynamic pricing approach is superior since it minimizes or eliminates the lost revenue opportunities (see Chap. 6, Fig. 6.13) by filling the gaps on the price–demand curve.

For fast response times during offer creation, it is impossible to execute a network optimization model in real time to capture the first order, second order, and third order effects in the airline network when the dynamic price is calculated for each request. The evaluation one market at a time meets the response time requirements for creating an offer but does not explicitly model the higher order network effects (see Sect. 3.5.4.1).

9.9.2 Pros and Cons of Dynamic Pricing

There are several benefits associated with dynamic pricing. First, it provides the ability to specify the revenue maximizing exact, continuous price point that is closer to a customer's true willingness to pay. Second, it maximizes the potential revenue benefit due to the round-trip control capability. This finer degree of control can drive additional revenue benefits. Third, by deploying dynamic pricing in a competitive revenue management framework with marketplace competitive shopping data, recommendations can be fine-tuned to reflect the selling fares offered by competitors.

Dynamic pricing also has two primary benefits over the traditional continuous nesting approach with ATPCO fare filings. First, it bridges the chasm between booking class availability and airline pricing by ensuring that the ticketed fare is greater than the total bid price. Second, the ticketed fare paid by a passenger will be the same as the dynamic price for the itinerary.

The potential revenue benefit from simulation studies of dynamic pricing with markups and markdowns can be up to 4% (Belobaba, 2019). Airlines always worry about the spiral down effect with dynamic pricing and dynamic availability. Simulation studies have shown that since dynamic pricing considers demand forecasts, the bid price, customer willingness to pay by segment and relative attractiveness of each itinerary, rational pricing recommendations will prevail, and it is not a zero-sum game as is reported by Belobaba (2019) and Wittman and Belobaba (2018). When four airlines were deploying dynamic pricing in the PODS (Passenger Origin-Destination Simulator), based on customer willingness to pay and not competitive shopping data, revenue gains ranged from 1.5% to 2.5%. However, the benefits can erode with irrational competitors.

The potential benefits of leveraging competitive shopping data can be even higher. The quality of the dynamic pricing recommendations can be further improved by looking beyond competitive selling fares; by considering the booking pace of competitors as observed in daily MIDT data.

An added benefit of continuous dynamic pricing is that it provides an infinite number of price points on the price demand curve that can be inventory controlled to generate incremental revenues.

Dynamic pricing will be disruptive to airline fare management processes, agency workflows, and GDS processing for comparison search. Another barrier to the adoption of dynamic pricing in the GDS channel is the absence of an industry standard for roundtrip inventory control (Isler & D' Souza, 2009).

Personalized offers and dynamic pricing continue to be an active area of research by both academia and industry (Szymański et al., 2021; Wang et al., 2021; Sznajder et al., 2023a, b; Wang et al., 2023).

9.10   An Omni-Channel Strategy: Myth or Reality

An omni-channel strategy is often discussed at industry forums. Many airlines have promoted their products across all channels of distribution. However, one of the biggest challenges with ancillary products and services in an omni-channel strategy is the ability to create a seamless customer experience through all channels with an identical display of offers. The challenge is for customers to experience the airline brand through direct and indirect channels.

The fundamental issue is that there is no standardization of branded fares across airlines. The airline websites showcase the airline branded fares exactly how the airline wants to display it. Through the agency channel, the GDSs must display content from many airlines and herein lies the problem. Hence, regardless of how the airline showcases their brands on their websites, the GDSs determine the *minimal set of relevant attributes* that agencies and their customers would like to see and uses this as a baseline to display nonhomogenous content across airlines.

This issue will continue to persist with NDC. Value scoring of itineraries across nonhomogenous brands using choice models or hedonic regression (Bacon et al., 2016) or other methods is therefore critical to normalize the content and determine the display order of the itineraries based on value on the agency desktop. Normalizing nonhomogenous content across airlines can be sophisticated, but it is not an exact science. Ultimately, it will continue to be difficult for customers to experience a true omni-channel experience through the agency channel. Further, while normalization of content is important, it is impossible to satisfy all airlines and can also trigger lawsuits like the Sabre New Storefront American Airlines lawsuit (American Airlines, Inc., 2021) where American claims that the search results have a systematic bias toward Delta Air Lines which is a violation of Sabre's contract with American. In November 2022, the Tarrant Country, Texas, court dismissed the case with prejudice.

9.11   Corporate Travel and Offer Management

Corporate travel is the dominant component of GDS bookings and is quite different from leisure travel. Leisure demand is B2C demand, and a supplier deals directly with a consumer and offers relevant product bundles for sale. Corporate bookings on the other hand are B2B demand where suppliers negotiate deals with a corporate buyer. The customers are employees of the organization that book travel based on negotiated deals. Corporate travel managers are responsible for managing the corporate travel program.

Corporate bookings must adhere to company policies and practices. Corporate bookings fall under the realm of managed travel with the active participation of a travel management company (TMC) through which bookings are made. Planned travel must conform to a corporate travel policy definition on fares, business versus coach based on distance traveled, preferred carriers, preferred hotels, and cars. Corporate travel governs elements such as leisure side trips as an add-on to a business trip, traveling with spouse to industry conferences, type of hotel room and size of the rental car. Corporate travelers make bookings using the corporate booking tool and non-compliance with the policy will be flagged as an exception and generate notifications up the management chain in the organization. Travel policy is also dictated by the employee profile based on the miles flown in a year. Allowances vary based on infrequent travel (e.g., one or two trips per year), average travel (e.g., six trips per year), and road warrior (e.g., 12 or more trips per year).

Companies negotiate corporate fares with airlines, which is a discount off the prevailing selling fare (ATPCO Category 25). Similar deals are negotiated with hotels and car rental companies. Business travel also requires pre-trip approval based on an estimate of the cost of the trip and the available travel budget. Further, all business travel expenses must be charged to a corporate card so that the company can review detailed reports on aggregate spend by category and take steps to reevaluate and fine tune travel policy.

Duty of care is an integral component of the travel management program. Companies have a legal obligation to ensure that employees are safe. Duty of care applies to employees on both domestic and international travel. Aspects of duty of care can be automated using machine learning. Duty of care includes educating employees on recommended inoculations before traveling to certain regions of the world, ensure access to adequate medical treatment while working overseas, business continuity plan in the event of unusual events, evacuation plans in the event of an accident or uprising, pre-trip briefings of political unrest and other risks where employees are sent. Central to duty of care is the ability to find and communicate with employees impacted by an event and allow employees to "check-in" periodically for a status update.

Over the past decade, airline contracts with corporations have evolved beyond schedules and fares. Corporate traveler preferences and how suppliers prefer to sell travel will influence the evolution of corporate contracts, which will allow corporations and suppliers to mutually achieve their objectives (Bradberry, 2013). What an airline promotes in the marketplace is not necessarily what a corporation wants. For example, an airline's pricing and revenue management function will focus on published fare levels, published ancillaries, airline branded fares, airline bundled fares, and airline branded bundles. In contrast, the corporate objective is to negotiate the corporate discounted airfares, discounted ancillaries, company branded fares, company bundled fares, and company branded bundles that are consistent with corporate policy. A related area is corporate air contracts, which are also on the verge of major changes with the incorporation of negotiated ancillary, merchandising, and personalization content, which will allow buyers and sellers to mutually achieve their objectives.

B2B personalization of offers has a greater number of constraints than B2C and should be balanced with corporate objectives.

9.12 Offer Management and the GDS Storefront

In the consumer direct channel airlines can showcase their branded fare products and support upsell across brands and à la carte pricing. The core issue with branded fares and ancillaries is that there are no standards for what constitutes a branded fare or an ancillary product or service. With the absence of standardization of branded fare products and ancillaries across airlines, how does the GDS support comparison shopping for a travel agency when they must display nonhomogenous content across airlines in the agency desktop display? This problem exists today prior to NDC and will continue after the adoption of NDC.

Value scoring of itineraries with nonhomogenous content from carriers is the first step toward normalizing content for display on an agency desktop. While the content displayed will never be identical to content on an airline website, the fundamental goal is to provide a comparison-shopping framework based on attributes associated with each "shelf" of the agency desktop. A shelf is screen real estate on the agency desktop that is designated with a selection of schedule, fare, and comfort attributes and their minimum levels for an airline offer to be classified for display in that shelf.

Estimating value of an itinerary is complicated by the fact that airlines promote branded fare products, and each branded fare product has a collection of attributes associated with it. To score the value of an itinerary, the marginal value of each ancillary product is required (Szymanski & Darrow, 2021).

Implicitly determining the value of each attribute from sales data is the preferred approach but sometimes difficult to calibrate due to the absence of sufficient data. Survey-based approaches are not perfect since there is no guarantee that the participant taking the survey is not giving a biased or faulty response. Survey participants should be preselected to represent the *typical* traveler based on criteria such as traveled for business at least n times in the last 12 months, take leisure trips, and ensure a good distribution of age, gender, and region. While crowd sourcing approaches such as Amazon Mechanical Turk (Mortensen & Hughes, 2018} and market research firms like Dynata and Qualtrics can be used in practice, a better approach is for travel entities to build the survey sampling feature directly into the airline website or agency desktop application so that surveys can be presented to every *n-th* customer to support continuous calibration.

An approach is to use a choice-based conjoint (CBC) survey analysis to determine the value of each itinerary based on a customer's willingness to pay. Survey participants are asked to book bundles virtually that include experimentally generated itineraries and sets of ancillaries. CBC is a specialized survey technique, which closely simulates the consumer selection process for products in competitive contexts (Orme & Chrzan, 2017). Survey respondents are shown a set of itinerary options and asked to make a purchase selection. Participants view products that consist of various attributes (e.g., seat comfort) made up of multiple levels of the product (extra leg room, standard seat, 127-degree reclining seat, etc.). Each attribute included in the product will have multiple levels associated with it. A price point is also associated with each level in the display. From the survey, based on how respondents evaluated products in response to changes in attribute levels, the impact of each attribute (the ancillary) on product performance can be estimated. The utility of the ancillary products can be determined using the hierarchical Bayes (HB) choice modeling technique as it models preferences heterogeneity at the individual survey participant level. By aggregating the HB model utilities, the significance of each ancillary can be determined on a normalized scale from 0 to 1. An easier way to interpret the ancillary utilities is to convert it into dollar equivalent terms which is the customer willingness to pay for the ancillary product or service.

9.13   Limitations of NDC Influenced Offers

"Personalized Travel" as defined by suppliers and IATA NDC is limited in scope within an airline to frequent flyers. Further, NDC does not manage the entire customer travel experience. An airline, for example, wants to maximize revenues with every customer interaction by selling the base fare and air ancillary bundle to a customer based on their preferences. While this is personalizing the offer from an airline perspective, it falls way short of managing a seamless, personalized travel experience for the entire journey that may include a hotel stay, car rental, and activities at the destination. This is because individual suppliers and travel entities only represent a component of the customer's journey.

An approach to support a true seamless customer experience in the future across various air and non-air travel entities that are part of a customer's journey requires the establishment of a universal profile and a decentralized id, and universal data exchange. This will be discussed in Chap. 13.

9.14   PNR, E-Ticket and EMD Transition to One Order

The electronic miscellaneous document (EMD) was introduced by IATA in 2010. It is a non-flight document that is used for the collection, financial settlement, and tracking of ancillary services offered by an airline thar are sold through travel agencies that subscribe to a GDS or through the airline's consumer direct channel. This requires the issuance of a separate electronic miscellaneous document (EMD) and messaging infrastructure to support it. There are two types of EMDs. The EMD-A is associated with an airline E-Ticket, and EMD-B is standalone. By 2014, over 180 airlines and all the major airline CRSs supported EMD processing.

The EMD also supports interline standards. The EMD can be transmitted to airline partners to enable revenue recognition and proration of fees. It also involves communicating ancillary information to operating carrier when not the validating carrier, receiving information from marketing carrier when they are the operating carrier, supporting associate and disassociate messages when the EMD and eTicket validating carriers differ and supporting ancillary sales and fulfillment when carrier both validates and operates.

Often referred to as legacy, order handling is managed by the PNR, ETicket, and EMD. The passenger name record is created by an airline or a GDS depending on where the reservation originates. It contains the passenger information, flight information, and ticket number. An electronic ticket is a digital equivalent of a paper ticket which became mandatory for ticket issuance in 2008. The E-Ticket that is issued is linked to an individual PNR. The EMDs work like an E-Ticket in the sense that an IATA standard document is issued for the ancillary services, a receipt is issued to the customer, and value coupons are stored in an electronic record in the airline CRS.

The new ONE Order was introduced in 2015 as part of the IATA NDC initiative. In the legacy environment, an airline's host CRS stores and processes reservation data and the PNR, E-Ticket, and EMD are generated and managed. ONE Order replaces the PNR, E-Ticket, and EMD. Consequently, the merged document or record will consist of

1. Passenger personal data.
2. PNR number.
3. Passenger contact information such an email, phone number, and passport.
4. Itinerary with flight details.
5. Payment status.
6. Ancillary services.
7. Ticketing information such as ticket issuer, issue date, expiration date.

This streamlines the sale of ancillary products with the elimination of multiple reservations records and the consolidation of the PNR/E-Ticket/EMD with a single reference order.

ONE Order standards were established in 2018 and Lufthansa was the first airline to be certified in 2019 following a yearlong pilot. By 2023, airlines have made some progress, but the number of airlines certified with ONE Order capabilities are few and far between.

ONE Order also has its benefits. It simplifies the order fulfillment process with a single reference record. It supports efficient billing and enables third parties involved in the order to edit purchases and check billing and payment status. The notion of a single reference order for all purchases such as airfare and air ancillaries simplify accounting and financial operations.

9.15 NDC Benefits for Revenue Management and Inventory Controls

If an airline reaches 100% adoption with NDC-enabled content, it no longer needs to publish booking class availability to the GDSs. With airline controlled offer creation and order management, and several airlines planning to adopt classless revenue management, there are several benefits that airlines can achieve in an NDC environment.

9.15.1 Classless Revenue Management

With NDC adoption, airlines do not have to distribute numeric availability by booking class since the airline is in control of all offers inclusive of the price for the itinerary. There is false excitement on classless revenue management.

Ultimately, airlines will continue to segment passengers, have restrictions on fares, and maintain multiple fare levels. The airline has the option to distribute the fares through ATPCO or not publish them. Regardless, booking classes will be internal to the airline pricing system since booking class availability is no longer distributed to intermediaries. The end product of traditional revenue management was booking class availability. With classless revenue management, the end product is the itinerary price. The availability by customer segment is internal to the airline. Airlines can deploy a rules-based or advanced decision support capability to determine the continuous price for an itinerary. Further, the quality of the inventory controls improves with classless revenue management since discrepancy in booking class availability caused by AVS transmissions from the host CRS to the GDSs no longer exists.

9.15.2   Continuous Dynamic Pricing

The primary benefit of continuous pricing is filling the gaps on the price–demand curve, thereby generating revenue that may otherwise have been lost. In the absence of discrete price points, the itinerary prices quoted by the airline for an offer request from a travel agent will be lower than how they are priced in the GDS today with discrete price points governed by RBDs.

A secondary benefit is that the continuous price for an itinerary is the ticketed fare. Before continuous pricing, the market class fare value (*MCFV*) and the total bid price for the itinerary are used to determine booking class availability followed by pricing the itinerary based on public fares, or applicable private (e.g., corporate fares) fares. Regardless of the level of accuracy of the market class fare values used, it will not be the same as the ticketed fare. Continuous pricing bridges this chasm.

A third benefit is the guarantee that the price quoted will never be below the total bid price for the itinerary. Traditional transaction processing relies on the bid price to determine the directional booking class availability only which is then used by airline pricing to price the itinerary. When the output of the revenue management process is the itinerary price and not booking class availability, the total bid price for the itinerary can serve as a lower bound during the calculation of the total itinerary price by the method of continuous dynamic pricing.

9.15.3   City Pair Availability

The calculation of city pair availability in the host CRS inventory system is no longer required to book NDC content in a GDS.

Since its inception in 1976, GDSs have used command line interfaces, commonly known as the green screen display. The most used command by travel agents is the city pair availability command line entry on the agency desktop. It is used to display flight times and availability by booking class for a city pair operated by various airlines. Agents use the three letter airport or city codes to make these entries. GDSs also provide encoding and decoding formats to determine the pertinent codes.

The syntax for city pair availability is similar for all GDSs. The Sabre syntax to request availability for a city-pair is shown below.

129NOVAUSSEA

The first character represents the command followed by the date and city pair.

For airlines that operate in a leg/segment environment, a direct connect availability (DCA) city pair availability request will return booking class availability for each leg of the request. Consider the city pair availability (CPA) request and response from Austin (AUS) to Seattle (SEA) shown below. Shown are the booking class availability for two hypothetical airlines, with IATA airline codes Z1 and ZZ. A sample response is shown below for a carrier on leg/segment inventory controls.

129NOVAUSSEA

```
1Z1 101 F7 Y7 B7 M7 H7 V7 Q3   AUSDFW 0815A 0920A 73H 0 DCA
2Z1 201 F4 Y7 B4 M0 H0 V0 Q0   DFWSEA 1030A 1230P 777 0 DCA
3ZZ 101 F7 Y7 B7 M2 H1 V0 Q3   AUSDFW 1115A 1220P 73H 0 DCA
4ZZ 201 F4 Y7 B7 M4 H1 V0 Q0   DFWSEA 0130P 0330P 777 0 DCA
```

For airlines on leg/segment revenue management controls displaying seats available by booking class is straightforward since numeric seat availability is stored on the inventory detail record on the airline host CRS. The seat availability is typically calculated based on net nesting or threshold nesting (Vinod, 2006) (see Sect. 3.5). It is a read-only look up from the inventory detail record for the corresponding legs of the host CRS.

When an airline employs O&D revenue management, the numeric availability by booking class must be the same for all segments of the outbound (and return) itineraries. However, to determine the numeric availability with continuous nesting (also known as bid price controls, see Sect. 3.5) is a computationally intensive task since the net contribution calculation must be applied iteratively to find the seats available by booking class.

With NDC and classless revenue management, the computational burden of supporting the city pair availability on airline reservations inventory systems is no longer required. Availability determination for O&D carriers is limited to determining the physical availability (overbooking limit—total seats sold) and continuous dynamic pricing to price the itinerary.

9.15.4 Round Trip Availability and Inventory Control

Issues that exist with GDS connectivity can be addressed with airline controlled offers in an NDC environment.

For round trip controls, there are two issues that can be addressed with NDC. They are availability and married segments.

Booking class availability in the GDS is directional. The availability on the inbound segments by virtue of being directional is not influenced by the outbound segment that has been booked. Continuous pricing with classless controls can address this problem.

If segments on an itinerary are married in the GDS, they are sold as a single unit and cannot be cancelled, priced, rebooked or issued individually. It is a feature that allows airlines to control inventory by combining two or more segments in the sell request and processing it as a single unit to avoid revenue leakage. Married segment indicators on the GDS can typically be applied for three segments (and in some cases up to six) on an outbound and return journey, but the indicator cannot be applied on the last outbound segment to the first inbound segment. Inconsistencies may exist between one-way pricing and round-trip pricing. For round-trip tickets, airlines require a capability to link all segments that are sold together as part of the round-trip fare during the booking and ticketing process. This link should only be broken with consent from the airline.

9.15.5   Out of Sequence Bookings

Travel agents who make bookings out of sequence may unintentionally reduce revenue. These bookings involve securing a reservation in a high-demand period and completing the itinerary at a later time. Let us consider the London-Bombay (LON-BOM) market as an example. If December is a peak season for travel from BOM to LHR, a travel agent may first book the return segment to secure a booking in a lower-priced booking class. Later, they may add the outbound segment from LHR to BOM for departure during an off-peak season, such as September. To determine availability with O&D controls, it is essential to focus on the true O&D directional market, which in this case is LON-BOM. This market fare value should be used when assessing availability, aligning with the fare filing process. The rules for fares, such as advance purchase, minimum stay, and seasonality, are always determined by the outbound departure date. It is crucial to adhere to these rules when retrieving the market value for availability assessment.

To compensate for the decrease in revenue, the inventory system must automatically close a booking class that is open on the outbound segment to balance out the potential revenue loss. This practice is known as revenue equalization. The main goal of O&D inventory controls is to ensure that SOTO (sold outside, ticketed outside) and SITI (sold inside, ticketed inside) bookings are priced identically. By utilizing journey data and conducting a thorough evaluation of the entire trip, any potential revenue loss caused by out-of-sequence bookings made by travel agents can be rectified. Only a small number of O&D carriers have integrated revenue equalization logic into their host CRS inventory control systems.

The problem of out of sequence bookings is largely eliminated since the travel agents submit a shopping request to the airline and have no knowledge of booking classes, which may not even exist if the airline is practicing classless revenue management. The airline response must be an itinerary price for every request based on the point of commencement of travel.

9.15.6   Branded Fare Families

Most airlines have three or more branded fare products. For fares associated with economy class, most airlines require a minimum of 10–15 fare levels. A fare level is a fare product identified by a unique RBD. For a given fare level, many airlines would like to maintain a fixed fare differential to the higher valued branded fare products.

For example, consider three branded fare products Basic, Standard, and Premium. For RBD M, the Standard branded fare can be $50 more than Basic and Premium can be $100 more than Standard. Booking class to branded fare product mapping is the norm. Hence, an RBD maps uniquely to a single branded fare product. If an airline has three branded fare products and wants to maintain 12 fare levels, then $3 \times 12 = 36$ RBDs are required which is not feasible. Market reality is that airlines run out of RBDs for a consistent representation of selling fares for the branded fares matrix. This can be overcome by mapping fare basis codes to booking classes by brand which is not a manageable process. Hence, airline inventory systems deploy a hybrid approach with few booking classes mapped to the more expensive and valuable branded fare products. Ideally, airlines want the capability to upsell to a higher valued branded fare product when a customer has selected a lower valued branded fare product. For simplicity, the "upsell" fare should be a derived fare, which is derived from the fare of the lower valued branded fare product, that is cheaper than the published fare for the higher valued branded fare product. GDSs cannot support this feature external to the airline host CRS due to the issue of the Fare Guarantee Policy where GDSs pay the debit memos resulting from fare or tax under collected tickets that meet the criteria for reimbursement.

The definition and deployment of branded fares and associated inventory controls are vastly simplified by NDC since the airline is in control of the offer creation process. The fundamental issue of not having enough booking classes is resolved since the price of an itinerary is controlled by the airline, regardless of the distribution channel. Consequently, content on airline websites is richer than what is offered in the GDS. While airlines publish fares to GDSs to be booked and ticketed, all the fares may not be reflected in the GDS that are available through the airline direct channel. For example, since the introduction of branded fare families in 2006, Air Canada does not distribute the lowest fares to the GDS but publish them only on the airline website for the Canada POS to compete in the domestic Canadian market.

In a world of airline controlled offers across all channels of distribution, an approach for airlines is to define reference fares for the lowest valued branded fare product. The number of fare levels is at the discretion of the airline, e.g., 12, 14, etc. The fares for the higher (or lower) valued branded fare products can be derived from the reference fares either with a fixed fare differential or a multiplication factor that is greater than one (or less than one).

An approach is to use the CAT25 discount calculation to derive all the fares for the lower valued branded fare products from the reference fares for the highest brand. For corporate discounts, since a discount cannot be applied over a discount (a CAT25 over an existing CAT25 fare), the negotiated fares capability with CAT35 together with CAT25 can be applied to calculate the corporate discount with the Security Table 983 and Fare Creator Table 979 located in CAT35.

To generate a price quote, continuous dynamic pricing can be applied to determine a derived fare value between two successive fare levels of a branded fare product. With this approach, inventory control can be synchronized by fare level across all branded fares, so that they are either open or closed.

Table 9.7 illustrates how an airline can support branded fare families across all channels of distribution without the issue of running out of RBDs. This is accomplished by using the same booking class across branded fare products, identified by a character in the fare basis code. This approach ensures consistency in content delivery across all channels of distribution, both direct and indirect.

The RBDs and fare basis codes shown in the table are internal to the airline and not published to intermediaries or third parties. The internal fare filing process will require a new standard that does not exist today. The first character of the fare basis code represents the booking class code, and the last character represents the branded fare product. Hence in this case Basic is "3", Standard is "2", and Premium is "1". Enforcing this standard eliminates the need to maintain a mapping of fare basis codes to branded fare products.

This approach supports synchronized nesting across all branded fares. This means that a booking class is either open or closed for all branded fare products. To avoid revenue dilution, there are two requirements:

1. The fare qualifications for fare basis codes associated with a row are roughly the same. Hence, the fare amount is the only difference in each row, besides the attributes bundled with the branded fare product.
2. Pricing analysts need to ensure that there is no fare overlap from one row to the next. In other words, for a given tier (row), the fare amount for premium should

Table 9.7 Branded fare family mapping example. Market: DFW-SFO

	Basic		Standard		Premium	
Booking Class (RBD)	Fare Basis Code	Fare	Fare Basis Code	Fare	Fare Basis Code	Fare
Y	Y......3	$700	Y......2	$750	Y......1	$825
B	B......3	$625	B......2	$655	B......1	$695
M	M......3	$550	M......2	$580	M......1	$620
...	...		...		...	...
Q	Q......3	$300	Q......2	$350	Q......1	$425

be less than the fare amount for basic for the next higher tier. In the example above, the Premium B fare of $695 should be less than the Basic Y fare of $700.

9.15.7 Ticketing Time Limit Churn

In the current environment (prior to NDC) for reservations processing, revenue integrity (RI) applications identify travel agents who circumvent ticketing time limits (TTL), which are enforced as a pricing rule. Travel agents can create churn by cancelling and rebooking the same passenger multiple times on the same flight to extend the ticketing time limits. This can now be avoided.

9.16 Decoupling the Value Chain

Thales Teixeira, a former Harvard professor, outlined three waves of digital disruption in his book *Unlocking the Value Chain* (Teixeira, 2019). He identifies these waves as unbundling, disintermediation, and decoupling.

The first wave was unbundling when PeopleExpress started charging $3 for each checked bag in 1981. This strategy generated incremental revenues to provide partial compensation for the deeply discounted fares. LCCs embraced unbundling. Though large-scale adoption by full-service carriers did not happen until the 2000s. In other industries, for example, Apple iTunes allowed customers to purchase songs individually instead of buying entire CDs and eBook publishers changed the way textbooks are sold by allowing chapters to be sold individually.

The second wave was disintermediation when airline websites started appearing in 1996. They bypassed the travel agencies to offer services directly to customers over the Internet. Dell with its direct build-to-order selling model achieved an edge over its competitors, Compaq and IBM, in the 1990s. Another example is the disruption of full-service stockbrokers.

The third wave, decoupling, involves entities splitting the customer value chain and delivering only a portion of the value, while avoiding the costs associated with sustaining the entire value chain. NDC aggregators can be viewed as disruptors that want to decouple the existing value chain by delivering NDC content to the point of sale. Other examples of decoupling are car sharing companies decoupling the link between purchasing and driving a car and Amazon decoupling the link between physically trying out a new TV (for example, at Best Buy) and purchasing it.

Chapter 10
Origins of Online Travel Agencies

10.1 Overview

The growth of the Internet since the early 1990s has had a major impact on supplier-direct bookings. A supplier-direct booking is one that is made without an intermediary. The Internet has also had a major impact on the indirect channels of product distribution.

Key milestones in the evolution of online travel that led to the creation of consumer direct channels and online travel agencies (OTA) are reviewed in this chapter.

10.2 Key Milestones in Personal Computing

The Kenbak-1 was released in 1971 by inventor John V. Blankenbaker of Kenbak Corporation. It is considered by the Computer History Museum to be the world's first personal computer. Unlike the modern era of personal computers that rely on microprocessors, the Kenbak-1 was built on small-scale integrated circuits. They originally sold for $750.

Apple Computer released Apple I in 1976. In 1977, the Apple II, Commodore PET 2001, and TRS-80 were released. The IBM PC revolutionized business computing by gaining widespread adoption with the Model 5150 that was released in 1981. Apple Computer introduced the Macintosh in 1984.

© The Author(s), under exclusive license to Springer Nature Switzerland AG 2024   329
B. Vinod, *Mastering the Travel Intermediaries*, Management for Professionals,
https://doi.org/10.1007/978-3-031-51524-8_10

10.3   Key Milestones in Online Travel

Three months after Neil Alden Armstrong and Edwin "Buzz" Aldrin landed on the moon, on October 29, 1969, the first message was sent successfully over the ARPANET. Computers from UCLA and Stanford University made the first host-to-host connection, which eventually evolved into the Internet.

In 1985, well before the first online travel agencies and supplier direct websites came into existence, American launched eAAsySabre, well before the idea of an interactive Internet had yet to take hold. eAAsySabre was a command line program with a green screen interface for a home user to access schedules, fares, and availability with any modem-equipped computer to make a booking without the aid of a travel agent. Allowing consumers to book airline, hotel, and rental car with a personal computer through the Sabre system was an industry-first.

eAAsySabre was a simple *request and reply* conversational system that allowed customers to make bookings, but the airline had to do the ticketing and mail the physical tickets. It was also the first time that an airline introduced channel conflict by allowing customers to book directly through the eAAsySabre interface, thereby bypassing the travel agent from the booking process. Consumer network service providers like CompuServe, Prodigy, Minitel in France, Nynex, AOL, and others typically used one or more of the three largest air fare services: eAAsySabre, the Official Airline Guide (OAG) electronic edition travel services and Travelshopper (the consumer version of PARS) from TWA and Northwest Airlines. The request and reply interface were upgraded to use templates where customers could fill in the blanks to initiate their booking. However, market penetration was low and even by 1989 fewer than 150,000 passengers had reserved and purchased tickets with a home computer through eAAsySabre (Gutis, 1989). By 1990, the three largest service providers were CompuServe with 550,000 members followed by Prodigy and Nynex (Lavin, 1990).

In 1985, Microsoft released Windows 1.0. Also in 1985, DATAS II was the first GDS to deploy a PC for the travel agent's desktop.

In 1984, Nick Lanyon founded his company Lanyon Ltd., which developed the Lanyon ALC Board for PCs. ALC is an acronym for Assembly Language Coding, a generic term for IBM mainframe assembly languages. He pioneered the use of a communications co-processor and positioned his company as a neutral vendor to connect PCs to airline reservations systems and the GDSs.

In March 1989, British computer scientist Tim Berners-Lee proposed an Internet-based hypermedia initiative for global information sharing, which would eventually become the World Wide Web. He wrote the first web client and server application in 1990. The World Wide Web was made public in 1993.

TravelWeb, owned by Pegasus Systems, Inc., was launched in 1994 with the first online catalog of hotel properties by a team under John Davis III of THISCO. In January 1996, TravelWeb was also the first website to provide online customers with access to 29,000 hotels to check room availability and make bookings interactively. In August 1996, TravelWeb enhanced its capability by adding air content for online bookings through the System One Amadeus GDS (Pegasus Solutions, 1996). It served as a one stop shop for hotels, air, hotel photos, maps, weather, currency conversion, and special discount programs.

Unrelated to travel, Daniel Dreilinger developed the first metasearch engine at Colorado State University called Search Savvy in 1994 that searched 20 websites to give back a consolidated result (Howe & Dreilinger, 1997). MetaCrawler was developed by Erik Selberg, a student at the University of Washington that used an updated version of Search Savvy. Developed in 1994, the site was operational in June 1995.

In 1995, a Palo Alto business called Internet Travel Network (ITN) claimed to have overseen the first airline ticket booking made over the web for a flight from San Francisco to Las Vegas. ITN was the forerunner of GetThere, which was acquired by Sabre in 2000 for $757 million.

Alaska Airlines was the first U.S. airline that allowed customers to ticket their reservations online. They issued a press release on December 27, 1995, entitled *"Alaska is first U.S. airline to make internet interactive; customers can now book themselves for travel"*. The homemade website was quite advanced for its time. It provided online users side-by-side comparison of Alaska and Horizon flight schedules, and lowest available fares in the markets they served.

Viator Systems was launched in 1995 for destination tours, activities, and excursions. There are a large number of small niche suppliers that promote activities and excursions, making it a challenge to uniformly aggregate accurate content and display to customers on a website. It was the first of its kind where there is immense fragmentation of content.

David Litman and Robert Diener co-founded Hotel Reservations Network (HRN) in 1991. They started by taking bookings over toll free telephone lines. Their most significant achievement was the invention of the hotel merchant model of buying hotel rooms at net rates and selling to customers at a gross rate. This is how Expedia and Booking.com came to dominate the business. In 1995, they migrated their hotel telephone booking service to the web with the launch of two websites, hoteldiscount.com and hoteldiscounts.com (Hoffmann, 2021). Hotel Reservations Network's first merchant model hotel was the Dorset Hotel on West 54th Street in Manhattan. The website connected hotel vacancies with customers who were searching for hotel rooms. It was one of the earliest examples of online hotel bookings. Subsequently, the merchant model transformed online travel and the hotel industry. Incidentally, the Dorset Hotel was torn down in 2002 and is an extension of MOMA, the Museum of Modern Art.

Larry Page and Sergey Brin, students at Stanford University, initiated the process of indexing web pages in 1996. This endeavor, focused on assessing popularity, eventually gave rise to the inception of Google search in 1998. Within a relatively brief timeframe, Google emerged as the dominant search engine, becoming the preferred choice for online travelers. Today, travel suppliers and OTAs invest substantial amounts of money to advertise alongside search results, thereby posing a significant challenge for advertisers as it elevates the cost of travel for customers.

OTAs emerged in 1996. Travelocity was the first to launch in March, followed by TravelWeb in August and Expedia in October. These platforms give travelers the ability to access and purchase air, hotel, and rental car options online. OTAs depend on GDSs for air shopping and booking API's. However, there are exceptions like Expedia, which has developed its own algorithm for domestic air shopping but still relies on GDSs for international flights due to the complexity of international fare rules. The older APIs are SOAP-based (Simple Object Access Protocol), while the newer ones are REST (Representational State Transfer) APIs. REST APIs offer a simpler way to access web services, are JSON-compliant, and can be advantageous for mobile applications. OTAs also rely on hotels to provide frequent ARI updates (availability, rates, and inventory) through their respective channel managers.

Sabre Interactive, a subsidiary of AMR Corporation (the former parent company of American Airlines until 2011), was the pioneer among global distribution systems (GDS) in recognizing the possibilities offered by the Internet. On March 12, 1996, Sabre Interactive introduced Travelocity, the world's first online travel booking system. Remarkably, Travelocity emerged even before airline websites under the leadership of Terry Jones. Expedia, on the other hand, originated as a division of Microsoft on October 22, 1996, under the leadership of Rich Barton. It eventually became an independent company in 1999.

Figure 10.1 highlights the important milestones in online travel through 1996.

Travelocity gained large public popularity by electing the Roaming Gnome, a lawn ornament figure, as its spokesperson to report on low fares across the world. The Travelocity Roaming Gnome had a distinctive British accent and his hops around the world portrayed him as a global citizen.

Priceline was founded in 1997 by Jay S. Walker. Priceline introduced the novel reverse auction "*name your price*" model in 1998 which was famously patented (Walker et al., 1998). This model reversed the buyer–seller relationship wherein the buyer submits a guaranteed price that they are willing to pay to travel to a destination. Priceline aggregates requests from buyers and submits them to sellers for a response. Airlines participate in the reverse auction model to get rid of surplus

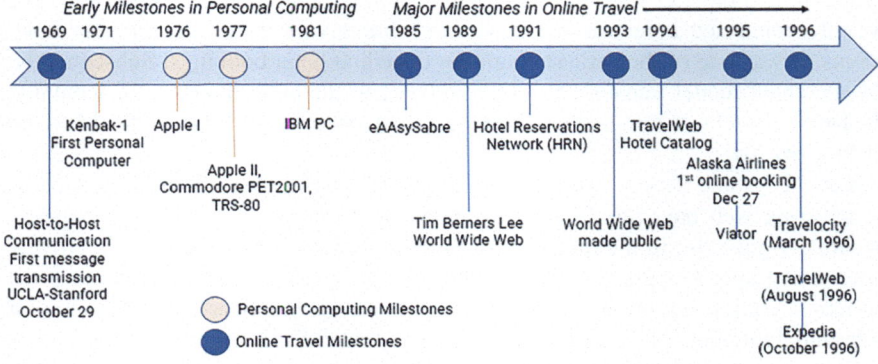

Fig. 10.1 Important milestones in online travel until 1996

inventory. However, the model has its flaws; it did not offer the identity of the seller or the travel schedule until after the purchase has been completed. The popularity of this model is limited to leisure travelers who are seeking deep discounts.

Codenamed Purple Demon before its launch, Karl Peterson, Eric Grosse, Gregg Brockway, and Spencer Rascoff launched the travel website Hotwire in 2000 with opaque airline tickets and expanded to opaque hotels and rental cars a few months later and packages in 2003.

Orbitz was founded in 2001 with investments from American, Continental, Delta, Northwest, and United to counter the power and dominance of the established OTAs. The secret code word for the startup was T2 (rumored to stand for Travelocity Terminator) (Hansell, 2002b). Launched in 2001, unlike the other OTAs, Orbitz supported a lower-cost model with direct supplier links, a percentage of their online air bookings bypassed the Worldspan GDS and were booked directly in the airline host CRSs. Travelocity's former Chief Executive Sam Gilliland (Hansell, 2002a) remarked on the direct link to suppliers *"If you have a direct connection, you are just pushing costs from one place to another. In the end, it will have very little economic benefit to the airlines."*

In 2002, Travelocity acquired the New York based Site59.com, a specialist in last minute travel booking deals, for $43 million (Sullivan, 2002). With the acquisition, the founder and CEO of Site59, Michelle Peluso joined Travelocity and was appointed CEO in 2003. In 1998, Brent Hoberman and Martha Lane Fox created an iconic web brand called lastminute.com, which was acquired by Travelocity (a Sabre company at that time) for £600 million in July 2005.

Barry Diller's USA Networks acquired Hotel Reservations Network and acquired Expedia in 2001. The HRN site was later rebranded as Hotels.com in 2002, to become one of the largest hotel reservations websites. Expedia and Hotels.com wield tremendous power over hotel distribution with the merchant model, much to the resentment of the hotel industry.

Zuji, the leading onsite travel agency in Asia Pacific, became a wholly owned brand of Travelocity in 2006. It was subsequently spun off to Webjet in 2012 for $25 million. Webjet Limited sold the Zuji businesses in Hong Kong and Singapore to a Hong Kong-based travel technology group in 2016.

The merger of Hospitality Franchise Systems (HFS) and CUC (Comp-U-Card) International created Cendant Corporation in 1997. Travelport was created in 2001 by Cendant following the acquisition of Galileo GDS and cheap tickets. Travelport acquired Orbitz in 2004. Travelport was sold to the Blackstone Group in 2006.

In 2006, Travelport, the travel distribution business of Cendant, was acquired by The Blackstone Group. Travelport acquired Worldspan in 2007. Orbitz was acquired by Expedia in 2015. Expedia also acquired Travelocity from Sabre in 2015. The Priceline Group, which owned booking.com, was renamed Booking Holdings in 2018. Today, the largest OTAs are Expedia (Expedia Holdings) and booking.com (Booking Holdings).

SideStep and FareChase appeared on the scene in 1999 as the first travel metasearch engines to generate revenue from advertising and referral fees. A metasearch site looks at several sites, supplier websites, as well as OTAs and produces a display based on itineraries returned from these individual sites and deep links to the various sites to facilitate a booking. They receive the referral fee even though the booking is not guaranteed since the customer can abandon the shopping cart after the deep link to the supplier site. FareChase was acquired by Yahoo. Tripadvisor, another metasearch engine, was launched in 2000, and this led to a revolution in the hotel industry. Guests left written reviews and user star ratings based on their experiences at the hotel where they stayed. This became a staple in the future for customers to read the reviews before making a hotel booking. In February 2020, the name was changed from TripAdvisor to Tripadvisor. The company has been plagued with criticism for allowing unsubstantiated reviews to be posted about hotels and restaurants. Competition from metasearch engines, Tripadvisor and Travelzoo, forced Yahoo to sunset FareChase in 2009 (Schaal, 2009). SideStep was acquired by Kayak for $200 million in 2007.

Metasearch engines Skyscanner emerged in 2003 and Kayak in 2004, followed by Google Flights in 2011. Metasearch sites have become mainstream. Today, many retailers advertise on Google, and a wide range of goods are available through Amazon. Platforms such as Google, Amazon, and Kayak are expensive for users of these systems. This new breed of intermediaries also threatens the existence of businesses since if they do not agree with their terms and fees, they would no longer be allowed to participate in the marketplace. Strategies to counter this threat are a high priority to control the cost of online advertising (Edelman, 2014).

Priceline acquired TravelWeb in 2004 (Limone, 2004), giving Priceline a foothold in the hotel merchant model distribution program. It also acquired U.K. based Active Hotels, a reservations service provider for independent hotels, for $161 million in 2004.

Priceline also acquired Bookings.nl, a Netherlands-based company (founded by Geert-Jan Bruinsma in 1996), in 2005, which is considered in travel industry circles as the greatest acquisition in the history of online travel (Schaal, 2014). Priceline melded Active Hotels and Bookings.nl to create the industry behemoth Booking.com, which reshaped the competitive landscape in online travel. The new undisputed leader dislodged competitors like Expedia, Orbitz, Travelocity, and Lastminute.com. It became the gold standard along two key dimensions: search engine marketing (SEM) to convert lookers to bookers and how customers make online hotel reservations.

To counter the OTA threat to their margins, Room Key was launched in 2012 by founding members Intercontinental Hotel Group, Hilton Hotels Worldwide, Hyatt, Marriott International, Choice Hotels, and Wyndham Worldwide. It was founded by John Davis as a successor to TravelWeb. Over 60,000 hotels from 60 brands were listed in Room Key. When a Room Key user conducted a search to make a booking for a hotel that belonged to a founding member, they were redirected to the hotel website. RoomKey also had access to other properties with links to other reservations systems like SynXis, Trust, and TravelClick. A fundamental issue is that hotels have very limited marketing budgets compared to OTAs. Another issue is that hotel chains prioritized their marketing dollars on growing their own brand and attract direct traffic rather than making a commitment to RoomKey. For these reasons, the site failed to attract traffic and was dissolved in 2022.

Facebook, the social network platform, was launched in 2004 by Mark Zuckerberg. This further cemented the role of online reviews and travel experiences to generate bookings. Today, many travel companies have their business page on Facebook.

Apple had a significant impact on the development of online travel, albeit indirectly. When the iPhone was introduced on June 29, 2007, it enabled users to access the Internet through mobile communication devices, which resulted in the emergence of numerous mobile applications related to travel.

In 2008, Air Bed and Breakfast brought individual renters to the web marketplace. It was later renamed Airbnb.

From its early origins, the OTAs transformed themselves into web supermarkets to serve as a one-stop shop for air, hotels, rental car, and cruise lines. Typically, an OTA uses a GDS as the back end to manage air bookings and a fulfilment agency for ticketing, customer service, and accounting. For hotels, the GDS model is different in the sense that an agent can shop and display hotels, but the hotel content including room rates and taxes are returned from a hotel CRS when a query is submitted. When a booking is made, a GDS encrypts the credit card information and sends it to the hotel; customers only pay when they check-out.

10.4   OTAs and Air Bookings

Online travel agencies (OTAs) acquire content from aggregators that compile schedules and fares, as well as various suppliers. This allows travelers to compare prices and make reservations for their preferred airline seat. OTAs have strategically positioned themselves as a convenient one-stop solution, offering a seamless booking experience as digital distribution technology has advanced. They provide a range of travel-related products, including a greater selection of flights compared to travel management companies (TMCs), including codeshare and interline flights, hotels, rental cars, cruises, rail, buses, and dynamic packages which include multiple components. Additionally, OTAs offer a variety of payment options and customer service. This has revolutionized the way air travel is sold, with OTAs now accounting for almost half of all online travel sales worldwide.

The proliferation of these services increased competition not only among OTAs but also with offline agencies, tour guides, and, most importantly, travel suppliers. For example, an airline website does not offer cheap interline flights with its competitors, but OTAs do to produce a lower total fare for leisure customers. Innovation with a wider choice set served as the primary differentiator to attract customers. Unlike airline websites where a ticket can be issued immediately after a purchase is made, with smaller OTAs there is usually a delay between the time of ticket purchase and the time of ticket issuance.

To increase revenue and strengthen their competitive positioning in the marketplace, OTAs started to offer solutions in the second decade of the twenty-first century that went beyond conventional air ticketing. Travelers were given the option to purchase ancillary services like seat selection and food choices through OTAs' interfaces with various airlines, including low-cost carriers. The OTAs also vastly improved the user experience. The second digital wave was introduced via apps with the introduction of smartphones and app stores.

On the NDC front, OTAs are more experienced than TMCs. OTAs first partnered with LCCs to access LCC content using carrier-specific APIs. With their experience with LCCs, OTAs are better positioned than TMCs to adapt NDC content into their booking workflow.

Besides the three largest OTA brands Booking Holdings, Expedia Group, and Trip.com, there are hundreds of smaller OTAs that make up the long tail. Examples are Ebookers, Kiwi, OneTravel, GoToGate, SkyBooker, Vayama, Webjet, and many more.

10.4.1   OTAs and the GDS

Larger OTAs have a dual or triple GDS strategy to access air content to provide redundancy in the event of a system failure, better access to certain markets, and, above all, it gives them negotiating power for GDS services and incentives.

OTAs rely on the web services provided by the GDSs for air shopping and booking. They also place a significant emphasis on the quantity of itineraries displayed across multiple pages for each shopping request. Unlike travel agencies, which typically show fifty to a hundred itineraries, OTAs expect a wider range of options that include non-stop flights, single and double connections, codeshare flights, interline flights, and carrier diversity. OTAs do not increase the prices of publicly published airfares set by airlines. In certain situations, such as in China, OTAs like CTRIP may offer lower prices to increase their presence in the market and capture more market share, compensating for the loss in revenue through service fees and additional services. The air shopping demands from OTAs have led to significant enhancements and flexibility of GDS air shopping algorithms.

10.4.2   OTA Revenue Streams

OTAs have multiple revenue streams. Their primary revenue streams are lodging, advertising, and media, air, and other lines of business such as car rentals, cruise lines, and destination activities. Of these revenue streams, lodging is the dominant component for many of the largest mainstream OTAs. Details of the hotel merchant model and pay at check-in (PAC) models are discussed in Chap. 11.

OTAs view air bookings as a loss leader. Selling airline tickets is like selling milk cartons in a grocery store; while it is not profitable it drives foot traffic. The cost of air distribution, paid by airlines to GDSs in the form of segment booking fees, ranges between 1.8% and 2.2%.

10.4.2.1   Advertising and Media Revenue

OTAs sell advertising space to businesses in the form of banner ads, email campaigns, and other forms of online marketing.

Pay per click (PPC) advertising is the second largest revenue stream for many OTAs after hotels. Hotels pay to be prominently placed above organic search results on an OTA results page. PPC-based advertising is a flexible way to maximize a supplier's visibility and improve bookings. For example, Expedia promotes TravelAds to drive booking demand with custom copy, images, precise traveler targeting, and real-time reporting.

10.4.2.2   GDS Incentives

GDSs pay incentives to OTAs based on performance. OTA incentives paid by GDSs are higher than the incentives paid to TMCs and can range from 70% to 85% of the segment booking fees paid by the airline to the GDS. From a GDS perspective, these are extremely low margin bookings due to the high incentives. Demand for even higher incentives also contributed to the fallout between Expedia and Sabre in November 2021 that resulted in Expedia shifting a significant portion of its GDS bookings in North America away from Sabre (Schaal, 2021). As a result of the Expedia initiative, Sabre raised the average booking fee globally to partially offset the losses and, more importantly, improve the booking mix.

This move by Expedia also resulted in a loss of market share for Sabre in North America, prompting Sabre CEO Sean Menke to make the statement on November 2, 2021, during the Q3 2021 Earnings Call,[1] *"Although we expect this shift to result in the loss of some volumes, it is important to remember that not all share is created equal in the GDS. As we have discussed, U.S. domestic leisure bookings are our lowest revenue and margin bookings. Therefore, this volume shift is anticipated to have a favorable impact on our overall average booking fee."*

Although we expect this shift will result in the loss of some volumes, it is important to remember that not all share is created equal in the GDS. As we have discussed, US domestic leisure bookings are our lowest revenue and margin bookings. Therefore, this volume shift is anticipated to have a favorable impact on our overall average booking fee.

10.4.2.3   Front-End Commissions

OTAs receive front-end commissions in selected markets, usually governed by a contract between the airline and the OTA. Front-end commissions from airlines are

---

[1] Sabre Q3 2021 Earnings Call, November 2, 2023, https://investors.sabre.com/static-files/167dea24-6fd0-4286-a6b0-9cea14e5d3ab

limited to markets that they want to grow. Individual contracts are signed between each airline and agency. A ticket can have one or more fare components, and ticketed commissions are applicable at a fare component level which can have one or more ticket coupons.

10.4.2.4   Back-End (Override) Commissions

Airlines negotiate performance-based back-end commissions, also called override commissions, with OTAs. These contracts are based on achieving specific goals, such as bookings in a particular market, market entity, or region. When the target metric is achieved, the OTA receives a lump sum payment from the airline. Override contracts are more complex than front-end commission contracts. A typical OTA manages multiple contracts with various airlines, each with its own complex rules for incentive payout. As an example, an OTA may have contracts with multiple airlines to sell $10,000,000 worth of tickets in North America in a predetermined timeframe, with a tiered incentive payout structure (e.g., 0.4%, 1%, and 1.5% for achieving sales of $4 million, $8 million, and $10 million). To maximize their payout, some OTAs have developed override commissions optimization models (Smith et al., 2007). The model recommends the sequence of thresholds by airline that should be achieved against a timeline to maximize the earnings potential.

10.4.2.5   Net Fare Markup

OTAs negotiate net fares with airlines. These negotiated net fares can be marked up by the OTA, and the markup is pure profit. The net fare can be marked up only when the prevailing selling public (ADT) fare is higher than the net fare. Major airlines do not like to distribute net fares to OTAs but do it in limited quantities for strategic reasons.

10.4.2.6   Bulk Fares and Packages

OTAs negotiate deeply discounted bulk fares with airlines known as JCBs. This refers to a passenger type code (PTC) for contract, bulk, adult. These deeply discounted fares can only be sold as part of a package. By incorporating these discounted fares that may include a hotel or car rental to produce a competitive package price, especially when the discounted fare is lower than the selling public (ADT) fare. However, this is not always the case, as there are instances where the discounted JCB fare can be more expensive than the public ADT fare. In markets where the OTA has negotiated JCB fares, the air shopping entry should provide both the public ADT fare and the JCB fare, enabling an automated decision regarding the inclusion of the JCB fare in the package that is promoted to the customer.

10.4.2.7   Service Fees

OTAs also make money by charging a nominal service fee for air bookings. Some OTAs also offer value-added services such as concierge services to VIP customers. In the future, OTAs can also expand into advisory services with tools like ChatGPT to plan a door-to-door trip that includes local transportation, air, hotel, and destination activities.

10.5   Growth and Consolidation of Online Travel Agencies

Online travel agencies constitute the largest source of hotel bookings. They specialize in sales of air, hotel, car, cruise, and packages. Travel websites like Booking.com and Expedia have millions of visitors every month. The OTAs make money by taking a commission for each booking. The actual commission amount varies based on the negotiated contract with hotel chains and independents.

Over the past two decades, the OTA segment has grown to be the dominant channel for hotel bookings. This is an issue for hotels since both the merchant model and the agency model have steep commissions, leading to higher distribution costs. However, hotels realize the importance of participating in all channels of distribution, and the OTAs cannot be ignored. Growth in bookings for a property frequently requires OTAs even with a higher distribution cost per booking. Larger chains have the greatest leverage when commissions are negotiated, and the smaller independent properties pay the highest commissions. OTAs also generate revenues through advertising, where hotels pay to be prominently displayed above organic search results during a search. Travelers use OTAs since it is a one stop shop where customers can compare various hotel brands and prices before making a booking.

In August 2000, Sabre acquired GetThere for $757 million, to bring together the top two players, at that time, in the online business-to-business corporate booking travel channel and the business-to-consumer E-commerce channel. At that time less than 1% of U.S. corporations actively used a business-to-business booking tool (Meehan, 2000). The acquisition surprised analysts and investors since GetThere was GDS independent and could be connected to any GDS. Bookings made on Sabre's Business Travel Solutions (BTS) were routed through the Sabe GDS. In the GetThere model at that time, suppliers were directly connected to corporate buyers. GetThere provided a more cost-effective solution where the GDS was bypassed. The initial plans for the integrated GetThere and Sabre BTS solution under the GetThere brand called for a hybrid model, giving corporate customers a choice between direct connectivity and GDS usage. The idea was that corporate customers who were looking for savings with a bare-bone service may opt for direct connectivity while customers that required core distribution services such as air shopping across airlines and PNR storage could opt for the GDS model. The GetThere supplier-direct model for corporate travel did not gain traction, and all GetThere bookings were made through the Sabre GDS, though the multi-GDS booking option was retained.

After a wave of consolidations between 2010 and 2020, the major OTAs were Booking Holdings, Expedia Group, and Trip.com. Booking Holdings operated its flagship booking.com and sites such as Priceline, Kayak, Agoda, Cheapflights, Momondo, Rentalcars.com, and Open Table. Expedia Group operates sites such as Hotels.com, Vrbo, Trivago, Orbitz, Travelocity, Hotwire, Wotif, CarRentals.com, and Expedia Cruises. Egencia, a corporate booking site for Expedia, was acquired by American Express Global Business Travel (Amex GBT) in 2021. Ctrip, founded in 1999, is the dominant OTA in China. It expanded globally with the acquisition of Skyscanner in 2006 and Trip.com in 2017, which serves as its global identity. The rebranded Ctrip has a global footprint in North America and Europe.

An emerging player in the OTA space is Airbnb which was launched in 2006 as a home sharing website. Airbnb partners with property owners, developers, and residents to provide hosted services worldwide. Their recent acquisition of Hotel Tonight and Urbandoor in 2019 and its investment in Experiences for one-of-a-kind experiences hosted by local experts augment its desire to compete globally with the major OTAs.

It is not clear if Airbnb and similar sites like FlipKey (owned by Tripadvisor), OneFineStay (owned by Accor Hotels), and Vrbo (originally known as Vacation Rental by Owner or VRBO were acquired by HomeAway in 2006. HomeAway was acquired by Expedia in 2015, and VRBO was rebranded Vrbo in 2019) will ever emerge as a viable major platform for hotels, given the fierce competition and projected declining margins in the future.

Figure 10.2 highlights the important milestones in online travel from 1997 to 2022.

Google Hotels operates on a pay for performance business model and is technically not classified as an OTA. By capturing customers at an earlier stage in the travel search, Google's search engine has a significant impact on hotel bookings and OTAs worldwide. Google and Facebook control a significant portion of internet advertising. The two largest OTAs, Booking Holdings and Expedia Group, are their most prominent advertisers. In 2019, before the pandemic, Booking Holdings and

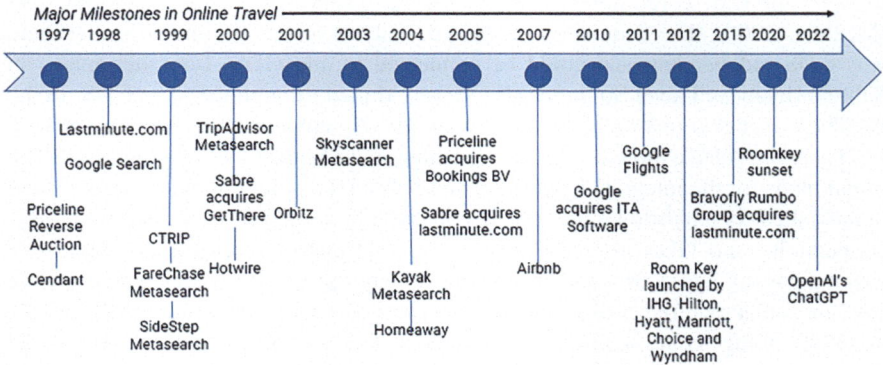

Fig. 10.2 Important milestones in online travel from 1997 to 2022

Expedia Group spent a record $11 billion on marketing. During the recovery period, they spent over a billion dollars in the first quarter of 2022 on marketing. Hotel companies face a dual challenge of high commissions from OTAs and advertising expenses from Google and Facebook. As a result, they are compelled to increase their marketing budgets to remain relevant in the online marketplace. Unfortunately, this is reflected in higher hotel rates for consumers.

10.6   The OTA Infrastructure

Figure 10.3 illustrates the core components of an online travel agency. The user experience is a façade of the OTA product. It identifies and defines the value of the brand and supports workflows for customers to book travel. OTAs invest in their own website and workflow design and do not use a GDS supplied travel agency desktop. The OTAs rely on the GDS web services for a variety of transactions such as shopping and booking. For hotels, they rely on a hotel or hotel chain's channel manager to update ARI content multiple times a day.

The supplier contracts team engages in contract negotiations with various suppliers such as airlines, hotels, rental car companies, cruise lines, rail providers, and activity providers. These contracts give the OTA a competitive edge from two perspectives: breadth of content and negotiated merchant rates or commissions. The marketing advertising campaigns receive the most scrutiny since it is a major driver of costs which can run into billions of dollars for an OTA to generate traffic to their websites.

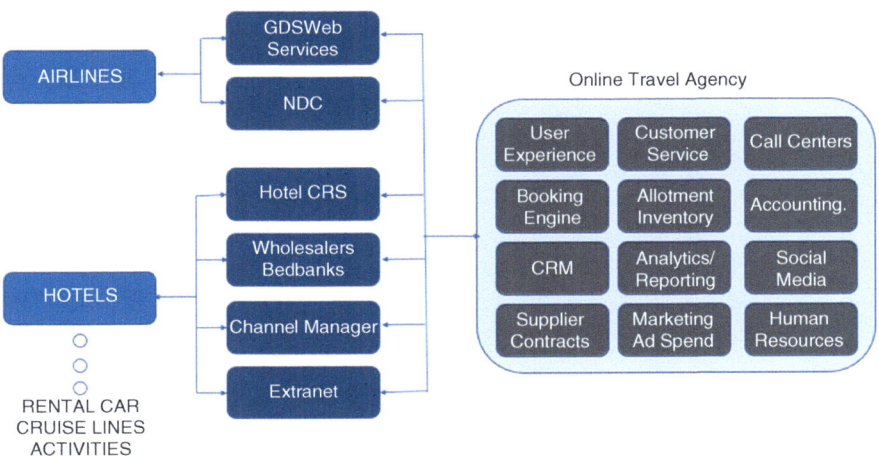

Fig. 10.3 Core components of an online travel agency

Chapter 11
Hotel Bookings and the Dominance of the OTAs

11.1 Introduction

Absent a recession in 2024, the global online travel agency market is expected to grow at a compound annual growth rate (CAGR) of approximately 6% over the next decade. The tourism sector was worth around $1.7 trillion, and the global online travel agency market was approximately $450 billion in 2022. Even with a modest CAGR of 8%, the online market will surpass a trillion dollars in a decade.

The growth of online bookings started in the late 1990s and continues unabated for several reasons including Internet access and strong economic growth in emerging nations, changing tech savvy demographic of travelers, solo travelers, growth in business leisure (bleisure) bookings, food tourism for authentic culinary experiences, and growth in middle class disposable income. In this transformation, mobile access is becoming increasingly important to shop for travel but not necessarily book at the expense of desktop users which has been dropping. Mobile conversion rates are significantly lower than desktop conversion rates. Before the Internet era, 80% of travel was booked through brick-and-mortar travel agencies.

With the growth of online direct and online indirect channels of distribution, the traditional travel agency has been in decline, measured as a percentage of total bookings. In 2022, it was estimated that only 25% of bookings are made by travel agencies.

Online travel agencies constitute the largest source of hotel bookings. They specialize in sales of air, hotel, car, cruise, and packages. Travel websites like Booking.com and Expedia have millions of visitors every month. The OTAs make money by taking a commission for each booking. The actual commission amount varies based on the negotiated contract with hotel chains and independents.

Over the past two decades, the OTA segment has grown to be the dominant channel for hotel bookings. Many small chains and independents are dependent on the OTAs to deliver bookings. It is the extreme fragmentation of the hotel industry with a long tail of independents that do not possess negotiating power that the OTAs exploit to generate revenues. Larger chains, though they participate in the OTA channel, have negotiating power and are not at the mercy of the OTAs. One of the first steps larger hotel chains took to counter the threat of reduced margins per booking transaction through the OTA channel because of their high commission rates was exempting bookings made by customers through OTAs to accrue loyalty points. This was a way to attract customers who were sensitive to accruing loyalty points directly to the hotel websites.

The OTAs have two distinct commission models by which they are compensated for each hotel booking. They are the merchant model and the agency model. Both have steep commissions, leading to higher distribution costs. However, hotels realize the importance of participating in all channels of distribution and the OTAs cannot be ignored. Growth in bookings, especially for small chains and independents, requires OTAs albeit with a higher distribution cost per booking. Larger chains have the greatest leverage when commissions are negotiated, and the smaller independent properties pay the highest commissions. OTAs also generate revenues through advertising, where hotels pay to be prominently displayed above organic search results during a search. Travelers use OTAs since it is a one stop shop where customers can compare various hotel brands and prices before making a booking.

For the air travel line of business, an OTA generates revenues from front-end and back-end (also known as override) commissions, incentive payouts from the global distribution systems (GDS), and service fees. Front-end commissions are either a fixed dollar amount for specific markets or a percentage of the ticketed revenue by fare type and market. Override commissions for air are performance-based lump sum payouts from the airlines at the end of a measurement period. Most OTAs have a multi-GDS strategy. The GDSs provide an incentive to OTAs to book on their system. These incentives are larger than the incentives given by the GDSs to the travel management companies (TMC). The service fees are paid by the retail customers.

For the hotel line of business, an OTA generates revenue from advertising, the merchant model, the agency (pay-at-check-out (PAC)) commission model, GDS incentives, front-end commissions, back-end commissions, markup of net fares, and service fees.

For example, Tripadvisor uses a cost-per-click (CPC) model, and the advertiser pays if the online user clicks on the link that leads to the advertiser's page away from Tripadvisor, regardless of whether it results in a booking or not. For banner ads, the pricing model is based on cost per 1000 impressions (CPM). There is also a subscription-based advertising model that gives hotels an opportunity to edit the profile, make special offers, and modify other content.

With the merchant model, the OTA negotiates a net rate with the hotel, usually at steep discounts, and marks up the hotel room for sale. This is a very lucrative model for OTAs and can generate revenues of 20%–40% of the value of the booking. The merchant model may be based on consignment inventory (where the hotel retains ownership of the rooms until sold) or risk inventory. Even with risk inventory, OTAs negotiate the release of unsold rooms at specific points in time before the check-in date. The OTA also has an incremental revenue stream since payment is collected from the customer when the booking is made, and the OTA enjoys the float until the customer checks out of the hotel when the negotiated merchant rate is paid to the hotel.

The pay at check-out (PAC) model is like the brick-and-mortar travel agency model, in the sense that the customer only pays at checkout, except for the booking fee payment structure from the hotel to the GDS and the commission payment from the hotel to the travel agency. With the GDS travel agency model, the GDS receives a booking fee from the hotel for each transaction regardless of the length of stay. This is usually in the $3–$6 range per transaction. In addition, the hotel pays a commission to the travel agency, usually 10%, based on the value of each booking.

The OTAs receive a commission for each hotel booking, and it can range from 10% to 30% based on the size of the hotel and placement on the OTA hotel search results display page. On Booking.com, hotels pay higher commissions to be ranked higher on a search results page. PAC commissions are much higher than CPC ads, and the key difference is that a hotel only pays the PAC commission for a confirmed booking and not for driving traffic to the hotel page.

While the merchant model produces higher margins, Booking.com has demonstrated faster growth. The PAC model is the dominant component of Booking.com's revenues, while the merchant model is the dominant component of Expedia's revenues.

A hotelier's view of revenue management is to maximize room revenues by opening and closing availability of rates or availability of rates by length of stay. The OTA has different business drivers and is a view from the outside of a hotel's enterprise revenue management business process, looking in. It is all about intermediaries on the outside looking in to maximize their revenues based on negotiated contracts with suppliers such as airlines, hotel chains, independents, and rental car companies. Managing the screen real estate to maximize conversions of shoppers to bookers is a fundamental objective of OTAs since 85% of online shopping carts are abandoned by desktop users and over 90% by mobile users. In the world of online leisure travel bookings, travel reviews play a pivotal role in securing a booking. Millennials prefer word of mouth feedback to traditional advertising. An issue with travel reviews is that sites that collect them do not verify that the customer took a flight or stayed in a hotel that promoted the review.

After a wave of consolidations between 2010 and 2020, the major OTAs are Booking Holdings, Expedia Group, and Trip.com. Booking Holdings operates its flagship booking.com and sites such as Priceline, Kayak, Agoda, Cheapflights, Momondo, Rentalcars.com, and Open Table. Expedia Group operates sites such as Hotels.com, Vrbo, Trivago, Orbitz, Travelocity, Hotwire, Wotif, CarRentals.com, Cheaptickets, and Expedia Cruises. Egencia, a corporate booking site for Expedia, was acquired by American Express Global Business Travel (Amex GBT) in 2021. Ctrip, founded in 1999, is the dominant OTA in China. It expanded globally with the acquisition of Skyscanner in 2006 and Trip.com in 2017, which serves as its new global identity. The rebranded Ctrip has a global footprint in North America and

Europe. Trip.com is the largest OTA in China followed by Qunar and Alibaba's Fliggy.

The COVID-19 pandemic has distorted results. Prior to the pandemic in 2019, the five largest OTAs (Prieto, 2020) on a standalone basis were Booking.com ($15.07 billion), Expedia ($12.07 billion), Trip.com ($5.10 billion), Tripadvisor ($1.56 billion), and Trivago ($0.84 billion). OTA dominance also varies by region. In North America, Expedia Group, and Booking Holdings had a 92% share of the OTA market in 2019. In South America, Despegar, BestDay.com, and Price Travel are the largest OTAs. In Europe, Booking Holdings had a dominant share with 67.7% followed by Expedia Group with 12.8%. In APAC, Trip.com has a 37% share, followed by Qunar (16.5%) and Fliggy (14%),

11.2   Full Content in the Lodging Industry

During the early 2000s, hotel chains implemented online rate party agreements to protect themselves from OTAs advertising lower rates than those offered by the chains on their websites. As hotels began to offer a variety of rates based on room type, OTAs introduced rate parity clauses into their contracts with hotels.

Hotels now have legal agreements with OTAs to ensure that the same rate is available for the same type of room across all distribution channels, including the hotel's own website. This has significant implications as the cost of distribution can vary greatly depending on the channel through which the booking is made. OTAs tend to be the most expensive, while direct bookings through the hotel's website are the least expensive. The more bookings a hotel receives through an OTA, the higher its exposure to increased customer acquisition costs. OTA commissions may often increase, or net merchant rates may decline, but the room rate remains the same for competitive reasons, further reducing the hotel's earnings. Many hoteliers see rate parity as an unfair practice, but a necessary evil to participate in OTA displays and bookings. Without rate parity agreements, hotels could potentially increase rates on indirect channels during periods of high demand to compensate for the higher cost of product distribution.

OTAs require rate parity clauses in their legal agreements with hotels to prevent the hotels from offering lower rates to customers who book directly through the hotel's website. This practice of rate parity is unjust as it results in revenue loss for hotels. Furthermore, it is seen as anticompetitive because customers are unable to distinguish between offers from the hotel's website and those from OTAs.

The agreements between airlines and GDSs, known as *full content*, agreements (see Section 3.3), ensure that the content available on an airline's website is also accessible through the GDS. This includes schedules, air fares, and availability. However, with the implementation of the IATA NDC messaging standard, these full content agreements are gradually losing their significance. Airlines now have more control over the content they promote to travel agents that subscribe to a GDS. Similarly, rate parity agreements in the hotel industry could be argued to impede true competition and channel efficiency. To protect their margins, hotels often have to offer higher rates through OTAs or GDS channels compared to their direct consumer channels.

Rate parity agreements between hotels and OTAs differ based on region and country. There are two main categories of rate parity agreements known as wide rate parity and narrow rate parity.

Wide rate parity is the more stringent type of rate parity agreement. In this scenario, a hotel agrees not to offer room prices lower than those available through all distribution channels. The narrow rate parity agreement emerged as a response to regulatory intervention in Europe. Typically, this agreement permits a hotel to provide lower rates to certain OTAs but not through their own websites. Additionally, hotels can offer lower direct rates through offline channels like email and call center bookings.

Emmanuel Macron became President of France in 2017 after he held the position of Minister of the Economy. During his tenure, he strongly supported the abolition of rate parity clauses. As a pioneer in this domain, France became the first country globally to prohibit rate parity agreements between hotels and OTAs. The objective behind this move was to enable hotels to regain authority over their offers to customers. In 2015, the French National Assembly passed the "Law Macron," which effectively eradicated rate parity clauses from contracts between hotels and OTAs. This legislation marked a significant milestone as it allowed hotels to reclaim control over their pricing strategies.

In addition to France, by 2019, Austria, Italy, and Belgium prohibit all OTA rate parity clauses outright. Austria ended the pricing parity clause in 2016. Italy and Belgium followed in 2017. The two largest OTAs, Booking.com and Expedia, have reached agreements with regulators in the European Union (EU) to adopt only the narrow rate parity clauses in 2015. Australia and New Zealand followed in 2016. Smaller OTAs that are under the radar continue to enforce wide rate parity clauses across all markets, except where they are banned.

India is unique because the OTAs only require a rate parity agreement across their competitors. Despite meta search parity, hotels can offer the best rate guarantee for consumer direct bookings.

In the United States, and Latin America, rate parity has not been regulated.

11.3   Impact of Online Travel Agencies and Revenue Dilution

Over the past two decades, OTAs have been progressively increasing their portion of online leisure hotel bookings. In comparison, when it comes to air and car rental, customers tend to favor booking directly on the supplier sites, whereas OTAs dominate the hotel segment and continue to outrun supplier websites. Considering online and offline bookings, OTAs accounted for 24% of gross bookings in 2021, up from a 20% share in 2020. According to Phocuswright, a travel research firm, before the pandemic in 2019, online travel agencies' share of bookings had dropped to 49% but has now rebounded to 52% of the hotel online market in 2022. In 2012, the OTAs had a 54% share of the online hotel market, though it is expected to decline to 48% by 2025 (Ko, 2022). One percent of online gross bookings could account for $1 billion worth of business, which would translate to commissions in the $100 million to $200 million range (King, 2023).

An often discussed topic is whether OTAs are *allies* or *adversaries.* The state of the travel industry after the terrorist attacks on September 11, 2001, and content fragmentation in the lodging industry led to their exploitation by the OTAs.

When negotiating with hotels, OTAs primarily use two models: the lucrative merchant model and the pay-at-check-out model, also called the travel agency model. Priceline.com introduced the third model, known as the opaque model.

11.3.1   The Merchant Model

David Litman and Robert Diener, of Hotel Reservations Network (HRN), created the merchant model during the mid-1990s. This model was later popularized by HRN's successor companies, such as Hotels.com, Expedia, and site59.com, along with other online travel agencies after the U.S. economy was impacted by the terrorist attacks on September 11, 2001. Under this model, the online travel agency (OTA) acts as the merchant of record and negotiates a net rate with the hotel. They then sell this rate for a profit through their online platform and collect payment from the customer when they make a reservation. Payment to the merchant hotel is made only after the customer has stayed, and the OTA has been invoiced. In addition to the markup of the net rate, which can vary from 12% to 40%, the OTA also generates a secondary source of revenue from the free loan. This float improves an OTA's cash flow since payment is collected at the time of booking. Larger hotel chains have become more vocal about the merchant model and have actively worked to switch to the agency model or limit the profit potential of the merchant model.

11.3.2   The Pay-at-Check-Out (PAC) Model

With the pay-at-check-out model, the OTA behaves like a brick-and-mortar travel agency that subscribes to a GDS. The OTA acts as an agent on behalf of the merchant to list prices and amenities offered on their site and facilitates bookings. The customer only makes a payment at the end of the stay. The OTA's are compensated with a commission after the customer makes a payment at the time of checkout.

Of the two models, the merchant model is far more lucrative for OTAs than the pay-at-check-out model. There are more properties today on the PAC commission

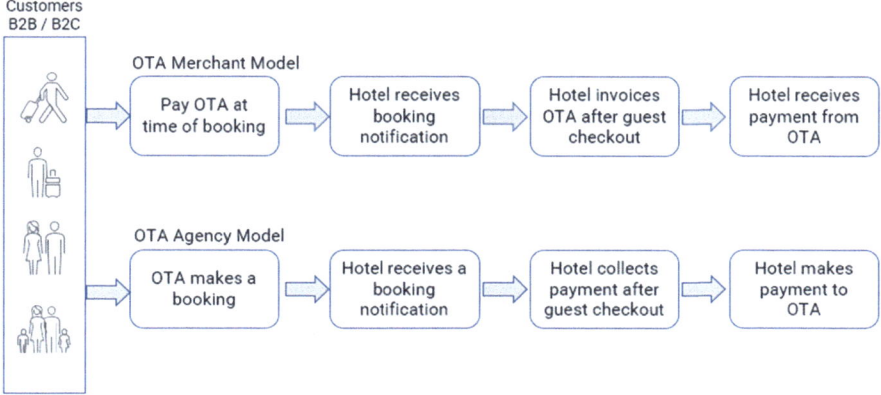

Fig. 11.1 OTA payment settlement process for merchant and agency models

model, and this model increases OTA booking rates and drives higher volumes, even though the profit per transaction is lower than the merchant model.

Figure 11.1 illustrates the differences in the OTA settlement process for the merchant and agency models.

There are third-party intermediaries for commissions management and reporting like Onyx CenterSource, Travel Agency Commission Payment (TACP) from NTT Data, and CTS Systems that enable payments of commissions from the hotel to the travel agencies. These entities can also be classified as intermediaries that provide a service to hotel chains and independents to pay commissions to travel agencies. Hotel chains facilitate the payment of travel agent commissions through a central payment system on behalf of hotels operating under the hotel brand. The hotel chain is not a guarantor of payment. The hotels are responsible for all payments.

Based on booking data received from the hotel, these commission payment intermediaries create a remittance notice to the hotel. After the hotel makes a deposit to their account, the commission payment processor makes a payment to each of the agencies that made a booking in the currency of their choice. Besides paper checks, virtual payments with virtual credit cards (VCC) are also available. The objective is to minimize delayed payments to travel agencies, thereby reducing the administrative overhead of collecting payments. Reconciliation and validation of commission payments can be accessed by the hotel and travel agencies using their portal.

Virtual credit cards are widely used to settle payments between hotels and OTAs. For example, with the merchant model, OTAs charge the customer through their standard online payment method such as credit cards or debit cards when a booking is made. The customer who makes the booking enters their CVV number, which is used for authentication but cannot be stored due to the payment card industry (PCI) data security standard and regulations. Therefore, when the booking details are sent from the OTA to the hotel, a virtual card replaces a customer's private data allowing the hotel to charge the OTA the negotiated merchant rate and taxes after the customer checks out of the property.

11.3.3   The Opaque Model

The Name Your Own Price (NYOP) model, famously introduced by priceline, was an opaque model. When Priceline.com first launched in 1989, the sole method for booking a flight on the website through a bidding process, which was sometimes referred to as a reverse auction. It was extremely popular with leisure travelers who were able to book flights up to 40% off retail prices. Using this model, customers would indicate their origin and destination and submit a fare that they were willing to pay. The identity of the airline was not disclosed to the customer until priceline accepted the bid. Priceline later extended this model to include hotels and rental cars.

In the hotel scenario, the bidding tool allowed customers to book hotel rooms up to 60% off retail rates. Priceline and the hotel would have an agreement where the OTA distributes the hotel's product at the requested price. The customer provides priceline with the desired neighborhood, hotel star level, and price they are willing to pay. Priceline receives and reviews bids from customers, accepting or declining them. Upon acceptance, priceline matches the bid with the lowest rate offered by a hotel. This approach enables priceline to profit from the price difference between the bid and the hotel's offered rate. Price-sensitive travelers without a brand preference preferred opaque sites like priceline.com, hotwire.com, and lastminute.com upon their launch.

The opaque model has undergone changes in the past two decades and has lost popularity. This led to the discontinuation of the Name Your Own Price (NYOP) feature for air travel on September 8, 2016 (Clarke, 2016) and for rental cars on March 26, 2018 (Trejos, 2018). Despite generating significant profit for Booking Holdings, the parent company of priceline.com, the bidding service for hotels was discontinued on February 25, 2020.

The bidding process is not popular in the current market as it was a few years ago. It is crucial to have firsthand knowledge of the experience to secure a booking. Priceline has replaced NYOP with Express Deals for air and car rental, which offers discounted rates without the need to bid. In the case of air travel, customers are provided with information regarding the departure and arrival airports, departure and return dates, and a list of airlines and fares. However, the exact airline, flight number, and layover times are only revealed after the booking has been made.

NYOP has been replaced with PriceBreakers and Express Deals for hotels. Pricebreakers is a platform that does not allow customers to name their own price. Instead, it groups three highly rated named properties that could be anywhere in the destination with the same star rating, offering a single, low discounted price for savings of up to 50%. The hotel being booked is unknown to the customer, but the discounted price is applied to the selected hotel. Express Deals is a slight variation where the actual neighborhood, star rating and three unnamed properties are shown on a map with the regular rates. Both Pricebreakers and Express Deals have limitations, such as promotional codes are not applicable since the price is already discounted, the quoted price excludes taxes and fees, cancellation policies vary by hotel, and refunds are unlikely without trip insurance.

11.4   Distribution Channel Effectiveness

Hotel operators require visibility into the value provided by each channel for distributing the hotel product. Direct variable costs associated with product distribution vary widely by channel which impacts the bottom line. A measure of channel effectiveness can be tracked by calculating the net yield (*Net Yield*) by channel (*c*).

$$Net\ Yield_c = \frac{Net\ Room\ Rate_c}{Average\ Rate_c} x100\%$$

and

$$Net\ Room\ Rate_c = Average\ Rate_c - Distribution\ Channel\ Costs_c$$

The higher the *Net Yield$_c$* expressed as a percentage, the more effective the distribution channel. Two measures need to be considered simultaneously to assess the effectiveness of a distribution channel. They are volume of rooms sold by the channel and the *Net Yield$_c$*.

11.5   Metasearch, Google Hotels and the OTAs

Metasearch engines operate at a level above the OTAs and GDSs. They have been growing at a rapid pace and are gaining ground on the OTAs. Examples of the leading hotel metasearch engines are Google Hotel Ads, Tripadvisor, Trivago, Kayak, Skyscanner, and WeGo. Various studies have reported that metasearch accounts for 40%–45% of global unique visitors.

The traditional metasearch model aggregates content from multiple sources such as supplier sites and OTAs to allow customers to compare airfares, hotel rates, and car rental rates. When a customer selects a hotel with the best price, the booking is completed at the source where the content originated. Metasearch engines make money when a user clicks on an offer, buys one of the offers, sponsored ad placement, and referral fees from travel insurance companies.

Metasearch allows a hotelier to compete with the larger travel agencies and deliver more direct bookings. Contrary to common belief, listing a hotel on a metasearch engine is not just about growing revenue with direct bookings, but serves as an online advertising platform to engage with an entirely new set of customers. Over the past decade (2012 to 2022), hotel metasearch has grown to generate $6 billion in advertising revenues.

Google Hotels is technically not an OTA, but it is the proverbial *elephant in the room*. It is a major competitor from the leading search engine where many travelers begin their dream and plan phase of travel. In terms of both growth in traffic and cost-per-click (CPC), Google Hotel Ads is surpassing its competitors in the meta space. By seamlessly incorporating hotel rates into Google Maps, the Google search engine has taken a commanding lead, making it the top choice among travelers. As an increasing number of vendors join the ranks of Google Hotel Ads, the company will gain even more influence in shaping booking preferences in the future. Hoteliers must fully leverage this channel to avoid a decline in direct bookings.

Google Hotels operates on a pay-for-performance business model, which technically sets it apart from other OTAs. Google's search engine is widely used by customers at the start of their travel search, which significantly changes the dynamic of bookings and OTAs worldwide. With this model, hotels and OTAs bid for the ad placement. The hotels displayed on the top of the page are the ones that receive the most clicks, but for a hotel with a limited budget, it is almost impossible to outbid OTAs that have advertising budgets that run into the billions of dollars. Booking intent varies from one metasearch engine to the next and hence the cost-per-click (CPC) plays an important role in ensuring that the channel is efficient. Experimentation is key to determine the CPC that is affordable and produces an acceptable conversion rate.

In response to COVID-19 in 2020, Google introduced a pay-per-stay (PPS) model that enables hotels to advertise directly on Google Hotel Ads and pay only for the clicks that result in bookings. This no-booking, no-fee model was established to ensure that hotels pay solely for the business they obtain, net of cancelled bookings, which increased during the pandemic's peak. PPS is also referred to as commission per stay (CPS). Through this model, hotels agree with Google on the commission to pay for a booking, which is paid net of cancellations and no-shows. This model has evolved into a one-stop search, with Tripadvisor and Google now offering instant booking and augmenting their revenue with a commission per booking model. The commission for a booking varies and can be 12% or more of the total price of the stay net of taxes. To appear higher in hotel ads booking links, the placement depends on the bid relative to competitive bids in the same auction. A bid may not be competitive enough to outrank other auction participants, resulting in a lower placement or no display at all. The commission rate must be raised for a property to increase its rank, resulting in even higher customer acquisition costs for the hotelier.

Pay-per-click (PPC) and PPS are distinct payment methods used in the hotel industry. PPC involves paying for each click, regardless of whether the customer books a room or not. PPS, on the other hand, involves payment per stay, with cancellations and no-shows factored in. While metasearch PPS is not a substitute for metasearch PPC, it does enable new properties to experiment with the platform without financial risk. Once a hotel has established a sufficient volume of bookings, it may want to consider transitioning to metasearch PPC for further growth. The cost of managing a PPS campaign is much lower than managing a PPC campaign.

Internet advertising is primarily dominated by Google and Facebook. Booking Holdings and the Expedia Group, the two biggest online travel agencies, are the largest advertisers on these platforms.

Expedia Group and Booking Holdings spent an astounding $11 billion on marketing in 2019, a pre-pandemic year. In the first quarter of 2022, during the recovery period, they spent more than $1 billion on marketing. Hotel companies are dealing with a duopoly on two fronts: high advertising costs and commissions. To stay relevant online, they must increase their marketing budgets. However, to compensate for commissions paid to OTAs and advertising costs on Google and Facebook, consumers are paying the price with higher hotel rates.

11.6   The Billboard Effect

The OTAs argue that they are not competing with hotel operators and instead claim that their relationship is a partnership that generates additional bookings. They refer to the increased visibility obtained by listing a hotel brand on an OTA as the "billboard effect" (Anderson, 2009). In the hospitality industry, the billboard effect describes a situation where online users see a hotel on an OTA website but ultimately choose to book directly through the hotel's own website. The study confirmed the long-standing claim by OTAs that they have a positive impact on generating more bookings and improving the average rate on both the OTA channel and the hotel's website and call center. This experiment was conducted in collaboration with Expedia and JHM Hotels, a national hotel company with 40 properties and 6500 rooms. JHM Hotels were alternately displayed and removed from the Expedia site. The study revealed that when the chain was displayed on the first page of hotel search results on Expedia, bookings through the JHM website increased by 7.5%–26%. Subsequent studies conducted by Anderson (2011) at InterContinental Brands produced similar results. Expedia and other OTAs have widely publicized these findings as evidence of the so-called billboard effect, where OTAs function as a search engine marketing tool for hotel websites.

Consumer behavior has undergone changes since Anderson's research. A study conducted in 2014 by Professor P.K. Kannan from the Robert H. Smith School of Business at the University of Maryland, on behalf of the American Hotel & Lodging Association's Consumer Innovation Forum, revealed that out of 50,000 online shoppers, the majority chose to book their hotel stays through intermediaries such as Expedia, Booking.com, and Tripadvisor (O'Neal, 2015). For independent hoteliers, it is crucial to focus on collaborating with the most suitable OTA (Online Travel Agency) to maximize their hotel and brand exposure and compete on an equal level with larger national and international chains. This becomes even more significant if the property caters to a niche market.

11.7   Hotel Strategy for OTA Participation

Hotel revenue analysts should periodically review their OTA relationships, both global and regional, and determine the value of doing business with each OTA. Visibility into commission costs per occupied room delivered by an OTA will influence future decisions on OTA participation.

Hotels typically experience wide variations in productivity, commissions paid, and the net room rate delivered by each OTA channel.

To understand the strategic importance of participating in an OTA, there are two metrics that should be monitored. They are OTA channel productivity and net room rate by OTA channel.

For OTA channel i

$$\text{Channel Productivity}_i \ (\%) = \frac{\text{Total Room Nights Sold}_i}{\text{Total Room Nights at Property}} \times 100$$

$$\text{Net Room Rate by Channel}_i = \text{Average Rate per Room Night}_i \\ - \text{Average Commissions per Room Night}_i$$

OTA channels that have the lowest productivity and deliver the lowest net room rate because of their higher commissions are candidates to be avoided. Participation in international and regional OTAs should be updated periodically based on channel productivity, the net room rate by channel and the target audience that the property attracts. If the audience delivered by an OTA is unique, such as demographics and point of commencement of travel, the quantitative measures may be overlooked.

Chapter 12
Customer Trust and NDC

12.1 The Importance of Trust in Travel

As the travel industry recovers from the COVID-19 pandemic, travel entities need to rebuild confidence across the value chain with suppliers to intermediaries to customers for a robust future. There are many dimensions to trust in bilateral relationships between entities in the value chain which are reviewed in this chapter. From an end consumer perspective, topics that should be addressed by suppliers and intermediaries are transparency and open communication regarding pricing, health and safety, and the handling of personal information.

12.2 The Importance of Trust to Accelerate Adoption of NDC

Ultimately, it is the customer who will dictate the success or failure of IATA NDC. What does the customer truly want? Suffice it to say that they want the right content at an acceptable price at the right time to make an informed decision on their next trip.

The long-term success of NDC depends on the extent to which travelers accept the new paradigm of continuous dynamic pricing, fragmented content, and targeted offers. This requires an understanding of trust from the point of view of an airline, a global distribution systems (GDS), and a travel agent. Since 2012, NDC has created chaos across organizations associated with travel. The crisis of trust requires a new type of leadership among executives in the travel value chain that is grounded in transparency, honesty, and effective communications both vertically and horizontally across organizations. Effective leaders are empathetic and build trust in chaotic environments that improves the long-term outlook of an organization with engaged employees.

© The Author(s), under exclusive license to Springer Nature Switzerland AG 2024   355
B. Vinod, *Mastering the Travel Intermediaries*, Management for Professionals,
https://doi.org/10.1007/978-3-031-51524-8_12

Fig. 12.1 Trust is a fundamental backbone for conducting business

So, what exactly is trust? It is a measure of confidence to engage in a transaction between business entities. Trust matters and hinges on expectations between entities in a business relationship.

It is important in all industries. Trust matters deeply in travel because a trip that is booked could be a journey into the unknown. The entire experience starting with making the booking is based on trust. Trust is also very fragile; it takes a lifetime to build and can be broken in an instant when mistakes are made. In the NDC context, it is a $n \times n$ (4 × 4 in this case) relationship between all the stakeholders: end consumers, travel agents, GDS, and the airline as shown in Fig. 12.1.

In 2019, Gartner, the travel research firm, conducted a brand trust survey among B2B and B2C customers (Ramaswami, 2020). Their results showed that 81% of customers refuse to do business with or buy from a brand that they do not trust. Further, 89% of buyers expect to disengage from a brand that breaches their trust.

The leading brands in every industry understand the importance of trust and strive to constantly meet customer expectations. From my perspective, companies like Costco and Carmax stand out because of the level of trust they have developed with customers. When you buy a product from these companies, customers have a very high degree of confidence and realize the value of an honest transaction because the customer is at the center of their universe. The same is true with Walmart and Target brick-and-mortar stores. Amazon was on a similar trajectory when they are responsible for the fulfillment. However, they are strictly an online store and many Amazon resellers have had a negative impact on customer trust. In the NDC context, with the absence of fare and schedule transparency, the idea is for an airline to build trust with customers with the promise that they do not need to shop around to get the best value for their money based on their preferences. Developing this trust is not a trivial task, cannot be accomplished overnight, and will take several years to develop (McMahon-Beattie et al., 2011).

Trust is important for the survival of airlines and intermediaries like GDSs and travel agencies. We shall now examine trust in the context of these key entities in the value chain for their survival and growth in an NDC world.

We shall now examine trust in the context of these key entities in the value chain for their survival and growth in an NDC world.

12.3   Airlines and Trust

While targeted offers to customers based on their preferences are at the heart of NDC, success of NDC adoption requires mutual trust between the buyer and the seller. Trust is at the center of effective customer relationships, whether it is the end consumer or a travel agent that controls the point of sale and is the single most important factor for the long-term sustainability of IATA NDC. It is also a fundamental challenge for an airline to convince customers when transparency regarding fares and schedules, as we know it today, has disappeared from the marketplace.

The transition to NDC has a significant impact on trust between the airline and travel agencies and their customers. American's decision to remove content from the standard EDIFACT channels in April 2023 resulted in customers paying higher fares (Davis, 2023). Besides the higher cost of air travel, it also created operational inefficiencies and negatively impacted TMCs and corporate travelers. American's actions negatively impact trust with intermediaries who unequivocally communicated that they would not be fully prepared to facilitate NDC transactions by the April deadline.

Price transparency is a fundamental requirement for airlines to build trust with customers. Authenticity and honesty are fundamental requirements to build trust in an airline brand. In an NDC world, the fundamental question is: *without brand transparency, how can an airline build trust with its customer base?* Perhaps an approach that can be adopted is social reassurance to ease customer concerns by providing visibility into flown trips and summaries of customer testimonials based on user generated content.

Ancillary revenues generated by airlines as a percentage of total revenue are significant as seen in Fig. 12.2.

Baggage, advanced seat selection, priority boarding, special baggage, change fees, Wi-Fi, pets on-board, and meals are the most widely supported air ancillaries by

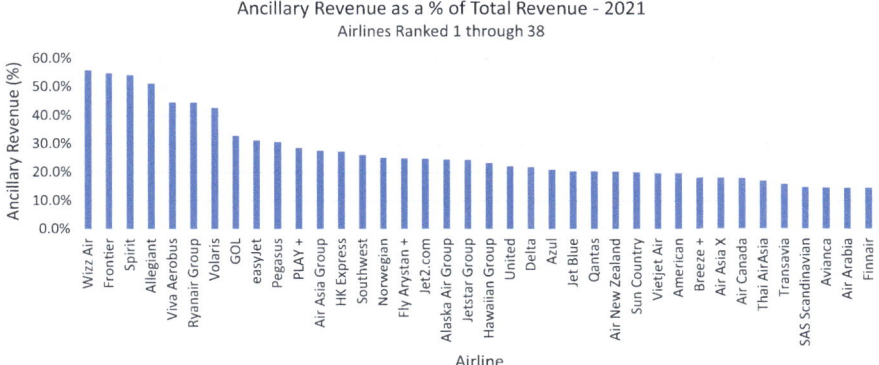

Fig. 12.2 Ancillary revenues as a percentage of total revenue. Source: 2022 Cartrawler Yearbook of Ancillary Revenue © IdeaWorksCompany.com LLC

airlines. While some airlines may only be selling a few ancillary products others boast of having several. With the large combination of ancillary products that can potentially be included in a bundle as a revenue generator, airlines need to also focus on fulfillment of these ancillaries. Management of infinite stock keeping unit (SKU) in the airline industry only makes sense if it can be fulfilled to a customer's satisfaction.

As part of the offer an airline can dynamically price the itinerary and ancillaries instead of relying on the traditional filed fares and booking classes for pricing. While the benefits of proactive pricing versus reactive pricing are well known, if the dynamic pricing offered by the airline is inconsistent with competitive market dynamics, it could result in loss of customer goodwill and lost customers. As a rule, with continuous pricing, since prices for an itinerary can be computed on the price-demand continuum, the dynamic prices should be lower than traditional filed fares with RBDs since the price jumps that exist with booking class closures can be avoided.

Another aspect of trust is full disclosure about ancillary services and fees by carriers as proposed by the DOT. Specifically:

1. Any fees associated with a first and second checked bag and a carry-on bag must be disclosed at all points of sale wherever schedule and fare information is provided to customers. Carriers are required to distribute usable, current, and accurate fee information to ticket agents that receive and distribute the carrier's fare and schedule information, including GDSs.
2. Information provided by carriers must be detailed enough to allow ticket agents to disclose fees as itinerary-specific or customer-specific charges.

12.4   GDS and Trust

In the world of NDC, a GDS has limited control over the environment since the airline controls the offer creation process—the base fare, ancillary products and services included in the bundle, and the price of the bundled offer.

For GDSs, trust is based on two factors:

1. Accuracy and availability of *all* content for a market from all airlines, regardless of whether it is traditional GDS content or NDC content.
2. The ability to rank and display nonhomogeneous airline offers on the agency desktop.

12.4.1   Comprehensive Content

The challenge for NDC content lies in the fact that are 600+ airlines in the world. Most GDSs today have 20 or fewer connections through which they aggregate NDC content. Many airlines are also not investing in NDC and their private NDC content will not be available on travel agency desktops for many years.

12.4.2   Shelf Definitions, Product Normalization, and Unbiased Display

Branded fare families were introduced by Air New Zealand, Qantas, and Air Canada in 2006. Branded fares allow airlines to differentiate their product. They are a bundling of popular ancillary products that are sold as an upsell during the sales process. Examples of ancillary products are first bag included, extra legroom, priority boarding, prereserved seats, refundability, extra loyalty miles, access to the frequent flyer lounge, etc. However, the display and sale of branded fare families on a travel agency desktop is difficult to accomplish since the content returned from individual airlines is not homogenous. The challenge to create a fair and neutral display is exacerbated by the absence of any standardization of branded fare families created by airlines. For each customer request, multiple airlines will respond through their own respective NDC gateways, and the content returned is not homogenous. To make a multisupplier experience consumable and scalable is probably the most difficult aspect of NDC for mass adoption.

From a travel agency desktop display perspective, the GDS is faced with two problems. First is the definition of the *shelves* based on a minimal set of attributes that matter to customers. Second is the task to *normalize the content* for ranking and display on the storefront on the travel agency desktop. The storefront display problem is by no means a perfect science. It is like fitting a square peg into a round hole. An airline may be satisfied, neutral, or dissatisfied with the approach taken by A GDS to classify and display an airline's offer.

A *shelf* defines the selected attributes, and their minimum levels for an airline's offer to qualify to be placed on that shelf. The shelves are ranked from the lowest set of qualifications to the highest levels. All the shelves above the lowest are upsell shelves since it entices a customer to purchase a branded fare product that is superior. For example, the lowest shelf may have a seat pitch of 32" and non-refundable fares, while the next shelf in the hierarchy may have a seat pitch of 34" and a fully refundable fare (no cancellation penalties). The GDSs work with airlines and standards organizations like IATA and ATPCO to define a nomenclature for these shelves. For example, star ratings can be used where a single star is the cheapest shelf followed with two stars, three stars, and so on that are considered more desirable by customers. An alternative approach is to define the icons for the attributes that are included in a shelf and virtual cabins.

The objective is to estimate the intrinsic value of each bundle and place them on shelves, thereby offering a level of clarity to the customer, the GDS is faced with the task of determining how to display this nonhomogeneous content returned by several airlines in a succinct way to support comparison shopping for travel agents to select and book travel. An approach is to focus on the most relevant features to include and create a minimal set to display on the travel agency desktop. To determine the value of each branded fare product returned by an airline three factors need to be considered: fare, schedule quality, and the value-added utility of the amenities included in the branded fares. Data scientists use a combination of techniques such as consumer choice models, hedonic regression, and other methods to normalize the content on a common scale for ranking in the display.

12.4.2.1   The New Airline Storefront (NAS) Controversy

The American Airlines lawsuit against Sabre serves as a prime illustration of the intricate process of managing diverse airline content for display on an agency desktop while ensuring the satisfaction of all suppliers. In April 2021, Sabre introduced the New Airline Storefront (NAS), designed to make comparison shopping for airline products easier by grouping the itineraries returned from shopping requests into virtual shelves based on similar product attributes. However, in June 2021, American Airlines sought a court order in Texas to prevent Sabre from utilizing the NAS display in GDS shopping results. American Airlines claimed that the new storefront exhibited a bias toward its competitor, Delta Air Lines (Silk, 2021; Skift, 2021).

   In addition, American asked the court to prohibit Sabre from paying higher incentives to travel agents for selling high-valued delta products than the incentives offered for selling American's high-valued products.

   American and Sabre have a tangled history. In November 2022 (Boehmer, 2022b), the Texas court that had overseen the breach-of-contract lawsuit by American dismissed the case with prejudice.

12.5   Travel Agencies and Trust

Probably, the biggest challenge faced by brick-and-mortar and online travel agencies is consumer confidence in IATA NDC. With the absence of transparency of schedules and fares in an NDC world, what can the travel agent do to convince the customer that all available options from multiple airlines have been reviewed and the best itinerary has been recommended? In the hybrid state of traditional GDS content and NDC content co-existing, this is a difficult task. It is difficult because the GDS that the travel agency subscribes to many not have connectivity to all the airlines who are publishing their NDC content. This may force TMCs and OTAs to supplement content from NDC aggregators.

   Travelers are also influenced by their experiences outside of travel. In the world of modern retailing, companies like Amazon changed consumer expectations forever. When a customer shops and makes a purchase on Amazon, they are confident that they have compared costs and brands with ease and are at peace with their actions. If airline retailing for fares and ancillaries across air suppliers falls short of these expectations, travelers notice and will ultimately lose confidence in the booking environment.

A very large percentage of corporate travel managers prioritize trust over everything else when selecting a travel provider. Hidden costs should be a thing of the past. Corporate travel managers expect transparency into the service fee structure and a product that is tailored to their needs. The importance of schedule and fare transparency into the fare products offered across airlines is more important to the corporate employee than anything else.

Chapter 13
The Impact of Emerging Technologies on Intermediaries

13.1 Emerging Technologies and Travel

The promise of artificial intelligence (AI) and machine learning (ML) in travel is to master complexity and vastly simplify the travel experience for the end consumer. Financial analysts estimate that by 2030, the AI industry will have an Internet and iPhone-scale economic impact of $15.7 trillion (Rosen, 2023).

There are several dimensions to travel complexity such as growth in air shopping volumes, growth in traffic, content fragmentation, IATA NDC, offer creation, dynamic pricing, payment systems, and many more. In an IATA NDC environment, the airline controls shopping and offer management (bundles) responses to requests from travel agencies and to end consumers through the direct channel. Therefore, it is the suppliers (airlines, hotels, rental cars cruise lines, and ferry lines) that are investing in this technology to gain a competitive edge. While AI can be used by GDSs to improve their traditional capabilities around air shopping, fraud prevention, payment processing, etc., their focus is diminished because of IATA NDC, and all eyes are now on the suppliers.

There is a second area in which the very foundations of the GDSs and OTAs are at risk of being dis-intermediated, and this is blockchain technology. It also has an impact on personalization in the future with the concept of a decentralized identifier for a customer. The scope of this chapter is limited to the potential future impacts on intermediaries with advances in technology around decentralized identifiers and blockchain.

B. Vinod, *Mastering the Travel Intermediaries*, Management for Professionals, https://doi.org/10.1007/978-3-031-51524-8_13

13.2   Overview of Artificial Intelligence

After airlines were deregulated in 1978, flight departures, filing fares, and passenger traffic all increased dramatically (Kahn, 1988a, b). Developments in decision support to enable airline planning and operations started in earnest to address the increasing complexity of running an airline. The developments were grounded in statistical modeling and analysis and utilized *operations research* approaches. The subject area of operations research solves complicated problems with many decision variables by applying mathematical methodologies. Substantial decision support advances were quickly adopted by other travel verticals, including hotels, auto rentals, cruise lines, and ferry lines, outside of the airline industry.

Throughout World War II (1939–1945), military planners faced logistical challenges. To address these, operations research (OR) was developed. Alan Turing, an English mathematician, produced his seminal paper "Computing Machinery and Intelligence" (Turing, 1950), a few years after he deciphered the Germans' "Enigma" code, which was employed as an enciphering machine to convey communications securely. The purpose of his essay was to address the query, *"Can machines think?"* This resulted in the creation of the Turing Test, which evaluated a machine's capacity for intelligent conduct on par with that of a human. Hector Levesque proposed the Winograd Schema Challenge (WSC) as an enhancement to the Turing Test (Levesque, 2011). To address the criticism that assessing a person's skill at a task alone is insufficient to gauge their intelligence, Francois Chollet proposed a new framework known as the Abstraction and Reasoning Corpus (ARC), which is based on Algorithmic Information Theory (AIT) (Chollet, 2019). The framework offers guidelines for creating a benchmark of general intelligence.

Dartmouth College math professor John McCarthy first used the term "*artificial intelligence*" in 1955. He influenced the creation of ALGOL (AlGOrithmic Language) and created LISP (LISt Processor), the second-oldest high-level programming language, after FORTRAN (FORmula TRANslation), in 1958. He is known as the "Father of Artificial Intelligence" because of his groundbreaking work in the realm of intelligent machines. Information processing language (IPL), the first artificial intelligence language, was created at the RAND Corporation with the help of economist Herbert Simon.

When Simon was employed at the Carnegie Institute in 1957, he boldly predicted that within ten years, computers would outsmart humans at chess. On February 10, 1996, IBM's Deep Blue, an artificial intelligence system created to play chess, defeated world chess champion Garry Kasparov, albeit it took nearly 40 years to achieve (Seirawan et al., 1997). When Google's Deep Mind AlphaGo defeated the greatest Go player in the world, Lee Sedol, in 2016, the news was widely reported (Gershgorn, 2016). Ross Quillian, Marvin Minsky, Edward Feigenbaum, and IBM (International Business Machines) are some of the early pioneers of AI.

AI, a field that is constantly changing, has become a common presence in the corporate world. It is no longer confined to universities and research labs but is now being embraced by various industries (Marr, 2018). Travel suppliers and intermediaries are adopting AI-based applications at a rapid pace to gain a competitive advantage related to productivity through robotic process automation and efficient interaction with customers.

AI, as a field, is extensive and encompasses the idea of machines performing tasks in an intelligent manner. It consists of three main areas: machine learning (ML), natural language processing, and deep learning. Machine learning and deep learning have direct relevance to advanced decision-making. Additionally, all three areas can be utilized throughout the travel industry's value chain.

Machine learning is a form of artificial intelligence that utilizes predictive modeling. Machines are provided with data to generate predictions based on the data. Machine learning facilitates ongoing learning and self-adjustment, resulting in improved recommendations over time. Unlike the conventional software development approach, which involves writing code for a predetermined result, machine learning learns from examples, making it fundamentally distinct.

Machine learning techniques can be grouped based on the types of problems the techniques are designed to solve.

1. *Unsupervised Learning*: Unsupervised learning uses algorithms to analyze and cluster unlabeled datasets by identifying hidden patterns in the data. It is used to solve a range of problems such as classifying incoming emails as spam or legitimate, clustering data for customer segmentation, etc. Methods include Clustering (e.g., k Means, k Medians, Fuzzy cMeans, Hierarchical), Gaussian Mixture, Hidden Markov Model, and dimensionality reduction (such as principal component analysis).

2. *Supervised Learning*: Supervised learning uses labeled datasets to train algorithms to classify data or predict outcomes. It is used for exploratory data analysis, customer segmentation, frequency of updates to an air shopping cache, image recognition, etc. Methods include Classification (e.g., Support Vector Machines, Discriminant Analysis, Naïve Bayes, K-Nearest Neighbor) and Regression (e.g., Linear, Generalized Linear Model (GLM), Decision Trees, Ensemble Methods, Neural Networks).

3. *Semi-supervised learning*: Semi-supervised learning is a bridge between supervised and unsupervised learning when availability of labeled data is limited. It is trained based on a combination of labeled and unlabeled data. It requires smaller labeled data sets to guide classification and determination of features from a larger unlabeled data set.

4. *Reinforcement learning*: Reinforcement learning is like supervised learning, but the algorithm is not trained using sample data, but learns over time by trial and error. It is used for predicting when airfares will go up or down, creation of bundles with experimentation, etc. Methods include Genetic Algorithms, Q Learning, Multiarmed Bandit test and learn, Approximate Dynamic Programming, and Markov Decision Processes.

5. *Deep Learning*: Deep learning can leverage labeled datasets, but it is not required. It can ingest unstructured data such as text and images and can automatically discern features to classify different categories of data. It is used for speech

recognition, computer vison and natural language processing (NLP). methods include multilayer neural networks, convolution neural networks (CNN), recurrent neural networks (RNN) with long short-term memory (LSTM).

13.3   Adoption of AI in Travel

AI adoption in the travel industry is still in the very early stages, unlike other industries such as automotive, life sciences, telecommunications, and education which have made significant advancements (Chui et al., 2018).

AI has the potential to serve two purposes within the travel sector: simplifying the travel experience for customers by reducing complexity and improving the operational efficiency and productivity of travel companies (Vinod, 2021c, 2023).

Commonly known applications of AI in the airline industry include the use of speech recognition, facial recognition for passenger identification at airports, chatbots for addressing basic customer queries, the utilization of biometric data to expedite passenger processes at airports, augmented reality, virtual reality, and AI-driven autonomous multi-lingual robots (BreakingTravelNews, 2019) to assist passengers in navigating airport terminals. However, AI holds further potential beyond these well-established use cases.

AI is a broad discipline that involves machines completing tasks intelligently. There are three primary domains in AI: machine learning, natural language processing, and deep learning. These domains can be used throughout the travel value chain. Machine learning is used for predictive modeling and learns from data examples to make better recommendations. It is particularly useful for processing large amounts of data and building predictive models.

Machine learning is a domain of AI which is used for predictive modeling. Machines are given access to data to make predictions based on that data. Machine learning is designed to learn continuously and recalibrate itself so that it can make better recommendations over time. Unlike traditional software development, machine learning learns from examples instead of being programmed for a specific outcome. It is used for processing large amounts of data and building predictive models.

Natural language processing is a domain of artificial intelligence that empowers computers to comprehend spoken words and text like humans. NLP is positioned at the crossroads of computational linguistics that utilizes customary rules-based modeling, statistical, machine learning, and deep learning models for understanding human language.

Understanding the travel intentions and sentiments of online customers is one of the greatest challenges in airline offer management. This information is crucial for recommendation engines to create offers that align with customers' preferences and expectations. The context for travel-based segmentation discussed in Chap. 8 examines consumer intent using input parameters. NLP takes this a step further by analyzing human language, either in voice or text, to interpret consumer intent and sentiment.

Deep learning is a type of machine learning method that utilizes neural networks with multiple layers to predict outcomes. It can be applied in various travel-related activities like forecasting demand, recognizing faces, creating conversational chatbots, and detecting fraud. This technique employs gradient descent and back propagation across the numerous hidden layers to ensure precise predictions. However, it should be noted that this process requires significant processing power (CPU).

13.3.1 *The Transition from Analog to Digital to Intelligent Digital*

Over the past two decades or more, the global community shifted from analog to digital and is now in the process of transitioning from digital to intelligent digital. This is illustrated in Fig. 13.1.

Every decision support application and software deployed in the future will be fused with an AI component that executes tasks intelligently, eliminates errors and improves the quality of recommendations over time through a process of self-learning. This will constitute one of the most profound changes in software development that enhances human productivity. In this new world, two types of companies will simplify and overcome the complexity in travel: Companies that sell AI software and robust established companies with a sound business model and a software platform. An AI-based landscape will also reduce or eliminate human manpower in the marketplace in repetitive functional areas. This will trigger retraining of human resources to adapt to the new reality.

13.4 Categories of Investment in AI

Travel industry players such as travel suppliers, GDSs, TMCs, OTAs, and tour operators have made significant investments in artificial intelligence (AI) and machine learning. To adopt AI in travel, successful organizations have taken an incremental approach with small, well-defined initiatives instead of tackling large, complex problems at once, according to Davenport and Ronanki (2018). The support of senior executives is crucial to the success and adoption of these solutions.

Fig. 13.1 Transition to the age of AI

AI applications fall into three categories based on increasing problem-solving complexity.

13.4.1 Robotic Process Automation

Robotic process automation (RPA) is a type of software technology that imitates defined human actions to interact with digital systems and software. RPA automates business processes that are structured, repetitive and can be encapsulated into rules. It is a quick and easy solution that delivers a high return on investment by enhancing productivity and eliminating human errors. Although RPA is not classified as a "smart" application, it is useful and can be adapted to multiple backend systems to improve productivity. There are numerous routine tasks that RPA can perform, including robotic shopping to aggregate competitor selling fares, processing payroll, financial portfolio planning, accounts payable, call center operations, processing orders on eCommerce sites, processing credit card applications, expense management, and human resource tasks like employee onboarding.

RPA is undergoing a transition from a programmed action tool to a smart application due to its utilization of AI and machine learning. The data processed by RPA can be utilized by AI tools to continually learn and determine what actions to take based on the processed data.

It is important to note that RPA and chatbots differ in functionality. Both use software robots, but RPA bots are meant for process automation, whereas chatbots have a conversational aspect. RPA bots lack a chat component, whereas chatbots utilize NLP to imitate human-like conversations. Chatbots were initially introduced as digital assistants that responded to simple questions in a Q&A format but have since transformed into intelligent conversational bots.

13.4.1.1 Generative AI

In November 2022, OpenAI's ChatGPT (Chat Generative Pre-Trained Transformer), an AI chatbot, made an immediate impact in the market. ChatGPT is a large language model (LLM) that has tremendous momentum and is a closely monitored field in the realm of natural language processing. LLMs have played a pivotal role in facilitating machines to comprehend, analyze, and produce human language in a manner that was previously unattainable. They have empowered various applications, ranging from machine translation and content generation to virtual assistants and chatbots. LLMs also facilitate the deployment of code generation for statistical modeling and data analytics.

The text data is processed by a neural network which consists of multiple nodes and layers. These neural networks learn continuously and refine the way they interpret data. These LLMs use a specific neural network architecture called a transformer. Prior to transformers, NLP models were trained using supervised learning for specific tasks and only worked for the data that they were trained on. Transformer-based libraries have been gaining popularity even prior to ChatGPT's emergence, which further brought LLMs into the limelight. The key difference with ChatGPT is that it is an AI chatbot that is optimized for dialogue.

Unlike traditional travel website chatbots, ChatGPT is generative, meaning it can create original content as part of its conversational response by analyzing and summarizing vast amounts of information on the Internet, including web pages, books, and blogs.

Expedia and Kayak were two of the first travel brands to launch plugins for ChatGPT.

What is a plugin? A plugin is a software component that links extensive language models, like the ones utilized by ChatGPT or Bing Chat to external data sources. This allows the plugin to obtain up-to-date data to address real-time queries from users and to help them with tasks, such as arranging travel. With plugins built into chat relevant recommendations are made based on the conversation. Many leading software vendors are adopting the same open plugin standard that OpenAI introduced for ChatGPT in March 2023, so plugins built on the standard are interoperable across multiple ecosystems.

In just a few months, artificial intelligence plugins such as those enabled in ChatGPT have become table stakes. New plugins are in development from travel brands such as Tripadvisor, Trip.com, Skyscanner, Fareportal, and Spotnana to name a few.

ChatGPT is particularly useful for trip planning, which is a core competency of travel agents. However, it is unlikely to replace travel agents but will complement the trip planning process as a productivity-enhancing capability that collects all the relevant information for a trip that a travel agent can edit and email to the customer. A question that is foremost in the minds of many travel executives is: *Will agencies continue to be the predominant way people shop and purchase travel?*

ChatGPT, like numerous statistical and machine learning predictive models, has a limitation in that its efficacy is dependent on the data it has processed. Consequently, it will always trail the market. At the time of ChatGPT's launch, the data it had processed only extended to 2021. For instance, if a volcanic ash eruption occurred in Iceland today, ChatGPT would still propose trips to Iceland. However, this problem is partly resolved for ChatGPT users who can opt for "Browse with Bing" within the chat platform to receive updated information from across the web.

13.4.2 Cognitive Insight

The capacity to discover patterns and concealed characteristics in extensive data sets and understand the implications of the findings is referred to as cognitive insight. It relies heavily on data and can identify patterns and forecast results.

Examples include tools that recommend offers, anticipate consumer purchasing habits, create tailored tactics for individual customers based on purchase history, target personalized ads, detect credit card fraud, assess insurance claims, detect safety and quality issues using warranty data, and actuarial models.

13.4.3   Cognitive Engagement

Cognitive engagement is a powerful instrument that can function as an expansion of cognitive perception. Its purpose is to immerse oneself in a learning process to tackle issues. These applications are data-intensive and undergo extensive training to enhance the quality of outcomes over time. Customer engagement is the primary focus, and innovative techniques can be employed, such as scrutinizing online search outcomes on a travel site to identify customers seeking beach destinations. A newsletter featuring curated beach destinations that match the customer profile can be dispatched to them via email without human intervention.

Examples of cognitive engagement involve the utilization of chatbots and intelligent agents to engage with employees and customers. Predictive analytics is utilized to process customer data and acquire insights into future customer behavior, which is then utilized to personalize customer engagement. Airline intelligent retailing depends on cognitive engagement to automate and create relevant customized offers based on context.

13.5   Future Outlook for Personalization

Personalization in the travel industry or any industry for that matter, begins with knowing the customer. In travel, every customer has a profile, but the customer may have different personas based on the context for travel. Knowing a customer's identity along with the context for travel enables personalization at a segment level of for a segment of ONE.

Most online users typically remain anonymous when they begin searching for an airline or hotel, only disclosing their identities once the search has led to a booking. To guarantee the relevance of the supplier products that are offered, recommendation engines must target both anonymous guests and stated (declared) guests. The context of the journey can be used to segment anonymous visitors and offer recommendations will be tailored to the personas revealed by the trip-purpose segment. However, when a customer is identified, sales history and individual preferences from the customer profile can be considered to ensure that the offer recommendations are precise and accurate. The importance of a traveler's personal identity for personalized recommendations cannot be understated.

Personal identity has been going through an evolution, and it is important to understand the future end state, which has far-reaching ramifications on travel intermediaries and the relationship between a customer and a supplier.

13.5.1 Components of Personalization and Frictionless Travel for the Future

In the current environment, suppliers, OTAs, and GDSs have fundamental gaps in addressing intelligent retailing, personalization, and frictionless travel from a customer point of view. The core components that need to come together with seamless inter-operability are:

1. Decentralized Digital Identity. Distributed and decentralized storage will allow authentication credentials to be verified independently by every participant without mutual trust between participants. This allows a user to control their own identity and manage their own risk.
2. Decentralized Universal Profile. A user owns the data on the universal profile and communicates directly and shares parts of the information on the profile with a travel/non-travel entity to provide the requested service for a seamless experience.
3. Universal Data Exchange. The data owner authorizes the exchange of data through a broker that securely exchanges data between travel and non-travel partners. The data are never centralized and stored but relevant data are exchanged in real time.

As an alternative to the Universal Data Exchange, the Digital Identity Foundation (DIF), and the World Wide Web Consortium (W3C) are promoting the concept of an Identity Hub (IH) to store profile data, sensitive documents and offers securely from suppliers that is controlled by the customer.

13.5.2 Evolution of Personal Identity

Throughout the modern computing era, the concept of personal identity has progressed from centralized identity to federated identity to decentralized identity. This progression is applicable to all websites that a customer may visit, rather than being specific to travel websites.

13.5.2.1 Centralized Identity

Using a username and password to enter a website is the hallmark of centralized identity. However, customers often have to access several websites, each with varying password requirements, leading to a proliferation of passwords. This proliferation poses a significant security risk if passwords are reused and compromised on one site, potentially exposing the user's other accounts. While password manager software offers the ability to manage multiple passwords, it does not directly address the underlying security issue.

13.5.2.2   Federated Identity

Federated identity management (FIM) emerged due to the widespread adoption of the web and software as a service (SaaS), coupled with the pressing issue of password management. FIM solves the problem of password proliferation by allowing users to use a single set of login credentials, called a single sign-on (SSO), to access multiple websites. In this system, a trusted identity provider (IDP), such as Google or Facebook, facilitates authentication for other websites known as service providers. The IDP provides login information in a token that is digitally signed to verify its authenticity. The service provider relies on this token and has a trust relationship with the IDP, hence the term "federation". However, a major drawback of this system is that it leaves out the customer, leading to concerns about privacy and data protection.

SSO and FIM are two separate concepts. Although SSO is part of FIM, it is specifically intended to verify one set of login credentials across several systems in a single organization. FIM, on the other hand, provides a unified and secure method for accessing multiple domains or organizations with the same set of credentials.

13.5.2.3   Decentralized Identity

By placing the customer in charge of their own identity, decentralized identity resolves the problems associated with federated identification. The customer determines the sharing and accessibility of their personal information through trusted interactions, all while maintaining privacy. This is accomplished by transferring the responsibility of identity information from websites to the customer. The decentralized identity ecosystem consists of three key components: the digital wallet, ledger, and credentials.

On their mobile device, the customer utilizes an identity wallet to gather verified information from certified issuers like the government, employer, and university. The customer has complete control over what information they choose to share with a third party from their digital wallet, guaranteeing their privacy.

The wallet can be installed on various devices used by the customer, ensuring that the data is always synchronized and easily accessible. In essence, the identity wallet serves as a manager for private keys used in the decentralized identity ecosystem, replacing the traditional password manager. The customer's private key is securely stored within their wallet, while their public key is published on a blockchain-based distributed ledger. The identity wallet is utilized to identify the customer within the decentralized identity ecosystem by generating a unique identifier, which is based on public key cryptography and establishes ownership of the identifier.

13.5.3   Customer Identity and Personalization

A digital identity is a collection of rich data that represents a unique entity such as a customer, an organization, an application, or device. It needs to be represented by a digital identifier to identify the subject and to be trusted by the receiving party.

To issue universally acceptable digital cards or credentials, we need digital identifiers that customers can own, that is, independent of any entity, organization, or institution. Today, travelers use email addresses and phone numbers as identifiers to access websites and apps. However, a customer's access to these identifiers may be revoked at any time by the service providers that issued them. Today, there are no universally acceptable standards for expressing, exchanging, and verifying digital credentials across organizational boundaries. This will change with the new digital identity based on emerging standards such as verifiable credentials and decentralized identifiers, which enables digital credentials to work anywhere, be trustworthy, and respect privacy.

Every travel supplier needs direct access to customers to establish a relationship and provide service based on a customer's unique preferences and desires. Figure 13.2 shows the ideal peer-to-peer interaction (P2P) between a travel supplier and the customer. This is what every supplier wants: to interact directly with their customers.

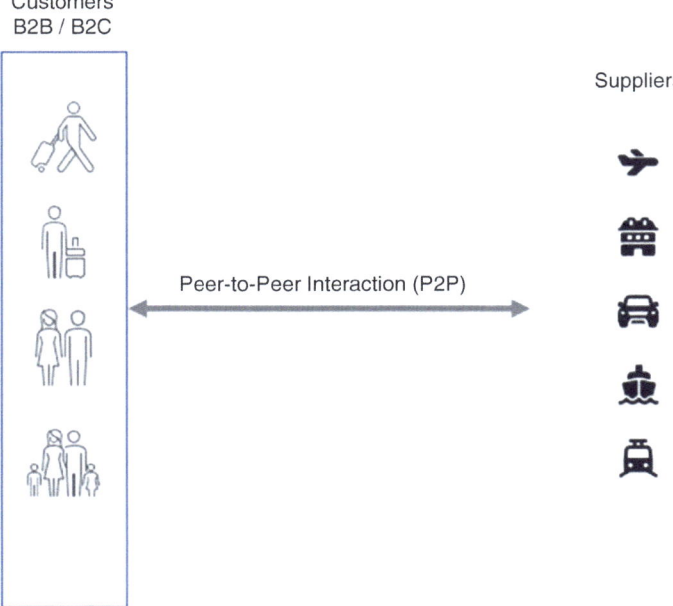

Fig. 13.2 Desired peer-to-peer interaction

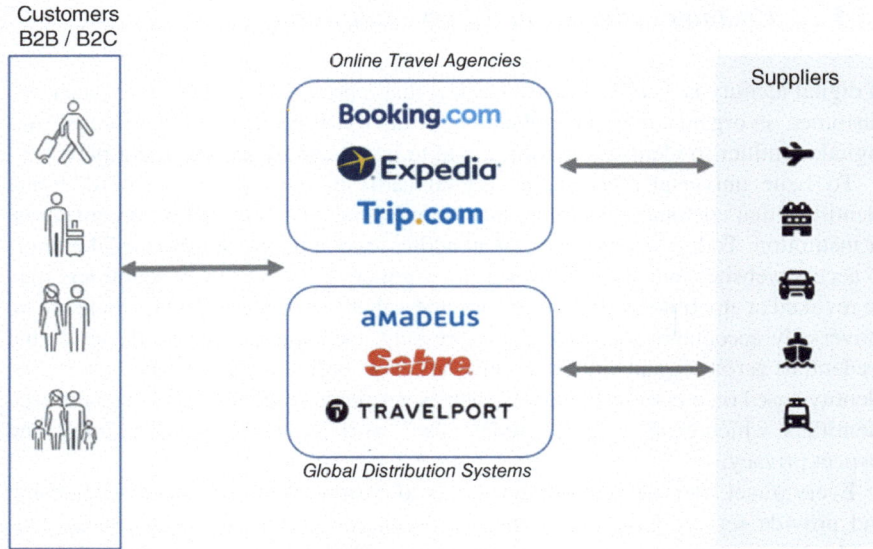

Fig. 13.3 Customers, intermediaries, and hotel suppliers

The presence of travel intermediaries presents a problem for a travel supplier. It is difficult for the supplier to have visibility into the arriving customers. An OTA stores limited information about a customer such as name, email address, address, and payment method. They serve as identity controllers. Customers see this as a convenience, and they keep coming back and the intermediary benefits by controlling the sales funnel. Figure 13.3 illustrates the two types of intermediaries that position themselves between a travel supplier and the customer. Leisure air bookings predominantly go through the OTAs while managed corporate air travel bookings and some leisure bookings go through the GDSs.

The issue of a supplier's ability to reach customers is compounded by the widespread utilization of federated identity mechanisms offered as complimentary services by social media and search companies such as Facebook and Google. This is illustrated in Fig. 13.4, leading to suppliers being even more disconnected from customers.

The consumer direct channel eliminates the intermediaries but uses the familiar federated identity models and the loyalty number. This is depicted in Fig. 13.5. Certain corporate clients (B2B) bypass company travel policy regulations and make reservations directly through hotel websites. Additionally, certain properties like the Fontainebleau Miami Beach offer exclusive discounted rates to corporations for conferences and special occasions. These rates can only be accessed through the hotel website, allowing them to offer tailored upgrades to their corporate clientele.

The customer is becoming more and more hidden from the travel provider's gaze as online reservations take over the market. This paradigm is ineffective for both travel suppliers and customers who may be looking for individualized service from

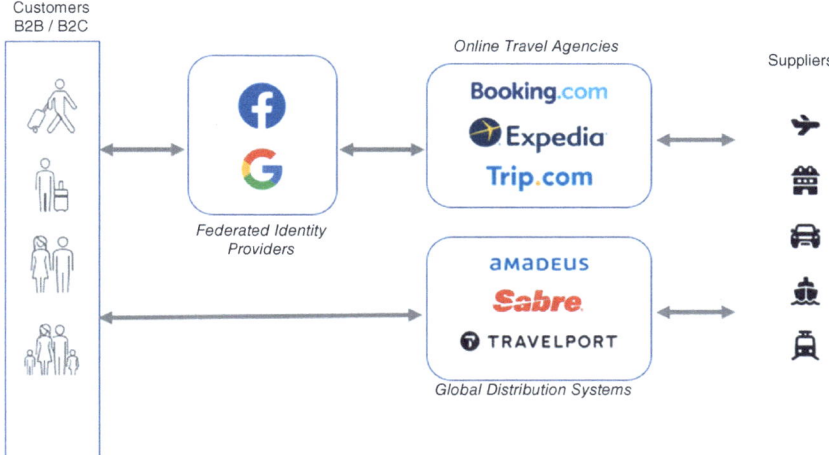

Fig. 13.4 Customers, federated identity providers, intermediaries, and hotel suppliers

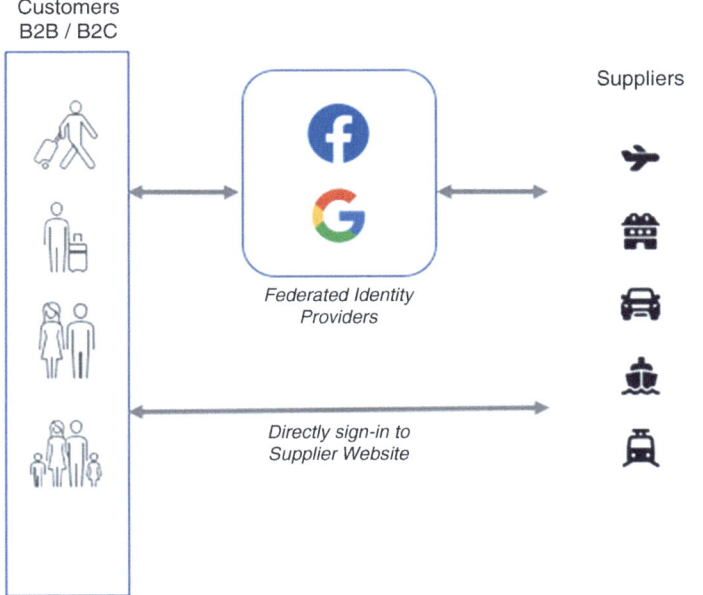

Fig. 13.5 Customer access to the direct channel

an airline or hotel, for example. To deliver the individualized service on the flight or during a hotel stay, travel providers need an identity toolkit to identify the customer. This offers an opportunity for self-sovereign digital identification leveraging decentralized technologies that are trusted, permanent, and owned by the consumer.

13.5.4   Role of the Universal Profile

Many travel suppliers, and intermediaries like OTAs and GDSs, believe that the heart of personalization is the systematic process of collecting data and storing it on the guest history database. For example, point of sale systems at a property can collect customer activity and send this data to a property management system that can serve as a repository to engage the guests. While personalization is an increasing area of focus in the travel industry, it should be balanced with individual privacy and not be intrusive. Invasion of privacy is not an option. The better option is for a customer to determine what data on their universal profile should be shared with a hotel for a better experience. In this scenario, the supplier is not in control, the customer is.

Hotels, GDSs and OTAs all believe that they "own" the customer and know the customer's preferences. This is so untrue. A customer typically has varied interests and any one supplier or intermediary will never have a 360° view of them. In fact, the only individual who knows it all is the customer. The role of the universal profile is a fundamental requirement for personalization, even within the constraints of GDPR. The customer owns the data and the customer profile and decides which entity should have selective access to parts of the data on the profile during a trip.

There is only a single, unique universal profile with travel-related and non-travel related content that is owned by the customer who owns the data stored on the profile. The individual customer determines who should have access to specific data elements in the profile for a seamless journey. Table 13.1 shows a sample universal profile that is owned by the customer. A customer's historical purchase behavior patterns, itinerary preferences, and future travel intent are critical components for travel suppliers to generate proactive offers if the customer provides access to their data.

The universal profile is stored in a private data store owned and controlled by the identity subject and referenced by the subject's identifier.

13.5.5   The Universal Data Exchange

A door-to-door seamless travel experience is talked about in travel but does not exist today. The end-to-end travel paradigm begins with a decision on what time to leave a residence to travel to the airport, the ride to the airport, going through the shortest security lines, boarding a flight, travel to the destination, renting a car, checking into a hotel, and engaging in local activities. The key to a seamless door-to-door travel experience will require the creation of a clearing house and universal data exchange between travel entities subject to fulfilling all the GDPR privacy laws. The data exchange must work with a decentralized universal profile.

P2P networks are systems that lack centralization and instead rely on direct communication and schema-less data exchange between nodes or peers. This type of architecture has gained popularity due to its efficient distribution of data and ability to handle high levels of traffic. The data exchange is a secure communications message routing platform between travel partners (e.g., brands such as United

Table 13.1 Sample universal profile owned by the customer

Name: Farah Cromwell			
Demographic	**Geographical**	**Psychographic**	**Socio-Economic**
Date of birth Gender Passport number, expiration date Marital status Parental status, number of children	Country, state, and city of residence Vacation homes: Country, state, and city	Habits Lifestyle Restaurants Values, options, attitudes Interests: Movies (genre), theatre, fine wine, scuba diving, parasailing	Education Banking, financial instruments Salary Occupation Corporation Home address, type of neighborhood
Travel preferences	**Itinerary preferences**	**Travel intent**	**Medical**
History of past trips Air: Seat type, frequent flyer, meal, etc. Hotel: Frequent stay, room type, room temperature, favorite TV channels, etc. Car: Frequent renter, car type, etc.)	Business: 4-star business hotel Leisure with spouse: Romantic hotels Leisure with family: 4-star resorts Visiting friends & relatives (VFR): Villa, boathouse	Leisure destinations Business destinations	Allergies Vaccination status

Airlines and Marriott Hotels) based on the permissions granted by the customer. Every customer has an encrypted electronic key that is used to identify the customer across travel entities such as airlines, hotels, cars, rideshare, and local activities. The data exchange should be schema-less, and not store any information including PII (personally identifiable information such as social security number, passport data, biometric data, and date of birth) data.

Figure 13.6 illustrates the concept of the universal profile and the universal data exchange to create a seamless customer experience. The customer maintains the data in the universal profile. When the customer makes a booking with an airline and the hotel independently for a future trip, the customer authorizes access to the data required to pre-populate content on the supplier websites to make a booking, avoiding duplication of data input. Further, the customer can authorize the airline to send the itinerary to the exchange and provides permission to the hotel to view the itinerary and status of the itinerary. Based on the status of the flight, the hotel can now plan for early and late arrivals and plan on additional product offers to recommend to the customer upon arrival at the hotel. All travel entities need to register with the universal data exchange to receive real-time notifications authorized by the customer.

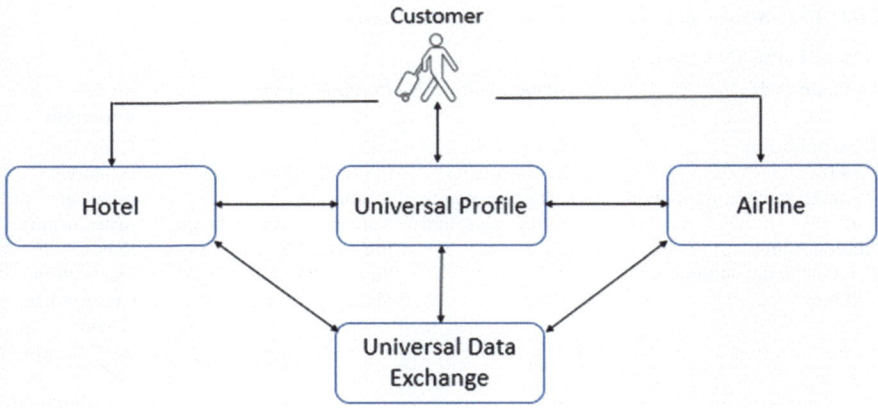

Fig. 13.6 Data sharing between travel entities

To fulfill a customer's travel experience, exchange of data authorized by the traveler between travel entities is required for frictionless travel. This concept can be easily extended to include other entities in the travel chain like rental car, rideshare, restaurants, and local activities for a true door-to-door seamless travel experience.

13.5.6   The Future of Decentralized Identity

The core architecture, data model, and representations for decentralized identifiers (DIDs) have been proposed by the W3C (W3C, 2021).

The future of decentralized identity that enables peer-to-peer commerce depends on unconstrained thinking and the availability of new technologies. These foundational technologies are a work in progress by many organizations such as DIF, W3C, Sovrin, and others and will take a few years to mature and gain mass adoption.

A shopping request issued by a digital agent that is controlled by the customer with access to information in the customer profile can be broadcast to hotel suppliers who can respond with offers. The interaction model enabled by the decentralized technologies and the trusted identities that they support is peer-to-peer.

The four principal DIF architectural components are:

13.5.6.1   Decentralized Systems

Decentralized systems are blockchain ledgers, which are decentralized and immutable. This means that once something is written to a chain, it cannot be undone. They are chronologically stored with time stamp actions against the record, and they need to be public.

13.5.6.2 The Universal Resolver

The W3C Universal Resolver serves a similar purpose to the Berkeley Internet domain name system (DNS) for the resolution of identifiers. Instead of resolving domain names the universal resolver resolves self-sovereign identifiers (DIDs) and retrieves against the identifiers information such as service end points and cryptographic keys. It enables secure, private, peer-to-peer connections between two entities such as a person or an organization. The universal resolver performs these tasks and constitutes the basic building block of a self-sovereign identity world.

13.5.6.3 Identity Hub

This is a mesh of replicated personal secure data stores controlled and owned by the customer. It has profiles, sensitive documents, and offers from suppliers. This data store is accessed by standardized APIs with permission.

Standardized APIs exist to access data from the secure data store through permission grants. Each piece of data self-contains all the metadata required for the interactions.

13.5.6.4 User Agent

It is a digital identity wallet commonly found today on mobile devices.

13.5.7 Digital Identity Adoption

Many governments worldwide support the creation of a decentralized digital identity, and some have achieved broad adoption (Domeyer et al., 2020; Price, 2021). For example, the European Union has an initiative to provide a digital id for 450 million people based on decentralized digital id technology and enabling Self-Sovereign identity (SSI). The U.S. has a digital identity research project sponsored by the Department of Homeland Security to replace the Social Security number. The German government has sponsored commercial projects across several sectors. One of these projects is a hotel check-in for corporate customers. The Indian government onboarded more than a billion people to its Aadhaar digital ID program by offering incentives to public and private sector entities for each successful registration.

Two centralized digital id initiatives that are available in the market are from Amadeus and Mastercard. Both Amadeus and Mastercard have identified digital identity as a commercial business opportunity and have publicly acknowledged their interest in migrating to a decentralized digital identity in the future. The Amadeus digital id solution is called Amadeus Traveler ID and provides a capability for travel suppliers and travel agencies to perform automatic identification and COVID-19 document verification at any online or biometric touchpoint such as check-in, boarding, and airport security. Mastercard provides a secure real time user identity verification service. They also provide a reusable digital identity service for trust providers to increase engagement with their customers.

The Digital Identity Foundation (DIF) organization has three primary work streams:

1. Engage in development of technical specifications, and emerging standards for protocols, components, and data formats, that implementers can deploy.
2. Produce reference implementations to develop open source reference implementations.
3. Coordinate and serve as the leading decentralized identity organization in the identity space.

Technologists view decentralized digital identification as a once-in-a-generation disruptive foundational technology like TCP/IP (1972–1982), HTTP and the browser (1989–1997), GSM (Global System for Mobile communication) and mobile (1993–2007) that were all brought to market by open source community development (Price, 2021).

13.5.8   Benefits of Decentralized Digital Identification

The benefits of decentralized digital identification for travel suppliers are far reaching.

Data received by a travel supplier from customers is authorized data that can be trusted for its accuracy and reliability. This authorized data, obtained with the customers' consent, ensures compliance with the regulatory requirements set forth by the General Data Protection Regulation (GDPR) and the California Consumer Protection Act of 2018 (CCPA). Additionally, it removes the necessity for a travel supplier to retain sensitive passport and credit card information.

Customers can submit their request, like a shopping query on an OTA website, without the need for an OTA. A travel supplier can then respond to the request using the information provided by the customers, such as their hotel product requirements, preferences, affiliations, and entitlements. This information is securely stored on the universal customer profile, which may be the Identity Hub. The communication channel between the travel supplier and the customer is a permissioned peer-to-peer (P2P) channel, without the involvement of a middleman. P2P also streamlines financial and administrative tasks, including refunds, exchanges, reconciliation of post-checkout data, expense reporting, and eliminates commissions and fees charged by intermediaries.

The future lies in digital decentralized identity. The global adoption of this ecosystem is crucial in the next ten years. This type of identity is a fundamental technology that paves the way for progress. In less than ten years, digital identifiers will surpass traditional forms of identification such as email addresses, phone numbers, social security numbers, and passport numbers, eventually replacing them altogether.

This innovative framework introduces a decentralized digital identity, universal profile, and universal data exchange (or Identity Hub). It is a transformative technology and business process that alters the flow of transactions in the travel industry. By enabling peer-to-peer (P2P) interaction without the need for intermediaries, it holds the potential to ultimately eliminate intermediaries such as GDSs and OTAs. An example of this is Winding Tree, which launched a decentralized travel marketplace in 2017 using an Ethereum blockchain to directly connect suppliers and buyers in a P2P environment. They have established agreements with various airlines and hotels (Chavez-Dreyfuss, 2021) to directly distribute content to customers, with a particular focus on corporate travel. The main benefit of this solution is that it reduces costs for consumers and increases profitability for travel suppliers by removing intermediaries. However, the scalability of the blockchain platform and its ability to achieve widespread adoption still need to be assessed.

The interaction between the supplier and the customer in a peer-to-peer setting creates fresh possibilities for selling hotel products, for example, that are limited on online travel agency (OTA) platforms. For instance, it allows for the sale of rooms for short corporate meetings (e.g., 4 h), the bundling of specific room numbers with their unique features, and the offering of highly personalized experiences for honeymooners that include champagne, roses, and chocolates. Additionally, the concerns regarding rate parity clauses and attribute-based room pricing, which are imposed by OTAs, are eliminated since the intermediary is no longer involved in the transaction, making it a direct peer-to-peer transaction.

13.6 Personalization and Corporate Travel

In the realm of corporate travel, personalization presents distinct challenges that differ from leisure travel. Rather than being individually organized, corporate travel is overseen by a travel management company (TMC) chosen by the corporation. The TMC receives instructions from the corporate travel manager regarding the corporate travel policy and objectives for optimizing travel expenses.

Corporate policy limits the extent of personalization in terms of what can be achieved. For personalization to become widely accepted in corporate travel, it must offer benefits to both the corporate buyer and the airline. This can be achieved through customized bundles that cater to the specific needs of corporate travelers. This is a crucial requirement to showcase the advantages of personalization and encourage corporate travelers to appreciate the benefits of being in the managed channel. In addition, corporate travel managers are faced with a dilemma with respect to bundled offers. Even if the bundle is compliant with corporate travel policy, is it competitive? How does the offer compare to what other corporations or travel management companies (TMCs) are receiving? This is an important question to answer to come to grips with the value of a managed corporate travel program.

The lodging industry encounters the same range of challenges as the airline. When dealing with business clients, personalized offers with a room attribute bundle must be compliant with corporate travel policy. Additionally, it is essential to illustrate how personalized offerings contribute to a well-managed corporate travel program.

13.7 Challenge of Interpretability

In the realm of advanced decision support for applications like offer management, flight scheduling, and fare optimization, a significant obstacle to the acceptance of machine learning techniques is the lack of interpretability. For instance, revenue management analysts often desire an understanding of how the system generates forecasts. Unlike statistical models, machine learning models are typically opaque producing an output that cannot be interpreted. It is of critical importance to grasp the reasoning behind a model's meaningful results and identify instances when they may fall short.

Machine learning models are utilized to forecast the anticipated result of a featured dataset. The loss function serves to assess the precision of the model's predictions. The accuracy increases as the cost function, an average of all loss function values, decreases. The accuracy of the model is influenced by the choice of the loss function. Loss functions can be categorized into two groups: classification and regression. Binary cross entropy loss/log loss and hinge loss are illustrations of classification loss functions. Mean squared error (MSE) and mean absolute error (MAE) are examples of regression loss functions.

Deployment of machine learning models is devoted to establishing the necessary environment and preparing the data for input into these opaque models. Once a machine learning model produces results, comprehending the expected behavior of a trained model becomes vital. Frequently, machine learning models can yield counterintuitive outcomes, which necessitates the intervention of data scientists who developed the model for debugging purposes.

Interpreting deep learning models presents an even greater challenge due to the multitude of parameters involved and how features are combined. Both academia and corporations are actively engaged in researching methods to address issues such as model bias, establishing connections between model inputs and predictions, and interpreting the model's response to gain insights. This area of research is commonly referred to as Explainable AI (XAI) (Arrieta et al., 2020).

13.8   Future of AI in Travel

Aspects of Artificial Intelligence have been a longstanding presence in the travel industry. For many years, travel suppliers have utilized operations research, advanced statistical modeling, and machine learning to their advantage. These technologies have been employed in various capacities, including demand modeling, schedule development, pricing, the creation of caching algorithms tailored to specific needs, as well as offer management. However, it is important to note that these applications represent specific solutions.

Today, we live in an AI-enabled landscape (Musser, 2019). The most significant transformation on the horizon for the travel and tourism industry lies in the increased integration of AI. This integration will become even more widespread in the coming decade, with the primary aim of simplifying the travel experience. Through stream-lined booking processes, personalized recommendations, and enhanced customer service, the industry will strive to create a more immersive travel experience for travelers.

The domain of operations research will not be substituted by AI, but rather augmented by AI to automate and enhance the value propositions (Vinod, 2021c). Numerous OR models will continue to be enhanced with artificial intelligence. AI exists at the convergence of technologies that rationalize, engage, and acquire knowledge. In a landscape empowered by AI, there exist prospects to introduce the notion of continuous learning without human intervention.

AI-powered applications such as ChatGPT hold great promise and have the potential to enhance the productivity of both travel agents and consumers when it comes to trip planning. Travel agents with a wealth of accumulated knowledge and expertise will still be required to personalize the travel plan.

Significant progress has been made in the field of AI and machine learning over the past decade. Notable advancements include speech recognition, image recognition, autonomous driving cars, robotic automation, recommendation engines, and personalization. Deep learning models have greatly improved pattern recognition capabilities, enabling them to identify patterns that traditional regression models cannot. However, these applications require a substantial amount of data for calibration. Additionally, there are various challenges that still need to be addressed, such as uncertainty, inference, decision-making, robustness, and scalability (Moritz et al., 2018). The question remains whether this trend will continue or if AI will undergo a transformative phase, becoming more widely distributed, relying on less data from multiple sources, exploring innovative solutions, and learning dynamically (Darrow, 2021).

The primary concern for corporations as they utilize AI to gain a competitive edge is how to expand its implementation throughout the entire organization (Leff & Lim, 2021). Typically, the obstacles that hinder the scaling of AI within these organizations include inadequate data architecture, underestimation of the data science lifecycle, lack of clarity in operationalizing models, insufficient involvement of business leaders, and a lack of executive sponsorship.

According to Michael I. Jordan, a professor at the University of California, Berkeley, the future of artificial intelligence lies in the marketplace (Jordan, 2018, 2019). His main point is to empower users to share their preferences (publish) and allow vendors (subscribers) to respond with offers. The current AI model operates differently, as it is a top-down push model. Jordan believes that the upcoming generation of AI will not consist of a single agent making decisions, but rather a vast interconnected network of data, agents, and decisions. From Jordan's perspective, self-driving cars can greatly benefit from a networked system. For example, if one self-driving car detects a cyclist on the right shoulder up ahead, it should notify all other self-driving cars, ensuring that the entire system is aware of the situation. The next generation of AI applications will continually interact with the environment and learn from these interactions. This will require a distributed system that meets new performance and flexibility requirements. It is a significant departure from the current focus of AI. In the future, recommendation engines for travel will need to adapt to this new reality.

Professor Jordan's proposal aligns closely with the direction in which NDC is heading. Under NDC, travel agents will transmit a message to airlines containing specific information about the market for a customer to book. Airlines, acting as subscribers, will then provide offers in response. Professor Jordan believes that Facebook and other social media platforms fail to grasp this concept and instead rely on a top-down (one-way) push model.

13.9   Blockchain in Travel

Blockchain is a disruptive emerging technology that has a wide range of applications and a unique value proposition across several industry verticals including travel (Vinod, 2020b). In various studies such as the IATA study "Future of the Airline Industry 2035" (IATA, 2018), blockchain has been identified as one of the technologies that may have a significant impact on the future of aviation.

Blockchain is a distributed network and has the potential to become a travel platform that is embraced by travelers worldwide. Central to the use of blockchain in travel to facilitate transactions between buyers and sellers is the deployment of a decentralized digital identifier.

So, what is a blockchain? A blockchain consists of a set of blocks that are linked to each other using cryptography, thereby ensuring data security, data integrity, confidentiality, authentication, and non-repudiation. Each block in a blockchain has a cryptographic hash of the previous block, a timestamp when it was created and transaction data, thereby enforcing chronological integrity. Transactions are stored in blocks with hash values and time stamps. Each block is connected to the previous one, creating a chain of blocks. When a block is created, it cannot be deleted or modified, known as immutability and data can only be modified with consensus from all parties in the blockchain. Hence, making a change to a block is an addendum to the block, making it a very secure environment. This technology makes it possible to build applications where multiple entities can execute transactions without the need for central authority. The power of blockchain lies in the internal mechanics of how it works.

A blockchain has three primary components: a ledger, cryptocurrency, and a mechanism to execute smart contracts. A smart contract is a self-executing program that automates the execution of the contract. Once the task is completed, the transactions are trackable and irreversible. The decentralized, distributed ledger allows entities to work together securely. Smart contracts are used to settle negotiated agreements among all parties automatically without an intermediary. A blockchain is ideal for storing smart contracts because security and immutability are built into the solution. All payments and settlements on the blockchain are made with a cryptocurrency. The most well-known cryptocurrency is Bitcoin, though there are over 1000 other cryptocurrencies in the market today. In addition to Bitcoin, the most widely traded are Ethereum, XRP, Litecoin, and EOS. Bitcoin is a virtual currency with no governing authority, such as a central bank. To avoid the volatility that most cryptocurrencies are famous for, some (such as True USD (TUSD), Gemini Dollar) are pegged to the U.S. dollar.

Blockchain miners are individuals possessing the necessary computer hardware and software to engage in the mining of digital currencies or the resolution of mathematical problems. A miner plays an important role in the execution of smart contracts. Consider the example when a buyer wants to buy a product or service from a seller. The transaction is inserted as a block on the blockchain which is verified by a miner to generate the final state followed by settlement. A miner solves a cryptographic puzzle with computer power by verifying that the transaction is legitimate. Miners are compensated with digital currency.

In the case of Ethereum, smart contracts are written in a programming language called Solidity and executed by the Ethereum Virtual Machine.

The distributed nodes in a blockchain operate peer-to-peer (P2P). P2P is not a new concept. It is a way for two or more computers to share files and devices such as printers without requiring a separate server computer or server software. For example, video sharing platforms like Skype operate on a P2P network. The unique aspect of blockchain is that it keeps a record of the actual transaction, including—the buyer, the seller, and the settlement—on every node. P2P has also eliminated the middleman.

13.9.1 Evolution of Blockchain

In 1994, U.S. computer scientist Nicholas Szabo proposed smart contracts. In 1998, he designed a mechanism for a decentralized digital currency called "Bit Gold". While "Bit Gold" was never productized, it was a precursor to the Bitcoin architecture, to which blockchain is commonly identified and associated. Satoshi Nakamoto is the pseudonym used by the creator or creators of Bitcoin during the financial crisis of 2008. The identity of Satoshi Nakamoto is not known. When Dorian Nakamoto

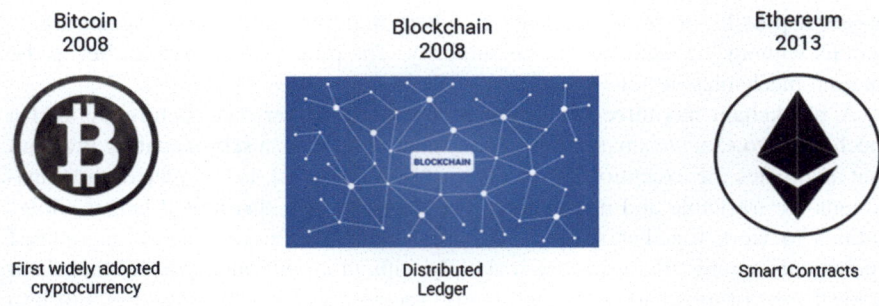

Bitcoin and Ethereum images from Pixabay
Blockchain Image by Maicon Fonseca Zanco from Pixabay

Fig. 13.7 The evolution of blockchain technology

was identified as Bitcoin's creator, he declined the claim. It is also rumored that Nicholas Szabo is the real Satoshi Nakamoto, which he has also denied.

Bitcoin has been the most successful crypto currency. Blockchain is the technology that was developed to support digital currencies, and the first decentralized blockchain was conceptualized in 2008. Ethereum was conceived in 2013 by programmer Vitalik Buterin as a decentralized open source blockchain with smart contracts. Ether is the native cryptocurrency of the platform. Figure 13.7 illustrates the evolution of blockchain.

13.9.2   Types of Blockchains

Content stored on a blockchain and its orchestration by the participants can be controlled depending on how the blockchain is configured and deployed. There are three types of blockchains in use. They are public blockchains, private blockchains, and permissioned blockchains.

13.9.2.1   Public Blockchain

A public blockchain provides general access to anyone who wishes to participate in the core activities of the network. It is also called a permissionless blockchain. Any participant can read, write, and audit the activities on the network. It is a decentralized operation that is democratized and authority-free. The Winding Tree platform is an example of a public decentralized travel ecosystem to distribute inventory for travel. Another example is Russian carrier S7 which has deployed a peer-to-peer network to issue and sell tickets using blockchain technology (Magas, 2020).

13.9.2.2   Private Blockchain

With private blockchains, as the name implies, is an invitation-only network limited
to only selected, verified participants governed by a single entity. A private
blockchain is not decentralized. It is a distributed ledger that operates as a closed
system. Only participants with permission can make transactions and validate the
changes in the blockchain. Private blockchains promote transparency since user
identities are known and promote efficiency and immutability. For example, TUI
Group which has 1600 travel agents, several airlines, over 300 hotels, and several
cruise liners that serves 20 million customers annually houses all of its contracts on a
private blockchain. If it were made public, travel agents and customers could use it to
book travel. Russian carrier S7 created a private blockchain based on the Ethereum
protocol and uses smart contracts to exchange data between entities to reduce the
settlement time between airlines and travel agents (Magas, 2020).

13.9.2.3   Permissioned Blockchain

Permissioned blockchain is a hybrid model where anyone can participate if they
have gone through the identity verification process. Permissioned blockchains have
grown in popularity because they are capable of allocating specific permissions, to
perform specific functions such as reading, accessing, or writing information, to
users on the network. Permissioned blockchains support blockchain-as-a-service
(BaaS). Travelers can benefit from BaaS when travel entities offer services such as
insurance against flight delays, accounting services, real-time financial reporting
services, which is accurate since financial data cannot be altered.

13.9.3   Peer-to-Peer and Disintermediation

"Eliminating the middleman" is a topic of interest in blockchain. In a private
blockchain with a distributed ledger the users are known, and it is controlled by a
known entity and the middleman is not eliminated. In a public blockchain, there is no
middleman, and the users are anonymous, each of whom has a copy of the ledger.
And this is what Bitcoin did with banks.

   At its core, blockchain is about decentralizing information. Blockchain does pose
a threat to transactional intermediaries that dominate the space today such as GDSs,
OTAs, and TMCs because they would no longer be needed. Like NDCs, new
XML-based messaging protocol that gave rise to new entrants called NDC
aggregators, blockchain also lowers the barrier for entry of new players in the
marketplace (Vinod, 2023). The question is: *What is the value provided by today's
leading intermediaries if a better form of technology comes along that supports peer*

to peer communications between a traveler and a supplier without a middleman, thereby eliminating commissions and booking fees?

This does not necessarily mean that this is the end of intermediaries. Consumer interfaces to interact with suppliers will still be required, and the business model will continue to evolve, and the roles of these intermediaries will be transformed over time. But to survive and serve customers, they must be willing to adapt and define a role of themselves in the new reality.

The telephone companies are a great example of adapting to changes in technology and the business landscape.

13.9.3.1   Telecommunication Industry Transformation Example

Take the telecommunication industry as an example of business and revenue model transformation because of competitive dynamics and technology changes since the Bell Telephone Company was established in 1877 by Alexander Graham Bell. Until the 1950s, AT&T was a "natural" monopoly. They owned Western Electric that manufactured and leased phones to customers. A second source of revenue was the talk time charged on landlines which had variable pricing; long distance calls within the United States were cheaper on weekends than weekdays and prices also varied during a business day. The first cracks in the monopoly occurred in 1968 when the Federal Communications Commission (FCC) granted third parties to be connected to the AT&T network. This brought about innovation with the introduction of numerous new products such as answering machines, fax machines, cordless phones, computer modems, and the early dial-up Internet.

The telephone continued to evolve from leased phones to owned phones, from telephone exchanges, landlines, and switchboards to VOIP, cell phones and cell towers, telephone leasing revenues vanished, and so did talk time revenue from AT&T land lines. Changes in technology forced telephone operators to adapt and change their business model to survive and serve their customer base with a new line of products.

It is expected that GDSs and OTAs will go through a similar change when blockchain gains large scale adoption for travel.

13.9.4   Self Sovereign Identity

With the growth in cross-border travel, an application of blockchain for safe and secure travel is the known traveler digital identity (KTDI). This is promoted by the World Economic Forum (World Economic Forum, 2020), working with Accenture, and public and private partners. This is a decentralized identity model where travelers can self-manage, control, and share their credentials selectively. The public blockchain ledger allows trusted organizations to issue credentials to individuals who in turn can present the credentials to verifying organizations on demand.

The Known Traveler Digital identity system is a joint platform developed by the World Economic Forum with Accenture to collect and host information from frequent international travelers.

13.9.5 Applications and Benefits of Blockchain in Travel

The value of blockchain in travel is in the very early stages of discovery that could evolve into new business models that increase resilience, reliability, transparency, and trust. There are several examples of the benefits of blockchain in travel. Here are a few examples.

13.9.5.1 Booking Without an Intermediary

Blockchain can be leveraged in travel for competitive advantage. Proponents of blockchain with its distributed ledger advocate the end of entrenched intermediaries such as global distribution systems (GDS), online travel agencies (OTA), and travel management companies (TMC). The value proposition is reducing the cost of distribution without intermediaries while simultaneously lowering the barrier for new entrants. A decentralized distributed ledger enables a traveler and an airline to securely work together to book a flight directly with an itinerary without the aid of an intermediary. This reduces the cost of distribution for the supplier which should in turn lower the cost of travel for the customer.

13.9.5.2 Reduce Airport Congestion

Blockchain also has the potential to reduce airport congestion. An airline can access all the documents of a traveler that is stored on a public ledger when they enter the airport. This minimizes the verification process of travelers thereby reducing long lines and potential flight delays.

13.9.5.3 Secure Payment Processing

Payments made with a decentralized ledger technology will be transparent and secure. Airlines can accept payments using cryptocurrencies that are pegged to the U.S. dollar to minimize volatility.

13.9.5.4 Decentralized Baggage Tracking

Blockchain technology has the potential to enhance the precision and transparency of mishandled baggage rate (MBR) calculations by establishing an unchangeable and distributed record of baggage handling occurrences. The immutable ledger can be created to document all baggage handling incidents such as lost, damaged, and delayed bags. Each incident can be stored in a block that is connected to the previous block, forming an unbreakable chain of blocks. This ensures the accuracy and visibility of MBR calculations.

13.9.5.5 Maximize Burn Rate of Loyalty Programs

In 2018, Singapore Airlines announced the launch of its blockchain-based airline loyalty digital wallet app for its KrisFlyer loyalty program (Shayon, 2018). This reward system, KrisPay, was a miles-based digital wallet of the frequent flyer program. Loyalty customers could convert their miles into KrisPay tokens, which could then be spent with multiple retailers.

In October 2020, Singapore Airlines extended its blockchain-based reward digital outlet KrisPay with Kris+, with 150 partners through 650 outlets to provide exclusive deals for frequent flyers. With Kris+ travelers can accrue miles from daily purchases, and the miles could be used as a payment vehicle (Ledger Insights, 2020).

The role of blockchain was to onboard new partners seamlessly and reconcile payments, accelerating productivity and efficiency. The added benefit of blockchain technology has been attributed to flexibility in redemption options for customers who can redeem the accrued digital tokens at several retail outlets. Thus, there is no depreciation of accrued miles over time since the digital tokens can be exchanged at will for other cryptocurrencies. This reduces the displacement of higher valued passengers closer to flight departure since customers may redeem the cryptocurrency for nontravel related purchases. For airlines, liability on the books for accrued miles is reduced, which is important in 10 K filings (IATA, 2018).

Critical to the success of blockchain-based loyalty programs is the ability of the airline to sign up retailers and credit card companies quickly who will accept the cryptocurrency as a form of payment when a customer wants to make a purchase. Added benefits are the instant accrual of mileage equivalent cryptocurrency when the travel segment is completed and the inherent security benefits that reduce fraud.

The use of blockchain with digital tokens also intrinsically increases the value of loyalty coalition programs by an order of magnitude.

13.9.5.6 Interline Ticketing

Airlines that use origin and destination (O&D) revenue management apply varying degrees of sophistication to determine availability for interline itineraries (Vinod,

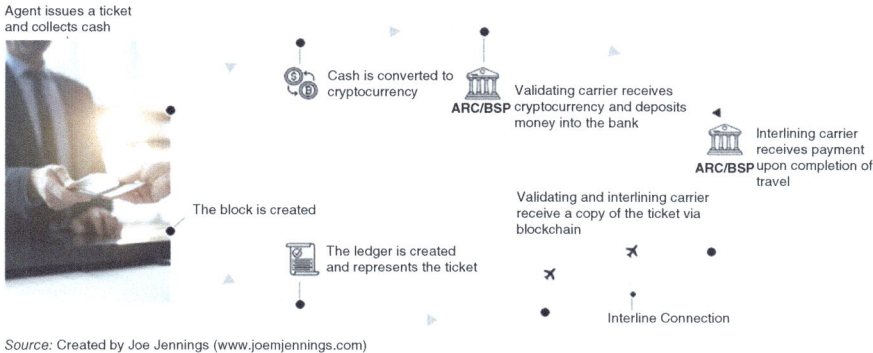

Agent issues a ticket and collects cash

Cash is converted to cryptocurrency

Validating carrier receives ARC/BSP cryptocurrency and deposits money into the bank

Interlining carrier receives payment ARC/BSP upon completion of travel

The block is created

Validating and interlining carrier receive a copy of the ticket via blockchain

The ledger is created and represents the ticket

Interline Connection

Source: Created by Joe Jennings (www.joemjennings.com)

Fig. 13.8 Interline ticketing using blockchain without an intermediary

1999, 2005b), which accounts for approximately 10% of the total traffic for network carriers. Dynamic interline proration is typically applied to determine the airline's revenue share of the total itinerary prior to determining booking class availability. The simpler approach is to prorate the itinerary based on the IATA standard cost weighted mileage factors (CWMF), and the more sophisticated proration method is based on the special prorate (bilateral) agreements, known as SPAs. After the segments have been flown, revenue accounting prorate engines are used to determine the revenue share for each airline, and the settlement is handled by intermediaries such as Airline Reporting Company (ARC) in North America and IATA Billing and Settlement Plan (BSP) worldwide.

When airlines migrate to the new IATA new distribution capability (NDC), the settlement will be handled directly by the airlines involved in the interline itinerary without an intermediary like ARC or BSP. The workflow for interline ticketing using blockchain without an intermediary is shown in Fig. 13.8.

13.9.5.7   Airline/Agency Contracts

In the realm of airline commissions, there exist two categories of contracts: front-end commission contracts and back-end commission contracts. Front-end commissions are remunerations provided by airlines to travel agents for the sale of tickets within designated booking classes (also known as reservations booking designators or RBDs) in specific markets or on certain routes. These commissions may take the form of either a flat dollar amount or a percentage of the ticket's value, with a maximum payout per ticket. On the other hand, back-end commissions are commonly referred to as override commissions. They are a type of performance-based compensation that airlines offer to travel agents for meeting specific targets stated in the contract. For instance, airlines may stipulate performance criteria for a particular market or market entity (a group of markets), such as a defined percentage increase in market share or bookings surpassing a certain threshold. Override commissions

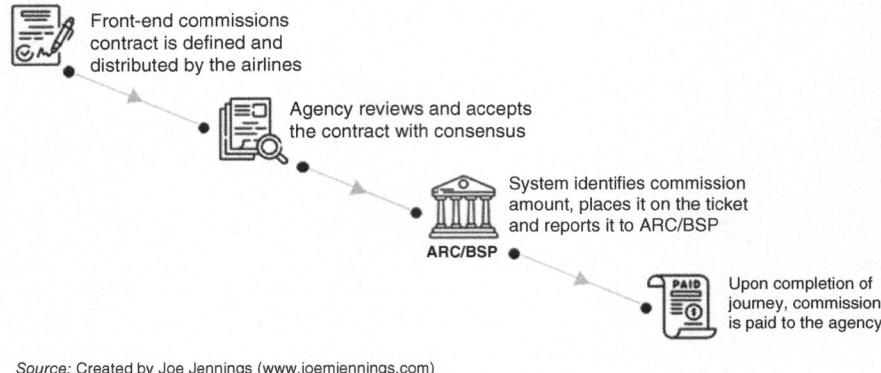

Front-end commissions
contract is defined and
distributed by the airlines

Agency reviews and accepts
the contract with consensus

System identifies commission
amount, places it on the ticket
and reports it to ARC/BSP

ARC/BSP

Upon completion of
journey, commission
is paid to the agency

Source: Created by Joe Jennings (www.joemjennings.com)

Fig. 13.9 Front-end commissions with blockchain

are disbursed as a lump sum payment by the airline only if the stipulated performance criteria are met according to the terms of the contract.

Blockchain offers a solution to the complexity of commission contracts in the
airline industry. These contracts involve various elements, including schedules,
fares, commission structures, and ticketing instructions. By utilizing a structured
data model, a digital contract can be created and represented on the blockchain. This
data model can then be used to execute smart contracts that calculate commissions
through a rules engine. Using blockchain's distributed model, contracts can easily be
distributed among parties. The benefits of using blockchain for airline/agency
commission contracts are significant. Traditional paper contracts can result in inefficiencies during the approval, implementation, and execution phases. These contracts typically remain effective for a quarter to a year, resulting in periodic renewals.
However, with a private blockchain, all parties can access a secure, digital version of
the contract, streamlining the process of automating approvals and payments. Plus,
since all changes are tracked, blockchain is a trusted source for all parties. Moreover,
contract addendums can be made near-real time, eliminating the legacy periodic
cycle of renewals. Blockchain also automates the secure payment process, making it
easier for parties to transact. Figure 13.9 outlines the workflow for front-end
commissions using blockchain.

13.9.5.8 Product Distribution with Winding Tree

Founded in Switzerland in 2017, Winding Tree is a nonprofit organization that uses
blockchain technology to connect travelers to suppliers and eliminate the intermediary in the process to lower the cost of product distribution. Winding Tree is an
Ethereum-based decentralized travel marketplace with the LIF cryptocurrency on the
platform. Winding Tree executes smart contracts and the ERC827 token standard to
deliver potential savings to the supplier and to the end consumer. Many airlines have

signed partnerships with Winding Tree based on the promise of lower distribution costs. American Airlines, the largest airline in the world, signed a partnership agreement in 2021 to enable corporate buyers to access American's flights directly, without an intermediary (Chavez-Dreyfuss, 2021).

13.9.5.9 Payment Disputes and Reconciliation

Webjet, the Australian online travel agency, utilizes blockchain technology to streamline payments and resolve conflicts through a reconciliation service. By incorporating blockchain, a reliable and transparent platform is established for both buyers and sellers within the hotel room marketplace. This innovative solution employs smart contracts to generate an indisputable and permanent record of hotel bookings between the parties involved with the transaction (Page, 2019). Once recorded, these records are stored across numerous decentralized nodes in the network, reducing the occurrence of errors between travel partners and hotel suppliers.

13.9.6 Proof of Concept and Scalability for the Future of Blockchain

Regardless of the promise of blockchain, there are many critics that question its viability. Every node in the blockchain network possesses a copy of all the data stored within the blockchain. When a new block is added, the system must update the copy on each node to maintain a single version of the truth across all participants. As the network grows with more participants and data, the speed of the update process decreases, resulting in concerns over latency.

For example, a core concern is that blockchain technology can only support low transaction volumes. IBM's Hyperledger boasts a maximum transaction rate of 3800 per second, which is well below what is required for the likes of GDSs and OTAs. There have been experimental studies on how a permissioned blockchain like Hyperledger Fabric can be scaled to higher volumes (Gorenflo et al., 2019) with a focus on I/O, caching, parallelism, and efficient data access. While promising, these results are preliminary.

In travel, a key consideration is the characteristic of the transactions. These include:

1. Transaction Path Length Complexity. A path length indicates the number of interactions required to achieve a conversion. It defines the sequence of customer interactions from the moment the booking process is initiated until the purchase is made. In blockchain terminology, it is also defined as the logic of the smart contract.

2. Data Access Patterns. Are the patterns of reads and writes consistent with production use? Further, do the transactions model the data dependencies?
3. Transaction Size. The size of the transactions has implications on elapsed times across the network. For example, there is a vast difference between a shopping transaction and a booking transaction.

13.10   Maturity of Blockchain in Travel

The travel industry is currently witnessing the early stages of blockchain adoption. It is expected that blockchain will gradually gain acceptance within travel, as both established and new use cases become a reality. The success of blockchain in travel ultimately depends on the customer experience and confidence in secure data sharing and traceability of transactions. If blockchain can enhance the travel experiences and security, its adoption rate will accelerate. Blockchain technology is disruptive and has the potential to revolutionize travel. Over the next decade, several hurdles in the areas of secure data exchange, high-volume transaction processing, airline planning, airline operations, and E-commerce must be overcome for rapid adoption of this technology. Future business processes must adapt to work with this technology. Multiple blockchains must be able to interoperate seamlessly, and regulatory constraints must be addressed to ensure mass adoption of blockchain technology.

Chapter 14
Future State

14.1 Overview

In 2019, prior to the COVID-19 pandemic, bookings for air travel through the global distribution system (GDS) were roughly the same as those made directly with the carriers. With the onset of the pandemic in 2020, corporate travel all but evaporated. In 2022, recovery in leisure travel bookings outpaced corporate bookings and the balance has not yet returned.

The question that is foremost on the minds of entities in the travel value chain is: *What is the future of travel intermediaries in the context of lowering the cost of distribution?* A related question is: *What is the future customer experience?* With rapid changes in the business and technology landscape, it is difficult to predict the precise future state. What is clear is that there are several core elements that need to come together over the next decade to be able to deliver on the promise of the future customer experience.

I will connect the dots from the previous chapters in this book to highlight a future state that begins with travel and extends to nontravel.

14.2 Future of Travel

What is foremost on the minds of industry analysts and airline executives is: *When will air traffic volumes return to 2019 levels?* The U.S. Travel Association, a nonprofit organization that represents all components of the U.S. travel industry estimates that U.S. domestic business travel will not recover to 2019 levels until 2024. In 2023, U.S. domestic corporate travel is expected to reach 76% of 2019 levels. In 2023, U.S. international corporate travel is expected to reach 72% of 2019 levels (French, 2023). A full recovery of international business travel to 2019 levels is only expected in 2025.

© The Author(s), under exclusive license to Springer Nature Switzerland AG 2024
B. Vinod, *Mastering the Travel Intermediaries*, Management for Professionals,
https://doi.org/10.1007/978-3-031-51524-8_14

While travel volumes will recover, the mix of business versus leisure traffic will be significantly different. While travel is expected to rebound, it does not necessarily translate into GDS performance.

This is not good news for airlines or intermediaries for several reasons. From an airline perspective, corporate yields are higher than leisure. Most leisure travelers book either on the airline consumer direct channel or through an online travel agency (OTA) and not a travel management company (TMC). OTA incentives are significantly higher than TMC incentives paid by the GDSs, and this will impair their performance. It is unlikely the GDSs will recover to the revenues they garnered in 2019.

Airlines will also realize that intermediaries, GDSs, TMCs, and OTAs, are still required to distribute and sell their products. With NDC, the airline focus should be on a governance structure and not bypass the GDSs.

The growing consensus among experts is that corporate travel will eventually return to pre-COVID 19 volumes, influenced by corporate travel spend and organic growth. However, as a percentage of total traffic, corporate travel will shrink in size and leisure travel will grow. This is not good news for the airline industry since corporate yields are higher than leisure. In 2020, corporate travel has become accustomed to transacting business with the latest video conferencing technologies. This will continue unabated in the years ahead as we come out of the COVID-19 pandemic.

What is the future of frictionless travel? A vision for a frictionless travel experience in the future where the savvy traveler traverses channels and brands has been articulated (Locke, 2009). Such an experience is attainable in the future.

Travel is experiential, and the question to ask is: *What has changed around the experience itself? Has technology made the experience of travel better?* (Klein, 2016).

When a customer travels, several purchases that are made do not relate directly to travel such as restaurants and destination activities. Over the next decade, the lines between travel and nontravel activities will blur because customers are strongly influenced by their purchase behaviors outside of travel and prefer a one-stop shopping experience. Offer management, the new incarnation of revenue management will create dynamic bundles that extend to other lines of business (e.g., hotel, car, etc.) and local activities at the destination. The universal profile and universal data exchange clearing house concept is not limited to travel. Instead, it holds the promise to promote a seamless customer experience that spans across travel and nontravel entities.

To promote a seamless customer experience across multiple travel entities, there are three components that must be inter-operable: decentralized digital identity verification from a trusted source in real time, a decentralized universal profile that is owned by the customer and stored on their mobile device or elsewhere, and a data exchange to transfer customer data to entities that the customer has authorized. It can be argued that the universal profile can be centralized or decentralized. If it is centralized, the identity verification can take place with the provider of the universal profile. However, for flexibility and security, the profile must be decentralized.

   The World Wide Web Consortium (W3C) has established standards for the decentralized identifier (DID). Digital identity solution for travel cannot be provided by a for-profit entity, travel or otherwise. It will require a vendor agnostic nonprofit organization that operates internationally. The Sovrin™ Network (Sovrin, 2018), a nonprofit company, is working with various travel entities including SITA to determine how to standardize the verification of digital signatures of credential issuers using blockchain for international travel. A public blockchain serves as a decentralized self-service registry for public keys. Every identity owner is their own self-sovereign identity provider, can issue a digitally signed credential, and any entity can verify it. All transactions have a digital signature that require a private key. Individuals (also known as identity owners) can register with the Sovrin™ blockchain ledger and are issued Sovrin™ verifiable credentials that is stored in the digital wallet. Entities that receive the credentials can authenticate the information provided.
   The customer owns the data in the universal profile and determines who has access to data elements in the profile. Data permissions will be at the atomic level, by individual data element, and access is provided to an entity at the discretion of the customer. Triggered by a customer, the universal profile will work with a data exchange clearing house to provide proactive notifications to travel and nontravel entities that subscribe for up-to-date information and are given access. Consider the following examples:

1. A universal data exchange (UDX) can be used to send advanced notification on status of checked luggage lost in transit. The hotel where the customer is staying can subscribe to the service to determine when the bags will be delivered. In the scenario, the customer authorizes the airline to provide the hotel (subscriber) access to the status of the lost bag.
2. The UDX can be used to send a message to the customer's airport limo service from the airline on the status of the arrival time of the flight for customer pick-up. In the scenario, the customer authorizes the airline to provide the limo service (subscriber) information on the precise flight arrival time at the airport.

   The same core components described for travel will be required for a seamless E-commerce experience: decentralized digital identity, universal profile, and universal data exchange. If the GDSs do not transform their current business model to an open marketplace to sell any type of content, someone else will own the consolidated order. The question is: *Who will own the order management system with an offer store that provides a single view of travel and nontravel purchases, like a bill of material, to a customer, which they can access on demand?*

14.3   "Coalition of the Willing" Is a Prerequisite for the Success of NDC

NDC was launched in 2012 promising enhanced traveler experiences rather than cost savings or disintermediation. Today, IATA NDC is also referred to as IATA Airline Retailing. However, to date, customers have seen minimal benefits as the focus has primarily been on restricting content from channels that the airlines do not like, resulting in higher fares paid by customers. Meeting and surpassing customer expectations is fundamental to the success of any initiative, including NDC, and we are still a long way from achieving this goal. A collaborative effort is necessary for NDC to succeed.

A *coalition of the willing* consisting of airlines, intermediaries, and NDC aggregators is required to address the dysfunctional nature of interactions between these entities to make NDC a success story. The term *coalition of the willing* typically refers to an international alliance that is focused on achieving a particular political or military objective. In the context of NDC, admittedly it is not a group of airlines imposing their will on intermediaries and NDC aggregators but finding the right technology partners and a governance model to achieve a common goal, which is to achieve shopping, booking, and post booking automation with NDC centric workflows at scale. Without scale, the promise of NDC to open the travel marketplace to a new breed of airline offers with rich content, and an abundance of ancillaries and differentiation cannot be fulfilled.

This urgency comes at a time when the economic outlook after the pandemic is uncertain, and many airlines have either stopped investing or have a limited investment in NDC and want to focus on flying airplanes. When the economy is robust, investment in distribution is discussed in airline board rooms, and otherwise it is usually an afterthought. Ultimately, in a two-sided marketplace, a technological and commercial solution is required that is not unilateral, but collaborative and addresses the demands of all stake holders.

14.4   Airline Offer Management

As revenue management transitions to offer management, offer creation will continue to play a key role in a post-COVID-19 world. The core components that will continue to evolve as an integrated workflow are customer segmentation, air shopping, itinerary pricing, offer management, and inventory control. With NDC adoption and the demise of full content agreements between airlines and GDSs, airlines can provide identical content across all channels of distribution, but they can also offer different bundles and prices for different channels if they choose. A horizontal enabler for these core functions is the universal profile (UP) where customer preferences and customer intent are stored.

Innovation will be strongly influenced by IATA NDC to generate offers to customers on demand across all channels of distribution. The promise of advanced data analytics and artificial intelligence to help airlines understand their customers better should over time positively transform the travel experience with highly targeted personalized offers. This will require customers to give information to suppliers and travel management companies from their universal profile that they manage, in real time, to ensure timely notification messages for a frictionless journey.

Customer segmentation will always be at the heart of offer management. How customers are segmented and what attributes need to be considered will change as personas evolve over time. The intent of segmentation is to understand the customer's context for travel and then determine how to price the product being offered and to display relevant content in an online setting. The so-called classless revenue management will be reality when the airline controls the dissemination of all offers to customers through their websites and through intermediaries on demand, but customer segmentation is a prerequisite to maximize revenues. Customer segmentation for air will extend to other lines of business to create dynamic packages with hotel, rental car, and local activities at the stop-over city or destination.

Customer travel begins with the dream and plan phase with air shopping to search for flights. While low fare efficacy will continue to be important as a reference fare, customer segmentation identifies the dominant shopping parameters (e.g., departure time window, return time window, elapsed time, etc.) that should be used by the shopping algorithm to return relevant itineraries with the *best fares* for customers based on their preferences. Customer segmentation used for the generation of the air ancillary bundle will also be leveraged to display itineraries during shopping consistent with a customer's preferences. Customer segmentation will also influence dynamic pricing of the base fare and air ancillaries and the composition of the targeted offers that are sent to customers. Boutique vendors and GDSs that provide enterprise IT solutions to airlines are working on the plumbing and the algorithms to sell or license this technology to remain relevant in the future.

In the absence of booking classes in the future, trip-purpose segmentation, defined by each airline individually depending on the customer base, will drive *demand forecasting* and *optimization of inventory controls* to create dynamically priced offers to individual customers. Airlines will manage the *recommendation engines* that generate offers on demand based on customer personas. Air ancillary products such as seat counts by seat type and section of the aircraft that are inventory controlled should be managed by the host CRS inventory system. Most of the innovation in airline offers will be directed toward leisure travel. Airline offers for corporate customers will be limited in scope, constrained by corporate travel policy. Besides the leisure agencies, corporate booking tools and large TMCs that service the corporate channel such as Amex GBT, BCD Travel, and CWT will have to adapt to a world with bundled offers.

Beyond customer segmentation, *personalization* of the offer with the *best fare* and air ancillaries for a segment of ONE based on individual preferences will be central to attract and retain profitable customers who make several trips a year. Determining the best fare based on customer preferences will continue to be an active area of research.

When NDC matures, all offers, base fares or base fares with ancillary bundles, will always be generated by the airline on demand. Rules-based offer creation systems will decline in favor of data science driven offer creation. Travel agencies that subscribe to a GDS and airline consumer direct websites can request schedules, fares, and air bundles which airlines will respond to in real time. This is like how hotel and car shopping, booking, and pricing works today through the agency channel. For example, GDSs do not price and calculate room rates and taxes for hotel content but rely on the hotel to provide this information on demand.

The benefits of ONE Order are process centric and not revenue centric. It is the modernization of the passenger name record (PNR) management process across the sale of the base fares and air ancillary product and services with the creation of a single integrated customer record for the airline industry. It is an XML-based standard that streamlines fulfillment, delivery, and accounting processes and phases out the traditional PNR bookings and ticketing records such as E-Tickets and electronic miscellaneous documents (EMD). The phaseout of traditional PNR bookings is contingent on the dependency of traditional PNR reservations processing on the host CRS.

14.5   Innovations in B2B and B2C Payment Processing Technology

Innovation in the B2C space lead by the Asian market has outpaced advances in B2B payment processing.

According to a recent research report published by the Economist and WEX, the business payments company (Economist, 2021), paper checks still account for more than one half of the $25 trillion in annual U.S. B2B payments. Travel is no exception. Digital solutions for B2B payments represent the future. For over a decade, travel agencies have been migrating to virtual credit cards (VCC), also known as virtual card numbers (VCN), to pay suppliers like hotels. These 16-digit VCCs are generated to allow for one-time purchases and give management the ability to have them available only at set amounts. Virtual cards have unique benefits such as secure authentication to reduce fraud, minimize the risk of default, are cheaper than paper checks, protect cashflow by preventing runaway spending by setting limits on transaction time frames and amounts, reducing, or eliminating cross-border fees, simplify back-office reconciliation, and improving the payment experience for travel agencies and travel suppliers.

To realize cost savings with virtual cards, travel suppliers and intermediaries must collaborate with a payment processing intermediary that supports multiple card schemes and access to multiple banking partners. This is required to minimize the cost of processing payments depending on the region of the world where the transaction is executed.

On the B2C front, global payment processing technology has been rapidly evolving. There have been several recent innovations in payment technology led by the Asian market to simplify the customer experience with payments.

The traditional forms of payment include credit cards (American Express, MasterCard, Visa), corporate cards, and later PayPal. There has been an explosion of new forms of payment such as Apple Pay, Google Pay, and Alipay to name a few. Intermediaries that control the point of sale realize the importance of supporting alternate forms of payment to stay competitive and minimize abandoned shopping carts. Mobile payment is gaining wide acceptance in various regions of the world and represents the future. For travelers, digital payments also support the new generation mobile wallets such as China's WeChat Pay and India's Paym. Giving instant credit to customers to pay back in installments is an evolving trend.

14.6 Future of NDC Aggregators

NDC aggregators came into existence shortly after the first NDC messaging standards were published. In December 2021, the IATA registry contained over forty certified NDC aggregators worldwide and the list keeps growing. It is unlikely that there is a future for stand-alone NDC aggregators. It is not a sustainable model for the long-term.

In a few years' time, it is unlikely there will be more than a handful of NDC aggregators due to scale economics, and it is unlikely that they will be independent entities. Their biggest competition is GDS vendors that also have reservation hosting services for airlines like Amadeus and Sabre. During the hybrid NDC era, large TMCs and OTAs will acquire NDC aggregators to ensure that they have access to NDC content from airlines that are not yet supported by the GDSs.

TMCs and OTAs have started acquiring NDC aggregators to ensure access to NDC content until the GDSs support them as well. Trip.com acquired a controlling stake in Travelfusion in January 2015. FlightCentre, one of the largest travel agencies in the world, increased its stake in NDC aggregator TP Connects to 70% in March 2022.

14.6.1 GDS Margins

GDS booking fees vary by point of commencement (POC) of travel. North America is the lowest, followed by APAC and EMEA. These imbalances in GDS booking fees have existed since the late 1980s. GDS margins represent the difference between the segment booking fees received from airlines and the incentives paid to travel agencies. Corporate bookings, the size of the airline, and the size of the travel agency influence GDS margins. Larger airlines require access to corporate customers that produce higher yields than leisure customers. While the GDSs control access to the TMCs, the TMCs control access to the corporate market. Smaller airlines tend to pay higher segment booking fees, leading to higher GDS margins for bookings involving smaller airlines. In addition, larger travel agencies command higher incentives from the GDS due to their higher volume performance. GDS margins for larger airlines also vary based on the type of booking, classified as "home markets" and "away markets".

GDSs lag behind NDC aggregators in airline NDC connections. As NDC aggregators attempt to capture a share of corporate travel bookings from the GDSs and NDC adoption increases, this dynamic will change in the future. NDC aggregators charge TMCs a booking fee that ranges between $0.75 to $1.50 per booking regardless of the number of segments (not per segment like the GDS model).

New contracts between airlines and GDSs have a two-tier revenue model specified as EDIFACT bookings and NDC bookings. The booking fees associated with NDC bookings are a fraction of the standard GDS EDIFACT booking fees. Further, there are no incentives paid to travel agencies for NDC bookings by the GDSs, who may also charge a fee for NDC content like NDC aggregators.

14.7   Future of Hotel Distribution

The hotel transaction flow model between the GDS agency channel and a hotel CRS is similar to the IATA NDC model. Pricing power has always been with the hotels. For example, when a travel agent that subscribes to a GDS submits a request for room rates with a GDS entry, the transaction is not processed internally by the GDS. Instead, the transaction is sent to the hotel's CRS which returns available selling rates and applicable taxes.

OTA distribution does not operate in this manner. The use of a channel manager is necessary to transmit availability, rates, and inventory (ARI) to different OTAs. The role of the channel manager, hotel CRS, hotel PMS, and hotel revenue management to manage the various points of sale is illustrated in Fig. 14.1 (Vinod, 2022a).

ARI refers to the transmission of data from a hotel's CRS (and sometimes the hotel property management system (PMS)) to intermediaries like OTAs, which then display and facilitate the booking of hotel rooms. ARI is also shared with Google, and the Hotel Finder app allows online users to explore, compare and book their preferred hotel. It is based on the availability of rooms and the corresponding rates at a specific property, informing intermediaries about which rate plans are available for purchase. To maintain accurate room rates and availability, a hotel CRS must provide numerous ARI updates daily to various channels, adapting to changes in bookings and demand patterns. ARI utilizes a portion of the OpenTravel Alliance (OTA) XML standards for availability and pricing.

ARI is inefficient as it generates updates en masse, many of which are for hotel products that may never be sold. Additionally, it is not feasible to have ARI updates

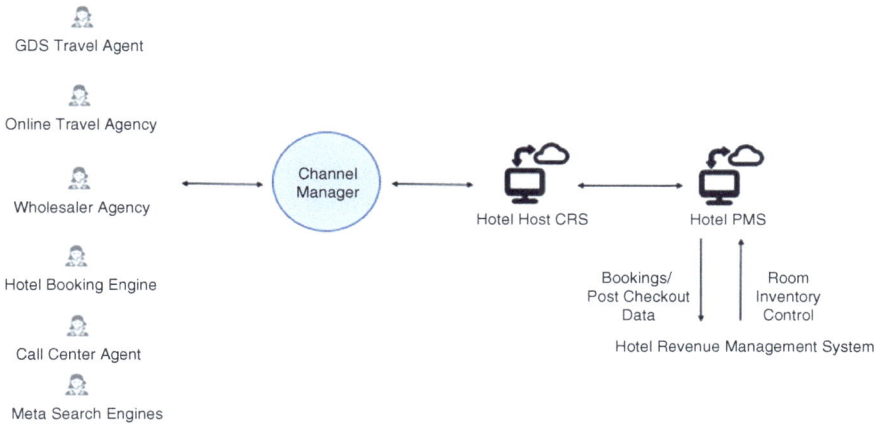

Fig. 14.1   Distribution channels, hotel CRS, PMS, and revenue management system

for every sale and cancellation. Instead, periodic updates on each channel's extranet are necessary, but this can lead to overbooking. The task of keeping all channels synchronized is a major burden and adds significant overhead. Furthermore, there has been an increase in ARI rate mapping errors for OTAs, which negatively impacts conversion rates and customer satisfaction. The speed and accuracy of ARI content is critical, with the ideal focus being on future performance rather than recent history.

The IATA NDC model for airlines is a request and response model, which is efficient compared to the ARI model used by hotels.

In contrast to the OTAs, GDSs utilize a pull method in which the hotel CRS receives the request and subsequently provides the available rates and applicable taxes. To prevent any instances of double bookings through OTA channels, a more effective approach is to request on demand room availability for each booking request from the hotel CRS in real time, like the NDC model used by airlines. However, implementing this method may overwhelm the hotel CRS with peak transaction volumes that it may not have the capacity to manage.

The question remains: *Is adopting an NDC-type standard for hotels an option to reduce the cost of distribution?*

Hotels were fast followers in adopting core concepts from airline revenue management to develop revenue management capabilities. However, the likelihood of hotels adopting the airline NDC initiative appears to be very low. The development of airline NDC has been ongoing since 2012, and achieving widespread adoption across all airlines worldwide will likely take another decade. The fragmented nature of the hotel industry further complicates this endeavor. In 2017, the Hospitality Technology Next Generation (HTNG), a global nonprofit trade association for hotel companies and technology providers, investigated the possibility of adopting NDC standards for hotels but ultimately decided against it (Sickel, 2018).

In 2023, Expedia and Booking Holdings were the dominant OTAs that command commissions between 15% and 35% of the total room rate per booking excluding taxes as the cost of customer acquisition. In contrast, GDSs charge between $4 and $8 per transaction, regardless of the length of stay. The OTA commissions are not sustainable over the long-term, from what we have seen with airline distribution over the past decade. OTAs exploit hotel suppliers because of content fragmentation. Hotels cannot ignore large OTAs that also function as aggregators of hotel content for GDSs, TMCs, and corporate booking tools, thereby expanding the options available for travelers. Investment in cloud-based hotel CRS and PMS and marketing to promote the direct channel is insufficient. It also contradicts corporate preferences for how corporate employees should book.

In the future, distributed ledger technology is expected to facilitate direct connections between hotels and customers, hotels and TMCs, as well as between hotels and corporates. Based on price economics, this may eliminate the necessity for intermediaries such as wholesalers, aggregators, switches, and GDSs.

14.8 Future of the GDS

The GDS business came into existence in 1976 and had a great run for 46 years (1976 until 1922). That is a long time for any business that has shown sustained, predictable profitability for most of those years. But there are significant changes coming and the GDS model is under siege and faced with the threat of extinction. Let us review the state of the union, of the GDS business.

When GDSs came into existence, they focused on technology to solve the complex problem of air shopping, booking, ticketing, refunds and exchanges, codeshare, and interline bookings at scale. The revenue model was simple: the GDS received segment booking fees from the airline as the cost of customer acquisition and the GDS in turn paid a portion of the fees collected in the form of incentives to travel agents for using their system.

When the GDS was created in 1976, owning the point of sale, the travel agencies, was never on their radar, which now threatens their very existence. Owning the point of sale provides continuity for the business by passing the segment booking fee costs to the customer when airlines are in open revolt over the cost of distribution.

The airline CFOs' view is GDS costs are one of the last controllable expenses for an airline. It is not just GDS booking fees but also credit card fees and agency commissions. So that is a definite factor in coming up with lower cost alternatives to product distribution. The long-term strategy of airlines has been to advance distribution from fare and schedule-led selling to merchandising; recommending an offer of base fare and ancillaries. It is no secret that airlines want to transform themselves from suppliers of a commodity, an airline seat, to product marketers of airline bundles. From their point of view, the GDS is not the vehicle to do it and they want to control the offer creation process of what is recommended to a customer through the indirect channel and all channels of distribution. Over time, even if GDSs offer NDC content exclusively to travel agents, the GDS or the travel agency will be levied a surcharge to offset the cost of product distribution. This makes the

GDS model untenable for the future. The TMCs on the other hand can pass this cost on to the customers.

For the first time, with the IATA NDC messaging standard and the move away from EDIFACT to XML, the threat of disintermediation is real. However, adoption of NDC has been slow and major issues such as scalability remain to be addressed. Rightfully, airlines see this as a technology problem that can be overcome with time and investment. The COVID-19 pandemic also set the brakes on NDC adoption when many airlines stopped investing in NDC since they did not have the free cash flow with reduced passenger demand.

In summary, with pricing power gradually shifting to the airline to control the offer creation process, the future of the GDS model looks uncertain. In the next 5–7 years, most airlines will have a direct-connect solution to acquire NDC content and the era of full content agreements between airlines and GDSs will be over. The lowest fares offered by an airline will only be available on the airline certified direct-connect NDC channels. According to Amtrav, a corporate travel agency, American's lowest fares were in the NDC channel 55% of the time (Silk, 2024a).

In 2023, ARC reported that 12% of airline bookings processed were NDC-enabled, showing an increase from 6.7% the previous year. However, adoption among corporate travel agencies and traditional leisure agencies was still low, with OTAs responsible for 95% of NDC-enabled transactions (Silk, 2024c).

Air consolidators depend on airlines for distressed inventory and low priced tickets. Ticket consolidators like Mondee, Picasso Travel, Downtown Travel, and Sky Bird Travel have incorporated NDC fares into their booking engines (Silk, 2024c). From an airline perspective, their product distribution costs through a GDS will continue to decline over the next decade. Airlines will also pay less front-end commissions to agencies.

14.8.1   Recovery of GDS Bookings from the Pandemic of 2020

GDS bookings and revenues will never match the revenue targets and margins achieved in 2019 before the onset of the COVID-19 pandemic. There are several factors that contribute to the decline in GDS revenues.

Business travel has been slow to rebound and will not reach 2019 levels until the third or fourth quarter of 2024 for domestic travel. In the meantime, leisure bookings have been growing at a fast pace, partly driven by the urge to travel after a 2 year hiatus. Travelers who book leisure travel either book them on the airline website or through an OTA like booking.com, Expedia and many others. Very few make a leisure booking through a TMC for domestic travel unless the TMC specializes in leisure bookings. Creative TMCs make leisure bookings for international travel to reduce the cost of travel. This has a direct impact on GDS margins since the incentives paid to OTAs are significantly higher than the incentives paid today to TMCs.

While aggregate airline bookings recovered to pre-pandemic levels in 2023, the GDS booking volumes did not. Channel shift contributes to lower GDS bookings and revenues. The presence of NDC aggregators, and corporate direct airline portals contribute to the channel shift.

Another aspects that contributes to lower revenues is that not all bookings are created equal. First, OTA leisure bookings through the GDS have very high incentives contributing to lower net GDS segment booking fees. Second, NDC bookings are far less profitable for GDSs than the traditional EDIFACT bookings.

14.8.2   Preserving Margins in a Shrinking Market

Since the GDSs do not own the point of sale, how can they fight back? GDSs have made significant investment in NDC adoption, typically about $25 million dollars a year or higher. So, how can they recoup their investments? There are two aspects for NDC bookings. First, a feature that has been gaining momentum is not pay incentives to travel agencies for NDC bookings made through the GDS. This is because with NDC bookings, there is a two-tier revenue model with lower segment booking fees for NDC bookings compared to the traditional GDS EDIFACT bookings. GDSs have initiated new subscriber agreements that ensure that NDC bookings will not be eligible segments for agency incentives. This approach questions the future loyalty of a TMC to a GDS. Second, the GDSs can add a markup fee to NDC content that is distributed to the TMCs. What this means is that the TMCs are paying a booking fee for NDC content from a GDS, as they would do with NDC content aggregators.

These steps give the GDSs a buffer to address declining segment booking fees, which are expected with the increase in NDC adoption. The operating economics of a GDS as an aggregator of NDC content is destined to change. This puts the burden on the travel agencies to collect additional service fees from their customers to offset the revenue shortfall.

14.8.3   Will GDSs Make an Exit?

As revenues shrink and margins decline with NDC adoption over the next 10–12 years, a fiercely debated topic is whether the GDSs, except for TravelSky that operates in a regulated market for China outbound bookings, will exit the distribution business. Or will they?

NDC aggregators today only support a few connections to airlines for NDC content, and they range from 15 to 25. GDSs as NDC aggregators support even less connections than NDC aggregators. Besides, the volume of bookings from NDC aggregators is not significant. There are over 600 airlines in the world and developing connectivity to all these airline systems will take time. There is also the issue of survival of the NDC aggregators. There is a large number of NDC aggregators in the market today, and it is likely that most will not survive unless they are able to attract a significant volume of bookings through their channel.

The future of the GDS is also dependent on TMC affinity to the GDSs in the absence of incentives for NDC bookings. The TMCs that have access to the corporate market are also under threat from self-service corporate booking tools that can enforce compliance to corporate travel policy and access to NDC content, but do not address duty of care.

To survive in the NDC era, GDSs need to rethink their investment in information technology. Core high investment functions such as GDS air shopping will not be required in the future. A lean organization with transparent cost controls is required for their survival as NDC aggregators.

14.8.4   GDS and the TMC

TMCs will continue to invest in the direct-connect solutions offered by the airline and NDC aggregators to address content shortfall if they rely on a GDS exclusively for all their content. In this new era, the role of NDC aggregators seem secure to feed content to the TMCs unless GDSs have a broad NDC aggregation strategy that addresses time to market.

What does all this mean for the customer? The era of total fare transparency for every market is over. It is like buying a new car. Consider a customer who wants to purchase a luxury car, preferably a BMW, Mercedes, or Audi. The various packages offered by the car manufacturers are nonhomogenous. Ultimately, the customer needs to decide by comparing the packages and the prices to select the automobile that maximizes their utility.

14.9   Can GDSs Embrace New Business Models for Survival?

Adoption of NDC by the GDSs is a prerequisite to receive air content in the future. Hence, GDSs need to focus on a multi-source content platform consisting of traditional GDS (EDIFACT itineraries) content, NDC (XML itineraries) content, and LCC content. For travel agent search queries, the content must be cleansed in real-time for duplicates and normalized using advanced statistical models and machine learning techniques to determine how the screen real estate for the agency desktop needs to be managed to display pertinent itineraries and offers to a travel agent. Beyond NDC adoption, the future of the GDS will be influenced by three key factors.

First, segment booking fees for full-service carriers should be at a price point that is acceptable to the airlines. What this means is that segment booking fees, if this revenue model survives, will continue to decline. However, this is going to make the survival of the GDS business a challenge since the unit economics of the GDS business, which requires high transaction volumes, is at risk because of channel shift through disintermediation which will increase the cost of sale of GDS processing.

GDS processing costs will decline with NDC since the airlines bear the burden of shopping, pricing, booking, and ticketing. Perhaps the survival of the GDS will be predicated on marking up the private NDC fares or charging a service fee before they are presented to travel agencies.

Second, is the ability to provide price points for segment booking fees that are economically viable for LCC participation in the GDS. The LCCs control over one-third of the market share worldwide (35% in 2020) and LCC content should be made available to travel agents for comparison shopping, booking, and fulfillment.

Third, is the ability for the GDS to gain access to *infinite content* and transform the legacy GDS platform into an *open marketplace* whose revenue streams extend beyond travel. Infinite content is not just air, hotel, car, and cruise line content, but the ability to sell and fulfill nontravel content such as destination activities and nontravel products customers wish to purchase at the destination. This will require a measure of risk and an appetite for transforming the business model as we see it today.

14.9.1 GDS Transformation to an Open Marketplace

The investment needed to transform the current GDS model into an open market-place for selling and fulfilling nontravel content is significant. This transformation requires the establishment of a publish and subscribe (Pub/Sub) environment. Pub/Sub is a messaging service that is both asynchronous and scalable. It helps in decoupling services that produce messages from the services that process those messages. This service enables services to communicate with each other asynchronously, with latencies of approximately 100 milliseconds.

In this scenario, the GDS will publish travel information, such as requests for flights, hotels, car rentals, and non-travel-related products such as theatre tickets, golf tee times, and tickets to professional sporting events. Travel and nontravel suppliers can subscribe to various queues, to view and respond to the various requests. This subscription also enables suppliers to promote their content proactively. For this to happen, suppliers must register and subscribe to the different queues. The GDS can simplify the process by providing an event calendar, which includes notable events like the All-Star Game, NFL Super Bowl, Consumer Electronics Show (CES), Balloon Festival in Albuquerque, New Mexico, the Pacific Air Show in Huntington Beach, California, and more.

In the case of booking requests, only the qualified airlines, based on their published schedules, that provide a service in the request market will receive a notification. This allows the airline to determine which ancillaries it wants to promote, with a continuous dynamic price for the itinerary through the NDC gateway. Beyond travel-related content, nontravel marketplace providers, such as theaters, professional or college sporting events, and local activity providers, can also subscribe and promote their content. Additionally, the GDS will need to share parts of a travel agency session (also known as the agent assembly area or AAA) with suppliers in real time. This sharing will enable suppliers to promote their

Transformation of the GDS Platform to an Open Marketplace

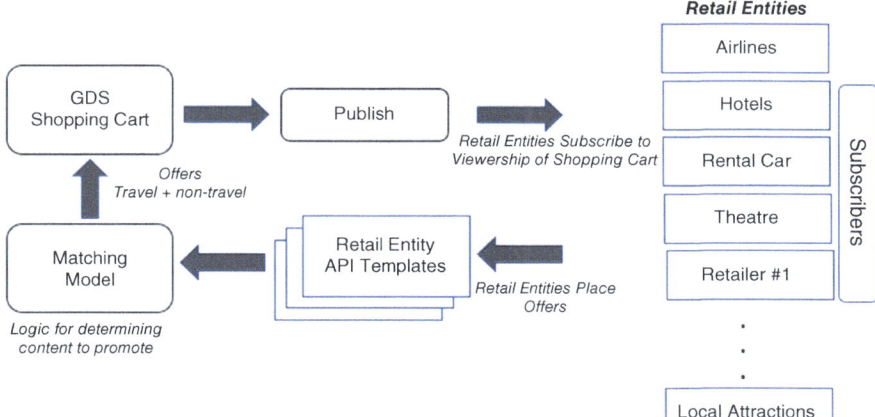

Fig. 14.2 GDS transformation to an open marketplace

content. Non-travel suppliers will require generic application program interface (API) templates to promote their content on the GDS open marketplace. The matching algorithm will determine which specific content should be promoted to travel agents for display. The matching model must consider travel agency preferences, GDS preferences, and customer preferences for non-air content. The economics of the revenue share model for non-travel content also plays a role in this determination.

Nontravel products can be sold through a GDS using a revenue share model, which will necessitate improvements to current billing systems. The revenue share agreement involves the nontravel retail entity, the GDS, and the travel agency. Figure 14.2 demonstrates the envisioned transformation of a GDS into an open marketplace.

14.9.2   Points to Ponder

Travel thrives in a technology-driven landscape. Travel entities that do not adapt to changes in technology and business will no doubt perish. Innovation is a prerequisite for survival of the fittest, and it requires a deep understanding of the business in all dimensions and the courage to explore new possibilities as consumer preferences and technology evolve. The urgency to paint a future desired end state and steps to transform the business to achieve the goal has never been greater for TMCs, GDSs, and OTAs.

Historically, this has been a fundamental shortcoming of the GDS business—thinking outside the box about new possibilities and not just defending the status quo. A fundamental question is: *Why did the GDS not own the point of sale?* This question is intriguing since it solves many of the problems that exist today such as segment booking fees that the airlines do not want to pay for. The larger question faced today is: *Does the GDS oligopoly have the courage and the will to embrace new business models with the technology and software that they have invested in since 1976?*

Here are some puzzles to ponder from history about companies that failed to adapt. In each of these cases, costly mistakes were made because the business entity with the first mover advantage was not in touch with the dynamics of the market-place, what opportunity could be captured, and did not understand what customers really wanted.

1. *How did CBS fail to see opportunities CNN saw?*
2. *Why did the Coca Cola Company launch New Coke in 1985?*
3. *Xerox invented the mouse and graphical user interface (GUI). Why did they not invest in its potential?*
4. *Why didn't IBM create the PC operating system?*
5. *How did General Motors miss the minivan in 1983?*
6. *Why didn't Blockbuster transform their dominant video rental business and invest in a streaming service to take on Netflix?*
7. *Why didn't Borders Books launch amazon.com?*
8. *Steven Sasson from Kodak invented the first Digital Camera. What happened to Kodak?*
9. *Why did Blockbuster turn down multiple offers to buy Netflix?*
10. *BlackBerry pioneered handheld devices and had 85 million subscribers in 2011. What happened?*
11. *Why didn't Excite, the search and directory services portal, buy Google for under a million dollars in 1999?*
12. *Why did the Ford Motor Company introduce the Edsel in 1957?*

Strategic agility in any line of business requires a deep understanding of the business across all dimensions, and the ability to stand in the future and see new possibilities.

In a changing business and technology landscape, where the odds are certainly not in favor of the GDS, how can the GDS adapt leveraging their assets? There are numerous nontravel-related applications where the core passenger name record and the new ONE Order capabilities used in a GDS can be leveraged to generate alternate new revenue streams.

The GDSs have no choice but to transform their business for survival. While there are endless possibilities to transform the business and leverage their existing infra-structure, they should focus on adjacencies which offer the path of least resistance. Two examples of adjacencies are discussed below.

14.9.3 Professional Sports Ticket Exchange

Consider the case of Ticketmaster Entertainment, Inc., a ticket sales and distribution company which dominates the primary market for the sale of season tickets and single event tickets for professional sports, the NBA, NFL, MLB, and NHL. They have exclusive rights for ticket sales and on average over 60% to 70% of all sales are to season ticket holders. Individual ticket purchasers may undergo a change in plan and may want to dispose of the tickets, which are sold through aftermarket sites, like StubHub, TicketLiquidator, TickPick and many more, or sold in person on the day of the event, sometimes for steep profits, depending on the popularity of the event. Today, there are several hundred aftermarket sites.

Reselling tickets is legal in all states in the United States though some states impose a limit on the amount above face value of the ticket that can be charged to a buyer. Consider the StubHub site, owned by eBay. Sellers can list their tickets on StubHub for free and are bound by the StubHub user agreement that includes local regulations. If the ticket sells, StubHub charges the seller a 15% fee. Buyers pay a 10% premium plus the cost of ticket delivery at $4.95/ticket. Hence, for a $100 ticket, the seller pays $15, and the buyer pays an additional $10 to the face value of the ticket or marked up price requested by the seller which could be higher than the face value of the ticket plus the ticket delivery fee. Hence, the seller receives $85, the buyer pays $100 + $10 + $4.95. The total cost of distribution is 30%! Even if the ticket delivery fee is not considered, the distribution fee is an astounding 25%! In contrast the average GDS bookings range between 1% and 2.2%. Note that on StubHub the buyer *always* pays an amount that is higher than the face value of the ticket.

GDSs can provide an alternative to the aftermarket sites with a unique value proposition that consists of three core components.

1. Facilitate a ticket exchange using the existing PNR infrastructure.
2. Verifying all name changes during a ticket exchange with TSA (Transportation Security Administration) for enhanced security to enter the venue.
3. Facilitate a revenue share model for the ticket exchange at a fraction of the surcharge paid by seller and buyer today to the active after marketplaces.

The GDS ticket exchange process will emulate how an airline ticket is purchased, but with some key differences. In the airline context name changes are not permitted. In this scenario, the name of the individuals attending an event should be printed on the ticket so there can be onsite validation to enter the stadium. The TSA no-fly list can be used to screen fans attending the event. This security check must be implemented at both Ticketmaster when the ticket is first purchased and the GDS ticket exchange when the ticket is being exchanged. When a consumer purchases multiple tickets, names need to be assigned to each ticket before it can be ticketed and issued. However, unlike airlines, name changes will be allowed for ticket resales and ticker transfers.

A key issue today is the presence of scalpers who benefit the most from sold out popular games by selling the seats at a premium outside the stadium for popular sporting events. Scalpers resell tickets directly to ticket seekers, typically in the vicinity of the sporting event. The ticket seeker also runs the risk of buying cash only fraudulent tickets. The GDS ticket exchange will also put an end to scalpers since the markup will be restricted to local laws at the time of ticket exchange with the prospective buyer.

14.9.4   Collaboration Between a GDS and Online Retailers for Destination Fulfillment

GDSs should consider partnering with large online retailers like Amazon.com to transform travel into a richer experience that looks very different from what it is today. The cost of travel is only a component of a traveler's budget for a trip. To enrich the experience, the online retailer can gain access to a customer's wallet during the journey to provide products and services at the destination for a competitive price. For example, consider a family that has booked travel to a beach destination. They wish to travel lightly and would like beach towels, suntan lotion, and snorkel gear delivered to the hotel where they are staying.

The GDS can share the customer itinerary with Amazon.com, subject to the customer accepting an opt-in clause. GDSs transact millions of dollars in revenue every day. In travel, scale matters and Amazon can leverage the GDS bookings to generate incremental revenues on an unprecedented scale. It also transforms the travel industry with a more complete view of fulfilling a customer's needs during a trip. A revenue share model for the sale of non-air ancillary products can provide a new revenue stream for the GDSs.

The GDS contribution to the partnership will be the traditional components of travel supplier connectivity, data management, booked itineraries, multiple forms of payment for non-airline products, and security related to authentication, authorization, and permissions for viewership. The Amazon contribution to the partnership is extensive fulfillment capability for non-air products at the destination, an advanced recommendation engine to recommend non-air product offers based on customer intent and preferences, and grow wallet share with the sale of non-air products to the travel destinations. The partnership also benefits the TMCs with a revenue share model, in the form of incentives from the GDS for the sale of non-air products.

14.10   The Future of Travel Management Companies and OTAs

The TMCs and OTAs will continue to thrive in an NDC world because they own the point of sale. To address declining margins from air bookings, the cost of air distribution will be passed on to the customer as a service fee. The changes in the competitive landscape will require the TMCs and OTAs to be less dependent on GDS content. This will require TMCs and OTAs to invest in direct connect technology to access prime airline NDC content either directly from the airlines or through NDC aggregators. Larger TMCs will invest in their own NDC connectivity technology and be more profitable than the smaller agencies that will have to rely on NDC aggregators. Some large travel management companies have already taken steps to acquire NDC aggregators. For example, Trip.com acquired Travelfusion in 2015 before the largest OTA in China, CTRIP, acquired Trip.com in 2017. Flight Centre Travel Group, based in Australia with a worldwide presence, acquired a controlling interest in Dubai-based NDC aggregator TPConnects (TPC), in March 2022 (Fox, 2022).

Hotel content is fragmented so OTAs and TMCs will continue to make revenue and be far more profitable than air.

In a decade, when peer-to-peer shopping and booking become a reality in a blockchain enabled world, the larger chains will benefit at the expense of the OTAs. The smaller chains and independents may not have the ability to invest in this technology and will continue to rely on the OTAs for bookings. A negative aspect of peer-to-peer payments between a buyer and a seller without a bank will lead to an exponential increase in new forms of payment that travel sellers will have to accept.

14.10.1   Reinvention of the TMCs and OTAs

Reinvention of the travel management company is a fundamental requirement for the long-term survival of a profession that was started by Thomas Cook in the nineteenth century.

Access to comprehensive travel content is a basic requirement from TMCs and OTAs. With the rise of a fragmented content landscape featuring independent hotels, local tours, and attractions, as well as NDC for air travel, TMCs need to prioritize securing content from many sources instead of depending on a few entities. For a travel agency, both brick-and-mortar and online travel agencies to be effective, it is vital to have access to content to meet the needs of their customer base.

An evolving trend is the investment in NDC aggregators by TMCs and OTAs. Trip.com acquired a controlling stake in Travelfusion in January 2015. FlightCentre, one of the largest travel agencies in the world, increased its stake in NDC aggregator TP Connects to 70% in March 2022.

14.10.1.1   Travel Management Companies

In the realm of managed travel, the core value proposition of a TMC should be centered around safety, value added services, travel spend optimization, monitoring active bookings for cancel and rebooking opportunities on the same itinerary with a lower fare to reduce the cost of travel, effective communications, and simplifying the travel experience for customers.

Safety and duty of care are crucial in corporate travel. The ability to provide 24 by 7 support, track employees and executives whether they are in the air, hotel, or a meeting is a valuable service to corporations. Travel advisors need to prioritize health and hygiene to simplify and avoid unexpected events during trips, especially after the pandemic of 2020. Customer feedback is valuable to enhance the travel experience. Customer feedback is a key input to refine and improve the travel experience over time. Travel advisors must have a laser focus on troubled hot spots, and supplier reliability such as airline dependability and on time performance, and guest feedback on hotel stays.

To optimize a corporation's travel expenses, partnering with its employees can be effective. By incentivizing employees to make better travel decisions and rewarding them for staying under budget and producing savings, behaviors can be modified. Rewards can be a powerful tool to encourage better decision-making, such as choosing cheaper hotels or booking in advance. Equally important, it increases employee awareness of the cost of travel to the company.

Understanding a traveler's intent for a trip is crucial for a travel advisor to ensure a good experience for the trip. With the trend toward combined business and leisure travel ("bleisure"), trip planning is more than just booking air and hotel but delivering the complete experience with local attractions to address the leisure component of the trip.

Virtual reality is an effective marketing tool for hotels and travel brands when planning a trip. Leisure focused TMCs should invest in 360-degree virtual reality tours categorized by theme, point of interest, or destination city. Virtual reality promotes bookings because leisure customers may book what they see.

Customizing itineraries is a necessary step for each trip, as it depends on various factors such as the duration of travel, dietary preferences, flight options, hotel choices, and the traveler's interests. Personalization and choice are highly valued by contemporary leisure travelers, making customized trips more attractive to them. Generative AI can complement the skills of a travel agent to create custom itineraries.

14.10.1.2   Anticipated Changes in the TMC Landscape

With the arrival of the NDC messaging standard, start-up corporate travel and expense management platforms see the opportunity to develop comprehensive corporate booking tools with pre-trip approval workflows, a rules engine to implement corporate travel policy, the ability to synchronize corporate bookings on calendars, custom reporting for true end-to-end transparency, and omni-channel deployment where corporate travelers see the same content (flights and fares) as travel agents inclusive of NDC content.

Many critics contend that the fee-per-transaction pricing model (see Sect. 2.13.1) will decline over the next decade, giving way to lower cost value proposition for corporate buyers. Today travelers are required to pay a fee every time they contact a TMC. For example, there are fees associated with air bookings, hotel bookings, change fees, agent touch fee, emergency contact fee, etc. A new breed of corporate travel and expense management platforms will create alternate cost-effecting pricing models such as a single consolidated fee per booked itinerary regardless of booking type (air, hotel, etc.) and touch. A second alternative is the all-you-can-book annual plan subscription model for small and mid-sized businesses. Only time will tell if they can move up from the mid-market to the Fortune 1000 companies. If they do, it will disrupt the traditional TMC model with new entrants that mimic a Walmart-like value model.

14.10.1.3   Online Travel Agencies

In 1996, the online travel agencies like Travelocity followed by Expedia were the original disruptors in the digital era.

The OTA's share of the total online market increased from 35% to 37% in 2021. However, supplier websites maintained their majority stake in the US online travel market, representing a 63% share of online gross bookings. OTAs continued to outrun supplier websites in the hotel segment, with a 52% share of the online market in 2021, but it is expected to decline to 48% by 2025. Although supplier websites are the preferred booking channel for air and car rentals, OTAs gained shares in both segments, capturing 20% of online air gross bookings in 2021 compared to 19% in 2020, and 35% of car rentals, up from 32% in 2020 (Jong, 2022).

Most travel experts believe that the time has come for the reinvention of the OTA model. The OTA approach of gathering digital content to assist customers in planning their travel on a budget can lead to their downfall. This model is not well-received by hotel suppliers and is an expensive way to acquire customers. From a supplier perspective, this method has become inefficient, and they continue to promote their consumer direct channel actively to avoid the extremely high cost of hotel distribution. Hoteliers know that they have been *held hostage* by the OTAs with incredibly high commissions and debilitating merchant rate margins, forcing them to cut direct variable costs to show a positive gross margin and in some cases even avoid bankruptcy. If history is any indication, it is only a matter of time before these sky high commissions and merchant rate margins reach levels that are acceptable to hoteliers.

The two largest OTAs, Booking Holdings and Expedia Group, spent a staggering $11 billion on customer acquisition in 2019, before the COVID-19 pandemic. Their primary competitor, Google, was the primary beneficiary since Booking Holdings and the Expedia Group are their biggest advertisers. The cost of search engine marketing (SEM) is increasing at an alarming rate, which presents a challenge for OTAs to maintain growth in the future. Over a billion dollars were spent on marketing during the first quarter of 2022 during the recovery period after the pandemic of 2020. Hotel companies must deal with a duopoly on two fronts: high commissions (Expedia Group, Booking Holdings) and advertising expenses

(Google, Facebook). They must increase their marketing expenses to stay relevant in the online marketplace. To balance the expenses from OTA commissions and advertising costs from Google and Facebook, consumers are paying more for hotel rates.

Google Hotels operates on a pay-for-performance business model, which technically sets it apart from traditional OTAs. As a result of Google's powerful search engine, customers are directed to the site earlier in their travel search process, impacting the booking dynamic of OTAs globally. Google, alongside Facebook, is a significant player in online advertising and dominates the world's two largest OTAs.

For OTAs to enhance their competitiveness, they need to reconsider their method and tactics for paid search. They must recognize the keywords linked to customer segments that yield greater profits per transaction to maintain an edge over increasing SEM expenses. Unless they transform their business model, they face the danger of being replaced and must gear up for their preservation.

To decrease dependence on SEM and minimize expenses associated with customer outreach, OTAs must rethink conventional forms of media, like television, radio, and outdoor advertising. Despite the challenges, the goal is to encourage customers to bypass search engines and directly access the OTA website or mobile app.

Besides higher SEM costs, the high commissions paid by travel suppliers are of concern and are not sustainable in the future. Airlines and hotels are actively promoting direct bookings. Unlike hotels, the airline model is different. For air bookings, the airlines pay a segment booking fee to the GDSs which in turn pays an incentive to the OTA that is based on established performance metrics. Fragmentation of the hotel industry with independent hotels helps the OTAs with higher commissions for both the merchant and agency models. Larger hotel chains have a laser focus on reducing OTA commissions and enticing customers to book direct to accumulate frequent stay points. Alternate lodging platforms like Airbnb that are not available through OTAs also puts them at a disadvantage.

To gain control of the advertising budget, OTAs need to revisit traditional media like television, radio, and outdoor advertising as a viable alternative to reach customers.

14.10.1.3.1 Members Only OTAs

The utilization of Member Only Online Travel Agencies is not a new idea, but it is becoming increasingly popular for more extended stays and high-end luxury vacation rentals. These agencies differ from the usual OTAs as they request an annual subscription fee from the property manager to list their properties. In return, they do not charge any commission when a booking is made. Various companies, such as Bidroom, Golightly, Voyage Privé, Travelzoo, and Secret Escapes, offer private sales to their subscribers. For instance, Bidroom provides direct bookings to properties without a commission and asks that the properties give members a minimum discount of 5% compared to prices available on public OTA platforms. While they evolve, these agencies prioritize a membership model that focuses on retaining guest loyalty. Their primary value proposition is the personalization of a unique end-to-end travel experience, which sets them apart from traditional OTAs' one-size-fits-all approach. Similarly, Golightly is an invitation-only club for women interested in home-sharing and vacation rentals.

Millennials are attracted to Member Only OTAs since they are open to the concept of paid loyalty in return for continued added value from the experience. Members Only OTAs typically maintain a strong partnership with property managers. This is because the OTAs expect an exceptional customer experience for their guests who arrive as members. As such, property managers have a unique opportunity to convert these guests into lifelong brand enthusiasts.

To access the opaque travel content, travelers must subscribe to this model. This is to solve rate parity issues and provide exclusive discounts to customers not available on traditional OTA platforms. Subscription services are becoming more popular with Tripadvisor launching its own subscription product, Tripadvisor Plus, in 2021. Members pay an annual fee to access savings that are better than traditional OTA channels. This membership model is more efficient for properties due to lower customer acquisition costs and zero percent commissions on bookings. It is a collaboration between travel intermediaries, hotels, and travelers who wish to participate in a private marketplace.

14.10.1.3.2   Loyalty Programs

Booking a flight through an OTA is different from booking a hotel room through an OTA in a very significant way. Generally, a customer will accrue loyalty points or miles when air travel is booked through these third party sites. This is not typically the case when a customer books hotel stays through OTA sites.

In the early 2000s when several new OTAs entered the market, hotel suppliers realized that more bookings were made through the OTAs than through their own branded consumer direct websites. These OTA bookings came with high commissions, usually in the 10%–30% range, unlike air where the average cost of distribution is less than 2%. To offset commissions and redemption of points accrued by members during stays booked across all points of sale, hotel suppliers eliminated accrual of points and status privileges when booked through an OTA. To attract customers to hotel websites, exclusive member rates and status perks were promoted to entice direct bookings.

The OTAs were very slow to react before they introduced their own loyalty programs. In 2021, Booking.com expanded its Genius loyalty and discount program to reach sister brands Priceline, Agoda, and Kayak. This strategic initiative aimed to expand Booking Holdings' customer base and shift consumer direct demand from major hotel chains. Properties that participate in the Genius program can offer additional discounts to customers and receive greater exposure on the various brands of Booking Holdings without impacting commission rates.

Expedia followed suit in May 2022 and announced One Key, a unified loyalty points program for customers to accrue and redeem points across their various brands starting in July 2023. This initiative fills a void in the OTA booking process since hotels do not permit accrual of hotel loyalty points when a booking is made on an OTA site. The new loyalty program is an expense categorized as marketing and customer acquisition to generate more bookings, strengthen customer affinity to the Expedia brands and redirect online users from supplier websites to the Expedia brands. Travelers can accrue Expedia loyalty points for purchases and redeem them toward hotel stays, flights, car rentals, and activities.

Trip.com and Yatra are online travel agencies that have implemented loyalty programs to foster customer loyalty and maintain retention. These programs, namely Trip Coins and eCash, aim to incentivize customers to remain loyal to their respective brands.

14.10.2 *Transforming OTAs and TMCs with New Capabilities for the Leisure Segment*

Historically, the leisure market segment has been larger than the corporate market segment. During the 2008 financial crisis, air travel bookings for leisure purposes were triple the amount of corporate travel bookings. Due to the COVID-19 pandemic in 2020, it is predicted that corporate air travel will not recover domestically until 2024, and internationally until 2025; and even when it does, it is not expected to experience substantial growth. As a result, it is projected that leisure bookings will surpass corporate bookings by five-fold in the coming years. Online travel agencies (OTAs) and leisure TMCs can attract both high-end and low-end leisure traffic.

OTAs and leisure TMCs must focus on innovation to add value to their customer base and create a competitive advantage. They need to think beyond the traditional booking workflow that is market (for air) and destination (for hotels) specific to attract leisure traffic. Here are a few examples where OTAs can solidify their position with capabilities that the individual property or property chain may not provide.

14.10.2.1 Optimizing Hotel Travel Spend Across Properties with Split Stays

OTAs are the go-to source for leisure travelers searching for competitive airfares and hotel rates across brands. However, there exist several deficiencies in the OTA workflow to cater to such travelers. While hotel chains require similar capabilities, OTAs enjoy an edge since they have access to hotel content across chains and individual hotels. Automated search capabilities are crucial for leisure travelers looking to optimize their total hotel expenses for a trip.

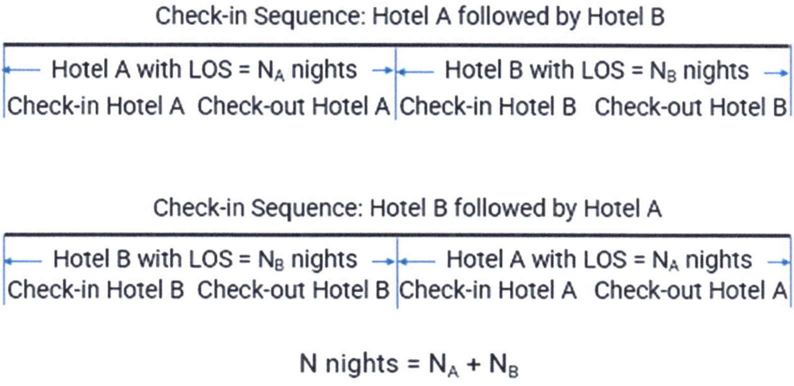

Check-in Sequence: Hotel A followed by Hotel B

← Hotel A with LOS = N_A nights →	← Hotel B with LOS = N_B nights →
Check-in Hotel A Check-out Hotel A	Check-in Hotel B Check-out Hotel B

Check-in Sequence: Hotel B followed by Hotel A

← Hotel B with LOS = N_B nights →	← Hotel A with LOS = N_A nights →
Check-in Hotel B Check-out Hotel B	Check-in Hotel A Check-out Hotel A

$$N \text{ nights} = N_A + N_B$$

Fig. 14.3 Split hotel alternatives

Leisure customers often face difficulty in booking their preferred hotel due to length of stay controls (restrictions) imposed by revenue management that result in the unavailability of rooms at lower rates. To bypass these room inventory control restrictions, customers can split their stay across multiple properties of equivalent or greater value, which also enables them to visit attractions situated in different neighborhoods. OTAs should incorporate an automated search function to assist leisure customers in optimizing their travel expenses by splitting their stay across multiple properties, prioritizing cost savings over the stay sequence which is less important (Blankenbaker & Peng, 2014).

When searching for a hotel in a destination that involves staying at two or more properties, either in the same city or in multiple cities that are nearby, the available properties should have a lower average rate and may be closer to local attractions. The number of split stays will always be less than or equal to the total duration for the trip.

When conducting a typical hotel search, rates are provided for a particular hotel and length of stay. As an illustration, the search results may display a 4-star property (such as Hotel A), available for a stay of N number of days.

When there are two split stays, the search optimization involves explicitly evaluating all the possible combinations of the trip duration if the customer will only make one stopover at the second hotel for convenience. In this situation, the customer can choose to start their stay at either Hotel A or Hotel B, as illustrated in Fig. 14.3.

In this scenario, the concept of splitting a stay between two hotels, Hotel A and Hotel B, is explored. The best available rate (BAR) is calculated for each valid length of stay (LOS) during the eligible check-in dates for both hotels. All possible combinations of nights spent at each hotel are evaluated, ensuring that the total number of nights (N) is equal to the sum of nights spent at each hotel ($N_A + N_B = N$). The least cost alternative is the recommended solution.

Table 14.1 Computational burden of split stays

Maximum number of split stays (hotels)	Total stays for the trip (length of stay)	Number of permutations to be evaluated
2	4	2 + 6 = 8
	6	2 + 10 = 12
	8	2 + 14 = 16
3	4	3 + 18 + 18 = 39
	6	3 + 30 + 60 = 93
	8	4 + 42 + 126 = 171
4	4	4 + 36 + 72 + 24 = 136
	6	4 + 60 + 240 + 240 = 544
	8	4 + 84 + 504 + 840 = 1432

Split hotel stays have benefits for hoteliers, OTAs, and customers, contrary to popular belief. Hotels can capture customers who would have been turned away due to the nonavailability of a room on a peak day or high price, by offering them a part of their stay at the destination. To enhance customer satisfaction, hotels that are part of the same hotel chain can provide free shuttle transportation for split stay customers to the next property. Alternately the OTA can provide free transfers for contiguous split hotel stays regardless of the hotel brand. Automation of the split stay process benefits OTAs, eliminating the need for manual intervention and leading to increased customer satisfaction, which can translate to repeat visits to the website. Additionally, OTAs can promote split stay opportunities, which can lead to cost savings for customers, and recommend properties that offer the highest margins.

Evaluating split stays involves assessing a considerable number of possible combinations. Therefore, it is advised to restrict the number of split stays to two or three at the maximum.

Table 14.1 illustrates the number of permutations that must be evaluated as a function of number of split stays m and total stays n for the trip. Note that m is the number of hotels and n is the total length of stay.

The formula for the number of evaluations when there are m hotels and n length of stay where P represents permutations and C represents combinations is shown below.

$$\sum_{i=1}^{m} P_i^m * C_{i-1}^{n-1}$$

The split hotel model can be used by leisure customers who reserve via a travel agency that subscribes to a GDS. Booking several hotel segments for one passenger name record (PNR) is supported by all GDSs. To generate a prompt split stay response, the TMCs and OTAs must support a hotel rate availability shopping cache using existing global hotel APIs to ensure a fast split stay response.

Airbnb customers often stay for extended periods ranging from ten to forty days. However, the longer the stay, fewer options are typically available. To combat this issue, Airbnb introduced a new feature in 2021 known as "split stays" (Sorrells,

2022). This feature enables customers to stay at multiple properties for the duration of their trip, thereby increasing the number of available options from which to choose.

14.10.2.2   Recommending Sequence of Stays for Multi-Destination Trips

For those who enjoy traveling for leisure and want to book hotel stays at multiple destinations, there is a way to search for the lowest cost across all destinations. For example, if someone wishes to spend three nights in San Francisco and three nights in the Napa Valley wine country, a sequence can be identified that minimizes the total cost of travel. This means that the Napa Valley stay can come after the San Francisco stay, or vice versa, to realize cost savings for the total stay of six nights. However, as the number of destinations increases, the number of sequences that must be evaluated also increases at a rapid rate. If there are N destinations to consider, then the number of sequences is N! For example, if there are only two destinations, there are only two sequences to consider, but if there are four destinations, there are 4! (24) sequences that must be explicitly evaluated. To speed up the shopping response, a rate availability shopping cache is necessary.

14.10.2.3   Theme-Based Shopping and Booking

During the dreaming and planning phase of a trip, a significant portion of leisure travelers prefer to engage in theme-based shopping on a budget rather than fixating on a particular destination. These themes may include but are not limited to nature parks, safaris, beaches, art history, gambling, skiing, and historical sites. When a customer selects a theme, the shopping response should display properties across various destinations that align with the selected theme, ultimately aiding in booking. The response should adhere to the customer's budget and present results in a profit-driven, rank-ordered format for the OTA, while also considering the customer's preferences.

14.10.2.4   The Multimodal (Virtual Interlining) Opportunity for OTAs

A potential area of business expansion focuses on leisure travel and is known as multi modal, or virtual interlining. However, it poses a complex challenge of optimizing travel costs by integrating various travel options, including full-service carriers (FSC), low-cost carriers (LCC), airport transfers, rail and bus transportation, hotel accommodations, and destination activities into a single package for customers. This challenge is compounded by the lack of travel agreements, such as interline ticketing agreements, between these travel entities to facilitate frictionless travel.

Interline travel refers to air travel that involves more than one airline. When two airlines have a ticketing agreement, they can show connections between them, allowing for automatic baggage transfer and a seamless travel experience. In this case, the passenger receives a single ticket. However, low-cost carriers (LCCs) typically do not have agreements with other LCCs or full-service carriers (FSCs). When a customer books an interline flight involving an LCC and an FSC or two LCCs, they must collect their baggage at the end of the first airline's journey and recheck it with the second airline. This process may require multiple tickets and longer connection times to collect and re-check bags. Despite the inconvenience, interlining can result in significant cost savings. To minimize the risk of missed connections, virtual interlining providers build connections using on-time arrival analytics and offer travel insurance to customers.

Virtual interlining offers a vast array of options beyond what is mentioned here. In addition to combining air bookings with scheduled rail services, bus services, and hotel stays, travelers can also choose to add a rental car or activities at their destination to their travel package. These activities vary depending on the location. For instance, in Maui, Hawaii, popular recreational activities include scuba diving, fishing, snorkeling, helicopter tour, and parasailing. On the other hand, in the United Arab Emirates, travelers might opt to visit the Burj Khalifa, attend the Abu Dhabi Grand Prix Formula 1, Atlantis Aquaventure Waterpark, experience the Dubai Desert 4x4 Safari and Dinner, and many more.

Virtual interlining involves optimizing routes, which can be a complex issue. When a traveler plans to visit n cities exactly once from an origin city, the number of potential routings is $n!$ This can create a challenge for leisure travelers who do not have a specific order in mind for visiting each city during their trip. The total number of possible route sequences when visiting n cities exactly once is given by:

$$n! = \Pi_{k=1}^{n} k; \quad n > 0$$

Route optimization aims to establish the most cost-effective sequence of cities to visit, considering airfare and accommodation expenses subject to airline preferences (if any) and hotel preferences (brand and star rating if any). As the number of cities in the itinerary grows, the quantity of possible sequences increases exponentially and for each sequence several total cost evaluations are required to determine the best price based on consumer preferences. To illustrate, let us assume a leisure traveler is departing from and returning to New York and wishes to explore Paris, Amsterdam, and Rome. In this case, there are three factorial permutations, or $3! = 3 \times 2 \times 1 =$ possible routings, as demonstrated in Table 14.2.

The problem resembles the Traveling Salesman Problem (TSP). In this problem, a group of cities and the distances between each pair of cities are given. The goal is to discover the shortest possible route, or cost, that visits each city exactly once and returns to the starting city. This is a problem that is considered NP-hard, meaning there is no solution that can be found in polynomial time. In travel, the sequencing problem of routes is more complex because the costs are not fixed but vary based on

Table 14.2 Routing alternatives for 3 cities

Routing alternative	Routing
1	New York—Amsterdam—Paris—Rome—New York
2	New York—Amsterdam—Rome—Paris—New York
3	New York—Paris—Amsterdam—Rome—New York
4	New York—Paris—Rome—Amsterdam—New York
5	New York—Rome—Amsterdam—Paris—New York
6	New York—Rome—Paris—Amsterdam—New York

the specific dates of air travel and hotel stays, based on the respective revenue management controls. The costs of transportation and lodging are affected by the availability of seats and rooms, which change as the flight departure date or hotel check-in date approaches. Therefore, the cost of traveling from Paris to Rome may differ depending on the routing sequence used. Moreover, the costs of traveling in the opposite direction, from Rome to Paris, can also vary depending on the routing sequence.

Figure 14.4 depicts how the complexity of the problem increases with the number of cities to be visited which determines the number of feasible routings. The data can be displayed on a graph with cities as nodes and air and hotel costs as arcs, the costs are not fixed but changes based on the actual day of travel and the hotel stay since air and room availability varies which impacts the price for each arc. As a result, we cannot establish fixed costs for these elements. While we may not be able to list all possible routing sequences, we can implicitly enumerate feasible alternatives based on the dominant cost component, either airfare or hotel expenses, and arrive at an acceptable feasible solution.

In addition to the problem complexity, it may be necessary to purchase multiple tickets for the trip. However, a single invoice can be created that presents all the information in chronological order, including the prices of individual components and the total cost of the journey.

14.10.2.5   Multi-Day Multi-Destination Personalized Tour

Due to its close connection with cross-border travel and detailed itinerary-based tour guides, the multi-day tours segment has been significantly impacted by the COVID-19 pandemic of 2020. As a result, tour operators that specialize in complex itineraries focused on single destination or single country trips to avoid multiple regulatory requirements for travel. However, the more profitable longer duration multi-destination trips have rebounded in 2023.

While dynamic packaging is effective for package tour sales, it is not suitable for personalized tours due to the complexity of merging customer activity preferences with traditional air, hotel, and car packages to create a complete itinerary. Examples of such activities may include visiting museums, dining at Michelin rated

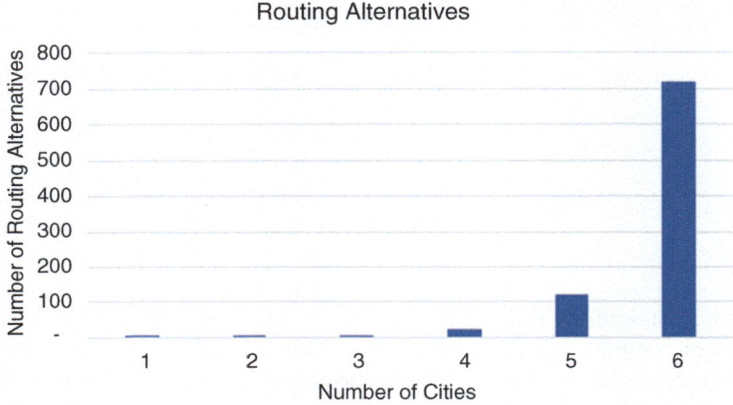

Fig. 14.4 Growth in feasible travel itineraries

restaurants, scuba diving, sailing, and helicopter trips. Multi-day, multi-destination personalized tours remain one of the last truly offline spaces in travel. However, to drive topline growth in the next decade, digital automation is required to streamline this complex and fragmented space. Customers have become accustomed to technology driven, self-serve, personalized shopping and booking experiences. Therefore, platforms that support personalized tours should have a B2C component for simple personalized tours and a B2B component with a local travel agent to support the shopping and booking process with localized knowledge of local attractions.

For an established OTA to be successful in this space is a formidable task due to the abundance of activity providers worldwide. These activity providers can be accessed through a consignment model or an allotment risk inventory model, with the latter offering higher margins and immediate confirmation for end-to-end tours. Additionally, access to destination specific travel agents is critical for personalized tours and to improve customer satisfaction.

Viator, which was established in 1995, was among the first companies to introduce online tours and activities. The firm collaborates with local operators to offer leisure travelers tours and activities that can be booked via their app or through a network of affiliate partners. Despite being acquired by Tripadvisor in 2014 for $200 million, the company failed to deliver on its promise of customized tours, instead focusing on prepackaged tours. It did not fundamentally change the structure of tours and activities and as a result traditional wholesalers, tour operators and retail agencies continued to be the preferred distribution channels. Today, boutique tour operators such as Evaneos, based in Paris, France, support over 2000 operators across 160 destinations. G Adventures, another leading operator, specializes in group tours and safaris. Both these specialized tour operators cater to both B2C and B2B clients (travel agencies) on their platform to offer personalized services.

14.10.2.6 Financial Ancillary Products

Travel fintech products are the financial ancillary products provided by TMCs. OTAs, and travel suppliers. These products include payment facilitation, lower cost of purchasing, and travel protection.

Offering travel protection is a risky venture for OTAs. TMCs, and travel suppliers as the use of predictive models can result in significant losses if the models are not well calibrated and poor recommendations are made. To mitigate these risks, machine learning models such as reinforcement learning are used to price fintech offerings at affordable prices dynamically. Intermediaries generate new revenue streams through two categories of travel protection for air, hotel, and rental car bookings: pre-travel purchases and consumer protection from unforeseeable events. Customers pay in advance for these services, which generates the new revenue stream. To support these fintech products, fare and rate prediction models that forecast when selling fares or rates will increase or decrease are required (Vinod, 2013a, 2016b).

Customers can receive information about potential increases in air fares or hotel rates before making a purchase. This is done through a "buy" or "wait" recommendation based on predicted future selling fares or rates. Additionally, customers can benefit from a "price drop" guarantee, which allows them to lock in the original price and receive a refund if the price drops after a "buy" recommendation. Another useful product is the "price freeze," which holds flight and hotel prices for 2 h to 21 days. If the price increases, the OTA or TMC covers the difference up to a certain limit. If the price decreases, the customer only pays the lower price.

OTAs such as Hopper are broadening their fintech offerings by providing "standalone trip protection" (Fox, 2022). This service permits customers to acquire protection against flight disruptions or cancellations for reservations that were not booked through Hopper's mobile app. Leading OTAs have singled out Hopper for unjustly exploiting customers, increasing customer angst, and confusion by forcing them to buy services that they do not need or understand. In July 2023, Expedia stopped supplying hotel and vacation rental inventory to Hopper (Biesiada, 2023a). In a preemptive move, Hopper terminated the relationship with Booking Holdings in September for hotel and vacation rental inventory (Schaal, 2023).

Another problem with companies that aggressively promote fintech services is their inadequate execution of reverse logistics. If a customer encounters an issue with a fintech service, like a price freeze, and requests compensation or a refund, they are left without online customer support. Instead, the dispute is addressed through email, which often involves combating false claims that are buried in the fine print of the financial service. As a result, customers may have to wait several weeks or even months to resolve the issue, causing dissatisfaction and discouraging future business. This frustration may lead customers to view these financial products as deceitful and fraudulent.

Fintech products have shown higher attachment rates during uncertain times of the COVID-19 pandemic. However, it is uncertain whether these products will maintain their popularity once travel returns to normal. Despite this, it is anticipated that fintech products will still be available as price fluctuations and weather conditions will continue to affect the travel plans of customers.

14.10.2.7 The Road Forward

Cracking the puzzle of multi modal travel can be challenging, but OTAs, first, and TMCs, second, are well-positioned to dominate this space in the future. Their entry into this arena could create a new revenue stream that currently doesn't exist.

  With the help of AI, including robotic process automation, generative AI and machine learning, OTAs could potentially offer advisory travel services. This innovative approach would provide automated expert advice tailored to individual traveler preferences, potentially becoming a major disruptor in the online travel industry.

Appendix Product Distribution Acronyms

Category	Acronym	Description
General	AHA	Airport handling agents
	ATI	Antitrust immunity
	CFR	Code of Federal Regulations
	CRS	Computerized reservations system used as a primary computer sales system by an airline. Synonyms include central reservations system and host CRS
	CRM	Customer relationship management
	DOT	Department of Transportation (U.S.)
	GSA	General sales agent
	GDS	Global distribution system used by travel agencies to buy/sell travel products, e.g., Sabre, Travelport, Amadeus
	FAR 121	Federal aviation regulation rules for scheduled air carriers
	KPI	Key performance indicator
	LCC	Low-cost carrier
	OTA	Online travel agency
	PSC	Passenger services conference
	PDG	Passenger distribution group
	SAGE	Semi-automatic ground environment
	SKU	Stock keeping unit
	TMC	Travel management company
	W3C	World wide web consortium

Category	Acronym	Description
GDS global id	1A	Amadeus
	1B	Abacus international
	1C	EDS information business
	1D	Radixx

(continued)

B. Vinod, *Mastering the Travel Intermediaries*, Management for Professionals, https://doi.org/10.1007/978-3-031-51524-8

Category	Acronym	Description
	1E	TravelSky
	1F	Infini
	1G	Galileo
	1J	Axess
	1K	Sutra
	1L	Open skies
	1M	Sirin
	1N	Navitaire
	1P	Worldspan
	1Q	Sirena
	1S	Sabre
	1U	ITA software
	1V	Apollo
	1W	Sabre (old)
	1X	GETS—Gabriel extended travel system
	1Y	EDS shares
	1Z	Fantasia
	F1	Farelogix

Category	Acronym	Description
Mainframe	ACP	Airline control program
	ALCS	Airline control system (running TPF in a MVS environment, now z/os)
	ATARS	Automated travel agency reservations system (based on PARS)
	CRT	Cathode ray tube
	DOARS	Donnelly official airline reservations system
	IPARS	International PARS
	MAARS	Multiaccess airline reservations system
	MVS	Multiple virtual storage
	PARS	Programmed airline reservations system
	SMART	Sabre multiaccess reservations terminal
	SODA	System one direct access
	TPF	Transaction processing facility
	TPFDF	TPF database facility
	z/os	An IBM operating system for mainframes that is suitable for continuous, high-volume operation with high security and stability.
	z/TPF	A high-volume, high-throughput transaction processor that can handle large continuous loads of complex transactions across large, geographically dispersed networks. Successor to TPF 4.1

Category	Acronym	Description
Messaging protocols	AIRIMP	A4A/IATA reservations interline messaging procedures
	API	Application program Interface

(continued)

Category	Acronym	Description
	EDI	Electronic data interchange—a format in which business data are represented using national or international standards.
	EDIFACT	Electronic data interchange for administration, commerce and transport, messages that are approved as standards for EDI.
	ICOT	The traditional piece of computer hardware for using network services. Terminals usually have minimal computing function, being completely dependent upon their host, and are often referred to as "dumb terminals."
	JSON	Java script object notation
	NDC	New distribution capability messaging standard for airlines to communicate to intermediaries.
	OTA	Open travel Alliance, a non-profit standards body that is creating messaging standards across all lines of business (air, hotel, car, etc.)
	PADIS	The passenger and airport data interchange standards (PADIS) board develops and maintains electronic data interchange and XML message standards for passenger travel and airport-related passenger service activities.
	REST API	Representational state transfer API
	SIPP	Standard interline passenger procedures
	SOAP	Simple object access protocol
	Type A	Immediate application response with timeout handling in place at the application level. EDIFACT follows a type A message format.
	Type B	Message guaranteed delivery, a slower response. The AIRIMP documents govern the type B message formats.
	TTY	Teletype is used to communicate reservations and messages between carriers using the AIRIMP standards. The airline industry uses teletype messages over ARINC or SITA networks to communicate between reservations systems.
	XML	eXtended markup language
	Zulu time	Greenwich mean time (GMT), due to "Z" used in CRSs to indicate GMT

Category	Acronym	Description
Schedules	ASM	Ad hoc schedule message
	Cirium	Schedule aggregator (like OAG and Cirium)
	DEI	Data element identifier in SSIM
	INNOVATA	Schedule aggregator, is now Cirium
	MCT	Minimum connect time
	OAG	Office airline guide (schedule aggregator, like Innovata and Cirium)
	QSI	Quality of service index, simpler linear form of CCM used in flight scheduling.
	SSIM	Standard schedule information manual
	SSM	Standard schedule message

Category	Acronym	Description
Emerging technologies	AI	Artificial intelligence
	AIT	Algorithmic information theory
	ARC	Abstraction and reasoning corpus
	CART	Classification and regression trees
	CBC	Choice-based conjoint
	CCM	Consumer choice model
	CLV	Customer lifetime value
	CNN	Convolution neural network
	DDS	Direct data solutions (from IATA in collaboration with ARC and Cirium)
	GBM	Gradient boosting machine
	GFS	Google file system
	GLM	Generalized linear model
	Hadoop	Java based open-source software that supports data intensive distributed applications on large clusters of commodity hardware
	HB	Hierarchical Bayes
	HBase	Open source nonrelational distributed database
	HDFS	Hadoop distributed file system based on GFS and MapReduceA data
	HIVE	Warehousing system that runs on top of Hadoop to allow SQL-like queries.
	LLM	Large language models
	LSTM	Long short-term memory
	MBR	Mishandled baggage rate
	MIDT	Marketing information data tapes
	ML	Machine learning
	MNL	Multinomial logit
	OR	Operations research
	Pig	Apache pig is a platform for analyzing large data sets
	PPC	Pay per click
	PPS	Pay per stay
	RFMTV	Recency, frequency, monetary value, tenure, variety
	RNN	Recurrent neural network
	RPA	Robotic process automation
	SVM	Support vector machines
	TOPSIS	Technique for ordering preference by similarity to ideal solution (multicriteria decision making)
	TPS	Trip purpose segmentation
	WTP	Willingness to pay
	XAI	Explainable AI

Category	Acronym	Description
Reservations/ CRS and GDS	AAA	Agent assembly area

(continued)

Category	Acronym	Description
	ACH	U.S.-based airline clearing house
	ACARS	Aircraft communications addressing and reporting
	ADM	Agency debit memo
	ADS	Agency data systems
	AHA	Airport handling agents
	ALTEA	Amadeus customer management suite (ALTEA RES, ALTEA INVENTORY and ALTEA DCS).
	APIS	Advanced passenger information system with pre-arrival and departure manifest data on passengers for border security agents
	ARNK	A surface sector segment in the PNR which means method of travel or arrival is unknown". These segments represent a break in journey. Pronounced ARUNK
	ASR	Agent sales report
	ATARS	Automated travel agency reservations system
	ATO	Airport ticket office
	ATSE	Air travel shopping engine
	BABS	British airways booking system
	BIDT	Billing information data tapes
	CPA	City pair availability
	CRS	Computer reservation system (used as the primary computer sales system by an airline); also called host CRS
	CTO	City ticket office
	DCS	Departure control system
	DOARS	Donnelly official airline reservations system
	ET	End transaction
	EMD	Electronic miscellaneous document, industry solution for collection and settlement of air ancillary fees (e.g., air extras) via ARC and BSP.
	ESV	Estimated seat value
	FPC	Fare pricing complex
	GDPR	General data protection regulation
	GSA	General sales agents
	JICRS	Joint industry computerized reservations system (JICRS)
	KTDI	Known traveler digital identity
	LNIATA	LiNe interchange address terminal address
	MAARS	Multi-access agent reservations system
	NAS	New airline storefront
	OLTP	Online transaction processing
	OPA	Offer partner airline
	ORA	Offer responsible airline
	PCA	Participating carrier agreement
	PDAS	Preference driven air shopping
	PII	Personally identifiable information
	PNR	Passenger name record

(continued)

Category	Acronym	Description
	PRS	Pre-reserved seats
	PSS	Passenger service system
	SABER	Semi-automatic business environment research
	SABRE	Semi-automated business research environment
	SabreSonic	*SabreSonic* CSS (customer sales and service) reservations system
	SSR	Special service request
	SSR-I	Special service request – Inventory
	STARS	Sabre traveler automation records
	TMC	Travel management company
	TOC	Total cost of ownership
	TTL	Ticketing time limit
	TVL	Travel segment
	UDX	Universal data exchange
	UC	Unable to confirm at sell
	UP	Universal profile
	VTCR	Vendor, tariff, carrier, rules
	VCR	Virtual coupon record

Category	Acronym	Description
Organizations	A4A	Airlines for America, a trade organization representing U.S. airlines
	AAA	American Automobile Association
	AARP	American Association of Retired Professionals
	AEA	Association of European Airlines
	AMR Corp.	Parent company of American Airlines, American Eagle, AmericanConnection and executive Airlines until 2013
	ARC	Airlines Reporting Corporation (ARC) is a technology solutions company providing transaction settlement and data information services.
	ARIG	Airline Revenue Integrity Group
	ARINC	Aeronautical Radio, Inc. was established in 1929, is a major provider of transport communications and systems engineering solutions for aviation and airports
	ASTA	American Society of Travel Agents (advisors)
	ATA	Air Transport Association
	ATPCO	Airline tariff publishing company (airline industry fare aggregator)
	BEA	Bureau of Economic Analysis (U.S.)
	BOAC	British Overseas Airways Corporation
	BSP	Billing and settlement plans (like ARC in the United States) for billing statements reflecting ticket sales made by each travel agent
	CAAC	Civil Aviation Administration of China
	CAB	Civil Aeronautics Board

(continued)

Category	Acronym	Description
	CASMA	Computerized Airline Sales and Marketing Association
	CDC	Control Data Corporation
	Cirium	Schedule aggregator, like OAG
	DDRS	Digital, data, and retailing symposium
	DOT	Department of Transportation (U.S.)
	Dynata	Online market research firm
	EU	European Union
	EEA	European Economic Area
	FAA	Federal Aviation Administration
	FCC	Federal Communications Commission
	IATA	International Air Transport Association
	IBM	International Business Machines Corporation
	ICAO	International Civil Aviation Organization
	ICH	IATA clearing house
	IFRS	International finance reporting standards
	ITT	International telephone and telegraph
	OAG	Official airline guide—schedule aggregator, like INNOVATA and Cirium
	Qualtrics	Online market research firm
	SABRE	Semi-automated business research environment
	SODA	System one direct access
	SITA	Société Internationale de Télécommunications Aéronautiques

Category	Acronym	Description
Connectivity	AVN	Numeric availability status sent from a host CRS to a GDS. AVN can use POS information to determine number of seats to be displayed at a specific location
	AVS	Availability status message sent from a host CRS to a GDS
	BBR	Based booking request
	DAI	Direct access interactive
	DCS	Direct connect sell (seamless sell)
	DCA	Direct connect availability (seamless availability)
	IDR	Inventory detail record, alternate term for IND
	IND	Inventory detail record, alternate term for IDR
	OAC	Office accounting code
	PCC	Pseudo city code
	POC	Point of commencement
	POS	Point of sale
	P2P	Processor to processor (P to P) communication (EDIFACT)

Category	Acronym	Description
Pricing, Revenue Management & Inventory Control	ADT	Adult fare
	AP	Advance purchase

(continued)

Category	Acronym	Description
	BRG	Best rate guarantee
	CASK	Cost per available seat kilometer
	CASM	Cost per available seat mile
	CLV	Customer lifetime value
	CN	Continuous nesting (also known as bid price controls)
	CWMF	Cost weighted mileage factors, published by IATA quarterly
	DACS	Dynamic availability calculation system
	EMD	Electronic miscellaneous document
	EMSR	Expected marginal seat revenue
	ePMP	Electronic prorate manual—passenger
	e-ticket	Ticketless
	E-ticket	Electronic ticket based on IATA specifications
	FBC	Fare basis code
	FBR	Fare by rule
	FC	Fare component
	FCU	Fare construction unit
	FET	Federal excise tax
	FQ	Fare quote
	IFRS	International finance reporting standards
	ITAREQ	IATADCS message request for sell transactions.
	ITARES	IATADCS message response for sell transaction request.
	LOS	Length of stay
	LUA	Last unit availability
	LLUA	Limited last unit availability
	MAF	Minimum acceptable fare
	MAT	Market adjustment table, synonymous with MCFA and MVT
	MCFA	Market class fare adjustment table, synonymous with MAT and MVT
	MPA	Multilateral prorate agreement
	MR-flights	Market restricted flights
	MVT	Market value table, also called a MAT or MVT
	MFEM	Multiflight expectation maximization
	NLF	Nominal load factor
	NSR	Net spill rate
	NUC	Neutral unit of construction (currency), superseded FCU
	OLF	Observed load factor
	PAOREQ	IATADCA message request for availability transactions
	PAORES	IATADCA message response for availability transaction request
	PFC	Passenger facility charge

(continued)

Category	Acronym	Description
	PFV	Potential future value
	PODS	Passenger origin-destination simulator
	PNI reconciliation	Passenger name index inventory count reconciliation request between host CRS and inventory to synchronize inventory counts on demand or during file maintenance
	PRM	Pricing and revenue management
	PTC	Passenger type code
	QSAP	Quality of service adjusted price
	RASK	Revenue per available seat kilometer
	RASM	Revenue per available seat mile
	RBD	The prime reservation booking designator is usually (though not required) the first character of the fare class code (fare class code is synonymous to fare basis code). The RBD is equivalent to the booking class code.
	RD	Reading day
	REVPAR	Revenue per available room
	RFDB	Resident fares database, same as MVT/MAT but on TPF/Sabre PSS to support O&D controls for American Airlines
	RFP	Restriction free pricing
	RFR	Recapture fare ratio
	RI	Revenue integrity
	ROM	Revenue opportunity model
	RS-13	Inventory control in BABS, called RS-13 when they had 13 booking classes, later expanded to 26
	SEM	Search engine marketing
	SEO	Search engine optimization
	SIFL	Standard industry fare levels
	SITI	Sold inside ticketed inside
	SOLO	Sum of locals
	SOTO	Sold outside ticketed outside
	SPA	Special prorate agreements
	SRP	Straight rate prorate
	TC	Traffic conference
	TPM	Ticketed point mileage
	TPMF	Total passenger miles flown
	UPA	Universal product attributes
	UTA	Universal ticketing attributes
	VFR	Visiting friends and relatives, a restricted leisure fare
	VN	Virtual nesting
	Yield	Revenue per revenue passenger mile (or kilometer)
	YQ/YR	Surcharges used in international markets.

Category	Acronym	Description
Emerging Technologies	DID	Decentralized Identifier
	TSP	Traveling Salesman Problem
	VCC	Virtual Credit Card
	VCN	Virtual Card Number

Category	Acronym	Description
Organizations	AT&T	American Telephone and Telegraph Company
	IATAN	International Airline Travel Agency Network

Glossary

Aeronautical Radio, Inc. Aeronautical Radio Inc. (ARINC) serves as the airline industry's single licensee and coordinator of radio communications outside of the government. The airline industry uses teletype messages over ARINC or SITA networks to communicate between reservations systems.

Agency Debit Memo An Agency Debit Memo (ADM) is a notification sent by an airline to a travel agency, requesting payment for a specific amount due to an error on the agency's part. Both ARC and IATA BSP also send debit memos to travel agencies on behalf of the airlines.

Agency Subscriber Desktop The customer facing product of a GDS is the travel agency subscriber desktop. It allows a travel agent to search and book travel. The names of the GDS desktops are Amadeus Selling Platform, Sabre Red 360, and Travelport Smartpoint.

Air Shopping Air shopping is the process of evaluating schedules, selecting, and pricing itineraries based on booking class availability to present to a customer during flight search. It is the single largest system and investment supported by a GDS.

Aircraft Communications Addressing and Reporting System It a digital datalink system for transmission of short, relatively simple messages between aircraft and ground stations via radio or satellite). The protocol was designed by Aeronautical Radio, Incorporated (ARINC) to replace their very high-frequency (VHF) voice service and deployed in 1978, uses telex formats.

Airline Code Designator A two-character code designated by IATA to identify an airline (e.g., Alaska Airlines (AS), Air France (AF), and United Airlines (UA)).

Airlines Reporting Corporation The airline reporting corporation (ARC) is an intermediary that provides ticket settlement services between airlines and travel agencies in the United States.

Airline Retailing Maturity Index The new airline retailing maturity (ARM) index is a replacement for the legacy IATA NDC certification process in 2022.

© The Author(s), under exclusive license to Springer Nature Switzerland AG 2024   437
B. Vinod, *Mastering the Travel Intermediaries*, Management for Professionals,
https://doi.org/10.1007/978-3-031-51524-8

Airport Code Designator A three-character code used to identify an airport (e.g., Dallas/Fort Worth (DFW), and London Heathrow (LHR)).

American Society of Travel Advisors The American Society of Travel Agents (ASTA) was founded in 1931. Its members represent 80% of all travel sold in the United States. through the travel agency distribution channel. They also have hundreds of internationally based travel agencies. In 2018, it was renamed the American Society of Travel Advisors.

Artificial Intelligence The development of computer-based methods able to mimic human-like processes such as learning, reasoning, and self-correction. Broad categories include machine learning, natural language processing, and deep learning.

Augmented Reality Augmented reality is an enhanced digital version of the real physical world. Augmented reality can be used to highlight room-specific attributes at a property to educate the guest and generate incremental bookings.

Authenticate Process of proving an assertion of an online user by verifying that identity. With decentralized identifiers (DIDs), it is the process of proving control of the cryptographic private key associated with a public key published in a DID document.

Availability Availability of seats by booking class observed at a point of sale.

Availability Cache Storage of availability data in cache, collected organically, frequently for a large number of airlines, to support air shopping. All GDSs have some form of availability cache; leg/segment based, or O&D based.

Availability Proxy A read-only version of the airline's reservations inventory system that replicates the availability processing logic resident in the airline's master inventory system. More accurate than the availability cache since it reflects the business rules resident in the airline host CRS for availability determination.

B2B Business to business.

B2C Business to consumer.

Bedbanks Bedbanks are B2B platforms that contract supply (rooms) from hotels and accommodation providers and make it available to travel sellers.

Bid Price A bid price is the opportunity cost of not having an incremental seat on a leg in the network. It can also be interpreted as the minimum acceptable fare for a reservation to be accepted on a flight leg.

Bid Price Vector Also called a bid price curve, it is constructed from the network optimization model that assumes primal feasibility is unchanged. It is a set of prices as a function of seats sold or seats available.

Big Data Structured and unstructured datasets so large and complex that it requires new tools and processes. Volume, velocity, and variety are the three dimensions of Big Data.

Billing and Settlement Plan The IATA Billing and Settlement Plan (BSP) is an intermediary that facilitates settlement, reporting and remitting procedures for IATA accredited Passenger Sales Agents. BSP operates in over 180 countries and territories and serves 370 participating airlines.

Blockchain as a Service This is the creation and management of cloud-based networks by third party service providers for companies who want to build blockchain applications.

Boarding Pass A boarding pass is issued by an airline at time of check-in for a flight to a passenger to board a flight.

Booking Class The booking class serves as an identifier for the fares of a specific type or value. It is used for the purposes of inventory control, selling, and ticketing. Booking classes are mapped into base compartments and are commonly known as reservations booking designators (RBD). Revenue management analysts frequently refer to booking classes and fare classes synonymously, which is incorrect. They are not interchangeable. The "owning carrier" of a fare determines the RBDs, which are exclusive to the airline.

Booking Limit The booking limit is the maximum number of seats that can be sold in a booking class or cabin. In a nested inventory control environment, the booking limit is the maximum bookings that can be sold to a given booking class and all booking classes that are nested into it.

Bucket Buckets are virtual and were introduced with virtual nestinginventory controls, where multiple service classes were mapped to a virtual bucket to control availability.

Buyer Refers to the role of the entity that purchases goods from an online merchant. Could be a single individual or could represent an organization.

Cabin or Base Compartment A cabin is a section of the flight leg that has a different seating configuration or is associated with a different level of inflight service. Examples are first class (F), business class (J), and coach class (Y) cabins.

Cabotage Cabotage is the transport of goods or passengers between two points in the same country by an aircraft registered in another country.

Capture The capture rate is the likelihood of a customer booking an alternate flight on any carrier if the first choice is unavailable.

Centralized Identity Management Identity and access management takes place in a single environment. In this environment a user signs into a single environment to access data and applications.

Central Reservations System Airlines, hotels, rental car, and cruise lines own a central reservations system where inventory is hosted (airline seats, hotel rooms, rental cars, and cruise line cabins). An entity's own CRS is commonly referred to as the host CRS.

Change of Gauge Airline marketing term used to designate that a specific flight number changes aircraft, part way through the direct flight.

City Code The three character designation for a city or airport. These codes are assigned by IATA.

City Pair Availability Six characters, a combination of 2 city or airport codes, the departure airport, and the arrival airport, served by direct or connecting flights. The display includes flights in the city pair with numeric booking class availability.

Classless Revenue Management A revenue management system that does not rely on booking classes to forecast demand and control seat inventory.

Code of Federal Regulations The Code of Federal Regulations (CFR) was made effective by the Civil Aeronautics Board on November 11, 1984, which prohibited anticompetitive behavior from airlines that owned the CRSs.

Codeshare Airlines enter partnerships to create marketing flights that are operated by a partner carrier to extend their reach.

Coefficient of Variation The coefficient of variation (CV) is the ratio of the standard deviation to the mean. This measure is used to understand demand uncertainty.

Communication Channel A *type of medium for communicating* marketing incentives. Examples include Web, Email, Call Center, Catalog, TV, etc.

Competition and Market Authority The Competition and Market Authority (CMA) is the competition regulator in the United Kingdom to strengthen business competition and prevent anticompetitive activities.

Competitive Revenue Management A revenue management capability to control what is available for sale based on prevailing market conditions such as competitor selling fares.

Competitive Revenue Management A revenue management capability to control what is available for sale based on prevailing market conditions such as competitor selling rates.

Conjoint Analysis A survey-based statistical technique used in market research to determine how consumers value features of a product or service.

Connecting Market A connecting market represents a service with predefined reasonable estimates of minimum and maximum connect time. Stop-overs do not constitute a connecting service.

Connectivity Connectivity defines the level of participation between the airline and the GDS.

Consolidator Fares Consolidators negotiate contracted rates with multiple airlines to sell tickets to qualified travel agents. These fares can be marked up by the travel agent. These consolidator fares are called net fares and bulk fares. Net fares (also called nett fares) can be marked up by the travel agent. Bulk fares normally can be sold as is and can only be sold as a package with a hotel, car, etc.

Consumer A consumer refers to the ultimate end-user of a product or service in a value chain.

Continuous Nesting Continuous nesting is also called bid-price controls. It is an inventory control technique for origin and destination control.

Continuous Pricing Continuous pricing is a dynamic pricing framework where the optimal price determined by the revenue management price optimizer is used *as is and will be the ticketed price*.

Corporate Fares These are private fares negotiated between an airline and a corporation.

Cost Weighted Mileage Factors IATA defined mileage factors weighted by traffic, used in some cases for interline proration. It is published quarterly.

Customer Relationship Management Customer Relationship Management (CRM) is a set of tools used by an organization to interact with current and future customers.

Customer A customer could refer to a single individual or could represent an organization in the value chain. For example, a travel agency is a customer of the GDS.

Customer Lifetime Value The customer lifetime value is an estimate of the future value of a customer combined with historical value provided to date.

Data Element Identifier The data element identifier (DEI) is an integral part of the schedule change information manual (SSIM). The DEI defines the mapping of the operating flights to the marketing flights.

Decentralized Identifier The decentralized identifier (DID) is a globally unique persistent identifier that does not require a centralized authority for registration. They are designed to enable individuals and organizations to generate their own identifiers using systems they trust. They can be generated and registered cryptographically.

Decentralized Identity Management Identity and access management is dispersed across multiple environments.

Decoy Offer A decoy offer, also known as the decoy effect, or asymmetric dominance, is an inferior offer, that when presented, increases the likelihood of the customer selecting the better offer.

Deep Learning A branch of AI that mimics the workings of a human brain in processing both unstructured and unlabeled data to make decisions by detecting objects, recognizing speech, and translating languages.

Department of Justice The Department of Justice (DOJ) is a federal executive department of the U.S. government that enforces federal law and administration of justice.

Department of Transportation The U.S. Department of Transportation is responsible for planning and coordinating federal transportation projects. It also sets safety regulations for all major modes of transportation. The FAA is under the DOT.

Digital Identity Foundation An organization focused on development of technical specifications and foundational elements necessary to establish an open ecosystem for decentralized identity for people, organizations, apps, and devices. Ensure inter-operability between all participants.

Dilution Revenue dilution occurs when adequate seats are not protected for higher valued passengers.

Direct Connect AvailabilityAgreement The DCA agreement is an addendum to the PCA and stipulates discounts to the rack rate for segment booking fees that a carrier will receive in exchange for full content from the airline. Usually this also includes the highest level of participation for connectivity between a GDS and an airline.

Distributed Availability Distributed availability is deployment of an availability proxy of an airline's inventory system in a public or private cloud to support the growing demands of air shopping worldwide. Deploying the solution in the cloud at multiple locations worldwide reduces network latency.

Distribution Cost Charge The distribution cost charge (DCC) was introduced by Lufthansa in 2015. It was €16 surcharge for GDS bookings.

Downsell Downsell is a special case of recapture to a lower fare on the same flight. Cross-flight recapture denotes recapture to other flight(s) on the host airline.

Dynamic Availability A competitive revenue management capability to respond to competitive market conditions by modifying inventory controls in real time.

Dynamic Pricing A competitive revenue management capability to respond to competitive market conditions by generating a dynamic fare in response to a customer request. There are two versions of dynamic pricing: laddered pricing and continuous pricing.

E-Ticketing E-Ticketing (ETKT) standards were defined by IATA in 1993. The coupons are produced in an electronic format and can be updated across different E-Ticketing databases and for the control of E-Tickets to be passed from one airline to the next.

e-ticketing (Ticketless) This is the ticketless solution invented by Morris Air and used by low-cost carriers. It does not allow the exchange of the ticket records among systems using IATA E-Ticketing standards.

EDI Electronic data interchange is a format for e-commerce in which business data are represented using national or international standards.

EDIFACT Electronic data interchange for administration, commerce, and transport. These are messages that are approved as standards for EDI and can be exchanged through a communication network.

Electronic Miscellaneous Document The electronic miscellaneous document (EMD) is an electronic document that may be issued and used for the collection of travel related ancillary services offered by an airline. EMDs replaced the traditional paper miscellaneous charges order (MCO) documents.

Expected Marginal Seat Revenue The expected marginal seat revenue is the expected revenue of an incremental seat based on a distribution of demand.

Explainable AI This is a field of research to develop tools and frameworks to understand and interpret predictions made by machine learning models. It is required to understand model behavior to improve model performance over time.

Extensible Markup Language Extensible markup language (XML) is a standard communications format in the Internet era.

Fare Basis Code The fare basis code appears on the ticket. It can include letters, numbers, and up to two slashes (/). A fare basis is a compilation of the fare class or ticketing code and one or two ticketing designators. Multiple fare basis codes map to a booking class code. The fare basis code is distinct from the fare class, which is associated to each fare and used in pricing.

Fare Breakpoint These are the terminal points of a fare component for fare construction. It is the destination point where a fare begins or ends.

Fare by Rule The creation of new fares using rules data to specify the market fares and the amounts. The fares can either be calculated from existing fares and rules in the market or specified to create a new fare using the rule provisions in ATPCO Category 25.

Fare Class Code A fare class specifies the rules of an airline's fare. Every fare has a fare class code, and this code appears on the ticket. It is synonymous with fare basis code. Revenue management analysts incorrectly refer to booking classes (RBDs) and fare classes inter-changeably, which should be avoided.

Fare Component The fare component is the most basic unit of fare construction and represents a specific fare between two city pairs.

Fare Construction Fare construction is the process by which various rules are applied to determine the final fare for a ticket.

Fare Guarantee Policy GDSs have a Fare Guarantee Policy where they are accountable for debit memos resulting from fare or tax under collected tickets that meet the criteria for reimbursement.

Fare Management Business process for managing all active fares by monitoring and responding to competitive fare activity.

Fast Response System A fast response system (FRS) has the capability to respond to a high volume of transaction requests, both online and offline, in less than a second. In the NDC context, airlines need to develop fast response systems to instantly respond to requests of offers (air itinerary and the airline bundle).

Federal Excise Tax A federal tax that is charged on airfare. The segment fee of U.S. $4.00 per segment, the September 11 security fee of U.S. $5.60 per one-way flight, and the Passenger Facility Charge (U.S. $4.50 per segment and U.S. $18.00 per round trip) are in additional to the federal excise tax.

Federated Identity Management Federated identity permits an authorized user to access multiple applications and domains using a single set of credentials.

Financial Availability The process of determining whether selling a seat to a customer over the requested origin and destination is financially viable using the net contribution calculation.

Flight Leg A flight leg is a nonstop flight departure from a board point and an off point with a specific flight number and departure time.

Flight Number A flight number is associated with a specific aircraft routing and can consist of one or more flight legs.

FullContent Access Fee The GDSs imposed an 80-cent full content fee per segment booked from participating airlines in 2006. This fee is deducted from the incentives paid by the GDSs to the travel agencies. These programs are called Content Plus Program, Efficient Access Solution, and Content Continuity Program by Amadeus, Sabre, and Travelport, respectively.

Full Content Agreements A full content agreement is a content parity clause in GDS and airline contracts. It stipulated that the same content (schedules, fares, and availability) that was available on the airline direct channel were also available on the indirect GDS channel. After GDS deregulation in 2004, airlines could publish web-only fares. The GDSs gave segment booking fee discounts to the airlines to secure all content, including web fares in the GDS. With IATANDC, full content agreements will eventually disappear.

Funnel Flights A funnel flight is an artificial direct flight that maps to operating flights.

Future Customer Value The future customer value is the net present value of a customer based on projections of future revenue and costs over the duration of the relationship.

GDS New Entrants In 2004, the GDS new entrants (GeNiEs) attempted to disintermediate the GDS model with their open systems platforms as an alternative to the legacy GDS platforms. They attempted to end the GDS oligopoly by promising significantly lower segment booking fees and flexible distribution technology. The prominent GDS new entrants were Trition Distribution Systems, ITA Software, G2 Switchworks, and Farelogix. Unable to penetrate the market, they abandoned their efforts in 2006.

GDS Passthrough This is a term used for NDC bookings made through GDS to the airline. In this scenario, the GDS is the NDC aggregator. NDC bookings, if stored in the GDS, are passive bookings.

Global Distribution System A global distribution system maintains airline schedules and accepts bookings for the requested itinerary. The major GDSs are Amadeus, Sabre, and Travelport.

Go Show Passengers who show up for a flight, without a confirmed reservation for a flight or those that show up with a confirmed reservation number for which no reference is found in the airline's host reservations system.

Gradient The gradient is the incremental bid price and is the adjustment that is made to the bid price for every sell and cancel over a specific leg. It can be interpreted as the rate of change in bid price for a unit change in capacity.

Gradient Boosting Machine A machine learning predictive modeling technique for regression and classification problems. Also known as multiple additive regression trees.

Group Bookings Bookings for a group of passengers that are negotiated with an airline sales representative.

Hidden City A hidden city is a O&D connection with a long layover. Leisure travelers search for cheap fares to their destination that is a hidden city of an O&D. Applicable when a local one-way fare to the destination is more expensive than the connecting O&D.

Hybrid State In this book the term "hybrid state" refers to the period when traditional GDS content that is booked via EDIFACT and NDC content that is booked via XML co-exist on the GDS platform.

IATA The International Air Transport Association was founded by member airlines in 1919. Today, IATA has over 270 airline members.

IATACarrier An airline that is a member of the International Air Transport Association.

IATAClearing House The IATA clearing house (ICH) provides billing and settlement services in multiple currencies for the air transportation industry.

Indexing This is the process of assigning a service class to a bucket on a flight/leg/date. Indexing can be static or dynamic.

Interline An itinerary where two or more airlines operate flights to complete the customer itinerary.

Intermediary An intermediary is a middleman between two entities and facilitates business transactions.

Inventory Control Ability to control seats sold by booking class in an airline's reservations inventory system. Inventory control can be by leg/segment or O&D.

Itinerary An itinerary is a complete trip for a passenger as determined from the passenger name record (PNR). Hence, an itinerary may be one-way or round trip.

Itinerary Pricing To determine the price of an itinerary subject to all the fare rules.

Large Language Model Large language models (LLM)are deep learning algorithms that can recognize, summarize, translate, predict, and generate content using very large datasets. They represent a class of deep learning architectures called transformer networks. They are neural networks that learn context and meaning by tracking relationships in sequential data.

Leg A leg is a nonstop board point and off point.

Leisure Fares Leisure fares are discounted fares with restrictions and sold to the public. They can be accessed from airline websites and through travel agencies.

Lifetime Customer Value The lifetime customer value is the net present value of a customer based on historical performance and projections of future revenue and costs over the duration of the relationship.

Load Factor The ratio of seats sold to the capacity expressed as a percentage. Load factor is post-departure. Predeparture load factor is called booked load factor. For reporting purposes, load factor is expressed as a ratio of RPK/ASK or RPM/ASM.

Local Availability Local availability is the availability for each RBD when an O&D is requested, and the O&D has only one segment. In an O&D inventory control environment, local RBD availability will be quite different from RBD availability for the same segment when it is part of a connecting O&D.

Logistic Regression Logistic regression is a predictive model that is used when the dependent variables are categorical variables.

Loss Function A loss function is a measure of the accuracy of the machine learning model to predict an expected outcome.

Machine Learning Machine learning uses statistical techniques to enable computer systems to "learn" (i.e., progressively improve on a specific task) with data, without being explicitly programmed. Broad categories include unsupervised learning, supervised learning, reinforcement learning, neural networks, and deep learning.

MapReduce Developed by Google, it is a data processing technique for distributed computing. The MapReduce algorithm contains two tasks, Map and Reduce Map, which converts a set of data where individual elements are broken down into tuples (key-value pairs).

Market A market describes a passenger's one-way origin and destination pair, regardless of connect points and time of day. An online market consists of one or more flight legs that are on a specific carrier. An interline market consists of one or more flight legs that are on a combination of two or more carriers.

Market Adjustment Table The market adjustment table (MAT) is required for O&D control. The conditioned fares are averaged by fare qualification rules and stored on the host CRS inventory system. This is called the market adjustment table which can be modified to increase or decrease availability by service class and POS. This is also called market class fare adjustment (MCFA) table or market value table (MVT).

Market Restricted Flight A market restricted (MR) flight is a specific flight posted on a GDS for which availability must be directly queried on the host CRS via a seamless request for true last seat availability.

Marketing Information Data Tapes Electronic records of a travel agency's sales history. The GDS records, owns, and market MIDT data based on booking transactions generated by travel agents.

Marketplace The electronic medium that brings together buyers and sellers to trade products and services online through fair and competitive means. A global distribution system (GDS) is a marketplace that brings suppliers (sellers) and travel agents (buyers) together to transact business.

Metasearch A metasearch engine is an online information retrieval tool that combines the results from various search engines into one and produces a consolidated result.

Micro Segmentation Process of segmenting customers at a lower level of detail based on unique attributes.

Minimum Connect Time The minimum connect time (MCT) is specified in the airline schedules published by schedule aggregators.

Most Favored Nation The most favored nation clause (MFN) was standard in airline-GDS contracts. In the former case, it is a provision in which the GDS agrees to give the airline (seller) the best terms it makes available to any other airline. MFN clauses also exist in some cases with TMC-corporation contracts.

Multimodal Transportation Multimodel transportation is a customer itinerary that includes two or more modes of transportation such as full-service carriers, low-coast carriers, rail, and bus.

Multinomial Logit Model A multinomial logit (MNL) model is a classification method that generalizes logistic regression to multiclass problems with more than two discrete outcomes.

Multiple Virtual Storage More commonly called MVS was the most used operating system on the System/370 and System/390 IBM mainframe computers for commercial applications.

NDC Aggregators An NDC aggregator is an intermediary that connects with an airline to access NDC content (priced itineraries) and deliver it to the point of sale. Besides third party new entrants in the marketplace, the GDSs are by default NDC aggregators.

NDCExchange Initiated by ATPCO, the NDC exchange was an industry owned solution that served as a neutral hub to connect airlines with sellers. The hub supported real time message translation to address the many versions of the XML-based messaging standard.

Nesting The hierarchy of booking classes determines the order in which the booking classes should be nested. The objective of nesting is to ensure that lower valued classes are not open for sale when higher valued classes are closed for sale. Net (standard) nesting and threshold (theft) nesting are the two methods for calculating availability. Applicable for both booking class controls and virtual nesting controls.

Net Contribution The net contribution is the difference between the market fare value for an O&D and the total bid price for the one-way (directional) itinerary. The net contribution calculation can also be extended to round trips on airline websites that exercise round trip control.

Network (airline) A network represents the schedule for an airline for a day or a subset of the schedule for the day.

Neural Networks Algorithms that attempt to recognize underlying relationships in data through a process that mimics how a human brain operates.

Next Generation Storefront The next-generation storefront (NGS) is a set of data standards established by ATPCO and adopted by GDSs with variations for their agency desktop to better represent, sort, and find the airline products and services for travel agents.

New Distribution Capability The new distribution capability (NDC) was established by IATA in 2021. The objective was to promote a new XML-based messaging standard to transform the way airline products are distributed and sold in the marketplace.

No Shows Booked passengers that do not show up for the flight at departure time.

NP-hard A problem in NP-hard if it can be translated into one for solving any NP-problem (nondeterministic polynomial time) problem. NP-hard problems are at least as hard as any NP problem. It is likely that there are no polynomial-time algorithms for NP-hard problems.

O&D Inventory Control The control of seat inventory by origin and destination of the request.

Offer and Order The NDC initiative is about offer and order management that is controlled by the airline. Besides airlines controlling the content of the offer (base fare + ancillaries) travelers will have a single record for their order.

Offer Management Offer management is an extension of traditional revenue management to offer a base fare and ancillary bundle to a customer.

Offer Partner Airline An "offer partner airline" (OPA) is the airline partner who responses to the ORA with the code share itinerary and offer for an itinerary that has code share segments.

Offer Responsible Airline An "offer responsible airline" (ORA) is the airline responsible for creating the offer for an itinerary that has code share segments.

Online Booking Engine An online booking engine is an application on airline websites to capture and process direct online reservations. Besides the direct channel, online booking engines are used by OTAs and self-service corporate booking tools.

Online Travel Agencies An online travel agency (OTA) is a web-based marketplace that allows online users to search and book travel products and services, including flights, hotels, cars, cruises, activities, etc.

Opaque Fares Opaque fares are frequently restricted and cannot be sold as is, and only as part of a package (with a hotel or car package).

Open Travel Alliance The Open Travel Alliance is a nonprofit standards body that is creating messaging standards across all lines of business (air, hotel, car, etc.). It was established in 1999.

Operation Research Techniques based on mathematical methods—discrete optimization, stochastic modeling, large-scale optimization modeling.

Optimization Group An optimization group represents a selection of flights from the schedule that can be grouped together based on the arrival departure pattern. For example, an arrival-departure complex from a hub can be an optimization group. The optimization group can also represent the entire schedule for a future departure date. Optimization groups are required for O&D revenue management to capture all interactions in network flow traffic.

Origin and Destination An origin and destination is a nonstop or connecting market with the departure time of day qualifier. It is synonymous with service.

Overbooking Authorizing more reservations than capacity to be accepted to compensate for the effects of cancellations and no shows.

Overlapping Flights Overlapping flights are also referred to as back-to-back ticketing. Passengers book overlapping flights to circumvent minimum stay restrictions.

PADIS Board The Passenger and Airport Data Interchange Standards (PADIS) Board develops and maintains Electronic Data Interchange and XML message standards for passenger travel and airport-related passenger service activities.

Participating Carrier Agreement The participating career agreement (PCA) is a common contract form between a GDS and an airline to distribute and sell the airline product. The PCA has a standard rack rate that governs the economics and participation of an airline in the GDS.

Passenger Facility Charge Commercial airports controlled by public agencies began imposing passenger facility charges on June 1, 1992 at $3.00 per passenger enplanement ($12.00 per round trip). The cap was raised to $4.50 ($18.00 per round trip) effective April 1, 2001. PFCs are federally authorized but levied by local airport operators, which set the amounts.

Passenger Type Code The passenger type codes are defined by ATPCO. It is a fare related classification for each passenger. For example, ADT is adult passenger, GVT is government travel, VFR is visit friends/relatives, AST is airline staff standby, etc.

Passive Segments A passive segment is a passive booking in the GDS for which a ticket will not be issued. It is an informational segment that has the complete itinerary of the passenger. NDC bookings are passive segments in the GDS.

Personalization Used in the context of offer management, to personalize the offer consisting of the base fare and ancillaries. Personalization is always 1:1 (for a segment of ONE).

Physical Availability This process determines whether a seat can be sold in a cabin by comparing the authorized capacity (including overbooking) against the seats sold count. If the authorized capacity is greater than seats sold, seats are physically available.

Point of Commencement The point at which a customer's journey originates.

Point of Sale Identifies the location of the travel agency where the booking was made.

Potential Future Value The potential future value is an estimate of the residual future value of a customer.

Preference Driven Air Shopping Preference driven air shopping (PDAS) display algorithms are based on trade-off analytics. PDAS is superior to traditional travel website filters, since a filter would exclude an itinerary based on one attribute, even though it would have been outweighed by the goodness in the other attributes. To deploy PDAS, there are two alternative techniques: TOPSIS (Technique for Ordering Preferences by Similarity to Ideal Solution) and VIKOR (VlseKriterijumska Optimiza-cija I Kompromisno Resenje).

Predictive Analytics A collection of techniques from statistical modeling and machine learning to process current and historical data to predict future events.

Preferred Channels These are product distribution channels that are preferred by airlines.

Price Elasticity A measure of a customer's willingness to pay for a product based on the price. The price elasticity of demand is the ratio of the percentage increase in demand to the percentage change in price.

Priceable Unit Also called a pricing unit (PU). One or more fare components make up a PU. One or more PU combinations produce a pricing solution for a trip.

Private Fares Private fares are only available through consolidators and travel agencies. Examples are corporate fares, net fares, and bulk fares.

Programmed Airline Reservations System The programmed airline reservations system (PARS) was developed by IBM as a generalized version of the Sabre reservations system they developed and launched with American Airlines in 1964. Within a few years, all IBM-based reservations systems were baselined on PARS. The international version of PARS, developed by IBM with functional requirements from BOAC and other airlines is called IPARS.

Protected Seats Seats that are protected in a booking class from lower valued booking classes.

Protection Levels This represents the seats protected for a booking class from lower valued booking classes.

Public Fares Also known as published fares, these fares are available for immediate purchase through any travel agency or airline website.

Publish and Subscribe Environment A publish and subscribe (Pub/Sub) environment is a messaging service that is asynchronous and scalable. It helps in decoupling services that produce messages from the services that process those messages. This service enables services to communicate with each other asynchronously.

Qantas Intelligent Keypad The Qantas Intelligent Keypad (QIK) is an intelligent airline agent application developed in the 1980s by Qadrant, a subsidiary of Qantas Airways, as a front end for mainframe airline reservations systems (host CRS).

Quality of Service Adjusted Price An airline's competitive fare response based on schedule and fare attributes. This is an improvement over traditional rule-based matching.

Quality of Service Index The quality of service index (QSI) has been used by airlines to predict their "fair share" based on relative attractiveness of their schedule versus competitors. QSI accounts for a range of schedule attributes such as aircraft type, frequency of service, and type of service (nonstop, single connect, double connect, interline, etc.).

Reading Day A reading day is a pre-departure snapshot when demand forecasts and inventory controls are updated. It is also referred to as data collection point (DCP).

Recapture Recapture is a special case of capture on to the same (host) airline.

Regression A predictive modeling technique that determines the relationship between a dependent variable and a set of independent variables.

Reinforcement Learning A machine learning technique to train machine learning models to make a sequence of decisions. With a goal to maximize the total reward, the agent learns to achieve the goal and receives either rewards or penalties for the actions it performs.

Request for Quotation Prior to negotiating the contract, a customer may submit a request for quotation (RFQ) to multiple entities.

Resolution 787 The purpose of Resolution 787 was to aid the growth of an open data exchange standard that uses extended markup language (XML). This new standard would be used alongside the teletype and EDIFACT data exchange standards that are already overseen by IATA for the airline industry. Also known as the new distribution capability (NDC), Resolution 787 aimed to replace the outdated messaging exchange standard with a modern XML standard for communication between airlines and intermediaries.

RestrictionFree Pricing Fare filings introduced by the LCC's where the fares have no restrictions (absence of fences such as advance purchase and minimum stay restrictions) and the fare amount is the only determinant of the market segment.

Revenue Per Available Seat Mile The revenue per available seat mile (kilometer) is defined as the passenger revenue per available seat mile (kilometer). RASM (RASK) is monitored at different levels such as route, market, market entity, and system.

Seamless Availability A seamless availability request is a direct connect availability request wherein a reservation request originating from a GDS will be queried for availability directly on a host CRS.

Seamless Sell A direct connect sell request is also called a seamless sell request. Interactive sells are made against the airline inventory system. Seamless sell allows full usage of host CRS O&D controls to compute availability before the booking is made.

Segment A segment is a sequence of one or more flight legs with the same flight number.

Segment Close Indicator An indicator on the airline's inventory control system that closes a booking class for future sales.

Segment Limit Sales A numeric limit on the airline's inventory control system that limits sales in a booking class. Usually used in conjunction with leg class controls.

Seller Refers to the role of the entity that is selling products online. Could be anywhere along the supply chain (e.g., supplier, manufacturer, distributor, and e-tailer).

Semi-Supervised Learning Situations arise when there is a large amount of input data and only some of the data are labeled.

Service A service describes the market plus the sequence of connections and time of day.

Sovrin The Sovrin Foundation is a nonprofit organization established to promote Internet identity for all and to administer the Governance Framework for the Sovrin Network, a decentralized global public network enabling self-sovereign identity on the Internet.

Spill Estimate of passengers who were turned away because their first choice was unavailable.

Spoilage Seats that are empty on a flight that was closed for sale before departure. This is called overbooking spoilage, which is different from discount allocation spoilage.

Station A station is equivalent to an airport.

Supervised Learning These techniques are designed to learn by example. The input data for this method is a labeled training dataset. The labeled data has the correct answers, and the algorithm learns from this dataset to make predictions.

Supplier The term "Supplier" in travel refers to airlines, hotels, rental car, rail, cruise lines, and ferry lines.

Support Vector Machines A set of supervised learning modes used for classification, regression, and outlier detection.

Thru Availability Thru (through) availability is the availability for each RBD when an O&D is requested, and the O&D has more than one segment. Pricing analysts frequently refer to connecting availability as "thru" availability since fare construction is based on "thru fares".

Thru Fare A thru (through) fare is a fare for the market with a fare class (fare basis code) regardless of the number of connections in the schedule.

Ticketed Point Mileage This is the shortest distance between any two points on an operating route, used in airline fare calculations regardless of the airports used.

Transaction Processing Facility Transaction processing facility (TPF) is a real time operating system for IBM mainframe computers. Key characteristics are high-volume transaction processing, supports many concurrent users, and fast response times.

Transactions Per Second Transactions per second (TPS) and peak transactions per second refers to the number of transactions a system (e.g., a GDS) can process in one second.

Travel Management Company Travel management companies are large travel agencies that pre-dominantly manage corporate business travel programs. Some large TMCs are also focused on leisure travel.

Trip-purpose segmentation A method to implicitly segment customers based on the context for travel.

Universal Product Attributes ATPCO Routehappy content type that provides access to airline products and services with images, messages, videos, and cabin tours. The content is specific to aircraft, cabin, route, time of day, and fare providing relevant content during air shopping by a customer.

Universal Ticketing Attributes ATPCO Routehappy content type provides descriptions of fare restrictions in simple language for customers during the shopping process. Data is sourced from ATPCO Fares, Branded Fares, and Optional Services.

Unsupervised Learning A collection of techniques used to draw inferences from data that does not have any labeled responses.

Upgrade An upgrade is an offer to a customer to sit in a higher class of service at no additional cost.

Upsell Upsell is a special case of recapture to a higher fare or bundle on the same flight. Cross-flight recapture denotes recapture to other flight(s) on the host airline.

Use Case A use case is a collection of workflows that together complete a particular business objective.

Value Pricing An approach to pricing products based on what the market would be willing to pay for the service level of the product.

Virtual Credit Card A *virtual credit card (VCC) is a digital version of a credit card with* a unique, disposable 16-digit card number with a CVV code and expiration date that can be generated instantly and used to make purchases.

Virtual Nesting An O&D inventory control technique where service classes are mapped into virtual buckets.

Virtual Reality Virtual reality is an enhanced digital version of an artificial environment to create immersive experiences. Luxury properties can leverage virtual reality to highlight unique features of the property such as free form pool, banquet rooms, signature restaurants, activities, etc.

Web Link Text that contains the address to other content.

Wholesaler In travel a wholesaler negotiates deeply discounted prices with suppliers and marks it up and delivers it to various points of sale.

Workflow A workflow is a collection of activities and events executed by systems and external agents participating in a business process.

Yield Yield is defined as the passenger revenue per revenue passenger mile (kilometer). Yield is monitored at different levels such as route, market, market entity, and system.

Yield Management The original name for revenue management, coined by Robert L Crandall of American Airlines. Yield Management was renamed revenue management in 1993.

YQ/YR Surcharges used in International Markets. Sometimes YQ/YR can be greater than the base fare since airlines do not pay commissions on YQ/YR (only pay commissions on the base fare). Fuel surcharges filed as YQF or YRF, insurance charges filed as YQI or YRI. YQ/YR fares are not a tax, but a validating carrier specific fee, not interline able, not commissionable. The Carrier-Imposed (YQ/YR) Fees solution provides marketing carriers (carriers that appear on the flight coupon) the ability to control and collect fees at the sector (coupon), at the portion of travel (multiple sectors), or on the journey.

References

Airoldi, D. M. (2022, December 5). American tells TMCs to be NDC-ready by April or lose some content access. *Business Travel News*. Retrieved from https://www.businesstravelnews.com/Distribution/American-Tells-TMCs-to-Be-NDC-Ready-by-April-or-Lose-Some-Content-Access

Alexander, K. L. (2002, March 15). Delta ending many travel agent commissions. *The Washington Post*. Retrieved from https://www.washingtonpost.com/archive/business/2002/03/15/delta-ending-many-travel-agent-commissions/aacf7a1f-5673-4b5e-aa84-a661cafb948e/

Alexander, K. L. (2006, January 31). Paying more for small extras. *The Washington Post*.

American Airlines, Inc Plaintiff vs Sabre. (2021). *Plaintiff's verified original petition and application for temporary injunction and permanent injunction, filed Tarrant country 6/29/2021, Thomas A Wilder, District Clerk*. Retrieved from https://skift.com/wp-content/uploads/2021/07/aa-sabre-lawsuit-july-1.pdf

Anderson, C. K. (2009, October). *The billboard effect: Online travel agent impact on non-OTA reservation volume*. Cornell University School of Hotel Administration, Cornell hospitality report, 9(16).

Anderson, C. (2011, April). Search, OTAs, and online booking: An expanded analysis of the billboard effect. *Cornell Hospitality Report, 11*(8), 1–10.

Ariely, D. (2010). *Predictably irrational, revised and expanded edition: The hidden forces that shape our decisions*. Harper Perennial. ISBN-10: 9780061353246, ISBN-13: 978-0061353246.

Arrieta, A. B., Diaz-Rodriguez, N., Del Ser, J., Bennetot, A., Tabik, S., Barbado, A., Garcia, S., Gil-Lopez, S., Molina, B., & R., Chatila, R., Herrera, F. (2020). Explainable artificial intelligence (XAI): Concepts, taxonomies, opportunities and challenges toward responsible AI. *Information Fusion, 58*, 82–115.

ASTA. (2006, July 13). *American Airlines announcement threatens to throw industry into chaos, says ASTA*.

Astrahan, M. M., & Jacobs, J. J. (1983). History of the design of the SAGE computer – The AN/FSQ-7. *Annals of the History of Computing, 5*, 340–349.

Aviation Week. (1997, February 03). DMR acquires Qantas's stake in Qadrant International. *Aviation Week*. Retrieved from https://aviationweek.com/dmr-acquires-qantass-stake-qadrant-international

Bacon, D. R., Besharat, A., Parsa, H. G., & Smith, S. J. (2016). Revenue management, hedonic pricing models and the effects of operational attributes. *International Journal of Revenue Management, 9*(2/3), 147–164.

Baker, M. B. (2023, June 05). Accelya enables NDC exchange for American Airlines EDIFACT tickets. *Business Travel News*. Retrieved from https://www.businesstravelnews.com/distribu tion/accelya-enables-ndc-exchange-for-american-airlines-edifact-tickets?utm_source= eNewsletter&utm_medium=email&utm_campaign=nstraveltoday&oly_enc_id=9607B24 85578C3T

Balcombe, K., Fraser, I., & Harris, L. (2009). Consumer willingness to pay for in-flight service and comfort levels: A choice experiment. *Journal of Air Transport Management, 15*, 221–226.

Bell, P. C., Anderson, C. K., & Kaiser, S. P. (2003) Strategic operations research and the Edelman Prize finalist applications 1989–1998 [Electronic version]. Operations Research, 51(1), 17–31. Retrieved [insert date], from Cornell University, School of Hospitality Administration site: http://scholarship.sha.cornell.edu/articles/218/

Belobaba, P. P. (1987). *Air travel demand and airline seat inventory management.* Massachusetts Institute of Technology, MIT Publication, MIT Flight Transportation Laboratory Report R87-7.

Belobaba, P. P. (1989). Application of a probabilistic decision support model to airline seat inventory control. *Operations Research, 37(2)*, 183–197.

Belobaba, P. P. (1992). Optimal vs heuristic methods for nested seat allocation. *AGIFORS Yield Management Study Group*, Brussels, Belgium, May 4.

Belobaba, P. P. (2019, March 19). *PODS consortium research update: Dynamic pricing mechanisms*. ATPCO Dynamic Pricing Working Group.

Ben-Akiva, M., & Lerman, S. (1985). *Discrete choice analysis: Theory and application to travel demand* (6th ed.). Massachusetts Institute of Technology Press.

Benzinger, M. A., Laohoo, L., Sandhu, J. S., Smith, B. C., Zhang, Y., & Zouaoui, F. (2008, July 31). *System and method for estimating seat value.* U.S. Patent Office Application Number 11/627684.

Biesiada, J. (2017, November 17). Private-channel airline distribution deals spark concerns. *Travel Weekly*. Retrieved from https://www.travelweekly.com/Travel-News/Travel-Agent-Issues/Trav elers-are-turning-to-travel-advisors-survey-reveals

Biesiada, J. (2021, May 05). Travelers are turning to travel advisors, survey from ASTA and Sandals reveals. *Travel Weekly*. Retrieved from https://www.travelweekly.com/Travel-News/ Travel-Agent-Issues/Travelers-are-turning-to-travel-advisors-survey-reveals

Biesiada, J. (2023a, July 24). Analysts say Expedia-Hopper split was bound to happen. *Travel Weekly*. Retrieved from https://www.travelweekly.com/Travel-News/Travel-Agent-Issues/Ana lysts-say-Expedia-Hopper-split-bound-to-happen

Biesiada, J. (2023b, October 2). In DOT complaint, ASTA requests punitive measures against American Airlines. *Travel Weekly*. Retrieved from https://www.travelweekly.com/Travel-News/Airline-News/ASTA-requests-punitive-measures-against-American-Airlines

Bilotkach, V., Rupp, N., & Pai, V. (2014). *Value of a platform to a seller: Case of American airlines and online travel agencies, working paper no. 13-08* (pp. 1–32). NET Institute. Retrieved from www.netinst.org

Blankenbaker, J., & Xie, P. (2014). Hotel Hopper, Innovation Garage Pitch, Sabre Research, June 12.

Boehmer, J. (2022a, October 5). Reinventing itself under bankruptcy, SAS readies a novel distribution model. *The Beat*. Retrieved from https://www.thebeat.travel/News/SAS-Readies-A-Novel-Distribution-Model

Boehmer, J. (2022b, November 3). American, Sabre resolve contract litigation over displays, alleged Delta bias. *The Beat*. Retrieved from https://www.thebeat.travel/News/American-Sabre-Resolve-Contract-Litigation

Boyd, E. A. (1998). Airline alliance revenue management: Global alliances within the airline industry add complexity to the yield management problem. *OR MS Today*.

Boyd, E. A. (2007). *The future of pricing: How airline ticket pricing has inspired a revolution.* Palgrave Macmillan. ISBN 10: 0230600190, ISBN-13: 978-0230600195.

Bradberry, R. (2013). A 'fare' deal: How to incorporate ancillaries, merchandising, and personalization into corporate air deals. *Ascend, 12*(1), 7–8.

Breaking Travel News! (2019, December 23). *Robots to guide British Airways passengers through Heathrow*. Retrieved from https://www.breakingtravelnews.com/news/article/robots-to-guide-british-airways-through-heathrow/

Bryan, J. A. (1989, January 16). Donald Burr may be ready to take to the skies again. *Business Week*.

BTS. (2012, September 10). *Table A-8 - U.S. Airline Industry Ancillary Fees: 2004–2010. Bureau of Transportation Statistics, U.S. Department of Transportation*. Retrieved from https://www.bts.gov/archive/publications/transportation_statistics_annual_report/2010/chapter_01/table_a_08

Burns, J. (2015, January 12). *Pegasus Solutions electronic distribution becomes DHISCO, hospitality upgrade*. Retrieved from https://www.hospitalityupgrade.com/techTalk/January-2015/Pegasus-Solutions-Electronic-Distribution-Becomes/

Business Travel News. (2004, November 29). *Amadeus delves deeper into new GDS pricing model*. Retrieved from https://www.businesstravelnews.com/More-News/Amadeus-Delves-Deeper-Into-New-GDS-Pricing-Model

Byrd, M., & Darrow, R. (2021). A note on the advantage of context in Thompson sampling. *Journal of Revenue and Pricing Management, 20*, 316. https://doi.org/10.1057/s41272-021-00314-1

CAAC. (2020). *Statistical Bulletin of Civil Aviation Industry Development in 2020*. Retrieved from http://www.caac.gov.cn/en/HYYJ/NDBG/202202/P020220222322799646163.pdf

Campbell, J. (2006, July 31). Interview: Travelport SVP Kurt Ekert. *The Beat*.

Carmichael, D. (1982, October 1). Two of nation' largest airlines reach agreement. *UPI Archives*. Retrieved from https://www.upi.com/Archives/1982/10/01/Two-of-nations-largest-airlines-reach-agreement/2350402292800/

Carroll, W. J., & Grimes, R. C. (1995). Evolutionary change in product management: Experiences in the car rental industry. *Interfaces, 25*, 84–104.

Castro, J., & Crandall, R. L. (1992, May 04). ROBERT CRANDALL: This industry is always in the grip of its dumbest competitors. *Time*.

Catron, F. J. (2024, March 25). AMEX GBT acquiring CWT in $570M deal. *Phocuswire*. https://www.phocuswire.com/Amex-GBT-acquiring-CWT-in-570-million-deal?utm_source=newsletter&utm_medium=email&utm_campaign=Daily&oly_enc_id=9607B2485578C3T

Chatwin, R. E. (1993). *Optimal airline overbooking*. Ph.D. thesis, Department of Operations Research, Stanford University.

Chavez-Dreyfuss, G. (2021, November 16). American Airlines, travel platform winding tree announce blockchain partnership. *Reuters*. Retrieved from https://www.reuters.com/business/finance/american-airlines-travel-platform-winding-tree-announce-blockchain-partnership-2021-11-16/

Chicago Convention. (1944). *Convention on International Civil Aviation done at Chicago on the 7th day of December 1944, Doc 7300*. Retrieved from https://www.icao.int/publications/documents/7300_orig.pdf

Chollet, F. (2019, November 5). *On the measure of intelligence*. Retrieved from https://arxiv.org/pdf/1911.01547.pdf

Choubert, L., Fiig, T., & Viale, V. (2015). Amadeus dynamic pricing. In *AGIFORS revenue management and distribution study group meeting, Shanghai, China*.

Christ, S. (2009). *Operationalizing dynamic pricing models: Bayesian demand forecasting and customer choice modeling for low cost carriers* (1st ed.). Ph.D. dissertation, University of Augsburg, Germany, Springer.

Chui, M., Manyika, J., Miremadi, M., Henke, N., Chung, R., Nel, P., & Malhotra, S. (2018, April). *Notes from the AI frontier: Applications and value of deep learning*. McKinsey & Company. Retrieved from https://www.mckinsey.com/featured-insights/artificial-intelligence/notes-from-the-ai-frontier-applications-and-value-of-deep-learning

Civil Aeronautics Board. (1967, January 10). *Economic Regulations Docket 16563*.

Clarke, P. (2016). Priceline eliminates 'Name Your Own Price' for airline tickets, Travel Pulse, September 07.

Cook, T. M. (1999, December 13). Creating competitive advantage using model-driven decision support systems. In *Presented at the international conference of information systems*.

Copeland, D. G. (1995). Sabre: The development of information-based competence and execution of information-based competition. *IEEE Annals of the History of Computing, 17*(3), 30–57.

Copeland, D. G., & McKenney, J. L. (1988, September). Airline reservations systems: Lessons from history. *MIS Quarterly*.

Crandall, R. L. (1991). How you benefit from overbooking. *American Way Magazine*.

Crandall, R. L. (1993, September 14). The airline in transition: Competitive challenges for the 1990s. In *Presented at the society of airline analysts*.

Crandall, R. L. (1995, September). The unique U.S. airline industry. In D. Jenkins (Ed.), *Aviation daily's handbook of airline economics*. The Aviation Weekly Group of the McGraw-Hill Companies.

Crandall, R. L. (1998, May 1). How airline pricing works. *American Way Magazine*.

Cross, R. G. (1995). An introduction to revenue management. In D. Jenkins (Ed.), *Aviation daily's handbook of airline economics* (pp. 443–458). The Aviation Group of the McGraw-Hill Companies.

Cross, R. G. (1997). Revenue Management: Hard Core Tactics for Market Domination, ISBN 0-553-06734-6. *Broadway Books*, New York, NY.

Cross, R. G. (1998). Trends in airline revenue management. In G. F. Butler & M. R. Keller (Eds.), *Aviation week group newsletters* (pp. 303–318). Edmund Pinto.

Croston, J. D. (1972). Forecasting and stock control for intermittent demands. *Journal of Operational Research Quarterly, 23*(3), 289–303.

Cummings, N. (2007, January). How Sabre invented YM. *British Operational Research Society, OR Newsletter*.

Curley, A., Garber, R., Krishnan, V., & Tellez, J. (2020, August 13). *For corporate travel a long recovery ahead*. McKinsey & Company.

Curry, R. E. (1990). Optimal airline seat allocation with fare classes nested by origins and destinations. *Transportation Science, 24*, 193–204.

Curry, R. (1995, October 17). A market-level pricing model for airlines. In *7th international revenue management conference, Toronto Ontario*.

Dadoun, A., Platel, M. D., Fiig, T., Landra, C., & Troncy, R. (2021). How recommender systems can transform airline offer construction and retailing. *Journal of Revenue and Pricing Management, 20*, 301. https://doi.org/10.1057/s41272-021-00313-2

Darrow, R. (2021). The future of AI is the market. *Journal of Revenue and Pricing Management, 20*, 381. https://doi.org/10.1057/s41272-021-00321-2

Daudel, S., & Vialle, G. (1989). *Le yield management: La face encore cachee du Marketing des services*. InterEditions.

Davenport, T., & Ronanki, R. (2018, January–February). Artificial intelligence for the real world. *Harvard Business Review*.

Davis III, J. F. (2002, June 25). *The formation of THISCO. HospitalityNet*. Retrieved from https://www.hospitalitynet.org/news/4012295.html

Davis, C. (2023, August 14). ASTA asks DOT to require AA to return fares to EDIFACT. *Business Travel News*. Retrieved from https://www.businesstravelnews.com/Distribution/ASTA-Asks-DOT-to-Require-AA-to-Return-Fares-to-EDIFACT

de Marcken, K., (2003). *Computational complexity of air travel planning*. ITA Software. Retrieved from http://www.demarcken.org/carl/papers/ITA-software-travel-complexity/ITA-software-travel-complexity.html

Dean, J., & Ghemawat, S. (2004, December). MapReduce: Simplified data processing on large clusters. In *Sixth symposium on operating system design and implementation* (Vol. 6). OSDI.

Dempsey, P. S., & Gesell, L. E. (1997). *Airline management: Strategies for the 21st century*. Coast Aire Publications.

Dezelak, M., & Ratliff, R. (2018). Towards new industry-standard specifications for air dynamic pricing engines. *Journal of Revenue Pricing Management, 17*(6), 394–402.

Domeyer, A., McCarthy, M., Pfeiffer, S., & Scherf, G. (2020, August 31). *How governments can deliver on the promise of digital ID*. McKinsey & Company. Retrieved from https://www.mckinsey.com/industries/public-sector/our-insights/how-governments-can-deliver-on-the-promise-of-digital-id#/

Donovan, A. W. (2005). Yield management in the airline industry. *Journal of Aviation/Aerospace Education and Research, 14*(3), 11–19.

Economist. (2021). The digital payments tipping point: Why the pandemic makes payments innovation mandatory, BrandConnect, The Economist Group and WEX.

Dudik, M., Langford, J., & Li, L. (2011). Doubly robust policy evaluation and learning. In *Proceedings of the 28th international conference on machine learning, Bellevue, WA, USA*.

Edelman, B. (2014, June 4–8). Mastering the intermediaries: Strategies of dealing with the likes of Google, Amazon and Kayak. *Harvard Business Review*.

Eklund, J. (1994). The reservisor automated airline reservation system: Combining communications and computing. *IEEE Annals of the History of Computing, 16*(1), 62–69.

Etzioni, O., Tuchinda, R., Knoblock, C. A., & Yates, A. (2003, August). To buy or not to buy: Mining airfare data to minimize ticket purchase price. In *Proceedings of the ninth ACM SIGKDD international conference on knowledge discovery and data mining* (pp. 119–128). ACM.

Fader, P. S., & Hardie, B. G. S. (2009, June 14–17). Probability models for customer-base analysis. In *20th annual advanced research techniques forum*.

Fader, P. S., Hardie, B. G. S., & Lee, K. L. (2005a). "Counting your customers" the easy way: An alternative to the Pareto/NBD model. *Marketing Science, 24*(2), 275–284.

Fader, P. S., Hardie, B. G. S., & Lee, K. L. (2005b, November). RFM and CLV: Using iso-value curves for customer base analysis. *Journal of Marketing Research, 42*, 415–430.

Feldman, J. M. (1995, August). Reclaiming control. *Air Transport World*.

Fielding, R. T. (2000). *Architectural styles and the design of network-based software architectures. PhD Dissertation, University of California, Irvine*. Retrieved from https://www.ics.uci.edu/~fielding/pubs/dissertation/top.htm

Fiig, T., Goyons, O., Adelving, R., & Smith, B. C. (2016). Dynamic pricing - The next revolution in RM? *Journal of Revenue and Pricing Management, 15*(5), 360–379.

Fiig, T., Guen, R. L., & Gauchet, M. (2018). Dynamic pricing of airline offers. *Journal of Revenue and Pricing Management, 17*, 281–293.

Fisher, J., & Mongalo, M. (1993). Integrating decision support. In *SAS users group international conference* (pp. 619–623).

FlightGlobal. (2003, February 28). *SITA opens up reservations system*. Retrieved from https://www.flightglobal.com/sita-opens-up-reservation-system/47135.article

Ford, G. R. (1976, September 08). *Statement on international air transportation policy*. Retrieved from https://www.presidency.ucsb.edu/documents/statement-international-air-transportation-policy

Fox, L. (2019, February 27). Sabre brings a dose of reality to artificial intelligence. *Phocuswire*. Retrieved from https://www.phocuswire.com/sabre-artificial-intelligence

Fox, L. (2022, March 14). Flight centre acquires majority stake in TPConnects. *Phocuswire*. Retrieved from https://www.phocuswire.com/Flight-Centre-acquires-majority-stake-in-TPConnects

Frary, M. (2022, December 9). Changes in the air. *Business travel News Europe*. Retrieved from https://www.businesstravelnewseurope.com/Powered-Up/Distribution

French, S. (2023, February 1). When will business travel return to normal? *NerdWallet*. Retrieved from https://www.nerdwallet.com/article/travel/how-long-until-business-travel-returns-to-normal#:~:text=When%20will%20business%20travel%20return%3F,its%202019%20levels%20this%20year

Gallacher, J. (1996, February). Pricing it right. *Airline Business*.

Gallego, G., & Hu, M. (2014). Dynamic pricing of perishable assets under competition. *Management Science, 60*(5), 1241–1259.

Gallego, G., & Topaloglu, H. (2019). *Revenue management and pricing analytics* (International series in operations research & management science). Springer. ISBN-10: 1493996045, ISBN-13: 978-1493996049.

Gallego, G., & van Ryzin, G. (1994). Optimal dynamic pricing of inventories with stochastic demand over finite horizons. *Management Science, 40*(8), 999–1020.

GAO. (2003, July). *Airline ticketing impact of changes in the airline ticket distribution industry: Report to congressional requesters.* U.S. General Accounting Office.

Garrow, L. A. (2016). *Discrete choice modeling and air travel demand: Theory and applications* (1st ed.). Routledge. eBook ISBN: 9781315577548.

Geraghty, M. K., & Johnson, E. (1997). Revenue management saves national car rental. *Interfaces, 27*.

Gershgorn, D. (2016, March 12). *Google's AlphaGo beats world champion in third match to win entire series.* Popular Science.

Ghemawat, S., Gobioff, H., & Leung, S.-T. (2003, October). The Google file system. In *19th ACM symposium on operating systems principles, Lake George, New York.*

Gitnux. (2021). *The most surprising travel agency statistics and trends in 2023.* Gitnux. Retrieved from https://blog.gitnux.com/travel-agency-statistics/#:~:text=In%202021%2C%20there%20 are%20105%2C000,the%20end%20of%20the%20year

Gittins, J., Glazebrook, K., & Weber, R. (2011). *Multi-armed bandit allocation indices* (2nd ed.). Wiley.

GlobalData. (2021a). *Top 10 online travel intermediaries in the world in 2021 by revenue.* GlobalData. Retrieved from https://www.globaldata.com/companies/top-companies-by-sector/ travel-and-tourism/global-online-travel-intermediaries-by-revenue/

GlobalData. (2021b). *Top 10 tour operators in the word in 2021 by revenue.* GlobalData. Retrieved from https://www.globaldata.com/companies/top-companies-by-sector/travel-and-tourism/ global-tour-operators-by-revenue/

Gorenflo, C., Lee, S., Golab, L., Keshav, S. (2019, March 4). *FastFabric: Scaling Hyperledger fabric to 20,000 transactions per second.* Retrieved from https://arxiv.org/pdf/1901.00910.pdf

Gottfredson, M. (2007, February 8). A new formula for airline profits. *Forbes.com.*

Green, P. E., Krieger, A. M., & Wind, Y. (2001). Thirty years of conjoint analysis: Reflections and prospects. *Interfaces, 31*, S56–S73.

Grimstad, H. (2015). IT and communications in SAS. In C. Gram, P. Rasmussen, & S. Østergaard (Eds.), *History of Nordic computing 4. HiNC 2014. IFIP advances in information and communication technology* (Vol. 447). Springer. https://doi.org/10.1007/978-3-319-17145-6_29

Guenther, D., Ratliff, R., & Sylla, A. (2012). Airline distribution, Chapter 4, Section 4.3.1. In C. Barnhart & B. Smith (Eds.), *Quantitative problem solving methods in the airline industry: A modeling methodology handbook.* Springer.

Gutis, P. S. (1989, December 23). More trips start at a home computer. *The New York Times.*

Hague, N. (2008). *The problem with price. B2B International.* Retrieved from http://www.b2 binternational.com/library/whitepapers/pdf/the_problem_with_ price.pdf

Hair, J. S., Anderson, R. E., & Tatham, R. T. (1984). *Multivariate data analysis with readings* (2nd ed.). Macmillan.

Hansell, S. (2002a, August 19). Technology: Orbitz can now book tickets on American Airlines directly. *The New York Times.*

Hansell, S. (2002b, October 27). Fare idea returns to haunt airlines. *The New York Times.*

Harrington, T. (1989a, April 18). Qantas to back fantasia. *Asian Financial Review.* Retrieved from https://www.afr.com/companies/qantas-to-back-fantasia-19890418-k3cnd

Harrington, T. (1989b, July 21). Qantas forced to limit breadth of fantasia system. *Asian Financial Review.* Retrieved from https://www.afr.com/companies/qantas-forced-to-limit-breadth-of-fantasia-system-19890721-k3if9

Harrington, T. (1989c, November 30). Loss of $3.7 million posted as fantasia takes off. *Asian Financial Review.* Retrieved from https://afr.com/companies/loss-of-3-7m-posted-as-fantasia-takes-off-19891130-k3pjs

Harris, B. (1993). *BEACON, and BOADICEA: A history of computing at British airways and its predecessor Airlines*. Speedwing Press. ISBN 0952226707.

Harris, S. (2007, August 12). BCD, HRG to deploy super-PNR systems. *Business Travel News*. Retrieved from https://www.businesstravelnews.com/More-News/BCD-HRG-To-Deploy-Super-PNR-Systems

Head, R. V. (2002). Getting Sabre off the ground. *IEEE Annals of the History of Computing, 24*, 32–39.

Hoare, M. B. (2010, June 19). *The hotel distribution workhorse, hospitality upgrade*. Retrieved from https://www.hospitalityupgrade.com/_magazine/magazine_Detail.asp/?ID=533

Hoffmann, J. (2021). I;s 1997 and you want to take a flight, The History of the Web, August 17. https://thehistoryoftheweb.com/its-1997-and-you-want-to-take-a-flight/

Hopper, M. D. (1990, May–June). Rattling SABRE – New ways to compete on information. *Harvard Business Review*.

Hoskins, A. (2023, March 22). SAS abandons planned wholesale model following new deal with Amadeus. *Business Travel News – Europe*. Retrieved from https://www.businesstravelnewseurope.com/Air-Travel/SAS-abandons-planned-wholesale-model-following-new-deal-with-Amadeus#:~:text=SAS%20abandons%20planned%20wholesale%20model%20following%20new%20deal%20with%20Amadeus&text=SAS%20Scandinavian%20Airlines%20will%20not,a%20new%20deal%20with%20Amadeus

Howe, A. E., & Dreilinger, D. (1997). SAVVYSEARCH: A metasearch engine that learns which search engines to query. *AI Magazine, 18*(2), 19–25.

IATA. (2012). Resolution 787 enhanced airline distribution, *Passenger Services Conference*, October 12.

IATA. (2018, December 2018–January 2019). *Airlines financial monitor*. Retrieved from https://www.iata.org/publications/economics/Reports/afm/Airlines-Financial-MonitorJan-2019.pdf

IBM. (2003). *Airline control system version 2.4.1, application programming guide*. Copyright International Business Machines Corporation 2003, 2019.

IdeaWorks and CarTrawler. (2018). *Airline ancillary revenue projected to be $92.9 billion worldwide in 2018*. Retrieved from https://www.ideaworkscompany.com/wp-content/uploads/2018/11/Press-Release133-Global-Estimate-2018.pdf

IdeaWorks and CarTrawler. (2019). *CarTrawler worldwide estimate of ancillary revenue for 2019*. Retrieved from https://www.cartrawler.com/ct/ancillary-revenue/worldwide-ancillary-revenue-2019

IdeaWorks and CarTrawler. (2020). *Airline ancillary revenues plummet to $58.2 billion in 2020, erasing 5-years of annual gains*. Retrieved from https://www.cartrawler.com/ct/ancillary-revenue/airline-ancillary-revenue-plummets-to-58-2-billion-in-2020-erasing-5-years-of-annual-gains/

IdeaWorks and CarTrawler. (2021). *Airline ancillary revenues begins recovery with a 13% increase to $65.8 billion for 2021*. Retrieved from https://ideaworkscompany.com/wp-content/uploads/2021/11/Press-Release-160-Global-Estimate-2021-version.pdf

Ideaworks and Cartrawler. (2023). *Airline Ancillary Revenues reach record $117.9 billion worldwide for 2023, October 31*. https://ideaworkscompany.com/wp-content/uploads/2023/10/Press-Release-178-Worldwide-Estimate-2023.pdf

Ingold, A., McMahon-Beattie, U., & Yeoman, I. (Eds.). (2000). *Yield management: Strategies for the service industries*. Continuum Books.

Isidore, C. (2002, March 15). Delta ends agent payments. *CNN/Money*. Retrieved from https://money.cnn.com/2002/03/15/news/companies/delta/

Isler, K. (2016, April 23). *Revenue management in a world without booking class availability*. ATPCO Workshop.

Isler, K., & D' Souza, E. (2009). GDS capabilities, OD control and dynamic pricing. *Journal of Revenue and Pricing Management, 8*(2/3), 255–266.

Jiang, H., Qi, X., & Sun, H. (2014). Choice-based recommender systems: A unified approach to achieving relevancy and diversity. *Operations Research, 62*(5). https://doi.org/10.1287/opre. 2014.1292

Jones, K. (1993, August 11). Big victory for American Airlines in Fare Suit. *The New York Times*.

Jong, A. (2022, March). *U.S. online travel agency market report 2021–2025*. Phocuswright.

Jordan, M. I. (2018, July 17). *Machine learning perspectives and challenges*. University of California, Berkeley.

Jordan, M. I. (2019). Artificial Intelligence—The revolution hasn't happened yet. *Harvard Data Science Review, 1*(1).

Kahn, A. E. (1988a). Airline deregulation. In *The concise encyclopedia of economics*.

Kahn, A. E. (1988b). Surprises of airline deregulation. *American Economic Review, Papers and Proceedings, 78*(2), 316–322.

Kärcher, K. (1996). *Reinventing the package holiday business: New information and communication Technologies in the British and German Tour Operator Sectors*. Ph.D. Thesis, University of Strathclyde.

Kavis, M. (2014). *Architecting the cloud: Design decisions for cloud computing service models*. Wiley. ISBN 978-1-118-61761-8.

King, D. (2023, April 10). In the recovery era, hotel bookings have tipped again toward online travel agencies. *CoStar*. Retrieved from https://www.costar.com/article/589850961/in-the-recovery-era-hotel-bookings-have-tipped-again-toward-online-travel-agencies

Klein, T. (2016). Tom Klein, CEO, Sabre Corp: The CEO perspective. *Travel Weekly*, Preview 2016.

Knight, M. (2008). Beacon 1963-7: A system design ahead of its time? *Computer Resurrection: The Bulletin of the Computer Conservation Society, 43*. Retrieved from http://vipclubmn.org/Articles/BEA-ReservationSystemEd1.pdf

Ko, E. (2022, April 8). *The state of online travel agencies*. Phocuswright contribution in *HospitalityNet*. Retrieved from https://www.hospitalitynet.org/news/4109835.html

Kothari, A., Madireddy, M., & Sundararajan, R. (2016). Discovering patterns in traveler behavior using segmentation. *Journal of Revenue and Pricing Management, 15*(5), 334–351.

Kretsch, S. S. (1995, September). Airline fare management and policy. In: D. Jenkins (Ed.), *Aviation daily's handbook of airline economics* (pp. 477–482). The Aviation Weekly Group of the McGraw-Hill Companies.

LA Times Archives. (1986, July 8). Northwest buys half of TWA's booking system. *Los Angeles Times*. Retrieved from https://www.latimes.com/archives/la-xpm-1986-07-08-fi-22674-story. html

Lammi, E. (1983, June 21). *A House subcommittee opened hearings Tuesday on allegations that...* UPI Archives.

Laney, D. (2001, February 6). *3D data management: Controlling data volume, velocity, variety. Application delivery strategies*. META Group.

Lavin, C. H. (1990, April 22). Practical traveler: Tour shopping with a computer. *The New York Times*.

Ledger Insights. (2020, October 15). Singapore Airlines extends its blockchain-based reward digital wallet. *Ledger Insights*. Retrieved from https://www.ledgerinsights.com/singapore-airlines-extends-its-blockchain-based-reward-digital-wallet/

Lee, A. (1990, September). *Airline reservations forecasting: Probabilistic and statistical models of the booking process*. Ph.D. dissertation in Transportation Systems, Massachusetts Institute of Technology.

Leff, D., & Lim, K. (2021). The key to leveraging AI at scale. *Journal of Revenue and Pricing Management, 20*, 376. https://doi.org/10.1057/s41272-021-00320-3

Leven, M. A. (1994, June). Superstar views on hotel technology. *CKC Report*, pp. 7–8.

Levenson, R. (1987). *Bill Bernbach's book: A history of the advertising that changed the history of advertising*. Villard Books. ISBN 10: 0394549201/ISBN 13: 9780394549200.

Levesque, H. (2011). *The Winograd schema challenge*. Retrieved from Commonsensereasoning. org

Li, L. (2008). *New heuristics for revenue management problem with customer choice models*. Ph.D. Dissertation, Graduate School of Arts and Sciences, Columbia University.

Lieberman, W. H. (2010). Revenue management in the travel industry. *Wiley Encyclopedia of Operations Research and Management Science*, edited by James J. Cochran.

Limone, J. (2004). Priceline to acquire 100% of TravelWeb, May 04. https://www.travelweekly. com/Travel-News/Hotel-News/Priceline-to-acquire-100-of-Travelweb

Littlewood, K. (1972, October). Forecasting and control of passenger bookings. In *AGIFORS 12th annual symposium proceedings, Nathanya, Israel*, pp. 193–204.

Locke, G. (2009). Consumer behavior trends and their impacts on airline product distribution. *Journal of Revenue and Pricing Management, 8*(2/3), 267–278.

Locke, E. (2020, June 23). *The strange way that airlines are actually central banks, issue 013, WTF is going on with the economy?! X abroaden*. Retrieved from https://abroaden.substack.com/p/the-strange-way-airlines-are-actually

Magas, J. (2020, July 01) Destination Blockchain: Shaking up travel industry and cutting costs. *Cointelegraph*. Retrieved from https://cointelegraph.com/news/destination-blockchain-shak ing-up-travel-industry-and-cutting-costs

Mantin, B., & Rubin, E. (2016, July–August). Fare prediction websites and transaction prices: Empirical evidence from the Airline Industry. *Marketing Science, 35*(4), 640–655.

Marr, B. (2018, December 7). The awesome ways TUI uses Blockchain to revolutionize the travel industry. *Forbes*.

Martin, J. C., Roman, C., & Espino, R. (2008). Willingness to pay for airline service quality. *Transport Reviews, 28*(2), 199–217.

Mauri, A. G. (2012). *Hotel revenue management: Principles and practices*. Pearson. ISBN: 9788865181461, 886518146X.

Mazareanu, E. (2020, June 10). Low cost carrier market – Global capacity share 2007–2019. *Statista*. Retrieved from https://www.statista.com/statistics/586677/global-low-cost-carrier-market-capacity-share/

McCarthy, D. (2023, August 14). ASTA lodges official complaint with DOT over NDC imple-mentation. *TravelMarket report*. Retrieved from https://www.travelmarketreport.com/Air/articles/ASTA-Lodges-Official-Complaint-with-DOT-Over-NDC-Implementation

McCartney, S. (1998, April 27). As he approaches retirement, AMR's Crandall is flying high. *The Wall Street Journal*.

McDonald, M. (2006, March). Yielding to the LCC's. *Air Transport World*, pp. 36–37.

McDowell, E. (1992, April 10). American air cuts most fares in simplification of rate system. *New York Times*.

McDowell, E. (1996, September 4). Lawsuits by travel agents against Airlines is settled. *New York Times*. Retrieved from https://www.nytimes.com/1996/09/04/business/lawsuit-by-travel-agents-against-airlines-is-settled.html

McKenna, B. (2017, June 02). Doug cutting 'father' of Hadoop talks about big data tech revolution. *Computer Weekly*. Retrieved from https://www.computerweekly.com/news/450420002/Doug-Cutting-father-of-Hadoop-talks-about-big-data-tech-evolution

McMahon-Beattie, U., Palmer, A., & Yeoman, I. (2011). Does the Customer Trust You? In I. Yeoman & U. McMahon-Beattie (Eds.), *Revenue management. A practical pricing perspective*. Palgrave Macmillan. https://doi.org/10.1057/9780230294776_6

Meehan, M. (2000, August 29). Sabre buys GetThere for $757 million, announces layoffs. *ComputerWorld*. Retrieved from https://www.computerworld.com/article/2596441/sabre-buys-getthere-for%2D%2D757-million%2D%2Dannounces-layoffs.html

Michael, S. C., & Silk, A. J. (1994, May 11). *American Airlines value pricing*. Harvard Business School.

Michaelis, L., & Menten, M. L. (1913). Die Kinetik der Invertinwirkung. *Biochemische Zeitschrift, 49*, 333–369.

Millet, H. (2023). Leveraging booking information for offer personalization and GDPR consequences. *Journal of Revenue and Pricing Management, 22*(2), 152–156. https://doi.org/10.1057/s41272-022-00401-x

Morello, G., & Lopatko, R. (2012). Airlines as retailers. *Ascend, 2.*

Moritz, P., Nishihara, R., Wang, S., Tumanov, A., Liaw, R., Liang, E., Elibol, M., Yang, Z., Paul, W., Jordan M. I., & Stoica, I. (2018). Ray: A distributed execution framework for emerging RL applications. *Research Faculty Summit, Microsoft.*

Mortensen, K., & Hughes, T. L. (2018). Comparing Amazon's mechanical Turk platform to conventional data collection methods in the health and medical research literature. *Journal of General Internal Medicine, 33*(4), 533–538.

Musser, G. (2019, May). Artificial imagination: How machines could learn creativity and common sense, among other human qualities. *Scientific American*, 59–63.

Moskowitz, C. (2008). Mind's limit found: 4 things at once, live science, April 27. https://www.livescience.com/2493-mind-limit-4.html

Nasiry, J., & Popescu, I. (2011). Dynamic pricing with loss-averse consumers and peak anchoring. *Operations Research, 59*(6), 1361–1368.

Neff, J. (2017, February 14). Study: Consumers get more fickle despite billions spent on loyalty. *Advertising Age*. Retrieved from http://adage.com/article/cmo-strategy/consumers-fickle-billions-spent-loyalty/307974/

New York Times. (1981, September 22). Company News; American to offer fixed-fare passes. *New York Times, section D, page 4.* Retrieved from https://www.nytimes.com/1981/09/22/business/company-news-american-to-offer-fixed-fare-passes.html

Ng, I. C. L. (2008). *The pricing and revenue management of services: A strategic approach* (Routledge advances in management and business studies). Routledge, Taylor & Francis Group.

O'Neal, S. (2015, July 15). The billboard effect is dead, says a study of hotels listed on OTAs. *Phocuswire.com*

Opricovic, S., & Tzeng, G.-H. (2004). The compromise solution by MCDM methods: A comparative analysis of VIKOR and TOPSIS. *European Journal of Operational Research, 156*(2), 445–455.

Orme, B. K., & Chrzan, K. (2017). *Becoming an expert in conjoint analysis choice modeling for pros. Sawtooth Software*. Retrieved from www.sawtoothsoftware.com

Page, A. (2019, June 18). *Webjet embraces Blockchain technology*. Retrieved from https://strawman.com/blog/webjet-asxweb-embraces-blockchain-technology

Palmatier, G. E., & Crum, C. (2003). *Enterprise sales and operations planning: Synchronizing demand, supply and resources for peak performance*. J. Ross Publishing. ISBN 1-932159-00-2.

Parsons, M. (2020, July 6). What a new surcharge from Singapore Airlines could mean for other carriers. *Skift.*

Pegasus Solutions, Inc. (1996, August 6). TravelWeb to offer both hotel and air bookings. *HospitalityNet.*

Pestronk, M. (2018, January 04). Can you reject your GDS' per-segment fee 'full content'? *Travel Weekly*. Retrieved from https://www.travelweekly.com/Mark-Pestronk/Can-you-reject-your-GDS-per-segment-fee-for-full-content

Pestronk, M. (2019, September 11). The issue with 'most favored nation' clauses. *Travel Weekly*. *Retrieved* from https://www.travelweekly.com/Mark-Pestronk/The-issue-with-most-favored-nation-clauses

Pestronk, M. (2023, March 31). What does NDC mean for full content fees? *Travel Weekly*. Retrieved from https://www.travelweekly.com/Mark-Pestronk/What-does-NDC-mean-for-full-content-fees

Peterson, R. M. (1986). *The penultimate hub airplane*. Internal Memo, Boeing Commercial Airplane Group.

Peterson, R. D. (1990, January). The CAB's struggle to establish price and route rivalry in world air transport. *American Journal of Economics and Sociology, 49*(1).

Phillips, R. L. (2005). *Pricing & revenue optimization*. Stanford University Press.

Pillai, K. G. J. (1969). *The air net: The case against the world aviation Cartel.* Grossman Publishers. ISBN-10: 067011118X, ISBN-13: 978-0670111183.

Price, N. (2021). *Decentralized identity for hospitality and travel, hospitality and travel SIG, Presented at DIF.* Retrieved from https://www.youtube.com/watch?v=zKKAcseqC_M&t=2 79s

Prieto, M. (2020, July 24). *The state of online travel agencies – 2020, travel tech media.* Retrieved from https://medium.com/traveltechmedia/the-state-of-online-travel-agencies-2020-f6acc899aca2

Quinby, D. (2005). *GeNiEs: Marketing or magic?.* The PhoCusWright Snapshot. PhoCusWright Market Research.

Quinby, D. (2006). *Un uneasy peace: Airlines, GDSs, GNEs and the outlook for air travel distribution in the North American marketplace.* The PhoCusWright snapshot. PhoCusWright market Research.

Ramaswami, R. (2020). Brand strategies focused on dependability score highest on customer trust, *Gartner*, December 1. https://www.gartner.com/en/marketing/insights/articles/brand-strategies-focused-dependability-score-highest

Ratliff, R., & Vinod, B. (2005). Airline pricing and revenue management: A future outlook. *Journal of Revenue and Pricing Management, 4*(3), 302–307.

Ratliff, R., & Vinod, B. (2016). An applied process for airline strategic fare optimization. *Journal of Revenue and Pricing Management, 15*(5), 320–333.

Reed, D. (2019, November 21). Airlines are earning more than ever from extra fees but are causing travelers more frustration and dissatisfaction. *Forbes.*

Reuters. (2012, October 31). *American Airlines, Sabre settle legal dispute.* Retrieved from https://www.reuters.com/article/us-americanair-sabre/american-airlines-sabre-settle-legal-dispute-idUSBRE89U1RY20121031

Robbins, J. (1952). Some aspects of the sequential design of experiments. *Bulletin of the American Mathematical Society, 58*(5), 527–535.

Rosen, P. (2023, March 1). Artificial Intelligence is on the bring of an 'iPhone moment' and can boost the world economy by $15.7 trillion in 7 years Bank of America says. *Markets Insider.* Retrieved from https://markets.businessinsider.com/news/stocks/artificial-intelligence-ai-chatgpt-iphone-moment-bank-america-bofa-markets-2023-3#:~:text=Ultimately%2C%20AI's%20ability%20to%20capitalize,use%20cases%20for%20the%20technology

Rothstein, M. (1985). O.R. and the airline overbooking problem. *Operations Research, 33*, 237–248.

Schaal, D. (2002, July 23). Study: 'Disintermediation' hasn't happened. *Travel Weekly.* Retrieved from https://www.travelweekly.com/Travel-News/Travel-Agent-Issues/Study-Disintermediation-hasn-t-happened

Schaal, D. (2006a, May 07). Sabre, Amadeus to share content if a carrier leaves either. *Travel Weekly.* https://www.travelweekly.com/Travel-News/Travel-Technology/Sabre-Amadeus-to-share-content-if-a-carrier-leaves-either

Schaal, D. (2006b, May 28). American moves forward with Farelogix deal. *Travel Weekly.* https://www.travelweekly.com/Travel-News/Travel-Technology/American-moves-forward-with-Farelogix-deal

Schaal, D. (2009, March 12). Yahoo to discontinue FareChase meta search engine. *Travel Weekly.* Retrieved from https://www.travelweekly.com/Travel-News/Online-Travel/Yahoo-to-discontinue-FareChase-metasearch-engine

Schaal, D. (2014, July 27). *The oral history of travel's greatest acquisition.* Skift.com. Retrieved from https://skift.com/oral-history-of-booking-acquisition/

Schaal, D. (2021, November 2). *Sabre loses a chunk of Expedia's North America business.* Retrieved from https://skift.com/2021/11/02/sabre-loses-a-chunk-of-expedias-north-america-business/

Schall, D. (2023, October 6). Hopper terminates Booking.com partnership in preemptive strike. *Skift*. Retrieved from https://skift.com/2023/10/06/exclusive-hopper-terminates-booking-com-partnership-in-preemptive-strike/

Schmitz, S. (2020). Tickets Please! ISBN-10: 1916039650, ISBN-13: 978-1916039650, *Astral Horizon Press*, (September 15, 2020).

Schmittlein, D. C., Morrison, D. G., & Colombo, R. (1987, January). Counting your customers: Who they are and what will they do next? *Management Science, 33*, 1–24.

Segmentify. (2021, June 17). *The 3 option decoy effect and relativity*. Retrieved from https://www.segmentify.com/blog/the-3-option-decoy-effect-and-relativity

Seirawan, Y., Simon, H., & Munakata, T. (1997). The implications of Kasparov vs. deep blue. *Communications of the ACM, 40*(8), 21–25.

Serling, R. J. (1985). *Eagle: The story of American Airlines*. St. Martin's/Marek.

Shao, S., & Kauermann, G. (2020). Understanding price elasticity for airline ancillary services. *Journal of Revenue and Pricing Management, 19*(1), 74–82.

Shayon, S. (2018). *6 reasons for Singapore Airlines' blockchain-based loyalty program*. Retrieved February 15, 2018, from www.brandchannel.com/2018/02/15/singaporeairlines-blockchain/

Sheivachman, A. (2017). Channel shock: The future of travel distribution, skift, August 7. https://skift.com/2017/08/07/channel-shock-the-future-of-travel-distribution/

Shifrin, C. (1982, July 2). Delta unveils unbiased air reservations system. *The Washington Post*. Retrieved from https://www.washingtonpost.com/archive/business/1982/07/02/delta-unveils-unbiased-air-reservations-system/1267b1c0-fd82-4117-bf23-7dada4bd25d6/

Sickel, J. (2018, October 28). Linking changes in hotel distribution to changes in customer expectations. *Business Travel News*. Retrieved from https://www.businesstravelnews.com/Research/Distribution/Linking-Changes-in-Hotel-Distribution-to-Changes-in-Customer-Expectations

Silk, R. (2021, July 1). American Airlines Sues Sabre over fare display. *Travel Weekly*. Retrieved from https://www.travelweekly.com/Travel-News/Airline-News/American-Airlines-sues-Sabre-over-fare-display

Silk, R. (2022, May 20). After 11 years, Sabre-US airways case results in Sabre paying $1 in damages. *Airline Weekly*. Retrieved from https://www.travelweekly.com/Travel-News/Airline-News/Sabre-US-Airways-verdict

Silk, R. (2023a, June 16). Singapore airways further boosts NDC and hinders legacy GDS bookings. *Travel Weekly*. Retrieved from https://www.travelweekly.com/Travel-News/Airline-News/Singapore-Airways-further-boosts-NDC

Silk, R. (2023b, July 10). Airline NDC integration varies, creating challenges for agencies. *Travel Weekly*. Retrieved from https://www.travelweekly.com/Travel-News/Airline-News/Airline-NDC-integration-varies-creating-challenges

Silk, R. (2024a). The public back-and-forth between ASTA and American airlines continues, *Travel Weekly*, February 08. https://www.travelweekly.com/Travel-News/Airline-News/ASTA-versus-American-Airlines-part-four

Silk, R. (2024b). American Airlines pressures travel agencies with new rules for loyalty points, *Travel Weekly*, February 20. https://www.travelweekly.com/Travel-News/Airline-News/American-new-rules-for-travel-agencies

Silk, R. (2024c, February 20). Consolidators are embracing NDC as they compete for the lowest airfares. *Travel Weekly*. https://www.travelweekly.com/Travel-News/Airline-News/Air-consolidators-embracing-NDC

Siwiec, J. E. (1977). A high performance DB/DC system. *IBM System Journal, 16*, 169–195.

Sjnajder, M., Ratliff, R., & Kaya, C. (2023a). A heuristic for incorporating ancillaries into air choice models with personalization (part 1: Estimating preferences using hedonic regression). *Journal of Revenue and Pricing Management, 22*, 122–139. https://doi.org/10.1057/s41272-022-00399-2

Sjnajder, M., Ratliff, R., & Kaya, C. (2023b). A heuristic for incorporating ancillaries into air choice models with personalization (part 2: Integrated multinomial logit and hedonic regression models). *Journal of Revenue and Pricing Management, 22*, 140–151. https://doi.org/10.1057/s41272-022-00400-y

Skift. (2021). *American Airlines Inc., Plaintiff, v. Sabre, Plaintiff's verified original petition and application for temporary injunction and permanent injunction.* Retrieved from https://skift. com/wp-content/uploads/2021/07/aa-sabre-lawsuit-july-1.pdf

Smith, B. C. (1982, August). Optimal departure booking levels. Internal technical report. *American Airlines.*

Smith, B. C. (1986). O&D Control with virtual nesting. Internal technical report. *American Airlines.*

Smith, B. C. (2007, May 14–16). Revenue management in the U.S. airline industry. In *AGIFORS passenger and cargo revenue management study group, Jeju Island, South Korea.*

Smith, B. C., Leimkuhler, J. L., & Darrow, R. M. (1992, January–February). Yield management at American Airlines. *Interfaces*, 22.

Smith, B. C., Vinod, B., & Green, R. (1997). *Apparatus and method of allocating flight inventory resources based on the current market value.* U.S. patent no. 6,085,164; filed 4th March, 1997, granted 4th July, 2000.

Smith, B. C., Darrow, R., Elieson, J., Guenther, D., Rao, B. V., & Zouaoui, F. (2007). Travelocity becomes a travel retailer. *Interfaces, 37*(1), 68–81.

Sorrells, M. (2022, May 11). Airbnb enables "split stays" to ease inventory woes. *PhocusWire.*

Sorrells, M. (2018a, July 11). Attribute-based selling comes to hotel reservation systems. *PhocusWire.*

Sorrells, M. (2018b, April 2). *AI analysis, part 1: Travel tech giants.* Retrieved from https://www. phocuswire.com/AI-series-part-1-GDSs

Sovrin. (2018, January). *Sovrin™: A protocol and token for self-sovereign identity and decentralized trust. A White Paper from the Sovrin Foundation. Version 1.*

Steeb, D., & Sohn, T. (2006, October 06). *Rule-based shopping.* U.S. Patent Office Patent Number 8126783.

Straus, B. (2008, April). Revenue window of opportunity. *Air Transport World*, pp. 37–40.

Sullivan, L. (2020, January 22). Google flights ends charges for airline booking, Referral Links. *Search and Performance Marketing Daily.* Retrieved from https://www.mediapost.com/ publications/article/346061/google-flights-ends-charges-for-airline-booking-r.html

Sullivan, B. (2002). Travelocity buys Site59.com for $43 million, Computerworld, March 26. https://www.computerworld.com/article/2588077/travelocity-buys-site59-com-for%2D%2 D43-million.html

Szende, P. (2020, November). *Hospitality revenue management: Concepts and practices* (1st ed.). Apple Academic Press. ISBN-10: 1771888881, ISBN-13: 978–1771888882.

Szymanski, T., & Darrow, R. (2021). Shelf placement optimization for air products. *Journal of Revenue and Pricing Management, 20*, 322. https://doi.org/10.1057/s41272-021-00315-0

Szymański, B., Belobaba, P., & Papen, A. (2021). Continuous pricing algorithms for airline RM: Revenue gains and competitive impacts. *Journal of Revenue and Pricing Management, 20*, 669–688. https://doi.org/10.1057/s41272-021-00350-x

Talluri, K. T., & van Ryzin, G. J. (2004). *The theory and practice of revenue management.* Kluwer Academic Publishers.

Taneja, N. K. (1976). *The commercial airline industry.* D. C. Heath.

Taneja, N. K. (2017). *21st century airlines: Connecting the dots* (1st ed.). Routledge. ISBN 1138093130.

Teixeira, T. S. (2019). *Unlocking the customer value chain, how decoupling drives consumer disruption.* Currency Publishers.

The Beat. (2007, May 24). *Interview: Amex on TravelBahn, the beat.* Retrieved from https://www. thebeat.travel/Interviews/INTERVIEW-Amex-on-TravelBahn?utm_source=website&utm_ medium=widget&utm_campaign=search

Touraine, S. (2021). The industry transformation to dynamic offering. *Journal of Revenue and Pricing Management, 20*, 611–614. https://doi.org/10.1057/s41272-021-00344-9

Train, K. E. (2003). *Discrete choice methods with simulation.* Cambridge University Press.

Travel Weekly. (2000, January 31). Galileo to buy its U.K. distributor. *Travel Weekly*. Retrieved from https://www.travelweekly.com/Travel-News/Travel-Technology/Galileo-to-buy-its-U-K-distributor

Travelport. (2023). Travelport acquires DEEM, furthering its investment in modern retailing and corporate travel, travelport press release. https://www.travelport.com/press-release/travelport-acquires-deem-furthering-its-investment-in-modern-retailing-and-corporate-travel

Trejos, N. (2018). Priceline ends 'Name Your Own Price' deals for rental cars, *USA Today*, March 26.

Turing, A. (1950). Computing machinery and intelligence. *Mind, 49*, 433–460.

Turow, J., Feldman, L., & Meltzer, K. (2005, June 1). *Open to exploitation: American shoppers online and offline*. University of Pennsylvania's Annenberg School for Communication.

van Westendorp, P. H. (1976). NSS – Price sensitivity meter (PSM) – A new approach to study consumer perception of Price. In *Proceedings of the ESOMAR Congress, Venice*.

Varian, H. (2014). Big data: New tricks for econometrics. *Journal of Economic Perspectives, 28*(2), 2–28.

Vellapalath, R. (2018, February 2). A viewpoint on GDS surcharges and the evolving airline distribution landscape. *PhocusWire*.

Vinod, B. (1989, March). *A partitioning algorithm for virtual nesting indexing using dynamic programming. Internal technical report*. Sabre Technology Solutions.

Vinod, B. (1995). Origin and destination yield management. In D. Jenkins (Ed.), *Aviation daily's handbook of airline economics* (pp. 459–468). The Aviation Group of the McGraw-Hill Companies.

Vinod, B. (1999). Airline alliances and its impact on pricing and revenue management. *IATA – the eleventh international airline revenue management conference proceedings*, Chicago, Illinois, October.

Vinod, B. (2004). Unlocking the value of revenue management in the hotel industry. *Journal of Revenue and Pricing Management, 3*(2), 178–190.

Vinod, B. (2005a). Retail revenue management and the new paradigm of merchandise optimization. *Journal of Revenue and Pricing Management, 3*(4), 358–368.

Vinod, B. (2005b). Alliance revenue management. *Journal of Revenue and Pricing Management, 4*(1), 66–82.

Vinod, B. (2006). Advances in inventory control. *Journal of Revenue and Pricing Management, 4*(4), 367–381.

Vinod, B. (2007). Seat availability: Alignment with the revenue management value proposition. *Journal of Revenue and Pricing Management, 4*(6), 315–230.

Vinod, B. (2008). The continuing evolution: Customer centric revenue management. *Journal of Revenue and Pricing Management, 7*(1), 27–39.

Vinod, B. (2009). Distribution and revenue management: Origins and value proposition. *Journal of Revenue and Pricing Management, 8*(2–3), 117–133.

Vinod, B. (2010). The complexities and challenges of the airline fare management process and alignment with revenue management. *Journal of Revenue and Pricing Management, 9*(1–2), 137–151.

Vinod, B. (2011a). The future of online travel. *Journal of Revenue and Pricing Management, 10*(1), 56–61.

Vinod, B. (2011b). Unleashing the power of loyalty programs: The next 30 years. *Journal of Revenue and Pricing Management, 10*(5), 471–476.

Vinod, B. (2013a). Leveraging big data for competitive advantage in travel. *Journal of Revenue and Pricing Management, 12*(1), 96–100.

Vinod, B. (2013b). Revenue management for the optimal control of group traffic. *Journal of Revenue and Pricing Management, 12*(4), 295–304.

Vinod, B. (2015a). The expanding role of revenue management in the airline industry. *Journal of Revenue and Pricing Management, 14*(6), 391–399.

Vinod, B. (2015b). *Sabre air shopping – Why it is important, why it's difficult, and how it works.* Sabre Research, August 20, 2013, revised August 15, 2015.

Vinod, B. (2016a). Evolution of yield management in travel. *Journal of Revenue and Pricing Management, 15*(3–4), 203–211.

Vinod, B. (2016b). Big data in the travel marketplace. *Journal of Revenue and Pricing Management, 15*(5), 352–359.

Vinod, B. (2017). The evolving paradigm of interactive selling based on consumer preferences. In N. Taneja (Ed.), *21st century airlines: Connecting the dots* (pp. 207–213). Routledge. ISBN 978-1-138-09313-3.

Vinod, B. (2019). Hotel retailing with attribute-based room pricing and inventory control. *Journal of Revenue and Pricing Management, 18*(6), 429–433.

Vinod, B. (2020a, February 25). Travel trends driving the paradigm shift of government travel. In *National Defense Transportation Association (NDTA) government travels symposium, Washington, DC.*

Vinod, B. (2020b). Blockchain in travel. *Journal of Revenue and Pricing Management, 19*(1), 2–6.

Vinod, B. (2020c). The Covid-19 pandemic and airline cash flow. *Journal of Revenue and Pricing Management, 19*, 228–229.

Vinod, B. (2021a, May). *The evolution of yield management in the airline industry: Origins to the last frontier.* Springer Nature. ISBN-13: 978-3030704230, ISBN-10: 3030704238.

Vinod, B. (2021b). An approach to adaptive robust revenue management with continuous demand management in a COVID-19 era. *Journal of Revenue and Pricing Management, 20*(1), 10–14.

Vinod, B. (2021c). Artificial intelligence in travel. *Journal of Revenue and Pricing Management, 20*, 368. https://doi.org/10.1057/s41272-021-00319-w

Vinod, B. (2021d). Advances in revenue management: The last frontier. *Journal of Revenue and Pricing Management, 20*(1), 15–20. https://doi.org/10.1057/s41272-020-00264-0

Vinod, B. (2021e). The age of intelligent retailing: Personalized offers in travel for a segment of ONE. *Journal of Revenue and Pricing Management, 20*, 473. https://doi.org/10.1057/s41272-020-00265-z

Vinod, B. (2021f). The influence of revenue management and inventory control on air shopping. *Journal of Revenue and Pricing Management, 20*, 490. https://doi.org/10.1057/s41272-020-00258-y

Vinod, B. (2022a, November). *Revenue management in the lodging industry: Origins to the last frontier.* Springer Nature. ISBN: 978-3-031-14304-5, ISBN: 978-3-031-14302-1.

Vinod, B. (2022b). Airline revenue planning and the COVID-19 pandemic. *Journal of Tourism Futures, 8*(2), 245–253.

Vinod, B. (2023, May 27). *Artificial intelligence and machine learning in the travel industry: Simplifying complex decision making.* Palgrave Macmillan. ISBN 978-3-031-25455-0, ISBN 978-3-031-25456-7.

Vinod, B., & Hobt, D. (2021). *Method, apparatus and computer program product for reservations, inventory control, shopping and booking with attribute based pricing.* U.S. patent number 11,017,326 filed May 6, 2015, granted May 25, 2021. Retrieved from https://patents.justia.com/patent/11017326

Vinod, B., & Huff, C. (2019, August 14). *How to scale NDC to GDS transaction volumes. Sabre research internal technical report.*

Vinod, B., & Moore, K. (2009). Promoting branded fare families and ancillary services: Merchandising and its impacts on the travel value chain. *Journal of Revenue and Pricing Management, 8*(2–3), 174–186.

Vinod, B., Nilson, V., & Hobt, D. A. (1997, October). Inventory control with the availability processor. In *Object Management Group (OMG) conference, OMG Awards Banquet Presentation, Frankfurt, Germany.*

Vinod, B., Xie, P., & Bellubbi, R. (2015). From shopper to customer: Preference driven air shopping with targeted one-to-one shopping responses. *Ascend*, pp. 11–13.

Vinod, B., Ratliff, R. M., & Jayaram, V. (2018). An approach to offer management: Maximizing sales with fare products and ancillaries. *Journal of Revenue and Pricing Management, 17*(2), 91–101.

Walker, J. S., Schneier, B., & Jorasch, A. (1998, August 11). *Method and apparatus for a cryptographically assisted commercial network system designed to facilitate buyer driven conditional purchase offers.* U.S. Patent Number 5,794,207.

Wang, K., Wittman, M. D., & Bockelie, A. (2021). *Journal of Revenue and Pricing Management, 20,* 654–668. https://doi.org/10.1057/s41272-021-00349-4

Wang, K., Wittman, M. D., & Fiig, T. (2023). Dynamic offer creation for airline ancillaries using a Markov chain choice model. *Journal of Revenue and Pricing Management, 22,* 103–121. https://doi.org/10.1057/s41272-022-00398-3

Watkins, C. J. C. H. (1989) *Learning from delayed rewards*, PhD. Dissertation, Cambridge University.

Weatherford, L. R. (1991). *Perishable asset revenue management in general business situations.* Ph.D. Dissertation, Darden Graduate School of Business Administration, University of Virginia.

Weatherford, L. R., & Bodily, S. E. (1992). A Taxonomy and research overview of perishable-asset revenue management: Yield management, overbooking and pricing. *Operations Research, 40* (5), 831–844.

West, E. (2021, December 3). Amex GBT makes SPAC deal to go public. *Business Travel News.* Retrieved from https://www.businesstravelnews.com/Procurement/Amex-GBT-Makes-SPAC-Deal-to-Go-Public

White, J. M. (2013, January 3). *Bandit algorithms for website optimization* (1st ed.). O'Reilly Media. ISBN-13: 978-1449341336.

Williamson, E. L. (1992). *Airline network seat inventory control: Methodologies and revenue impacts.* PhD Thesis. Massachusetts Institute of Technology.

Wittman, M. D., & Belobaba. P. (2018, May). *The implications of dynamic pricing for airline revenue management.* In *PODS meeting, Hong Kong.*

World Economic Forum. (2020, March). Known traveler digital identity: Specifications guide. In *World Economic Forum in collaboration with Accenture.* Retrieved from https://ktdi.org/

Wright, C. P., Groenevelt, H., & Shumsky, R. A. (2010). Dynamic revenue management in airline alliances. *Transportation Science, 44*(1), 15–37.

Yanofsky, D. (2015, October 27). *Half of American's revenues comes from 13% of its customers. Quartz.* Retrieved from https://qz.com/533501/half-of-american-airlines-revenue-came-from-13-of-its-customers

Yeoman, I., & McMahon-Beattie, U. (Eds.). (2004). *Revenue management and pricing: Case studies and applications.* Cengage Learning EMEA Higher Education.

Yeoman, I., & McMahon-Beattie, U. (Eds.). (2011). *Revenue management: A practical pricing perspective.* Palgrave Macmillan.

Yuen, B. B. (2002). Group revenue management: Redefining the business process – Part I. *Journal of Revenue & Pricing Management, 1,* 267–274.

Yuen, B. B. (2003). Group revenue management: Redefining the business process – Part II. *Journal of Revenue & Pricing Management, 1,* 345–354.

Zhang, B. (2016, October 22). *A single statistic shows why all airline passengers are not created equal.* Business Insider.

Zhang, Y., Bradlow, E. T., & Small, D. S. (2015). Predicting customer value using clumpiness: From RFM to RFMC. *Marketing Science, 34*(2), 195–208.

Zouaoui, F., & Rao, B. V. (2009). Dynamic pricing of opaque airline tickets. *Journal of Revenue and Pricing Management, 8*(2–3), 148–154.

Index

© The Author(s), under exclusive license to Springer Nature Switzerland AG 2024

B. Vinod, *Mastering the Travel Intermediaries*, Management for Professionals,

https://doi.org/10.1007/978-3-031-51524-8

The manufacturer's authorised representative in the EU is Springer
Nature Customer Service Centre GmbH, Europaplatz 3, 69115 Heidelberg,
Germany. If you have any concerns regarding our products, please
contact ProductSafety@springernature.com

Printed and bound by CPI Group (UK) Ltd, Croydon, CR0 4YY

29/04/2026

02099522-0006